Prognostics and Remaining Useful Life (RUL) Estimation

Prognostics and Remaining Useful Life (RUL) Estimation

Predicting with Confidence

Diego Galar
Kai Goebel
Peter Sandborn
Uday Kumar

CRC Press
Taylor & Francis Group
Boca Raton London New York

CRC Press is an imprint of the
Taylor & Francis Group, an **informa** business

First edition published 2022
by CRC Press
6000 Broken Sound Parkway NW, Suite 300, Boca Raton, FL 33487-2742

and by CRC Press
4 Park Square, Milton Park, Abingdon, Oxon, OX14 4RN

© 2022 Diego Galar, Kai Goebel, Peter Sandborn and Uday Kumar

First edition published by CRC Press 2022

CRC Press is an imprint of Taylor & Francis Group, LLC

ISBN: 978-0-367-56306-6 (hbk)
ISBN: 978-0-367-56309-7 (pbk)
ISBN: 978-1-003-09724-2 (ebk)

DOI: 10.1201/9781003097242

Typeset in Times
by Newgen Publishing UK

Contents

Preface

Reliability, availability, and safety are determining factors in the effectiveness of industrial performance, thus calling for enhanced maintenance support. Traditional concepts like preventive and corrective strategies are being replaced by new ones like predictive and proactive maintenance.

Fault detection, failure diagnosis, and response development are basic maintenance concepts, corresponding to the need to "see" phenomena, to "understand" them, and to "act" appropriately. However, rather than simply understanding a phenomenon which has just appeared, like a failure, it is useful to anticipate it to take action as soon as possible. This can be defined as the "prognostic process" – the topic of this book. Prognostics are defined by the International Organization for Standardization (ISO) as "the estimation of time to failure and risk for one or more existing and future failure modes". Another way to put this is the prediction of a system's lifetime or its remaining useful life (RUL) before failure, given the current condition and past history. Prognostic approaches to maintenance have the potential to reduce costs and increase reliability.

Many techniques have been proposed to support the prediction of RUL and prognosis, but choosing a technique depends on classical constraints: data availability, system complexity, implementation requirements, and available monitoring devices. There is no general agreement on an appropriate and acceptable set of metrics that can be employed in prognostics, and researchers and condition-based maintenance (CBM) practitioners are still working on this. Building an effective tool to measure RUL is not a trivial task: prognostics is an intrinsically uncertain process, and prognostic tools must take this into account when performing decisional metrics to select preventive actions.

Methods of data analysis can be:

1. Physical-model-driven.
2. Experience-based.
3. Data-driven.

Experience and expert opinion are still important, but the output generated from real monitoring data tends to give more precise information. Today, experience-based analysis tends to supplement and/or complement other types of analysis.

Data-driven methods rely on previously observed data to predict the system's future state or match similar patterns in the history to infer RUL. These include statistical methods, reliability functions, and artificial intelligence (AI) methods. The following data-driven techniques are discussed in this book:

- Use of run-to-failure (RTF) data.
- Neural networks (NNs), fuzzy logic, decision trees, support vector machines (SVMs), anomaly detection algorithms, reinforcement learning, classification, clustering and Bayesian methods, and data mining algorithms.
- Conventional numerical techniques, including wavelets, Kalman filters, particle filters, regression, demodulation, and statistical methods.
- Statistical approaches like the gamma process, hidden Markov model, regression-based model, and relevance vector machine, and least square stochastic models, such as the autoregressive (AR) model, the threshold AR model, the bilinear model, the projection pursuit, the multivariate adaptive regression splines, and the Volterra series expansion.

Model-based or physics-based methods involve the knowledge of a system's failure mechanisms (e.g., crack growth) to build a mathematical description of the system's degradation process to estimate the RUL. However, model-based methods may not be applicable to complex systems due to the lack of understanding of all failure modes and behaviors under a range of operating conditions. Model parameter identification also requires extensive experiments. In addition, a model-based method is often built case by case. Hence, it is not generally applicable to a different system without a significant amount of effort. One of the main advantages of the approach is that it is applicable even if the data are scarce, taking advantage of the knowledge gained.

Each method has its advantages and disadvantages, and the "right" model can sometimes be case specific. In many cases, neither a pure model-based approach nor a pure data-driven approach can provide the desired prognostic capability. In these situations, hybrid approaches can help. It is intuitive to use a hybrid approach to leverage their respective strengths and improve the RUL prediction. There are five types of hybrid approaches:

- Hybrid approach incorporating experience-based models and data-driven models.
- Hybrid approach incorporating experience-based models and physics-based models.
- Hybrid approach incorporating multiple data-driven models.
- Hybrid approach incorporating data-driven models and physics-based models.
- Hybrid approach incorporating experience-based models, data-driven models, and physics-based models.

For the last decade, the development of prognostics has been a popular research topic from both an academic and an operational point of view. The vast majority of research has focused on the prediction of the RUL of individual components. Moreover, much of this research has focused on the propagation of a single mechanism leading to a single failure mode. However, industrial equipment is complex and can have concurrent multi-failure modes and multi-failure mechanisms involving several components and sub-components. The propagation of failure mechanisms may also involve several components, and various diagnostic tools are used to detect and track them at different system scales. Once predetermined degradation thresholds are reached, specific maintenance actions should be taken to avoid a system failure. Depending on the types of active failure mechanisms and their progression, maintenance actions may not have the same effect to stop or slow down their propagation toward their related failure modes. It is therefore important to understand how the mechanisms have and will propagate when we want to apply specific maintenance tasks to extend the RUL of complex equipment.

One approach to predictive maintenance of complex equipment is physics of failure (PoF). A PoF model uses knowledge of a product's lifecycle loading and failure mechanisms to perform reliability assessment. It provides insight into life and reliability aspects and addresses the root causes of failure, such as fatigue, fracture, wear, or corrosion. PoF-based prognostics permit the assessment of system reliability under its actual application conditions. The method integrates sensor data with models that enable the *in situ* assessment of the deviation or degradation of a system from the expected normal operating condition and the prediction of its RUL.

Since the prediction of RUL is critical to operations and maintenance decision-making, it is imperative that it be estimated accurately. Prognostics deal with predicting future behavior. Several sources of uncertainty influence such future prediction; therefore, it is rarely feasible to obtain an estimate of the RUL with complete precision. In fact, it is not even meaningful to make such predictions without computing the uncertainty associated with RUL. Because we obviously can't really make "future measurements", predicting RUL necessarily entails propagating uncertainty.

Methods for quantifying uncertainty in prognostics and RUL prediction can be broadly classified as applicable to two types of situations: offline prognostics and online prognostics. Methods for offline prognostics are based on rigorous testing before and/or after operating an engineering system;

methods for online prognostics are based on monitoring the performance of the engineering system during operation.

In general, uncertainty management is the most significant challenge faced by monitoring systems in an era of ever-increasing system autonomy. As decision-making in complex engineered systems shifts toward highly evolved algorithms, simple threshold-based diagnostic decisions or single-valued model or data-driven prognostics are insufficient. The various sources of uncertainty inherent to diagnostics and prognostics must be accounted for in a probabilistic fashion for the approach to make sense. The Bayesian tracking framework allows the estimation of state of health parameters in prognostics, making use of available measurements from the system under consideration. It models uncertainty in the measurement process and in the evolution of degradation.

Context awareness is also desirable for prognostics and can be extended to support monitoring of multiple asset types. Context information in health monitoring can be used to improve the quality of prognostics, use limited maintenance resources more efficiently, and better match the maintenance services to the current asset conditions and needs of the machines under health monitoring.

For complex assets, a great deal of information must be captured and mined to assess the overall condition of the whole system. Various types of information must be integrated to get an accurate health assessment and determine the probability of a failure or slowdown. The data collected are not only huge but are often dispersed across independent systems that are difficult to access, fuse, and mine because of their disparate nature and granularity. If the data from these independent systems were combined, this new set of information could add value to the individual data sources through data mining.

System maintenance is obviously important to users of complex systems, with the potential to improve reliability, cost, and safety. To improve the maintenance capabilities of a complex system, an onboard health monitoring system must be deployed; it must provide enough informative data to determine which type of maintenance operation is required. Complex systems are an assemblage of heterogeneous components (continuous, discrete, and hybrid); each may require a different technique to monitor its health. A global health monitoring system would be useful but presents certain difficulties.

Fortunately, monitoring systems continue to increase in complexity and spatial coverage. Methods of digital signal processing, machine learning, knowledge representation, machine vision, and machine reasoning have been adopted in architectures and system implementations. In addition, modern integrated data communications systems include smartphones, allowing workers to communicate data immediately and permitting flexible production management for on-demand products.

Achieving high reliability and availability of complex systems is a crucial task. The goal is to detect and correct problems before they become severe and shut down the system. One recent technology that has the potential to transform the monitoring landscape is digital twin. Acting as a mirror of the real world, digital twin provides a means of simulating, predicting, and optimizing physical manufacturing systems and processes. Using digital twin, together with intelligent algorithms, organizations can achieve data-driven operation monitoring and optimization, develop innovative products and services, and diversify value creation and business models.

The explosion of Internet of Things (IoT) sensors makes digital twins possible. And as IoT devices are refined, digital twin scenarios can include smaller and less complex objects, giving additional benefits to companies. With additional software and data analytics, digital twins will optimize an IoT deployment for maximum efficiency and help designers decide where things should go or how they operate before they are physically deployed.

While the benefits that digital twins will bring to the built environment are clear, they can also play a role in achieving sustainability and solving other societal challenges in the design and engineering of civil infrastructure. The ultimate vision is the generalized development and adoption of a higher-level digital twin able to autonomously learn from and reason about its environments. The evolution of digital twin will help asset owners, managers, and larger society make informed decisions on the basis of real-time data. While we currently have widespread autonomy of warning systems, this will turn into widespread predictive systems, and finally into reasoning twins.

Not surprisingly, given the increasing complexity of assets, organizations also face increasingly complex problems. For example, there is a need to recognize the possibility of unlikely critical situations – this is not easy, but organizations have access to huge amounts of data. This vast knowledge should be managed systematically to identify and eliminate unpredictable events or reduce the undesirable consequences. Operations and maintenance may experience cases of extreme natural degradation and man-made accidental or malevolent intentional hazards in critical facilities; these can be handled by taking a risk-based approach, where risk is a function of the likelihood of event occurrence and the resulting consequences. However, a "black swan effect", an event that comes as a surprise, has a major effect, is often inappropriately rationalized after the fact, and is not foreseeable by the usual calculations of correlation, regression, standard deviation, reliability estimation, or prediction. In addition, expert opinion has minimal use, as experience is inevitably tainted by bias. The inability to estimate the likelihood of a black swan precludes the effective application of asset management and risk calculation, making the development of strategies to manage their consequences extremely important. In the not-so-distant future, we believe Industrial AI-empowered digitalization will be able to cope with the potential or actual consequences of these unforeseen, large-impact, and hard-to-predict events.

The increased complexity of systems calls for the development of integrated systems for fault detection, diagnostics, failure prognostics, maintenance planning, operation decision support, and decision-making as Industry 4.0 and smart manufacturing become a reality. Extracting useful knowledge from Big Data is one of the major challenges of smart manufacturing. In this sense, advanced data analytics is a crucial enabler of Industry 4.0. Advanced analytics focus on performance improvement in the context of smart manufacturing in the information age. To succeed, this evolution requires organizational commitment and work process adaptation on all levels, from the board of directors to the field operators.

Tomorrow's smart manufacturing will include the use of deep machine learning, prescriptive analytics in industrial plants, and analytics-based decision support in manufacturing operations. Advanced analytics will be able to answer an organization's questions about its past activities (descriptive analytics), probable future outcomes (predictive analytics/prognosis), and how to capitalize on short-, medium-, and long-term activities (prescriptive analytics).

Prescriptive analytics is an integral part of advanced analytics. It goes beyond predictive analytics (i.e., prognostics) to include predictions to estimate the effect of possible actions. Prescriptive analytics aim at going beyond the question "What just happened and why?" to answer the questions "What should I do?" and "Why should I do it?" These analytics yield business value through adaptive, time-dependent, and optimal decisions on the basis of accurate predictions about future events. Prescriptive analytics are part of smart manufacturing and are considered to be the next evolutionary step in increasing data analytics maturity for optimized decision-making ahead of time.

The most important challenges of prescriptive analytics include:

1. Addressing the uncertainty introduced by predictions, incomplete and noisy data, and subjectivity in human judgment.
2. Combining the knowledge acquired by machine learning and data mining methods with the knowledge of domain experts.
3. Developing generic prescriptive analytics methods and algorithms utilizing AI and machine learning instead of problem-specific optimization models.
4. Incorporating adaptation mechanisms capable of processing data and human feedback to continuously improve decision-making processes over time and generate non-intrusive prescriptions.
5. Recommending optimal plans from a list of alternative actions.

Although automated decisions with the use of prescriptive analytics may potentially yield more extensive benefits than other advanced analytics, this possibility has only recently become of significant interest. As the method becomes ubiquitous, it will undoubtedly lead to changes in the workplace. It may lead to the replacement or extinction of certain work positions, for example, teams of decision-makers searching for feasible solutions using trial-and-error procedures.

Advanced analytics have the potential to reduce costs, increase profit, and improve customer services. For this, a mixed model-based and data-driven architecture for solving complex problems in the industry must be combined with comprehensive information and communication technologies, high-performance computing, and auxiliary mechatronics. The major challenge is to find and integrate good quality data.

The massive volume, variety, and velocity of data are changing work processes. The timeliness, consistency, and, above all, the integration of these Big Data are key considerations of advanced analytics.

Good data are the smart manufacturing basis for better data reconciliation and better parameter estimation to be used for real-time optimization and control. If a company's data are out-of-date or non-integrated, there will be inconsistencies in both offline and online applications, and the company will not benefit from the value of advanced analytics. The bottom line is the data: well-polished and timely data seamlessly integrated into the decision-making body are the foundation of smart manufacturing in the Industry 4.0 context and the secret to future success.

Authors

 **Diego Galar** is Full Professor of Condition Monitoring in the Division of Operation and Maintenance Engineering at Luleå University of Technology (LTU), where he is coordinating several H2020 projects related to different aspects of cyber-physical systems, Industry 4.0, IoT or Industrial AI, and Big Data. He was involved in the SKF UTC located in Luleå focused on SMART bearings and actively involved in national projects with Swedish industry or funded by Swedish national agencies like Vinnova.

He is also principal researcher in Tecnalia (Spain), heading the Maintenance and Reliability research group within the Division of Industry and Transport.

He has authored more than 500 journal and conference papers, books, and technical reports in the field of maintenance, working also as member of editorial boards and scientific committees, chaired international journals and conferences, and actively participated in national and international committees for standardization and R&D on the topics of reliability and maintenance.

In the international arena, he has been visiting Professor at the Polytechnic of Braganza (Portugal), University of Valencia and NIU (USA), and the Universidad Pontificia Católica de Chile. Currently, he is visiting professor at the University of Sunderland (UK), University of Maryland (USA), and Chongqing University in China.

Kai Goebel is a Principal Scientist in the System Sciences Lab at the Palo Alto Research Center (PARC). His interests are broadly in condition-based maintenance and system health management for a broad spectrum of cyber-physical systems in the transportation, energy, aerospace, defense, and manufacturing sectors. Prior to joining PARC, He worked at the NASA Ames Research Center, and General Electric Corporate Research and Development Center. At NASA, he was a branch chief leading the Discovery and Systems Health Tech area which included groups for machine learning, quantum computing, physics modeling, and diagnostics and prognostics. He founded and directed the Prognostics Center of Excellence, which advanced our understanding of the fundamental aspects of prognostics. He holds 18 patents and has published more than 350 papers, including a book on prognostics. He was an adjunct professor at Rensselaer Polytechnic Institute and is now adjunct professor at Luleå Technical University. He is a co-founder of the Prognostics and Health Management (PHM) Society, and associate editor of the International Journal of PHM.

Peter Sandborn is a Professor in the CALCE Electronic Products and Systems Center and the Director of the Maryland Technology Enterprise Institute at the University of Maryland. His group develops lifecycle cost models and business case support for long field life systems. This work includes obsolescence forecasting algorithms, strategic design refresh planning, lifetime buy quantity optimization, return on investment models for maintenance planning and system health management, and outcome-based contract design and optimization. He is the developer of the MOCA refresh planning tool. He is an Associate Editor of the IEEE Transactions on Electronics Packaging Manufacturing and a member of the Board of Directors of the PHM Society. He is the author of over 200 technical publications and several books on electronic packaging and electronic systems cost analysis. He was the winner of the 2004 SOLE Proceedings, the 2006 Eugene L. Grant, the 2017 ASME Kos Ishii-Toshiba, and the 2018 Jacques S. Gansler awards. He has a B.S. in engineering physics from the University of Colorado, Boulder, in 1982, and an M.S. in electrical science and a Ph.D. in electrical engineering, both from the University of Michigan, Ann Arbor, in 1983 and 1987, respectively. He is a Fellow of the IEEE, the ASME, and the PHM Society.

Uday Kumar is the Chair Professor of Operation and Maintenance Engineering, Director of Research and Innovation (Sustainable Transport) at Luleå University of Technology and Director of Luleå Railway Research Center.

His teaching, research, and consulting interests are equipment maintenance, reliability and maintainability analysis, product support, lifecycle costing (LCC), risk analysis, system analysis, eMaintenance, and asset management.

He is visiting faculty at the Center of Intelligent Maintenance System (IMS), a center sponsored by the National Science Foundation, Cincinnati, USA since 2011, External Examiner and Program Reviewer for the Reliability and Asset Management Program of the University of Manchester, Distinguished Visiting Professor at Tsinghua University Beijing, and Honorary Professor at Beijing Jiaotong University, Beijing. Earlier, he was visiting faculty at Imperial College London, Helsinki University of Technology, Helsinki, and University of Stavanger, Norway.

He has more than 30 years' experience in consulting and finding solutions to industrial problems directly or indirectly related to maintenance of engineering asserts. He has published more than 300 papers in international journals and conference proceedings dealing with various aspects of maintenance of engineering systems and has co-authored four books on maintenance engineering and contributed to World Encyclopaedia on Risk Management.

He is an elected member of the Royal Swedish Academy of Engineering Sciences.

1 Information in Maintenance

1.1 TRADITIONAL MAINTENANCE: CORRECTIVE AND PREVENTIVE

1.1.1 TRADITIONAL CORRECTIVE MAINTENANCE

Traditionally, the main tool of maintenance was the reactive response to equipment failures; that is, repairing damage as quickly as possible. A good traditional maintenance plan included the identification of critical parts of the machinery, a stock of spares, and the availability of specialized personnel to solve breakdowns efficiently.

For example, if an engine suffered an irreparable breakdown, it was replaced by a new engine (Sigma Industrial Precision, 2018).

1.1.1.1 Corrective Maintenance

Corrective maintenance (also called breakdown maintenance) comprises maintenance tasks to rectify and repair faulty systems and equipment (UpKeep, 2020(a)).

1.1.1.1.1 Corrective Maintenance Workflow

Corrective maintenance is initiated when an additional problem is discovered during a separate work order, as shown in Figure 1.1. For example, during an emergency repair, as part of a routine inspection, or in the process of conducting preventive maintenance (PM), a technician might spot another issue that needs correction before other problems occur. When this problem is noticed, corrective maintenance is planned and scheduled for a future time. During the execution of corrective maintenance work, the asset is repaired, restored, or replaced (UpKeep, 2020(a)).

1.1.1.1.2 Types of Corrective Maintenance

Corrective maintenance can be planned or unplanned (Fiix, 2020). Corrective maintenance is part of planned maintenance when:

1. A run-to-failure (RTF) maintenance strategy is used. In RTF, equipment is allowed to run until it breaks down; at that point, it is repaired or replaced. This type of corrective maintenance should be used on non-critical assets that are easily and cheaply repaired or replaced.
2. A PM or condition-based monitoring maintenance strategy is used. Preventive and condition-based maintenance (CBM) try to identify problems early in the process before failure occurs. If a problem is found, corrective maintenance is scheduled.

Corrective maintenance is unplanned when:

1. A PM schedule is in place, but equipment breaks down between scheduled maintenance periods.
2. Equipment shows signs of failure or fails unexpectedly.

Corrective maintenance may be grouped under the following five categories (Dhillon, 2006):

1. Fail repair: Restoring the failed item or equipment to its operational state.
2. Overhaul: Repairing or restoring an item or equipment to its complete serviceable state, meeting requirements outlined in maintenance service ability standards.

DOI: 10.1201/9781003097242-1

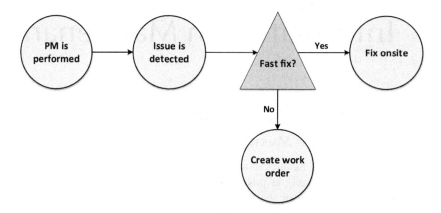

FIGURE 1.1   Corrective maintenance workflow (UpKeep, 2020(a)).

3. Salvage: Disposing of non-repairable materials and using salvaged materials from items that cannot be repaired in the overhaul, repair, or rebuild programs.
4. Servicing: Performing maintenance required after a corrective maintenance action; for example, engine repair can result in a requirement for crankcase refill, welding, and so on.
5. Rebuild: Restoring an item or equipment to a standard as close as possible to its original state in appearance, performance, and life expectancy. This is accomplished through actions such as complete disassembly, examination of all parts, replacement or repair of unserviceable or worn components according to original specifications and manufacturing tolerances, and reassembly and testing to original production requirements.

1.1.1.1.3 Corrective Maintenance Steps, Downtime Components, and Time-Reduction Strategies at System Level

Corrective maintenance comprises the following five steps (Dhillon, 2006):

- Failure recognition: Recognizing the existence of a failure.
- Failure localization: Localizing the failure within the system to a specific piece of equipment item.
- Diagnosis within the equipment or item: Identifying the specific failed part or component.
- Failed part replacement or repair: Replacing or repairing failed parts or components.
- Return system to service: Ending maintenance and returning the system back to service.

Corrective maintenance downtime is made up of three major components as shown in Figure 1.2. Active repair time has six sub-components: checkout time, preparation time, fault correction time, fault location time, adjustment and calibration time, and spare part obtainment time (Dhillon, 2006).

To improve corrective maintenance effectiveness, it is important to reduce corrective maintenance time. Useful strategies for reducing system-level corrective maintenance time include the following (Dhillon, 2006):

1. Improve accessibility: Past experience indicates a significant amount of time is spent accessing failed parts. Careful attention to accessibility during design can help to lower the accessibility time of parts and, consequently, reduce the corrective maintenance time.
2. Improve interchangeability: Effective functional and physical interchangeability is important when removing and replacing parts or components, as it lowers corrective maintenance time.

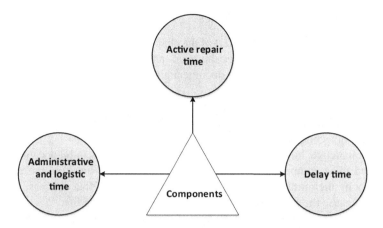

FIGURE 1.2 Major corrective maintenance downtime components (Dhillon, 2006).

3. Improve fault recognition, location, and isolation: Experience shows that within a corrective maintenance activity, fault recognition, location, and isolation consume the most time. Factors reducing corrective maintenance time are good maintenance procedures, well-trained maintenance personnel, well-designed fault indicators, and unambiguous fault isolation capability.

4. Consider human factors: During design, careful attention should be paid to human factors, such as selection and placement of indicators and dials; size, shape, and weight of components; readability of instructions; information processing aids; and size and placement of access and gates. This can lower corrective maintenance time significantly.

5. Employ redundancy: Redundant parts or components should be designed to be switched in during the repair of faulty parts so that the equipment or system continues to operate. In this case, although the overall maintenance workload may not be reduced, the downtime of the equipment could be impacted significantly.

1.1.1.1.4   How Corrective Maintenance Decreases Downtime

Corrective maintenance is something that gets caught just in time. Within the maintenance arena, corrective maintenance is triggered when a technician sees something that is about to break or will affect the overall performance of a piece of equipment. It can still be repaired or restored without incurring downtime.

If corrective maintenance is not scheduled, the problem may become an emergency maintenance work order down the road and result in halted production lines, interruption in service, or unhappy customers (UpKeep, 2020(a)).

1.1.1.1.5   Benefits of Corrective Maintenance

Since corrective maintenance is performed "just in time", the main benefits are reduced emergency maintenance orders and increased employee safety. Corrective maintenance work orders are scheduled and prioritized in a computerized maintenance management system (CMMS). This helps maintenance teams resolve problems before delays in production or service interruptions occur.

Corrective maintenance, coupled with good PM, helps a business extend the lifetime of its assets, reduce employee injury, and optimize resource planning. Corrective maintenance work orders are often less expensive to implement than emergency maintenance work orders, as the latter may need to be completed during overtime hours (UpKeep, 2020(a)).

1.1.2   Traditional Preventive Maintenance

Traditional maintenance mentioned at the start of the chapter has evolved to become the well-known PM. The knowledge of the machinery over the years and the historical behavior that both companies and manufacturers have been able to document allow us to estimate the useful life of certain machines.

Accordingly, it is possible to plan maintenance actions to replace parts or mechanisms that are likely to fail in the short term. Since the cost of replacing a particular piece tends to be more economical than repairing a complete machine, PM is cost-effective.

PM has a clear challenge, however. A number of different variables influence the wear of an item or mechanism, and most companies invest in PM without really exhausting the entire life of the item, going simply by the numbers. The consequence, of course, is that it is assuming a cost that could be avoided because the item has not reached the end of its useful life.

Consider the following example: Given the load to which it is subjected, its useful life, and other factors, it is anticipated that a certain motor will fail at 10 years. Therefore, at age 9, it is replaced. The intention is to avoid a hypothetical future flaw. However, this engine may be able to operate for 12 years, but we will never know (Sigma Industrial Precision, 2018).

1.1.2.1   Preventive Maintenance

PM refers to replacing or overhauling assets or equipment at fixed intervals, regardless of its condition at the time. These long-term maintenance policies do not consider the present equipment state. Scheduled restoration and scheduled discard are examples of PM tasks.

PM relies on historical data. Failures are tracked and recorded in a database, and these data suggest general preventive actions and time lines. A PM plan can be based on the age of equipment. It involves replacing or repairing equipment at age T or failure, whichever is first. Commonly used equipment reliability indices such as mean time between failure (MTBF) and mean time to repair (MTTR) are extracted from the historical data on equipment behavior. These measures give a rough estimate of the time between failures and the time required to restore a system after breakdown. Although processes of degradation and the causes of failure can vary, even for the same type of equipment, MTBF and MTTR are still informative. Other useful indices include the mean lifetime, mean time to first failure, and mean operational life.

Another PM policy focuses on periodic maintenance. In this policy, equipment is repaired or replaced at fixed time intervals independent of previous equipment failures.

In sum, PM is time-based. It does not consider the present health state of the asset. It can easily happen that equipment in good operating condition is removed for overhaul, thus wasting both time and money (Lee & Wang, 2008).

1.1.2.1.1   Types of Preventive Maintenance

Any maintenance that is not reactive maintenance is PM. There are many different types of PM that require different types of technology and expertise (UpKeep, 2020(a)).

Four common types of PM include (UpKeep, 2020(a)):

1. Calendar-based maintenance:

A recurring work order is scheduled in the CMMS when a specified time interval is reached.

2. Predictive maintenance (PdM):

When work order data are logged in the CMMS, maintenance managers can predict when an asset will fail based on historical events and create specific PMs to prevent this from happening again.

3. Usage-based maintenance:

Meter readings are used and logged in the CMMS. When a specific unit is reached, a work order is created for routine maintenance.

4. Prescriptive maintenance:

This is similar to PdM, but in this case, machine learning software assists the maintenance manager prescribing PM.

1.1.2.1.2 Preventive Maintenance Components and Principles for Choosing Items for Preventive Maintenance

There are seven elements of PM (Dhillon, 2006):

- Inspection: Periodically inspecting items to determine their serviceability by comparing their physical, mechanical, electrical, and other characteristics to established standards.
- Calibration: Detecting and adjusting any discrepancy in the accuracy of the material or parameter being compared to the established standard value.
- Testing: Periodically testing to determine serviceability and detect mechanical or electrical degradation.
- Adjustment: Periodically making adjustments to specified elements to achieve optimum performance.
- Servicing: Periodically lubricating, charging, and cleaning materials or items to prevent incipient failures.
- Installation: Periodically replacing limited-life items or items experiencing time cycle or wear degradation to maintain the specified tolerance level.
- Alignment: Making changes to an item's specified elements to achieve optimum performance

The following formula can used to decide whether to implement a PM program for an item or system (Dhillon, 2006):

$$(n)(C_a)(\theta) > C_{pm} \tag{1}$$

where n is the total number of breakdowns, θ is 70% of the total cost of breakdowns, C_a is the average cost per breakdown, and C_{pm} is the total cost of the PM system.

1.1.2.1.3 Steps for Developing a Preventive Maintenance Program

Development of an effective PM program requires the availability of such items as test instruments and tools, accurate historical records of equipment, skilled personnel, service manuals, manufacturer's recommendations, past data from similar equipment, and management support and user cooperation (Dhillon, 2006).

A highly effective PM program can be developed in a short time by following the steps listed below (Dhillon, 2006):

- Identify and select areas: Identify and select one or two important areas on which to concentrate the initial PM effort. The main objective of this step is to obtain good results in highly visible areas.
- Highlight PM requirements: Define the PM needs and develop a schedule for two types of tasks: daily PM inspections and periodic PM assignments.

- Determine assignment frequency: Establish the frequency of assignments and review the item or equipment records and conditions. The frequency depends on factors such as vendor recommendations, the experience of personnel familiar with the equipment or item under consideration, and engineers' recommendations.
- Prepare PM assignments: Prepare the daily and periodic assignments in an effective manner and get them approved.
- Schedule PM assignments: Schedule the defined PM assignments over a 12-month period.
- Expand the PM program as appropriate: Expand the PM program to other areas on the basis of experience gained from the pilot PM projects.

1.1.2.1.4   How Preventive Maintenance Decreases Downtime

Oil changes and regular servicing are part of a PM schedule that ensures a car runs properly and without unexpected failure. If people ignore the maintenance schedule and miss service intervals, their cars will depreciate in value and utility. The same goes for machinery and equipment.

With a PM schedule in place, maintenance managers can decrease downtime. This schedule is usually automated with a CMMS that comes with PM scheduling software. However, managers are always cautious of over-maintaining assets. There's a point where PM starts costing too much, as shown in Figure 1.3, in relation to the amount of downtime it prevents (UpKeep, 2020(b)).

1.1.2.1.5   Preventive Maintenance Optimization: The Right Approach

In optimal PM, the maintenance costs will decrease at the same time as the costs of production losses decrease. Optimal PM has the following traits (IDCOM Inc., 2020):

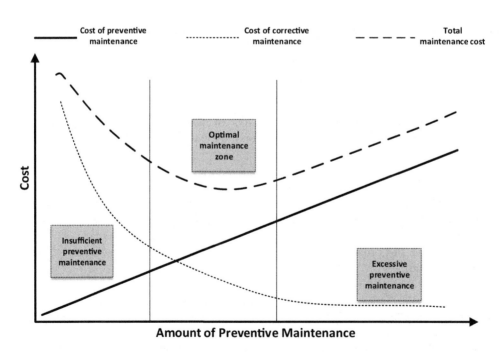

FIGURE 1.3   Optimal amount of preventive maintenance occurs when the cost of corrective and preventive maintenance meet (UpKeep, 2020(b)).

- Perform PM only when the cost of maintenance is less than the cost of the failure. A consequence of failure analysis can be used to determine this.
- Perform PM based on the condition (i.e., condition monitoring (CM)) of the equipment; however, RTF is acceptable when it is the most cost-effective maintenance procedure.
- Develop and document a standard corrective maintenance procedure.

1.2 CMMS AND IT SYSTEMS SUPPORTING MAINTENANCE FUNCTION

1.2.1 COMPUTER MAINTENANCE MANAGEMENT SYSTEMS (CMMS)

Many organizations use specialized software, a CMMS, to support management and to track operations and maintenance (O&M) activities.

1.2.1.1 CMMS Needs Assessment

To determine the need for a CMMS, managers should assess their current mode of operation. Key questions to ask include the following:

- Do you have an effective way to generate and track work orders? How do you verify the work was done efficiently and correctly? What is the notification function upon completion?
- Are you able to access historical information on the last time a system was serviced, by whom, and for what condition?
- How are your spare parts inventories managed and controlled? Do you have excess inventory or are you consistently waiting for parts to arrive?
- Do you have an organized system to store documents (electronically) related to O&M procedures, equipment manuals, and warranty information?
- When service staff are in the field, what assurances do you have that they are compliant with health and safety issues and are using the right tools/equipment for the task?
- How are your assets, that is, equipment and systems, tracked for reporting and planning?

If the answers to these questions are not well defined or even entirely lacking, it may be worth investigating the benefits of a well implemented CMMS (Sullivan et al., 2010).

1.2.1.2 CMMS Capabilities

CMMS automate most of the logistical functions performed by maintenance staff and management. They come with many options and have numerous advantages over manual maintenance tracking systems. Depending on the complexity of the system, typical CMMS functions include the following:

- Work order generation, prioritization, and tracking by equipment/component.
- Historical tracking of all work orders generated, sorted by equipment, date, person responding, and so on.
- Tracking of scheduled and unscheduled maintenance activities.
- Storing of maintenance procedures, as well as all warranty information, by component.
- Storing of all technical documentation or procedures by component.
- Availability of real-time reports of ongoing work activity.
- Generation of calendar or run-time-based PM work orders.
- Capital and labor cost tracking by component, as well as shortest, median, and longest times to close a work order by component.
- Complete parts and materials inventory control with automated reorder capability.
- Digital interfaces to streamline input and work order generation.
- Outside service calls/dispatch capabilities.

Many CMMS can now interface with existing energy management and control systems, as well as enterprise resource planning or even supervisory control and data acquisition (SCADA) systems for interaction from the managerial to the shop-floor level in maintenance decisions. Coupling these capabilities allows condition-based monitoring and the creation of component use profiles (Sullivan et al., 2010).

While CMMS can go a long way toward automating and improving the efficiency of most maintenance programs, there are some common pitfalls (Sullivan et al., 2010). These include the following:

- Improper selection of a CMMS: This is a site-specific decision. Time should be taken to evaluate needs and look for the proper match.
- Inadequate training of staff on proper use of the CMMS: Staff need dedicated training on input, function, and maintenance of the CMMS.
- Lack of commitment to properly implement the CMMS: A commitment needs to be in place for the start-up/implementation of the CMMS.
- Lack of commitment to persist in CMMS use and integration: A CMMS provides significant advantages, but it needs to be maintained.

CMMS is an information repository whose outcomes strongly depend on the quality of the data stored in the system. The data in the CMMS could be wrong, or there may be no data at all. If managers do not have the right kind of data or enough of it, they must make decisions without all the facts. The wrong decision could be costly.

To maximize CMMS use, managers need to continually revisit the system's features and functions, review and upgrade training for users, and reinforce among users the need for a steady stream of reliable maintenance and repair data (Hounsell, 2008).

1.2.1.3   CMMS Benefits

One of the greatest benefits of a CMMS is the elimination of paperwork and manual tracking activities, permitting maintenance staff to become more productive. In addition, the maintenance cost can be tracked, so the tool can be used as a cost control. A CMMS should be able to collect and store information in an easily retrievable format so the manager has access to the information required to make optimal maintenance decisions (Sullivan et al., 2010).

Benefits of implementing a CMMS include the following:

- Can detect problems before failure.
- Can plan optimal maintenance activities to use staff more efficiently.
- Can improve inventory control by better forecasting the need for spares, thus eliminating shortages while also minimizing existing inventory.
- Can ensure optimal equipment performance, thus reducing downtime and increasing equipment life.
- Can keep a detailed record of maintenance costs along with investments. This is useful at the business level, for example, to stay within budget constraints and predict future expenses. Although a CMMS primarily assists managers by providing information, some can perform lifecycle cost (LCC) calculations for asset renovation, overhaul, and replacement.

1.2.1.4   Finding a CMMS

The Internet is a great resource to find CMMS vendors. A simple search under "CMMS" provides links to the many vendors and the resources they offer.

1.2.1.5   Role of CMMS

CMMS have evolved over the last three decades, starting as elementary asset trackers with some PM functionality to becoming maintenance information systems. Hundreds of vendors provide solutions on a variety of platforms. State-of-the-art systems give users the comprehensive ability to facilitate the flow of maintenance information and to check the health of the maintenance organization at a glance.

As the maintenance, repair, and operations software market continues to expand, vendors have developed solutions that focus on specific segments of asset and work management. Systems described as enterprise asset management, asset life cycle management, asset performance management, asset or enterprise reliability management, and CM are all focused on achieving the same goals: increasing equipment availability and performance, increasing product quality, and reducing maintenance expense. When a CMMS is implemented to facilitate an established process and standards, those goals can be realized (Crain, 2003).

1.2.1.6 CMMS Implementation

Once the requirements have been established and a system selected, the implementation process begins. Unfortunately, it is not as easy as flicking a switch. To get the CMMS to provide useful information, it must be configured, key data collected and entered, and employees motivated to use the system.

The team who identified the business processes and requirements is a good choice for the initial configuration. The best first step is to have them participate in an extensive overview of the selected application in consultation with the vendor or a consultant familiar with the selected CMMS. During the overview, the team needs to ensure the system meets all requirements. The team should also identify all the components of the system that are required to achieve the planned level of functionality (e.g., labor, value list, cost/general ledger codes, assets, bins, and inventory).

As the data collection begins, the system needs to be "sold" to end users:

- Engage users in the process of collecting and "owning" the information entered into the system.
- Take advantage of assemblies like safety or tailgate meetings to provide updates on the progress of the implementation.
- Explain how the system will benefit maintenance users, management, and the company as a whole.
- Ask senior management to address the employees on the significance of the system and the importance of their contributions.
- Promote question and answer sessions, possibly even short product demonstrations.
- Emphasize that the goal of the system is not to monitor employees, but to move the organization from a reactive to a proactive mode (Crain, 2003).

1.2.1.6.1 Consulting Services

Many vendors and consultants provide maintenance management system implementation support. Some focus on the business side of the system, while others are systems integrators with an information technology focus. Before hiring a consultant, companies should ask these questions:

1. Can staff remove themselves enough from the existing way of business to evaluate ways to make the old processes more efficient?
2. Does someone on the maintenance staff have enterprise software implementation experience? If not, does the project schedule allow for the learning curve?
3. Do maintenance and information technology (IT) staff members have the time to add this project to their existing workload?

If the answer is no to any of these questions, or if the system needs to be fully functional in 6–12 months, a consultant may be helpful. As well as contributing years of lessons learned to the implementation team, a seasoned consultant will help develop a project plan and guide teams through the implementation process.

For a turnkey type implementation, it is critical to evaluate and approve key deliverables, such as business processes, nomenclature, customizations, data configuration, and the final system configuration. Regardless of the consultant's role, during testing, training, and going live, the existing resources should visibly lead the effort (Crain, 2003).

1.2.1.6.2   System Configuration

The collection of data, data migration, system configuration, and testing comprise the bulk of the work that goes into a CMMS implementation. The data entered in the system must follow consistent naming standards. The identification numbers should be recognizable. If a new naming convention is implemented, label the assets in the field and provide hierarchy or plot plan diagrams at maintenance workstations. The data should be as complete and accurate as possible. Incorrect or inconsistent data represent the quickest road to frustration for maintenance users.

The system should be configured to provide the most intuitive user interface available. Work request and work order entry screens should take advantage of as many "value lists" as possible. Use default values if supported by the system. If the CMMS supports hiding unused fields, take advantage of that feature as well. To ease future upgrades, keep customizations to the application and custom database triggers to a minimum. When customizations can't be avoided, make sure they are well documented.

The system should be well tested before training. To this end, develop test scripts that model the work process and recruit computer savvy end-users to participate in testing. They will also be able perform data quality spot checks. Do not start training until you are confident the system will perform "as advertised" (Crain, 2003).

1.2.1.6.3   Training

Training is where the "rubber meets the road". End users are getting their first hands-on experience with the system and developing first impressions. "Canned" data should never be used for training. Instead, the training environment should mirror the production database. The format of training should be role and process based. A work requestor is going to be overly confused if he/she is trained on all the functionalities of a work order instead of on the one screen and few fields required to create a work request. The following techniques are also useful:

- Make sure that one or two implementation team members are available to help users who fall behind during training.
- Follow up with those users during the "go-live" period and provide auxiliary training.
- Provide a "play" environment where users can practice and reinforce what they learned in training without corrupting production data (Crain, 2003).

1.2.1.6.4   "Go Live"

"Go live" should be scheduled with plenty of support for the users and lower than average work order volume. Have a back-up plan to manage the flow of information in case the system goes down unexpectedly. If users have to fall back to the paper system for a few hours, have them create work orders in the CMMS at the end of their shift. Have quick reference or "how to" sheets available, as well as diagrams, to help them enter the proper values into the system.

At the end of every day, review what went right and where changes are required. Reviewing the work orders daily is an excellent way of seeing how successfully each maintenance user is interacting with the system, or who needs extra help. Update and re-post the work processes the following day. Track and prioritize configuration issues and system "bugs". Provide a daily update to vendors or consultants, if applicable (Crain, 2003).

1.2.1.6.5 *Post-Implementation*

Following "go live" but before the transition to normal operation, review all the defined requirements and determine whether they have been fulfilled. Ensure that the mechanisms to facilitate and measure business processes are in place. If a vendor or consultant has been used, make sure all his/her deliverables have been completed.

Develop a support and issues escalation path. Have users go to their supervisor or resident "power user" with issues. If the issue can't be resolved, escalate it to a site or company administrator. The company administrator should be the sole point of contact between the organization and the vendor. The administrator should be responsible for logging and tracking all open issues and enhancement requests (Crain, 2003).

1.2.2 WHAT IS IT SYSTEM SUPPORT AND MAINTENANCE?

With the digital era on the rise, the strength of a company's IT systems may make or break it. Organizations dominating the market have a limited user downtime and a high rate of staff productivity; therefore, many invest in IT system support and maintenance.

Managing the company's IT assets may be difficult, with frequent problems and fixes. As the repair process takes place, there will be a diversion from primary tasks and priorities. This damages the flow of the operations and affects the quality of products and services. IT systems enable companies to control and monitor different business processes. Updating and maintaining IT assets avoids service disruption, thereby strengthening the quality, reliability, and consistency of the operations without compromising the needs of the customers. A smooth and efficient IT system can produce a quicker turnaround service.

Business operations involve various processes. It is impossible to know the improvements necessary to speed them up if the company fails to check and review the systems it uses on a daily basis. Anchored processes slow things down and lessen productivity. It is necessary for every establishment to have a quality and up-to-date IT system that can deliver accurate and relevant information from managers to employees and vice versa. A good IT system plays a huge role in generating information, communicating instructions, identifying problems, and making decisions. Designing a computerized management information system for maintenance aids in routine decisions for maximum uptime and restoration (Stīpnieks, 2018).

1.2.2.1 Importance of Quality IT Assets and Maintenance

Quality IT systems aid business operations by decreasing downtime, maximizing resources, lowering costs for maintenance, and maintaining or increasing key performance indicators at a certain level. A good IT system performs many functions: it collects, manipulates, and distributes data the business needs to meet its clients' expectations. Maintaining IT systems ensures the quality and functionality of the systems used in the organization. The following must be considered when planning IT maintenance (Stīpnieks, 2018):

- Whether it involves modification of IT peripherals, correcting faults, or improving attributes, the maintenance of systems plays a huge role in the technical and managerial aspects of the company.
- Solving a company's IT issues involves staffing, reengineering, estimating costs, and prioritizing customers' needs.
- Software may smooth the flow of information by correcting errors and optimizing entries, thus paving the way to modify operations and evaluate mechanisms. It also lengthens the lifespan and usability of the programs that accompany it.

- System updates allow companies to conduct the necessary changes to stay at the top of their game – not only technological aspects but also legal ones, including the privacy and security of users, as business patterns and government rules may change.

1.3 SENSORS FOR HEALTH MONITORING AND SCADA SYSTEMS: OPERATIONAL TECHNOLOGIES (OTS)

1.3.1 HEALTH MONITORING

Health monitoring is a non-destructive technique that allows the integrity of systems or structures to be actively monitored during operation and/or throughout their lives to prevent failure and reduce maintenance costs. Health monitoring tracks an aspect of a structure's health using reliably measured data and analytical simulations in conjunction with heuristic experience so that the current and expected future performance of the composite part, for at least the most critical limit events, can be described in a proactive manner. Minimum standards are required for analytical modeling to ensure reliable computer simulations, and measurements, loads, and tests must be designed and implemented in conjunction with the analytical simulations (Koncar, 2019).

Health monitoring systems were introduced in the 1970s in such applications as computer integrated coal monitoring systems, a production monitoring system for the weaving industry, the MINOS long-distance monitoring system, a wireless interceptor with channel scan and a priority channel monitoring system, an electric network in-time monitoring system, and a diesel engine CM system.

Current health monitoring technologies include intelligent monitoring, wireless monitoring, long-distance monitoring, real-time monitoring, and embedded monitoring (Xu & Xu, 2017), both online monitoring techniques and off-line non-destructive detection and inspection techniques. In general, online monitoring keeps track of the key parameters of the products or systems. For example, in an aero engine, the vibration parameters, high and low compression rotation speeds, turbine exhaust temperatures, fuel flux, and metal content in the oil need to be monitored, and baseline methods are commonly used to determine the threshold values. Off-line non-destructive abnormal detection or inspection techniques include hole detection, fluoroscopy, X-ray isotopes, borescope with remote control, ultrasonic detection, oil physics/chemical analysis, and oil debris analysis techniques such as iron spectrum, scanning electron microscopy energy spectrum, automated granule counts, and spectroscopic analysis.

1.3.1.1 Wireless Standards for Health Monitoring

The data collected from biosensors can be transferred to a system's central node, directly to a distant medical station, or directly to the cell phone of a physician using wireless health monitoring systems. Data can be transmitted through a wired medium in which the movement and comfort of the user are hindered or through a wireless medium. In wireless transmission, a collection of sensor nodes known as a body area network (BAN) is formed to facilitate the flow of data to a central node, which can be a smartphone, a microcontroller-based device, or a personal digital assistant (PDA).

The network should be cost effective, allow for extra nodes, consume less power, have a flexible configuration, and so on. The most widely preferred networks for BANs are ZigBee (IEEE 802.15.4) and Bluetooth (IEEE 802.15.1). ZigBee is usually preferred in applications which require security in networking and long battery life. If data are transmitted long distances, a mesh network of intermediary devices can be used. Bluetooth is commonly used for transmission of data at short range. Problems of synchronization are solved by Bluetooth, as it can connect several devices (Bhelkar & Shedge, 2016).

1.3.1.2 Sensors Facilitate Health Monitoring

Sensors are used in electronics-based medical equipment to convert various forms of stimuli into electrical signals for analysis. Sensors can increase the intelligence of medical equipment, such as

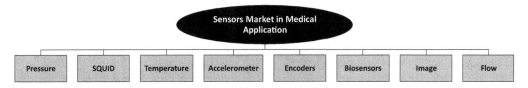

FIGURE 1.4 Product types in the medical market (Thusu, 2011).

life-supporting implants, and can enable bedside and remote monitoring of vital signs and other health factors. New and different types of medical equipment being developed include sensors used inside both equipment and patients' bodies. Healthcare organizations want real-time, reliable, and accurate diagnostic results provided by devices that can be monitored remotely, whether the patient is in a hospital, clinic, or at home.

Figure 1.4 shows pressure, temperature, flow and image sensors; accelerometers; biosensors; superconducting quantum interference devices (SQUIDs); and encoders used in medical applications (Thusu, 2011). These are discussed in the following sections.

1.3.1.3 Different Types of Sensors

Physiological parameters should be monitored with precision over longer periods of time. Sensors are the core of these types of monitoring systems. Different types of sensors are available for human activities. Progress in the fields of micromechanics, microelectronics, and other similar technologies has enabled the development of acute sensors with faster and more efficient measurement of data, along with lower consumption of power.

Body temperature is one of the vital signs for assessment of a patient's health. Fluctuations in body temperature occur with infection, inflammation, heart attack, shock, and so on. Thermometers were traditionally used to measure body temperature but they are being replaced by temperature sensors and other electrical and electronic methods.

When there is a need for measurement of acceleration over a certain axis within a specific range of frequencies, accelerometers are used. They are also used to detect falls in health monitoring.

Another important physiological parameter is heart rate. Fluctuations in heart rate can be caused by illness, injury, exercise, and so on. Monitoring heart rate can help identify health-related issues. Chronic heart problems are a cause of concern and may require monitoring. Electrocardiogram (ECG) sensors give information on the consistency and rate of the heart beat and can be used to detect cardiovascular diseases.

Table 1.1 shows some biosensors and the signals obtained from them (Bhelkar & Shedge, 2016).

1.3.1.4 Key Sensors and Applications

Pressure sensors are used in anesthesia delivery machines, oxygen concentrators, sleep apnea machines, ventilators, kidney dialysis machines, infusion and insulin pumps, blood analyzers, respiratory monitoring and blood pressure monitoring equipment, hospital beds, surgical fluid management systems, and pressure-operated dental instruments.

Temperature sensors are used in anesthesia delivery machines, sleep apnea machines, ventilators, kidney dialysis machines, blood analyzers, medical incubators, humidified oxygen heater temperature monitoring and control equipment, neonatal intensive care units to monitor patient temperature, digital thermometers, and organ transplant system temperature monitoring and control.

Applications for flow sensors include anesthesia delivery machines, oxygen concentrators, sleep apnea machines, ventilators, respiratory monitoring, gas mixing, and electro-surgery, in which high-frequency electric current is applied to tissue to cut, cause coagulation, desiccation, or destroy tissue such as tumors.

TABLE 1.1
Biosensors and Their Corresponding Biosignals

Type of Sensor	Type of Sensor	Measured Data
Temperature	Body or skin temperature	Ability of body to generate and get rid of heat
Pulse oximeter	Oxygen saturation	Quantity of oxygen carried by human blood
Accelerometer	Movements of body	Measurement of acceleration forces in 3D spaces
Chest or skin electrodes	Electrocardiogram (ECG)	Measurement of waveform showing contraction and relaxation of cardiac cycle
Piezoresistive or piezoelectric sensor	Respiration rate	Rate of breathing per unit time
Phonocardiograph	Heart sound	Measurement of heart sound by use of stethoscope

Source: Bhelkar and Shedge (2016).

Image sensor applications include radiography, fluoroscopy, cardiology, mammography, dental imaging, endoscopy, external observation, minimally invasive surgery, laboratory equipment, ocular surgery and observation, and artificial retinas.

Accelerometers are used in heart pacemakers and defibrillators, patient monitoring equipment, blood pressure monitors, and other integrated health monitoring equipment.

Biosensors find applications in blood glucose and cholesterol testing, testing for drug abuse, infectious diseases, and pregnancy.

Magnetoencephalography and magnetocardiography systems use SQUIDs. These highly sensitive magnetometers measure extremely weak magnetic fields and are used to analyze neural activity inside the brain.

Encoders can be found in X-ray machines, magnetic resonance imaging machines, computer-assisted tomography equipment, medical imaging systems, blood analyzers, surgical robotics, laboratory sample-handling equipment, sports and healthcare equipment, and other non-critical medical devices (Thusu, 2011).

1.3.2 WEARABLE HEALTH MONITORING SYSTEMS (WHMS)

Non-invasive, non-intrusive sensors are indispensable elements of ambulatory and long-term health monitoring systems. Wearable sensors, being progressively more comfortable and less obtrusive, are appropriate for monitoring an individual's health or wellness without interrupting his/her daily activities. The sensors can measure several physiological signals/parameters, as well as activity and movement of an individual, by being placed at different locations of the body. The advancement in low-power, compact wearables (sensors, actuators, antennas, and smart textiles), inexpensive computing, and storage devices, coupled with modern communication technologies, pave the way for low-cost, unobtrusive, and long-term health monitoring systems (Majumder et al., 2017).

Wearable health monitoring systems (WHMSs) attracted the attention of the research community and industry during the last decade. As healthcare costs are increasing and the world population is aging, there is a need to monitor patients' health status outside the hospital. A variety of system prototypes and commercial products have been produced to provide real-time feedback information on health condition, either to the user or to a medical center or a physician, with the ability to alert the individual of threatening conditions. WHMSs also constitute a new way to manage and monitor chronic diseases, elderly people, postoperative rehabilitation patients, and persons with disabilities.

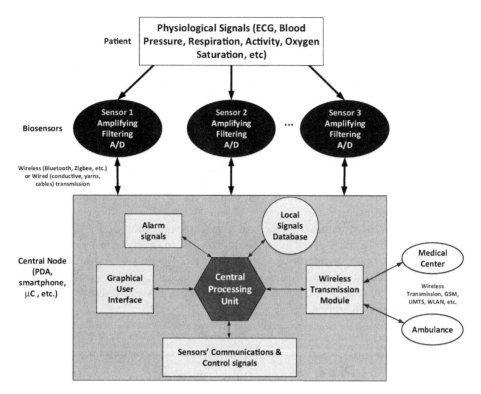

FIGURE 1.5   Architecture of a wearable health-monitoring system (Pantelopoulos & Bourbakis, 2010).

Wearable systems for health monitoring may comprise various types of miniature sensors, wearable or even implantable. These biosensors are capable of measuring significant physiological parameters like heart rate, blood pressure, body and skin temperature, oxygen saturation, respiration rate, and ECG. The obtained measurements are communicated either via a wireless or a wired link to a central node, for example, a PDA or a microcontroller board, which may display the relevant information on a user interface or transmit the aggregated vital signs to a medical center. A wearable medical system includes a wide variety of components: sensors, wearable materials, smart textiles, actuators, power supplies, wireless communication modules and links, control and processing units, interface for the user, software, and advanced algorithms for data extraction and decision-making.

A general WHMS architecture is depicted in Figure 1.5, including the system's functionality and components. However, this should not be perceived as the standard system design, as different systems may adopt significantly different architectural approaches (e.g., biosignals may be transmitted in analog form and without pre-processing to the central node, and bi-directional communication between sensors and central node may not exist) (Pantelopoulos & Bourbakis, 2010).

Wearable systems for health monitoring need to satisfy strict medical criteria while operating under several ergonomic constraints and significant hardware resource limitations. More specifically, a wearable health-monitoring system design needs to take wearability criteria into account, for instance, the weight and the size need to be small, and the system should not hinder any of the user's movements or actions. Radiation concerns and possible aesthetic issues also need to be accounted for. In addition, the security and privacy of the collected personal medical data must be guaranteed by the system, while power consumption needs to be minimized to increase the system's operational lifetime. Finally, such systems need to be affordable to ensure wide public access to low-cost ubiquitous health-monitoring services.

Designing such a system is a very challenging task, as many highly constraining and often conflicting requirements have to be considered by designers. It is understandable that there is no single ideal design for such systems; rather, the various "antagonizing" parameters should be balanced based on the specific area of application (Pantelopoulos & Bourbakis, 2010).

1.3.3 SCADA SYSTEMS

SCADA is a software system that works with coded signals over communication channels. These coded signals are combined with a data acquisition system via communication channels to get information on the status of remote equipment for display or for recording functions (Boys, 2009). The SCADA system is mainly used to control the production of power, infrastructure processes, and facility processes, including water treatment, distribution space stations, energy consumption, and siren systems in civil defense.

Shalini and Kumar (2013) reported on the real time implementation of SCADA in power distribution networks. The electrical distribution system consists of several substations with multiple controllers, sensors, and operator-interface points, as shown in Figure 1.6 (Arora, 2016). The issues faced by SCADA include integration, functionality, communication, network technology, security, and reliability. Having SCADA in the power system network increases the system's reliability and stability for integrated grid operation.

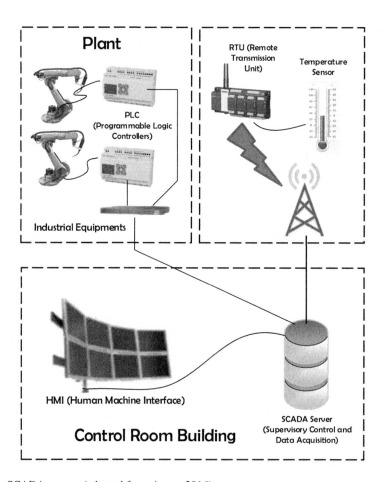

FIGURE 1.6   SCADA system (adapted from Arora, 2016).

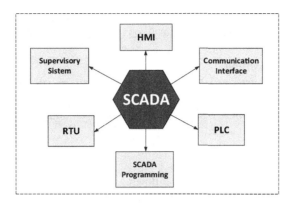

FIGURE 1.7   Basics of the SCADA system (adapted from Arora, 2016).

In general, programmable logic controllers (PLCs), circuit breakers, and power monitors are used to control and monitor a substation in real time. Data are transmitted from the PLCs and other devices to a computer-based SCADA node located at each substation. One or more computers are located at different centralized control and monitoring points. The SCADA system, as shown in Figure 1.7, consists of a number of blocks: a human-machine interface (HMI), supervisory system, remote terminal units (RTUs), PLCs, communication infrastructure, and SCADA programming (Arora, 2016).

1.3.3.1   Basics of SCADA

1.3.3.1.1   Human-Machine Interface (HMI)
HMI is an input-output device that presents the process data to be controlled by a human operator. It is linked to the SCADA system's software programs and databases to provide management information, including the scheduled maintenance procedures, detailed schematics, logistic information, and trending and diagnostic data for a specific sensor or machine. HMI systems allow the operating personnel to see the information graphically (Arora, 2016).

1.3.3.1.2   Supervisory System
A supervisory system is used as a server for communicating between the equipment of the SCADA system, such as RTUs, PLCs, and sensors, and the HMI software used in the control room workstations. A master station or supervisory station comprises a single personal computer (PC) in smaller SCADA systems; in larger SCADA systems, a supervisory system comprises distributed software applications, disaster recovery sites, and multiple servers.

1.3.3.1.3   Remote Terminal Units (RTUs)
Physical objects in SCADA systems are interfaced with microprocessor controlled electronic devices called RTUs. These units are used to transmit telemetry data to the supervisory system and receive messages on controlling the connected objects from the master system. They are also called remote telemetry units.

1.3.3.1.4   Programmable Logic Controller (PLC)
In SCADA systems, PLCs are connected to the sensors collecting the sensor output signals. They convert the sensor signals into digital data. PLCs are used instead of RTUs because of their advantages, including flexibility, configuration, versatility, and affordability.

1.3.3.1.5   Communication Infrastructure
A combination of radio and direct wired connections is generally used for SCADA systems, but in large systems like power stations and railways, SONET/SDH is frequently used. A few

communication protocols which are standardized and recognized by SCADA vendors send information only when the supervisory station polls the RTUs.

1.3.3.1.6   SCADA Programming

SCADA programming in a master or HMI is used to create maps and diagrams with important situational information in case of an event failure or process failure. Standard interfaces are used for programming most commercial SCADA systems. SCADA programming can be done using derived programming language or C language.

1.3.3.2   Architecture of SCADA

The SCADA system usually includes the following components: local processors, operating equipment, PLCs, instruments, RTUs, intelligent electronic devices, and a master terminal unit or host computers.

The block diagram in Figure 1.8 represents the basic SCADA architecture. The figure depicts an integrated SCADA architecture which supports TCP/IP, UDP, and other IP-based communication protocols, as well as industrial protocols like Modbus TCP, Modbus over TCP, or Modbus over UDP. These all work over cellular, private radio, or satellite networks.

Complex SCADA architectures have a variety of wired and wireless media and protocols to get data back to the monitoring site. This allows the implementation of powerful IP-based SCADA networks over landline, mixed cellular, and satellite systems. SCADA communications can use a diverse range of wired and wireless media.

The choice of communication depends on a number of factors: remoteness, available communications at the remote sites, existing communication infrastructure, polling frequency, and data rates. These factors impact the final decision on SCADA architecture.

1.3.3.3   Types of SCADA Systems

There are many different types of SCADA systems, and SCADA architectures have evolved over four generations:

1. First generation: Monolithic or early SCADA systems.
2. Second generation: Distributed SCADA systems.
3. Third generation: Networked SCADA systems.
4. Fourth generation: Internet of Things (IoT) technology SCADA systems.

1.3.3.4   Applications of SCADA

SCADA systems are used to monitor a variety of data, like flows, currents, voltages, pressures, temperatures, and water levels, in various industries. If the system detects any abnormal conditions from any monitoring data, the alarms at central or remote sites will be triggered to alert the operators through HMI.

There are numerous applications of SCADA systems; some frequently used ones include:

- Manufacturing industries.
- Wastewater treatment and distribution plants.
- Power systems.

1.3.3.4.1   SCADA in Manufacturing Industries

In manufacturing industries, the SCADA system can be applied to regular processes, like running production systems to meet productivity targets, checking the number of units produced, and counting the completed stages of operations at various stages of the manufacturing process.

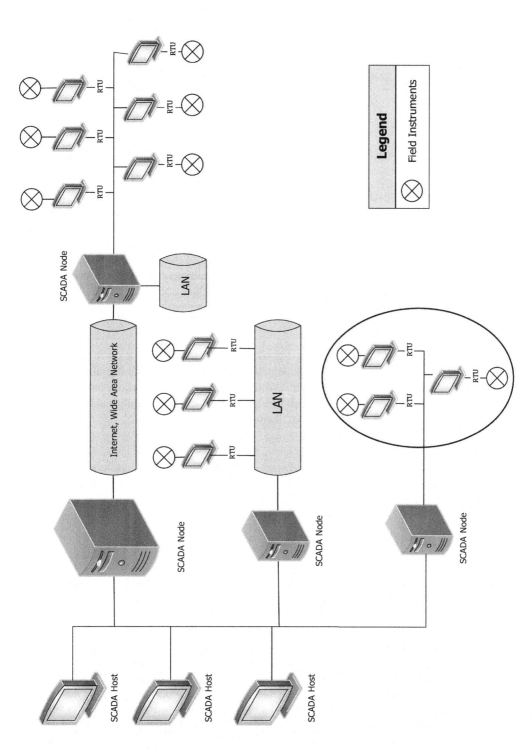

FIGURE 1.8 Architecture of SCADA (adapted from Arora, 2016).

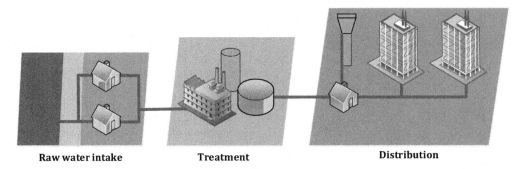

Raw water intake   **Treatment**   **Distribution**

FIGURE 1.9   Wastewater treatment and distribution plants (adapted from Arora, 2016).

1.3.3.4.2   SCADA in Wastewater Treatment and Distribution Plants

Wastewater treatment plants include surface water treatment and well water treatment systems. SCADA systems are used to control the automatic operation of equipment, such as backwashing filters based on the hours worked or amount of water flow through the filters. This is shown in Figure 1.9. In distribution plants, the water tank levels, pressure of the system, temperature of the plant, sedimentation, filtration, chemical treatment, and other parameters or processes are controlled using SCADA applications, such as PLCs or PC-based workstations connected to each other using local area networks such as Ethernet.

1.3.3.4.3   SCADA in Power Systems

Power systems include power generation, transmission, and distribution. All need to be monitored regularly to ensure system efficiency. The application of SCADA improves the overall efficiency of the system by providing supervision and control of the generation, transmission, and distribution systems. SCADA in the power system network increases the system's reliability and stability for integrated grid operation; see Figure 1.10 for an illustration.

1.3.3.4.4   Wireless SCADA

In large-scale industries, like power plants and steel plants, many processes and operations, such as the movement of conveyor belts for coal or product transport, and boiler heat temperature, must be constantly monitored, and it is necessary to control the factors affecting these parameters. Wireless SCADA can provide better control over the required control systems and operations; see Figure 1.11.

1.3.3.5   Understanding SCADA

To sum up the preceding discussion, SCADA refers to a class of industrial control systems (ICSs). Simply stated, these computer systems monitor and control industrial, infrastructure, or facility-based processes. A simplified ICS schematic is shown in Figure 1.12. Industrial processes include those of manufacturing, production, power generation, fabrication, and refining. Infrastructure processes may be public or private utilities, water or waste treatment, oil or gas distribution, and the power grid. Facility processes generally monitor and control heating, ventilation, and air-conditioning (HVAC) and lighting, particularly in building energy management systems.

A SCADA system comprises the following elements: an HMI, a supervisory system, and RTUs, all of which interface with particular sensors and PLCs, which, in turn, interface with sensors, actuators, and indicators. The communication infrastructure, which may be wireless, connects the supervisory system to the RTUs or PLCs (Greenland, 2012). Figure 1.13 gives a schematic representation of a SCADA system.

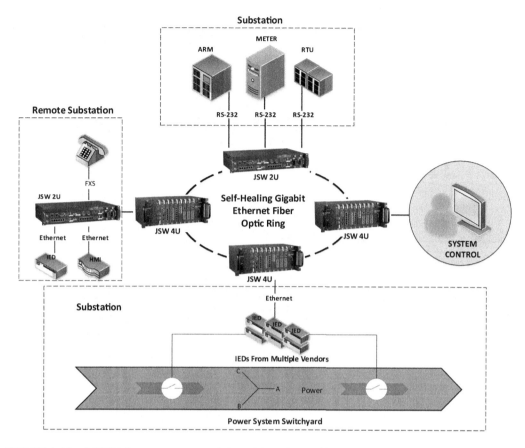

FIGURE 1.10 SCADA in a power system (adapted from Arora, 2016).

1.4 SENSOR FUSION, DATA FUSION, AND INFORMATION FUSION FOR MAINTENANCE

1.4.1 Sensor Fusion

Sensor fusion (sometimes called multi-sensor fusion) combines various sensory data or data derived from disparate sources, such that the resulting information has less uncertainty than would be possible if these sources were used individually. The term uncertainty reduction in this case can mean making conclusions more accurate, more complete, or more dependable. It can be done, for example, by combining two-dimensional images from two cameras at slightly different viewpoints.

Direct fusion is the fusion of sensor data from a set of heterogeneous or homogeneous sensors or soft sensors; data can include historical values. Indirect fusion uses human input (i.e., experience) and *a priori* knowledge about the environment (Wikipedia, 2020).

The concept of sensor fusion attempts to replicate the ability of the central nervous system to process sensory inputs from multiple sensors simultaneously. For robotic devices, feedback from one sensor is typically not enough, particularly to implement control algorithms. Sensor fusion can be used to compensate for deficiencies in information by utilizing feedback from multiple sensors. The deficiencies associated with individual sensors in calculating particular types of information can be compensated for by combining the data from multiple sensors. The net effect of sensor fusion is that the resulting information should have less uncertainty than would have occurred if the sensors were used individually. Sensor fusion can also help to compensate for sensor noise, limited accuracy,

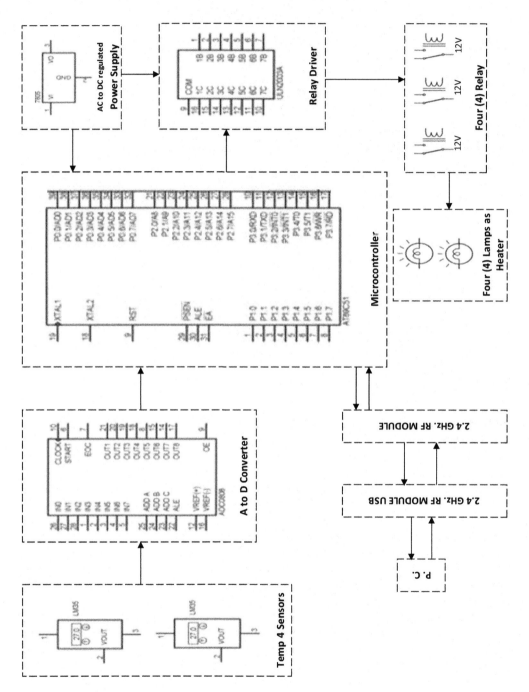

FIGURE 1.11 Wireless SCADA system (adapted from Arora, 2016).

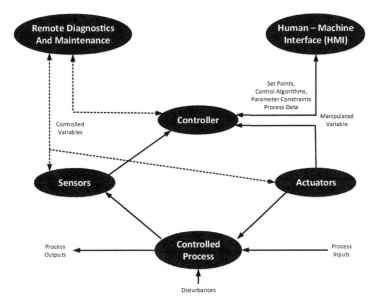

FIGURE 1.12 Simplified ICS block schematic (adapted from Greenland, 2012).

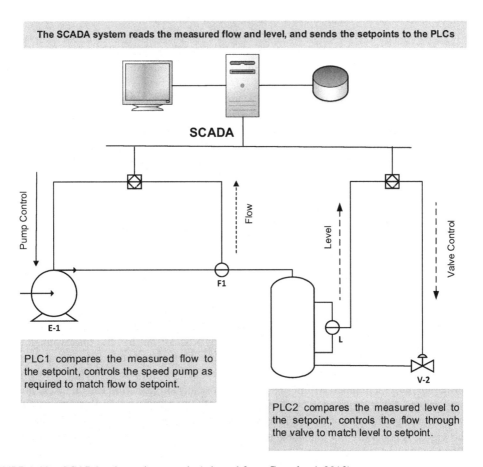

FIGURE 1.13 SCADA schematic example (adapted from Greenland, 2012).

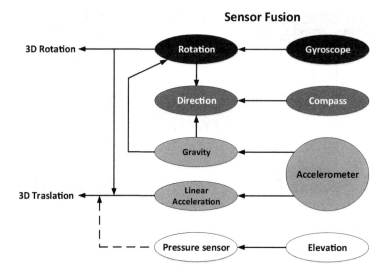

FIGURE 1.14 Fusion of data from multiple sensors to get information about 3D rotation and translation (Lamkin-Kennard & Popovic, 2019).

failure, or a lack of information about a particular aspect of the environment. In addition, sensor fusion can be used when the ideal sensor of choice is cost prohibitive.

Sensor fusion can use information directly from sensors or historic sensor data or use indirect information from prior knowledge about the system inputs. Figure 1.14 illustrates the concept of sensor fusion to get information about three-dimensional (3D) rotations and translations (Lamkin-Kennard & Popovic, 2019)

Sensor fusion combines data from disparate sources to create coherent information and give an enhanced description of the surrounding environment (Tirindelli, 2016). As mentioned previously, the resulting information is more certain than would be possible if these sources were used individually. This is especially important when different kinds of information are combined. For example, autonomous vehicles need a camera to mimic human vision, but information on obstacle distance is best gained through light detection and ranging (LiDAR) or radar sensors. The fusion of camera data with LiDAR or radar data is very important, as they are complementary. Combining information from LiDAR and radar will provide more certain information about the distance of an obstacle ahead of the vehicle or the general distance of objects in the environment (Kocić et al., 2018).

1.4.1.1 Sensor Fusion Architecture

Architectures performing sensor fusion construct a single unified world model based on all available sensory data, such as an occupancy grid which represents objects within a map, or a symbolic model containing assertions about world states. This world model is then used for planning actions, as illustrated in Figure 1.15. This approach has the advantage of being able to combine evidence to overcome ambiguities and noise inherent in the sensing process, but has the disadvantage of creating a computationally expensive sensory bottleneck. In such architectures, all sensor data must be collected and fused into a single monolithic world model, and a complete path considering all relevant information must be planned within this model before any single action can be taken, no matter how simple or urgent that action may be (Rosenblatt & Hendler, 1999).

Another difficulty with sensor fusion is that information from disparate sources, such as maps, sonar, and video, is generally not amenable to combination within a single representational framework suitable for planning such dissimilar tasks as following roads and avoiding obstacles. For example, although the Autonomous Land Vehicle in a Neural Network (ALVINN) system uses an

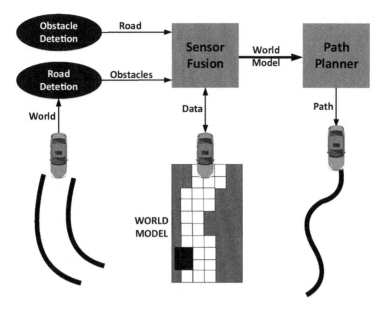

FIGURE 1.15 Sensor fusion creates a centralized world model for planning (Rosenblatt & Hendler, 1999).

artificial neural network to associate video images of roads with appropriate steering directions, and has been one of the most successful road-following systems to date, it has been less successful than other systems using range data for obstacle avoidance. Thus, by requiring a single representation for all sensor and map data, a centralized architecture does not allow specialized modules to use other representations and algorithms best suited to the task at hand. A single monolithic world model is also more difficult to develop, maintain, and extend (Rosenblatt & Hendler, 1999).

1.4.1.2   How Sensor Fusion Works

Sensor fusion provides myriad capabilities; it makes life easier in many ways and enables a variety of services. A very basic example of sensor fusion is an e-compass, in which data from a 3D magnetometer and a 3D accelerometer are combined to improve compass functionality. More complex technologies leverage and combine data from more sensors, thus giving users an enhanced experience. The combination of 3D accelerometers, 3D gyroscopes, and 3D magnetometers is called a nine-axis system, as it affords the user nine degrees of freedom (Karimi, 2020). Fusion is beneficial in this case because each of these sensor types provides unique data but has limitations:

- Accelerometer: Offers x-, y-, and z-axis linear motion sensing but is sensitive to vibration.
- Gyroscope: Offers pitch, roll, and yaw rotational sensing but has zero bias drift.
- Magnetometer: Offers x-, y-, and z-axis magnetic field sensing but is sensitive to magnetic interference.

By using special algorithms and filtering techniques, sensor fusion reduces or eliminates the deficiencies of each individual sensor. In other words, it takes the simultaneous input from the various sensors, processes it, and creates an output that is greater than the sum of its parts.

A remaining difficulty in sensor fusion is the lack of standardization across operating systems (OSs). This limits the use of the full capabilities of sensor fusion. With more development, this will change. For example, Windows® 8 OS by Microsoft® supports sensors in a cohesive manner, using sensor-class drivers based on industry standards developed in collaboration with Microsoft's ecosystem partners.

1.4.1.3   Sensor Fusion Levels

There are several categories or levels of sensor fusion (Wikipedia, 2020).

- Level 0: Data alignment.
- Level 1: Entity assessment (e.g., signal/feature/object); tracking and object detection/recognition/identification.
- Level 2: Situation assessment.
- Level 3: Impact assessment.
- Level 4: Process refinement (i.e., sensor management).
- Level 5: User refinement.

Sensor fusion level can also be defined based on the kind of information used to feed the fusion algorithm. More precisely, sensor fusion includes fusing raw data coming from different sources, fusing extrapolated features, and fusing decisions made by single nodes (Wikipedia, 2020).

- Data level: Data-level (or early) fusion aims to fuse raw data from multiple sources and represents the fusion technique at the lowest level of abstraction. It is the most common sensor fusion technique in many fields of application. Data-level fusion algorithms usually aim to combine multiple homogeneous sources of sensory data to achieve more accurate and synthetic readings. When portable devices are employed, data compression is an important factor, as collecting raw information from multiple sources generates huge information spaces that could define an issue in terms of memory or communication bandwidth for portable systems. Data-level information fusion tends to generate big input spaces that slow down the decision-making procedure. Data-level fusion often cannot handle incomplete measurements; if one sensor modality becomes useless because of malfunctions, breakdowns, or other reasons, the whole system could have an ambiguous outcome.
- Feature level: Features represent information computed by each sensing node. These features are then sent to a fusion node to feed the fusion algorithm. This procedure generates smaller information spaces for data level fusion, and this is better in terms of computational load. Obviously, it is important to select the best features for classification: choosing the most efficient feature set should be a main aspect of method design. Using feature selection algorithms that properly detect correlated features and feature subsets improves the recognition accuracy, but large training sets are usually required to find the most significant feature subset.
- Decision level: Decision-level (or late) fusion requires selecting a hypothesis from a set of hypotheses generated by individual (usually weaker) decisions of multiple nodes. It is the highest level of abstraction and uses information already elaborated through preliminary data- or feature-level processing. The main goal in decision fusion is to use meta-level classifiers while data from nodes are pre-processed by extracting features from them. Typically, decision-level sensor fusion is used in classification and recognition activities; the two most common approaches are majority voting and naïve Bayes. Advantages of decision-level fusion include communication bandwidth and improved decision accuracy. It also allows the combination of heterogeneous sensors (Wikipedia, 2020).

1.4.1.4   Leveraging Sensor Fusion for the Internet of Things (IoT)

The IoT comprises smart machines interacting and communicating with other machines, objects, environments, and infrastructures, gathering volumes of data, and processing those data to derive actions (see Figure 1.16). Examples include connected homes and cities, connected cars and roads, and devices that track behavior (Karimi, 2020). The basic idea is to make life easier for humans.

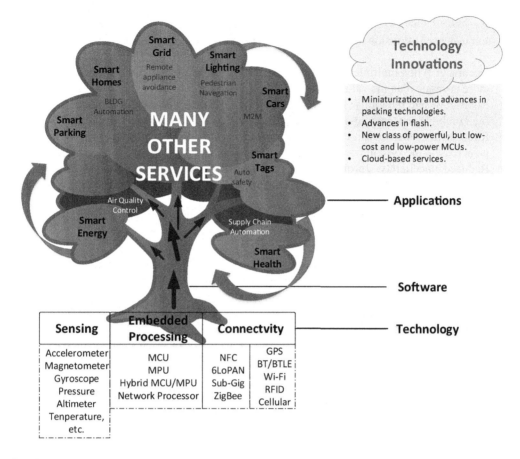

FIGURE 1.16   Internet of Things (Karimi, 2020).

All connected objects contain the following:

- Sensing nodes to collect data.
- Local embedded nodes to process data.
- Connectivity nodes with wired and/or wireless communication capability.
- Software to automate tasks and enable new classes of services.
- Remote embedded processing nodes with either network or cloud-based processing capability.
- Full security across the signal path.

Sensing nodes vary quite widely, depending on the particular application. For example, a camera system could be used for image monitoring, water or gas flow meters may be applied for smart energy, radar is useful for safety applications, radio-frequency identification (RFID) readers sense the presence of an object or person, and doors and locks may be equipped with open/close circuits to indicate a building intrusion. Each of these sensors does something quite different. Each will be controlled separately via a remote command and control topology. At some point, they will be able to communicate with each other. For example, in some cases, a smartphone with RFID and/or near field communication (NFC) capability and Global Positioning System can approach individual RFID/NFC-enabled objects in a building, communicate with them, and register their physical location.

With time, the further development of sensor fusion platforms and remote emotive computing will improve the capability of sensing nodes (Karimi, 2020).

1.4.1.5 Sensor Fusion Advantages

A system using sensor fusion has a series of advantages over a single sensor system:

- Accuracy: When multiple independent measurements are fused, the resulting value is much more accurate than the value achieved with a single sensor.
- Robustness and reliability: Even if there is a partial failure, the system's inherent redundancy (i.e., multiple sensor suites) enables it to continue providing information.
- Robustness against measurement outliers: When the dimensionality of the measurement space is increased, the system becomes more robust to outliers (no coherent measurements).
- Reduced uncertainty: Information from more than one source reduces the possibility of ambiguous interpretations of the measured value. A measurement of one sensor is confirmed by others covering the same domain.
- Extended coverage: Different sensors can look in different places and take different measurements.

Sensor fusion is not a new concept, but the advances in sensor technology and processing techniques, combined with improved hardware, make real-time fusion of data possible in complex applications, such as simultaneous localization and mapping (Tirindelli, 2016).

1.4.1.6 Challenges to Sensor Fusion

Figure 1.17 illustrates the challenge of sensor fusion, using a self-driving car as an example. The objective is to determine the environment around the vehicle trajectory with enough resolution, confidence, and latency to navigate the vehicle safely.

Figure 1.17 row 1 shows the ideal case when two sensors agree on an object, and the object is detected early enough to navigate the car. Figure 1.17 row 2 shows a case where each of the sensors classifies the object differently. In this case, the best option may be to just agree that it is a big enough object to avoid if possible.

Figure 1.17 row 3 shows a similar situation where a person on a bicycle may be identified as either a person or a bicycle. Nonetheless, we can agree it is an unidentified large moving object that needs to be avoided (Thakur, 2017).

The last two rows show smaller objects that pose difficult questions. Is it better to run over a small dog than to risk braking and getting rear-ended? Can the pothole be detected and classified early enough to navigate? Is the pothole or object small enough to run over?

These questions will take a longer time to resolve as they require improved technology in sensing, better computing, more public acceptance, and the development of legislation. The 80/20 Pareto

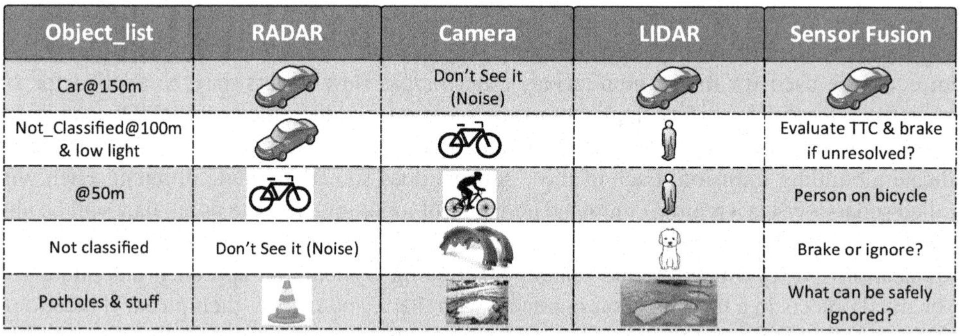

Object_list	RADAR	Camera	LIDAR	Sensor Fusion
Car@150m		Don't See it (Noise)		
Not_Classified@100m & low light				Evaluate TTC & brake if unresolved?
@50m				Person on bicycle
Not classified	Don't See it (Noise)			Brake or ignore?
Potholes & stuff				What can be safely ignored?

FIGURE 1.17 Challenges of sensor fusion (Thakur, 2017).

principle suggests solving the last 20% of the problems for self-driving cars will take 80% of the time it takes to bring them to mass market (Thakur, 2017).

1.4.2 DATA FUSION

Data fusion may be defined as the process of combining data and knowledge from different sources with the aim of maximizing the useful information content, for improved reliability or discriminant capability, while minimizing the quantity of data ultimately retained.

Most data fusion users find the field is wider than they thought. The sensor and signal processing communities have been using fusion to synthesize the results of two or more sensors for many years. This simple step recognizes the limitations of a single sensor but exploits the capability of another similar or dissimilar sensor to calibrate, add dimensionality, or simply to increase statistical significance or robustness to cope with sensor uncertainty. In many such applications, the fusion process is necessary to gain sufficient detail in the required domain.

Crucially for CM, data fusion allows maintainers to cross boundaries where recent applications have faltered:

- It is possible to deal with the selection of data processing methods based on problem characteristics, for example, data or knowledge density; the relationships are becoming clearer.
- Qualitative and quantitative information can be merged, for example, data and expert knowledge.

There are many problems to be overcome, however. There are several proposed frameworks, but each needs work before it could be called generic. There is still much to be learned from existing methods for technique selection.

Ultimately, once the bugs are worked out, integrated approaches will involve multilevel fusion of sensor data, novelty detection, feature classification, diagnosis, and decision-making (Starr & Ball, 2000).

1.4.2.1 Structures in Data Fusion

Several framework architectures have been proposed. Their layout depends on the field of application. The US Department of Defense (DoD) Joint Directors of Laboratories (JDL) architecture (Waltz & Llinas, 1990) assumes a level distribution for the fusion process, characterizing the data from the source signal level to a refinement level, where the fusion of information takes place in terms of data association, state estimation, or object classification (see Figure 1.18). Situation assessment, at a higher level of inference, fuses the object representations provided by the refinement and draws a course of action. This architecture can be adapted, in principle, to CM problems (Starr et al., 2002).

Three stages are commonly identified, but it is not always necessary to apply all three in data fusion (Starr et al., 2002):

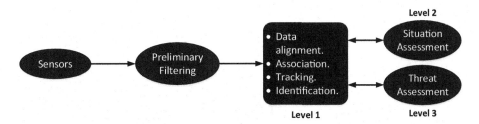

FIGURE 1.18 JDL data fusion architecture (Starr et al., 2002).

- Pre-processing: Reduction of the quantity of data while retaining useful information and improving its quality, with minimal loss of detail. This may include feature extraction and sensor validation. Techniques include dimension reduction, gating, thresholding, Fourier transform, averaging, and image processing.
- Data alignment: Techniques must fuse the results of multiple sensors or features already extracted in pre-processing. These include association metrics, batch and sequential estimation processes, grouping techniques, and model-based methods.
- Post-processing: Combines mathematical data with knowledge and decision-making. Techniques can be classified as knowledge-based, cognitive-based, heuristic, and statistical. A method map of these techniques appears in Hannah, Starr, and Bryanston-Cross (2001). Much research focuses on methods applied to particular problems, or aspects of the architecture. Examples include architectural issues dealing with multiple sensors in similar or dissimilar domains (Grime & Durrant-Whyte, 1994).

1.4.2.2 Classification of Data Fusion Techniques

Data fusion is a multidisciplinary area, making it difficult to classify data fusion techniques (Castanedo, 2013). However, the following are some of the current suggestions.

1.4.2.2.1 Classification Based on Relations between Input Data Sources

Durrant-Whyte (1988) proposed three types of data fusion based on relations between the data input sources: complementary, redundant, and cooperative (see Figure 1.19).

1. Complementary: Data provided by the input sources represent different parts of a single scene and when combined can give more complete global information. For example, in visual sensor networks, the data from two cameras with different fields of view are considered complementary because one provides information that the other cannot.
2. Redundant: Two or more input sources provide data on the same target. For example, in visual sensor networks, the data coming from overlapped areas are considered redundant because both sources give the same information.
3. Cooperative: Data are combined to create new information that is typically more complex than the original. For example, audio and video data fusion (i.e., multi-modal) is considered cooperative.

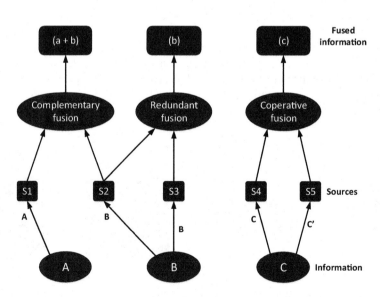

FIGURE 1.19   Whyte's classification based on relations between data sources (Castanedo, 2013).

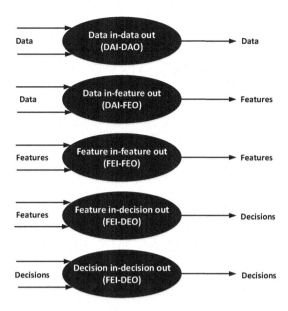

FIGURE 1.20 Dasarathy's classification (Dasarathy, 2001).

1.4.2.2.2 Dasarathy's Classification

Dasarathy (2001) offered one of the best-known data fusion classification systems. Dasarathy's classification is based on input/output data types and comprises five categories (see Figure 1.20):

1. Data in-data out (DAI-DAO): The most basic data fusion method inputs and outputs raw data; the purpose is to ensure accuracy of analysis. Data fusion is conducted immediately after the data are gathered from the sensors. The method uses signal and image processing algorithms.
2. Data in-feature out (DAI-FEO): The process extracts features or characteristics of an entity of interest by fusing the raw data from various sources.
3. Feature in-feature out (FEI-FEO): Both input and output are features. The fusion process, also called feature fusion, information fusion, symbolic fusion, or intermediate level fusion, fuses a set of features to improve or refine them or to discover new ones.
4. Feature in-decision out (FEI-DEO): A set of features is the input and a set of decisions is the output. In other words, the fusion process makes a decision based on a sensor's inputs.
5. Decision in-decision out (DEI-DEO): The process, also called decision fusion, fuses input decisions to get better decisions or make new decisions.

Importantly, Dasarathy's data fusion classification system specifies the abstraction level either as an input or an output, thus providing a useful framework to classify methods and techniques (Castanedo, 2013).

1.4.2.2.3 Classification Based on Abstraction Levels

Information fusion typically addresses three levels of abstraction:

1. Measurements (lowest level of abstraction).
2. Characteristics.
3. Decisions (highest level of abstraction).

Luo et al. (2002) added one level, the pixel level, and proposed a classification system based on these four abstraction levels:

1. Signal level: Fusion of signals acquired from the sensors.
2. Pixel level: Fusion at the image level, used to improve image processing tasks.
3. Characteristic level: Fusion of features extracted from images or signals (i.e., shape and velocity).
4. Symbol level: Also called the decision level; information is represented as symbols.

A similar classification of data fusion based on abstraction levels is the following:

1. Low-level fusion: Raw data are the input in the data fusion process; fusion of several sources provides more accurate data (e.g., a lower signal-to-noise ratio) than each source individually.
2. Medium-level fusion: Also called the feature or the characteristic level. At this level, characteristics or features (e.g., shape, texture, and position) are fused to obtain features to use for other tasks.
3. High-level fusion: Also called decision fusion. At this level, symbolic representations are the input; they are fused to reach more accurate decisions. Bayesian methods are typically employed at this level.
4. Multiple-level fusion: Input data come from different levels of abstraction and are fused; for example, a measurement can be combined with a feature.

1.4.2.2.4   JDL Data Fusion Classification

Arguably the most popular classification system was proposed by the JDL and the US DoD (JDL, 1991). Their configuration includes an information bus connecting five levels of data fusion (see Figure 1.21) and a database (Castanedo, 2013):

1. Sources: Various types of sources provide data, including sensors, *a priori* information (e.g., references and geographic data), databases, and human input (i.e., experience).
2. Human-computer interaction (HCI): The HCI interface accepts inputs from the operators and produces outputs for the operators. It includes commands, queries, and information on the results.
3. Database management system: The database management system stores a large amount of highly diverse information, including both raw data and the fused results.

JDL's five levels of data processing are the following (also see Section 1.4.1.3):

1. Level 0, source pre-processing: This is the lowest level of the data fusion process. It includes fusion at the signal and pixel levels. For text sources, this level also includes the information extraction process. This level reduces the amount of data and retains useful information for the high-level processes.
2. Level 1, object refinement. Inputs at this level are the processed data from the previous level. Input information is transformed into consistent data structures. The outputs are object discrimination (i.e., object classification and identification) and object tracking (i.e., object state and orientation). Common procedures used at this level are spatio-temporal alignment, association, correlation, state estimation, clustering or grouping techniques, false positive removal, identity fusion, and the combining of features extracted from images.
3. Level 2, situation assessment: This level aims to identify likely situations given the observed events and obtained data and establishes relations (i.e., proximity and communication) between the objects. The aim is to make high-level inferences and identify patterns. Thus, the output is a set of high-level inferences.

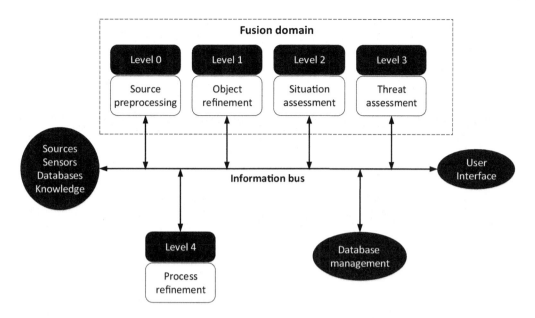

FIGURE 1.21 JDL data fusion framework (Castanedo, 2013).

4. Level 3, impact assessment: This level evaluates the impact (e.g., risk, threat, and opportunity) of the activities and patterns identified in level 2 and predicts the logical outcome.
5. Level 4, process refinement: This level adds resource and sensor management. The aim is to achieve efficient resource management while accounting for task priorities, scheduling, and available resources.

1.4.2.2.5 Classification Based on Type of Architecture
Data fusion architecture (i.e., where fusion takes place) can be centralized, decentralized, distributed, or hierarchical. Classification can be based on the type of architecture (Castanedo, 2013):

1. Centralized architecture: In this architecture, the various sources obtain observations as measurements and transmit them to a central processor, where data fusion occurs.
2. Decentralized architecture: This architecture features a network of nodes; each node has its own processing capabilities, and there is no single point of data fusion. Instead, each node fuses its local information with the information received from other nodes. Decentralized data fusion algorithms typically communicate information using Fisher and Shannon measurements instead of the object's state (Durrant-Whyte & Stevens, 2001). The main disadvantage of this architecture is the communication cost. In addition, as each node communicates with all of its peers, the architecture may have scalability problems if the number of nodes is increased.
3. Distributed architecture: In this architecture, measurements from each source node are processed independently before the information is sent to the fusion node. In other words, each node provides an estimation of the object state based on its local views. This information becomes the input to the fusion process. The output is a fused global view. The architecture may have one or several fusion nodes.
4. Hierarchical architecture: It is possible to have a combination of decentralized and distributed nodes, thus creating a hierarchical architecture, with data fusion performed at different levels in the hierarchy.

1.4.3 Information Fusion

1.4.3.1 An Introduction to Information Fusion

Information fusion encompasses the theory, techniques, and tools used to exploit the information acquired from multiple sources (from sensors, databases, or humans) to reach a decision that is somehow better than a decision reached if these sources were used individually. Improvements can be qualitative or quantitative, in terms of accuracy, robustness, and so on (Dasarathy, 2001).

An example of information fusion in manufacturing is the fusion of information from multiple sensors (Dasarathy, 2003). Information comes from the past operation of a manufacturing system (e.g., stored in databases), from the present operation (e.g., sensor signals, machine status), and from the future (e.g., predictions obtained through simulations) (De Vin et al., 2006).

In information fusion research and applications, the scope is often limited, for instance, in application area (e.g., military applications only) or in sources (e.g., from sensors only). Information fusion has a background in the defense sector but has become more generic and has been introduced to many other areas.

Information fusion theory is based on sensor fusion approaches, where large data quantities need to be fused, and on data mining, where the problem is to identify and fuse huge amounts of existing data, possibly from several sources.

1.4.3.2 Information Fusion Model

A well-known model for information fusion is the JDL model; see Section 1.4.2.1, Section 1.4.2.2.4, and Figure 1.22.

The model has five levels (see Section 1.4.2.2.4 and Section 1.4.1.3), which form a hierarchy of processing (Hilletofth et al., 2007).

- On Level 0, "sub-object pre-processing" is performed. This is an iterative process of fusing data to determine the identity and other attributes of sub-entities (signals and features).
- On Level 1, "object refinement" is performed. This is an iterative process of fusing data to determine the identity and other attributes of entities (objects). This level also builds a situation picture.
- On Level 2, "situation refinement" is performed. This is an iterative process of fusing spatial and temporal relationships between entities. This results in a situation assessment.

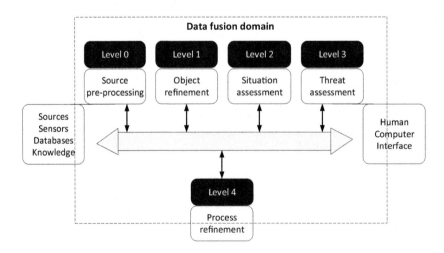

FIGURE 1.22   JDL model (Llinas et al., 2004).

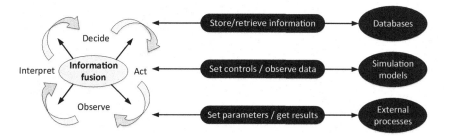

FIGURE 1.23 Information fusion process and interaction of information fusion with databases, external processes, and simulations (De Vin et al., 2006).

- On Level 3, "threat refinement" is performed. This is an iterative process of fusing the combined activity and capability of enemy forces to infer their intentions and assess the threat that they pose. The result from this level is called the threat assessment.
- On Level 4, "process refinement" is performed. This is an ongoing monitoring and assessment of the fusion process to refine the process itself, including optimization of data acquisition.

There can be problems applying the JDL model. For instance, in many cases, analogies for Levels 2 and 3 may be absent. Furthermore, acquiring data from external sources (e.g., geographical databases as provided by commercial companies) means optimization of data acquisition is hampered (De Vin et al., 2006).

An adjacent concept used in the information fusion community is agent-based modeling, used specifically for modeling, simulating, and evaluating future scenarios (Hilletofth et al., 2007).

A military-specific model is the OODA loop (Llinas et al., 2004; Bass, 2000) where OODA stands for "Observe, Orient, Decide and Act". However, in this model, the only military-specific term is "Orient". If this term is replaced with the more generic expression "Interpret" as in Figure 1.23, the model becomes suitably generic. The whole loop is part of the information fusion process, even though intuitively, the "Interpret" phase is a phase in which a large part of the information fusion takes place. However, observing, that is, gathering certain information, implies that certain information is going to be fused, and this is the result of deciding and acting (i.e., deciding which information is needed and executing that decision). The information fusion process does not just import and fuse information; it interacts with databases, external processes, and simulation models in various ways (Figure 1.23). For instance, results from the information fusion process can be used to control processes or to improve simulations by tuning parameters in the simulation model. Furthermore, the information fusion process can result in the design of a new set of simulation experiments to be carried out using the simulation model.

1.5 CONDITION MONITORING AND THE END OF TRADITIONAL PREVENTIVE MAINTENANCE

1.5.1 CONDITION MONITORING

The core of CBM is CM. Most equipment failures are preceded by certain signs, conditions, or indications, and if those signs and conditions are recognized, more timely maintenance can be performed (Bloch & Geitner, 1983). An asset's condition can be monitored in a variety of ways, using various approaches and employing different types and levels of technology (Jardine et al., 2006), as presented in Figure 1.24.

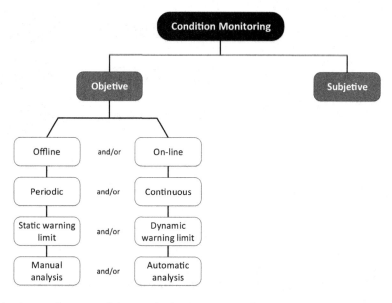

FIGURE 1.24 Approaches to condition monitoring (Rastegari, 2015).

An asset's condition can be monitored periodically or continuously:

- Periodic monitoring is conducted at certain regular intervals, such as every hour or at the end of every working shift, using portable indicators, for example, hand-held meters, acoustic emission units, or vibration pens. The process includes subjective monitoring (i.e., evaluations based on human senses) to measure or evaluate asset conditions, such as the degree of dirtiness or abnormal color. Periodic monitoring is low cost and provides accurate diagnostics using filtered and/or processed data, but it is possible to miss failure events that occur between inspections (Rastegari, 2015).
- In online or real time monitoring, an asset is continuously monitored, and a warning alarm is triggered if an error or incipient failure is detected. Online monitoring is often expensive, and the presence of noise in the raw signals may produce inaccurate diagnostic information (Jardine et al., 2006).

In general, the purpose of the CM process is two-fold. First, it collects data on the condition of the equipment. Second, it improves the knowledge of the causes and effects of equipment failure and deterioration patterns.

In CM, information on the equipment's internal condition must be obtained externally while the equipment is operating (Randall, 2011). The two principal techniques for obtaining information on internal conditions are vibration analysis and oil analysis (Rastegari, 2015).

- Vibration monitoring: The most popular CM technique, especially for rotating equipment (e.g., bearings and gearboxes), is vibration monitoring. A machine in standard condition has a certain vibration signature, and fault growth changes that signature in a way that can be linked to the fault (Rastegari, 2015). Machines constantly generate vibrations. Many are linked to periodic events in operation, such as rotating shafts, meshing gear teeth, or rotating electric fields. Some are due to events that are not phase-locked to shaft rotations, such as engine combustion. Others are linked to fluid flow, as in pumps and gas turbines. Each type has unique characteristics (Randall, 2011).

- Oil monitoring: Oil analysis (or lubricant monitoring) is mainly used in circulating-oil lubricating systems to determine the wear of internal components, such as engine shafts (Rosmaini & Kamaruddin, 2012). The lubricant transmits information in the form of wear particles or chemical contaminants. Grease lubricants can also be analyzed in this fashion (Randall, 2011).

Vibration and oil analysis may be the most common, but there are other types of CM:

- Performance monitoring: For certain types of equipment, performance analysis is an effective way to determine whether the equipment is functioning correctly.
- Thermography monitoring: In this CM technique, sensitive instruments are employed to remotely measure temperature changes. The method is mainly used in quasi-static situations, such as electrical switchboards to detect local hot spots or in hot fluid containers to find faulty refractory linings (Randall, 2011).
- Sound or acoustic monitoring: This method of analysis is closely related to vibration analysis, but there is a fundamental difference. Vibration sensors are rigidly mounted on a component to register local motions, but acoustic sensors pick up sounds. Like vibration analysis, sound or acoustic analysis is executed online, either periodically or continuously (Rosmaini & Kamaruddin, 2012).

1.5.1.1   Condition Monitoring as Tool of Preventive Maintenance

CM is used to detect changes or trends in controlling parameters or in the normal operating conditions which indicate the onset of failure. By providing an early problem diagnosis, CM can organize in advance the intervention for replacement of components whose failure is imminent, thereby avoiding more serious consequences. CM is particularly important in cases when the time required to mobilize resources for repair is significant. Early problem diagnosis helps to reduce significantly downtime associated with unplanned intervention for repair. A planned or opportune intervention is considerably less expensive than unplanned intervention initiated when a critical failure occurs. Ordering spare components immediately after the onset of failure is indicated significantly reduces the downtime for repair and the associated cost of lost production. The earlier the warning, the larger the response time, the more valuable the CM technique. Early identification of an incipient failure reduces the following (Todinov, 2007):

1. Risk of environmental pollution.
2. Number of fatalities.
3. Loss of production assets.
4. Cost of repair (damaged components require the mobilization of specialized repair resources).
5. Losses caused by dependent failures.
6. Loss of production associated with uncontrolled shutdowns.
7. Loss of production due to the time spent on troubleshooting.

By using the laws of probability, information obtained from CM can be used to obtain new information.

In CM, the reliability of the monitoring system must exceed the reliability of the monitored equipment. Furthermore, the cost of the CM should be considerably lower than the losses accruing from failure. An infrastructure making it possible to react to the data stream delivered from the CM devices must be in place if the CM technique is to be of any use. There is little use of a CM system giving an early warning of an incipient failure if the infrastructure for intervening with preventive actions is missing (Todinov, 2007).

CM is based on measuring the values of specific parameters critical to the failure-free operation of the monitored equipment. Measured parameters include the following (Todinov, 2007):

- Temperature measurements to detect increased heat generation (usually an indication of intensive wear, poor lubrication, failure of the cooling system, overloading, or inappropriate tolerances).
- Measurements of pressure and pressure differences to detect leaks in a hydraulic control system, dangerous pressure levels, and so on. Increased pressure difference before and after a filter, for example, is an indication of clogging or blockage of the filter, while a pressure drop in a hydraulic line is an indication of leakage.
- Measurement of the displacement of components and parts to determine excessive deformations, unstable fixtures, material degradation, and design faults.
- Vibration monitoring to detect incipient failures in bearings, increased wear, out-of-balance rotating components, and so on.
- Monitoring the cleanliness of lubricants and hydraulic fluids. Periodic sampling of lubricants and hydraulic fluids for debris is often used to determine the extent of wear of components. Another useful outcome of sampling hydraulic fluids for debris and their timely replacement is reducing the possibility for jamming of control valves. Such failures could render inoperative large sections of a control system and could be very expensive if the cost of intervention is high. This type of CM also reveals the degree of deterioration of the lubricants and hydraulic fluids because of poor cooling or increased heat generation due to wear.
- Ultrasonic inspection, radiographic examination, magnetic particles, and penetrating liquids to detect various flaws (cracks, pores, voids, and inclusions) in components, welded joints, and castings.
- Measurement of the degree of wear, erosion, and corrosion. Since most mechanical systems undergo some form of degradation caused by wear, corrosion, and erosion, they benefit from periodic monitoring of the extent of degradation. Deteriorated components can be replaced in time by opportunity maintenance.
- Measurement of fracture toughness and deterioration of materials due to aging, corrosion, and irradiation.
- Monitoring of manufacturing and process parameters. Controlling the tolerances during manufacturing by monitoring the wear rate of cutting tools reduces the possibility of failures caused by misfit during assembly; there is less possibility of jamming, poor lubrication, or accelerated wear out.
- Monitoring of electrical parameters such as current, voltage, and resistance. Increased current in electric motors indicates increased resistance from the powered equipment. This is an indication of jamming due to a build-up of debris or corrosion, misalignment, increased viscosity of the lubricant or lack of lubrication, damage of the bearings, clogged or blocked filters (in pumps), and so on. Decreased conductivity of working fluids (hydraulic fluid) often indicates contamination which may induce poor heat dissipation, increased energy losses, and accelerated erosion and corrosion.

CM is different from status monitoring which determines whether a component is in working or failed state. CM can improve the availability of the system immensely if combined with immediate intervention for repair/replacement of failed components. This can be illustrated on a simple system with active redundancy (Figure 1.25(a)) consisting of two identical components logically arranged in parallel. Without status monitoring, repair is initiated only if the system stops production (no path between nodes 1 and 2). The result of this breakdown-induced intervention will be a sequence of uptime corresponding to the case where at least one component works, followed by a downtime when both components have failed and a mobilization of resources for repair has been initiated (Figure 1.25(a)). Status monitoring changes this availability pattern dramatically. If the

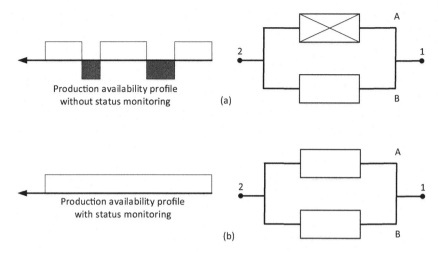

FIGURE 1.25 Production availability profile if (a) a breakdown policy is adopted and (b) constant status monitoring and immediate replacement of a failed redundant component is adopted (Todinov, 2007).

status of both components is constantly monitored and repair is initiated whenever any component fails, the downtime associated with a critical (system) failure could be avoided almost completely (Figure 1.25(b)). Therefore, for large systems, downtime status monitoring can help increase availability (Todinov, 2007).

Another example, related to dependent failures, is a common coupling fan-cooled device, where failure of the fan causes an overheating failure of the cooled device. Such a failure can be prevented if the status of the fan is monitored and, in the event of failure, operation is discontinued until the fan is replaced.

It should be pointed out that while status monitoring benefits systems with built-in redundancy, CM of the extent of degradation of components and structures benefits systems without any built-in redundancy.

CM can be used to predict the approximate time of failure; this provides a basis for planning the necessary resources for repair. The number of required spare parts, and the cost of their preservation and storage can also be reduced significantly. Orders for components whose failure is imminent can be placed, thereby reducing delays associated with their delivery. In addition, CM provides the basis for improved designs by feeding back information on actual times to failure, vulnerable components, root causes of failures, rate of material degradation, and the impact of the operating conditions (Todinov, 2007).

1.5.2 OPTIMIZING PREVENTIVE MAINTENANCE

PM can always be improved. The following are ways to perfect maintenance (Odesie, 2020):

1. Eliminate low-value tasks (waste).
2. Review PM-related failures shortly after they occur.
3. Replace intrusive PM tasks with non-intrusive CBM or PdM.
4. Reduces the ambiguity of maintenance tasks that are not clearly written.
5. Align environmental health safety requirements with specific PM tasks.
6. Review PM procedures regularly to improve them.

1.6 PREDICTIVE MAINTENANCE AS THE EVOLUTION OF CBM PROGRAMS

1.6.1 CONDITION-BASED MAINTENANCE (CBM)

CBM is popular because it reduces the uncertainty of maintenance activities. The equipment's operating condition is measured using monitoring parameters such as vibration, temperature, lubricating oil, contaminants, and noise levels. Maintenance decisions are based on the equipment's condition, so it allows maintainers to identify and solve problems before failure, making it very attractive for organizations with high-value assets. CBM can ensure equipment health management, lower LCC, and avoid catastrophic failure (Jardine, Lin, & Banjevic, 2006).

1.6.1.1 CBM Elements and Techniques

CBM is a failure management strategy for a particular failure mode that meets criteria such as:

- Potential failure is clearly defined.
- Failure interval is identifiable.
- Maintenance task interval is less than the failure interval and physically possible.
- Time between the discovery of the potential failure and the occurrence of the function failure is long enough for the maintenance action to be taken to avoid, eliminate, or minimize the consequences of the failure mode (Lawler & Felker, 2020).

Whereas PM addresses age-dependent failure probabilities, CBM addresses failures that can be measured by one or several indicators (Rausand & Vatn, 1998). When applying maintenance efforts (people, processes, and tools) in a CBM environment, maintenance is based on the actual condition of the equipment versus the age of the equipment, so that equipment in good condition does not need to be maintained as frequently as equipment that has reached the predicted age of deterioration. The core of CBM is using test equipment or statistically modeling data to predict the condition of equipment (Cadick & Traugott, 2009). The vision of CBM is to enable equipment to achieve nearly zero breakdowns, transforming traditional maintenance practices from RTF to PdM and prevention of failures. CBM utilizes failure history to predict breakdowns before they happen to prevent future failures from occurring in the field where repairs are costly and operations are impacted (Gillespie, 2015).

1.6.1.2 Requirements for CBM Implementation

CBM requires the following (Christiansen, 2018):

1. CM sensors.
2. CM tools.
3. Properly trained staff who can read sensors and use CM tools.
4. Staff who are able to analyze the data and schedule maintenance work based on the analysis.
5. Specialized software (i.e., CMMS) to collect, track, and analyze data coming from CM sensors.
6. A maintenance team willing to learn and adapt to change.

If this seems too complicated and expensive, PM is a good first step. The results can be comparable to those of CBM but it is easier to introduce (Christiansen, 2018).

1.6.2 PREDICTIVE MAINTENANCE VS. CONDITION-BASED MAINTENANCE (CBM)

PdM and CBM occur before breakdowns happen. As such, they are both forms of proactive maintenance and are designed to increase reliability and decrease downtime.

The primary difference between them is the way maintenance is measured. PdM relies on precise formulas in addition to sensor measurements (temperature, vibration, and noise), and maintenance work is performed based on the analysis of these parameters. In this way, PdM is a very exact form of maintenance because it predicts future maintenance events.

CBM relies on real-time sensor measurements. Once a parameter reaches an unacceptable level, maintenance workers are dispatched. This means CBM systems perform work only in the moment it is needed.

Both PdM and CBM can be expensive to initiate, though they both justify their upfront cost by saving money on downtime and equipment maintenance. CBM in particular can be expensive because of the cost of maintaining sensor devices, so it's best used on critically important equipment (UpKeep, 2020(c)).

Some people use CBM and PdM interchangeably. While they overlap in some ways, and the confusion is understandable, they are *not* the same thing. PdM combines condition-based diagnostics measuring variables with predictable patterns with complex predictive formulas to more precisely predict when a piece of equipment will fail. CBM relies on measurements at set intervals and lacks the predictive formulas needed to interpret trends. Seen from a certain angle, PdM is a more accurate version of CBM (Christiansen, 2018).

1.6.3 DIFFERENCES BETWEEN PREDICTIVE MAINTENANCE AND CONDITION-BASED MAINTENANCE (CBM)

Table 1.2 shows the differences between PdM and CBM (UpKeep, 2020(c)):

TABLE 1.2

Differences between Predictive Maintenance and Condition-Based Maintenance (CBM)

	Predictive Maintenance (PdM)	**Condition-Based Maintenance (CBM)**
Definition	Predictive maintenance (PdM) is work scheduled in the future based on analysis of sensor measurements and formulas.	Condition-based Maintenance (CBM) is work performed at the exact moment when measured parameters reach unacceptable levels.
Workflow		
Trigger	Predicted date	Measured parameter level
Cost	Medium/high	Medium/high (startup cost)
Cost Savings	25% to 30%	Dependent on amount of equipment using CBM
Resources Needed	• Maintenance software for scheduling • Maintenance scheduler (for larger organizations) • Condition monitoring software • Condition monitoring tools and sensors • PdM training	• Maintenance software for dispatching technicians and creating work orders • Condition monitoring software • Condition monitoring tools and sensors • CBM training

(continued)

TABLE 1.2 (Continued)
Differences between Predictive Maintenance and Condition-Based Maintenance (CBM)

	Predictive Maintenance (PdM)	Condition-Based Maintenance (CBM)
Pros	• Maintenance need is predicted in advance • Reduces maximum amount of downtime • Improved automation of maintenance tasks	• Maintenance work is performed only as needed • Fewer unplanned downtime events • Improved prioritization of maintenance time
Cons	• Expensive to implement and maintain • Time-intensive	• High cost of installation, training, and maintenance • Difficult to choose proper sensor equipment
Use Case	An organization has assets with slow-speed bearings that frequently fail. Preventive maintenance is already in place but the organization suspects its assets are being over-greased. To perform maintenance with more precision, they use ultrasound analysis (good for slow-speed bearings). Now, work orders for greasing are only scheduled when certain ultrasound measurements are reached.	An organization needs to make sure that a critical piece of safety equipment can be maintained while it is running at load. One critical equipment parameter is the amount of vibration produced. The organization decides to implement CBM so that maintenance is performed only when vibrations begin to reach unsafe levels. This way, the safety equipment can be run constantly during maintenance and only receives maintenance when necessary.

Source: UpKeep (2020(c)).

1.6.3.1 Key Conclusions

1. PdM is performed when needed based on measurements and calculations; CBM is performed at the exact moment a parameter reaches an unsatisfactory level.
2. Both types of maintenance use CM and measurements.
3. PdM can result in cost-savings of 25% to 30%; CBM's savings depend on how much equipment runs on the system.
4. PdM and CBM are both somewhat expensive, but PdM is much more precise (UpKeep, 2020(c)).

REFERENCES

Arora N., 2016. Supervisory control and data acquisition (SCADA): Back bone of modern industry – A review supervisory control and data acquisition (SCADA). Conference Proceeding of 4th International Conference on Recent Innovation in Science, Technology and Management (ICRISTM-16) at YMCA, Jai Singh Road, Delhi, India on December 3, 2016, ISBN:978-81-932712-9-2.

Bass T., 2000. Intrusion detection systems and multisensor data fusion. Communications of the ACM, 43(4), 99–105.

Bhelkar V., Shedge D. K., 2016. Different types of wearable sensors and health monitoring systems: A survey. 978-1-5090-2399-8/16. ©2016 IEEE.

Bloch H. P., Geitner F. K., 1983. Machinery failure analysis and troubleshooting. Gulf, Houston, TX.

Boys W., 2009. Back to basics: SCADA. Automation TV: Control Global – Control Design, August 18, 2009.

Cadick J., Traugott G., 2009. Condition-based maintenance: A white paper review of CBM analysis techniques, Cadick Corporation Tech Bulletin TB-017.

Castanedo F., 2013. A review of data fusion techniques. The Scientific World Journal. 2013, 704504.

Crain M., 2003. The role of CMMS: A white paper on the selection and implementation of computerized maintenance management systems (CMMS). Industrial Technologies. Northern Digital, Inc.

Christiansen B., 2018, October 2. A complete guide to condition-based maintenance. https://limblecmms.com/blog/condition-based-maintenance. Viewed: May 18, 2020.

Dasarathy B. V., 2001. Information fusion: What, where, why, when, and how? Information Fusion, 2, 75–76.

Dasarathy B. V., 2003. Information fusion as a tool in condition monitoring. Information Fusion, 4, 71–73.

De Vin L. J., Amos H. C., Oscarsson J., Andler S. F., 2006. Information fusion for simulation based decision support in manufacturing. Robotics and Computer Integrated Manufacturing, 22, 429–436.

Dhillon B. S., 2006. Reliability, quality, and safety for engineers. CRC Press.

Durrant-Whyte H. F., 1988. Sensor models and multisensor integration. International Journal of Robotics Research, 7(6), 97–113.

Durrant-Whyte H. F., Stevens M., 2001. Data fusion in decentralized sensing networks. Proceedings of the 4th International Conference on Information Fusion (pp. 302–307).

Fiix, 2020. Corrective maintenance: Free guide to preventive maintenance. www.fiixsoftware.com/corrective-maintenance. Viewed: May 17, 2020.

Gillespie A., 2015. Condition-based maintenance: Theory, methodology, & application. Science Applications International Corporation (SAIC).

Greenland P., 2012. Understanding industrial system structures is key to powering them. Exclusive Technology Feature.

Grime S., Durrant-Whyte H. F., 1994. Data fusion in decentralized sensor networks. Control Engineering Practice, 2(5), 849–863.

Hannah P., Starr A. G., Bryanston-Cross P., 2001. Condition monitoring and diagnostic engineering: – A data fusion approach, Proc. 14th International Congress on Condition Monitoring and Diagnostic Engineering Management (Comadem 2001) (pp. 275–282).

Hilletofth P., Ujvari S., Hilmola O-P., 2007. Information fusion in maintenance planning. Logistics Research Group, University of Skövde. Swedish Production Symposium 2007.

Hounsell D., 2008. Justify budgets with a CMMS. Facility Maintenance Decisions. September 2008. www.facilitiesnet.com/software/article/Justify-Budgets-with-a-CMMS-Facility-Management-Software-Feature--9600. Viewed: May 13, 2020.

IDCOM Inc., 2020. Preventive maintenance optimization. www.idcon.com/resource-library/preventive-maintenance/preventive-maintenance-optimization. Viewed: May 17, 2020.

Jardine A. K. S., Lin D., Banjevic D., 2006. A review on machinery diagnostics and prognostics implementing condition-based maintenance. Mechanical Systems and Signal Processing, 20(7), 1483–1510.

JDL, 1991. Data fusion lexicon. Technical Panel for C3, F.E., Code 420, 1991.

Karimi K., 2020. The role of sensor fusion in the Internet of Things. www.mouser.es/applications/sensor-fusion-iot. Viewed: May 14, 2020.

Kocić J., Jovičić N., Drndarević V., 2018. Sensors and sensor fusion in autonomous vehicles. School of Electrical Engineering, University of Belgrade. November 2018.

Koncar V., 2019. Smart textiles for in situ monitoring of composites. The Textile Institute Book Series 2019 (pp. 153–215).

Lamkin-Kennard K. A., Popovic M. B., 2019. Sensors: Natural and synthetic sensors, in biomechatronics.

Lawler J., Felker D., 2020. Condition-based maintenance gap analysis. Reliability Web.Com. US Army AMRDEC. https://reliabilityweb.com/articles/entry/condition_based_maintenance_gap_analysis, Viewed: May 18, 2020.

Lee J., Wang H., 2008. New technologies for maintenance. Foxconn Technology Group.

Llinas J., Bowman C., Rogova G., Steinberg A., 2004. Revisiting the JDL data fusion model II. Proceedings of the 7th International Conference on Information Fusion (pp. 1218–1230).

Luo R.C., Yih C.-C., Su K.L., 2002. Multisensor fusion and integration: Approaches, applications, and future research directions. IEEE Sensors Journal, 2(2), 107–119.

Majumder S., Mondal T., Deen J. M., 2017. Wearable sensors for remote health monitoring. Sensors, 17(1), 130.

Pantelopoulos A., Bourbakis N. G., 2010. A survey on wearable sensor-based systems for health monitoring and prognosis. Transactions on Systems, Man, And Cybernetics – Part C: Applications and Reviews, 40(1).

Odesie, 2020. Preventive and predictive maintenance concepts. Technology transfer services. www.myodesie.com/wiki/index/returnEntry/id/2965#Optimizing%20Preventive%20Maintenance. Viewed: May 18, 2020.

Randall R. B., 2011. Vibration-based condition monitoring: Industrial, aerospace and automotive applications. John Wiley & Sons.

Rastegari A., 2015. Strategic maintenance development focusing on use of condition based maintenance in manufacturing industry. Mälardalen University Press Licentiate Theses, No. 213.

Rausand M., Vatn J., 1998. Reliability centered maintenance. Risk and Reliability in Marine Technology, 60(2), 121–132.

Rosenblatt J. K., Hendler J. A., 1999. Distributed information resources. In Advances in Computers, 48, 315–353.

Rosmaini A., Kamaruddin Sh., 2012. An overview of time-based and condition-based maintenance in industrial application. Computers & Industrial Engineering, 63(1), 135–149.

Shalini J., Kumar S. J., 2013. Birtukan Teshome, Samrawit Bitewlgn Muluneh, Bitseat Tadesse Aragaw. International Journal of Advanced Research in Electrical, Electronics and Instrumental Engineering, 2.

Sigma Industrial Precision, 2018. Differences between predictive and traditional maintenance. November 2018. www.predictive-sigma.com/en/2018/11/27/blog-diferencias. Viewed: May 16, 2020.

Starr A. G., Ball A. D., 2000. Systems integration in maintenance engineering, Proceedings of the Institution of Mechanical Engineers Part E – Journal of Process Mechanical Engineering, 214 (pp. 79–95).

Starr A., Willetts R., Hannah P., Hu W., Banjevic D., Jardine A.K.S., 2002. Data fusion applications in intelligent condition monitoring. January 2002.

Stīpnieks K., 2018. What is IT system support and maintenance? NETCORE with passion for perfection. November 5, 2018. https://netcore.agency/hub/it-system-support-and-maintenance. Viewed: May 17, 2020.

Sullivan G. P., Pugh R., Melendez A. P., Hunt W. D., 2010. Operations & maintenance best practices. A Guide to Achieving Operational Efficiency, Release 3.0. Chapter 4: Computerized Maintenance Management System.

Thakur R., 2017. Infrared sensors for autonomous vehicles. IntechOpen. December 20, 2017. DOI: 10.5772/intechopen.70577.

Thusu R., 2011. Sensors facilitate health monitoring. Fierce Electronics. April 1, 2011. www.fierceelectronics.com/components/sensors-facilitate-health-monitoring. Viewed: May 19, 2020.

Tirindelli P., 2016. Sensor fusion of raw GPS measurements for autonomous vehicle localization. Master in Artificial Intelligence. April 21, 2016.

Todinov M. T., 2007. Risk- based reliability analysis and generic principles for risk reduction. Elsevier Science.

UpKeep, 2020(a). Corrective maintenance examples and definition: Information for building effective maintenance programs. www.onupkeep.com/learning/maintenance-types/corrective-maintenance. Viewed: May 17, 2020.

UpKeep, 2020(b). What is preventive maintenance? www.onupkeep.com/preventive-maintenance. Viewed: May 17, 2020.

UpKeep, 2020(c). Predictive vs. condition-based maintenance. www.onupkeep.com/learning/maintenance-types/predictive-condition-based. Viewed: May 18, 2020.

Xu J., Xu L., 2017. Integrated system health management. Perspectives on Systems Engineering Techniques. Academic Press.

Waltz E. L., Llinas J., 1990. Multisensor data fusion, Artech.

Wikipedia, 2020. Sensor fusion. https://en.wikipedia.org/wiki/Sensor_fusion. Viewed: May 16, 2020.

2 Predictive Maintenance Programs and Servitization Maintenance as a Service (MaaS) Creating Value through Prognosis Capabilities

2.1 INDUSTRY 4.0 AND SERVITIZATION

2.1.1 INDUSTRY 4.0

The term Industry 4.0 is used to designate the fourth industrial revolution which is now well under way. It was preceded by three other industrial revolutions. The first began with the introduction of mechanical production facilities in the latter half of the 18th century. At the turn of the 20th century, electrification and the division of labor tasks (i.e., Taylorism) led to the second industrial revolution. The third industrial revolution, often called "the digital revolution", started in the 1970s, when advanced electronics and information technology (IT) were introduced into production processes (Hermann et al., 2015).

The idea of a fourth industrial revolution was formulated by a group of representatives from business, politics, and academia seeking to strengthen the competitiveness of the German manufacturing industry (Kagermann et al., 2011). The German government supported the idea and incorporated Industry 4.0 into its "High-Tech Strategy 2020 for Germany" initiative. An Industry 4.0 Working Group subsequently developed the following vision:

> In the future, businesses will establish global networks that incorporate their machinery, warehousing systems and production facilities in the shape of Cyber-Physical Systems (CPS). In the manufacturing environment, these Cyber-Physical Systems comprise smart machines, storage systems and production facilities capable of autonomously exchanging information, triggering actions and controlling each other independently. This facilitates fundamental improvements to the industrial processes involved in manufacturing, engineering, material usage and supply chain and lifecycle management. The Smart Factories that are already beginning to appear employ a completely new approach to production. Smart products are uniquely identifiable, may be located at all times and know their own history, current status and alternative routes to achieving their target state. The embedded manufacturing systems are vertically networked with business processes within factories and enterprises and horizontally connected to dispersed value networks that can be managed in real time – from the moment an order is placed right through to outbound logistics. In addition, they both enable and require end-to-end engineering across the entire value chain.
>
> *Kagermann et al., 2013*

DOI: 10.1201/9781003097242-2

These ideas were widely adopted in Germany, including by the German National Academy of Science and Engineering who produced a manifesto on the topic (Acatech, 2013), and outside Germany as well. At the European level, the Public-Private Partnership for Factories of the Future works to develop Industry 4.0 topics and the Industrial Internet Consortium (IIC) promotes Industry 4.0 in the USA (Stock & Seliger, 2016).

The paradigm of Industry 4.0 has three dimensions (Industry 4.0: Whitepaper FuE-Themen, 2015; Acatech, 2015; VDI/VDE-GMA, 2015(a)):

1. Horizontal integration: Cross-company and internal intelligent cross-linking and digitaliza-tion of value creation modules throughout the value chain of a product's lifecycle and between value chains of adjoining products' lifecycles.
2. End-to-end engineering: Intelligent cross-linking and digitalization throughout all phases of a product's lifecycle, from the raw material acquisition to manufacturing systems, product use, and product end of life (EoL).
3. Vertical integration and networked manufacturing systems: Intelligent cross-linking and digit-alization within the different aggregation and hierarchical levels of a value creation module from manufacturing stations via manufacturing cells, lines, and factories to integrating the associated value chain activities, such as marketing and sales or technology development.

In a manufacturing system, intelligent cross-linking is realized by the application of CPSs operating in a self-organized and decentralized manner (Acatech, 2015; Gausemeier et al., 2015; Spath et al., 2013). These systems use embedded mechatronic components (i.e., sensor systems) to collect data and actuator systems to influence physical processes (Gausemeier et al., 2015).

CPSs are part of a socio-technical system. They use human-machine interfaces to interact with operators (Hirsch-Kreinsen & Weyer, 2014) and they are intelligently linked with each other and continuously interchange data via virtual networks (e.g., a cloud) in real time. The cloud itself is implemented in the Internet of Things (IoT) and Internet of Services (IoS) (Acatech, 2015).

2.1.1.1 Industry 4.0 Definition

Industry 4.0 is inclusive of all the technologies and concepts of the entire value chain. Once Industry 4.0 is fully materialized, CPSs will monitor physical processes, create virtual copies of the physical world, and make decentralized decisions. They will communicate and cooperate with each other and humans in real time via IoT. They will offer internal and cross-organizational services to be used by all participants in the value chain (Hermann et al., 2015).

2.1.1.2 What Is Industry 4.0?

Industry 4.0 is all about communication: screws communicate with assembly robots, self-driving forklifts stock high shelves with goods, intelligent machines coordinate independently running pro-duction processes – people, machines, and products are directly connected with each other (Plattform Industrie 4.0), creating an intelligent networking of machines and processes for industry with the help of information and communication technology.

There are many ways for companies to use intelligent networking. The possibilities include, for example:

• Flexible production: In manufacturing a product, many companies are involved in a step-by-step process. By being digitally networked, these steps can be better coordinated and the machine load better planned.
• Convertible factory: Future production lines can be built in modules and quickly assembled for tasks. Productivity and efficiency will be improved; individualized products can be produced in small quantities at affordable prices.

- Customer-oriented solutions: Consumers and producers will move closer together. The customers themselves will design products according to their wishes – for example, sneakers designed and tailored to the customer's unique foot shape. At the same time, smart products that are already being delivered and in use can send data to the manufacturer. With these usage data, the manufacturer can improve products and offer the customer novel services.
- Optimized logistics: Algorithms will calculate ideal delivery routes. Machines will independently report when they need new material. This type of smart networking will enable an optimal flow of goods.
- Use of data: Data on the production process and the condition of a product will be combined and analyzed, providing guidance on how to make a product more efficiently. More importantly, new business models and services will be created. For example, elevator manufacturers can offer their customers predictive maintenance (PdM): elevators equipped with sensors that continuously send data about their condition. Product wear will be detected and corrected before system failure.
- Resource-efficient circular economy: The entire lifecycle of a product can be considered with the support of data. The design phase will already be able to determine which materials can be recycled (Plattform Industrie 4.0).

2.1.1.2.1 Revolution: What Is New in Industry 4.0?

Since the 1970s, IT has been incorporated into business. Desktop personal computers, the use of office IT, and the first computer-aided automation revolutionized the industry. For Industry 4.0, the core technology is the Internet. Digitalizing production is gaining a new level of quality with global networking across corporate and national borders. The IoT, machine-to-machine communication, and intelligent manufacturing facilities are heralding the fourth industrial revolution, Industry 4.0 (Plattform Industrie 4.0).

2.1.1.2.2 On the Path to Industry 4.0: What Needs to Be Done?

Implementing Industry 4.0 is a complex project: the more process companies digitalize and network, the more interfaces are created between different actors. Uniform norms and standards for different industrial sectors, IT security, and data protection play equally central roles as the legal framework, changes in education and jobs, the development of new business models, and corresponding research (Plattform Industrie 4.0).

2.1.1.2.3 Digital Transformation "Made in Germany": Platform Industry 4.0

The overarching goal of Platform Industry 4.0 is to secure Germany's leading international position in the manufacturing industry. For this, the participants of the platform discuss appropriate and reliable framework conditions. As an initiator and moderator of various interests and messages, Platform Industry 4.0 ensures a space for pre-competitive exchange between relevant stakeholders from politics, business, academia, trade unions, and associations. The platform is one of the world's leading Industry 4.0 networks (Plattform Industrie 4.0).

Platform Industry 4.0:

- Develops core concepts in working groups on how to tackle challenges on the road to Industry 4.0.
- Provides concrete recommendations for academics, companies, and politicians.
- Supports small and medium-sized enterprises with specific service offerings.
- Drives national and international exchanges through bilateral and multilateral cooperation – particularly in the areas of IT security and standardization (Plattform Industrie 4.0).

2.1.1.3 Industry 4.0 Conception

In the 21st century, products' lifecycles are shorter, and consumers are demanding more complex, unique products in larger quantities. This reality poses many challenges to production, and current practices are not sustainable. The industrial sector is going through a paradigm shift, which will change production drastically. Traditional centrally controlled and monitored processes will be replaced by decentralized control built on the self-regulating ability of products and work pieces that communicate with each other.

The essence of Industry 4.0 conception is the introduction of network-linked intelligent systems, which realize self-regulating production: people, machines, equipment, and products will communicate to one another. The goal is to make flexible, custom production economical and to use resources efficiently.

Virtual and actual reality will merge during production. Products will schedule their own production, while factories will be self-regulating and optimize their own operation (Gubán & Kovás, 2017).

2.1.1.3.1 Five Components of Networked Production

The five main elements of networked production are the following:

- Digital work pieces: The dimensions, quality requirements, and the order of technological processing will be supplied by the work piece itself.
- Intelligent machines: Intelligent machines will communicate simultaneously with the production control system and the product being processed, so the machines will coordinate, control, and optimize themselves.
- Vertical network connections: When processing the unique specifications given by the customer for the product to be manufactured, the production control system forwards the design of the digital product created by automated rules to the equipment. The products control their own manufacturing process, as they communicate with the equipment and devices on the conditions of production.
- Horizontal network connections: Communication is realized not only within a factory, but in the whole supply chain; between the suppliers, manufacturers, and service providers. The main purpose is to improve the efficiency of production and use resources in a more economical way.
- Smart work pieces: The product to be manufactured senses the production environment with internal sensors and controls and monitors its own production processes to meet production standards, as it is able to communicate with the equipment and the components already incorporated and about to be incorporated.

Industry 4.0 is not a technology from the far-distant future. In July 2015, Changing Precision Technology (Dongguan, China) became the first factory where only robots work. Each labor process is executed by machines: the production is done by computer operated robots, the transport uses self-driven vehicles, and even the storage process is completely automatic (Gubán & Kovás, 2017).

The importance of production arranged in a global network is that the manufacturing process can flexibly adapt to unique customer demands, to the activity of the other parties in the supply chain, and to the rapidly changing economic environment.

Industry 4.0 affects the corporate world in the following ways:

- Integration and digitalization of horizontal and vertical value chains.
- Digitalization of products and services.
- Formation of digital business models and customer relations.

The connected technologies are shown in Figure 2.1.

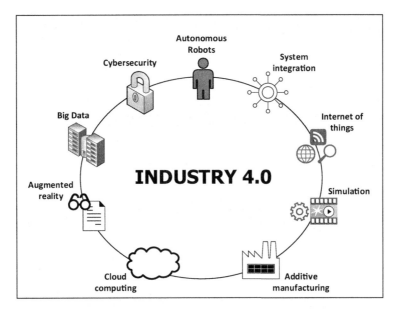

FIGURE 2.1 Main technologies of Industry 4.0 (Gubán & Kovás, 2017).

2.1.1.4 Industry 4.0 Components

2.1.1.4.1 Cyber-Physical Systems

Industry 4.0 fuses physical and virtual worlds (Kagermann, 2014) via CPSs. Simply stated, CPSs integrate computer technology and physical processes. More specifically, embedded computers and networks monitor and control the physical processes. At the same time, feedback loops mean the physical processes affect computations (Lee & Suh, 2009).

CPSs have seen three phases of development. First iterations of CPS included identification technologies like radio-frequency identification (RFID) tags. The second included sensors and actuators performing specific functions. Now, third-generation CPSs can store and analyze data, are equipped with multiple sensors and actuators, and are network compatible (Bauernhansl et al., 2014).

An increasing number of companies are using intelligent CPS. For example, Würth Canada uses a system to manage small parts, its intelligent bin (iBin). The iBin contains a built-in infrared camera module. If the quantity of small parts (or C-parts) falls below a threshold number, the iBin automatically orders new parts via RFID. This allows real-time consumption-based parts management (Günthner et al., 2014).

2.1.1.4.2 Internet of Things

As part of Industry 4.0, the IoT is being incorporated into the manufacturing process (Kagermann et al., 2013). IoT allows things and objects (e.g., sensors, mobile phones, actuators, and RFID) to interact and cooperate to reach common goals (Giusto et al., 2010). Each of these things and objects can be understood as a CPS. Therefore, the IoT can be defined as a network in which CPSs cooperate with each other. Examples include Smart Homes, Smart Factories, and Smart Grids (Bauernhansl et al., 2014).

2.1.1.4.3 Internet of Services

Another development of Internet possibilities is the Internet of Services (IoS). Basically, IoS permits service vendors to offer their services via the Internet. IoS includes users of the services, the organization offering services, and the services themselves (Buxmann et al., 2009), creating a dynamic variation of how individual value chain activities are distributed (Industry 4.0: Whitepaper FuE-Themen, 2015).

In the future, this concept will likely apply to entire value-added networks. For example, production technologies may be offered over the IoS and used to manufacture products (Scheer, 2013). In fact, in Germany, the IoS has already been implemented in a project like this. The project, SMART FACE, is developing a service-oriented architecture for a new distributed production control for the automotive industry. The system will permit the use of modular assembly stations that can be flexibly modified or expanded. Automated vehicles (AVs) serve as modes of transportation between the assembly stations. The assembly stations and AVs both supply their services through the IoS (Fraunhofer-Institut für Materialfluss und Logistik (IML, 2014).

2.1.1.4.4 Smart Factory

Smart factories are a main component of Industry 4.0 (Kagermann et al., 2013). A smart factory is context-aware and thus is able to assist people and machines to carry out their respective tasks (Lucke et al., 2008).

For example, a smart factory in Germany uses intelligent work piece carriers to report when a work piece is ready for pick-up. This relieves employees of unnecessary work (Schlick et al., 2014).

2.1.1.5 Industry 4.0: Design Principles

Six design principles can be derived for the main Industry 4.0 components discussed above: CPSs, IoT, Internet of Systems, and Smart Factories. These design principles are interoperability, virtualization, decentralization, real-time capability, service orientation, and modularity. These are shown in Table 2.1 (Hermann et al., 2015) and explained below using the example of SmartFactory[KL] a technology initiative at the German Research Center for Artificial Intelligence. In this initiative, a demonstration plant processes parts for key finders and assembles them. Each key finder has an RFID tag that provides relevant production data (Schlick et al., 2014).

2.1.1.5.1 Interoperability

Interoperability is one of the pillars of Industry 4.0. All CPSs in the SmartFactory[KL] plant communicate with each other "through open nets and semantic descriptions" (Hermann et al., 2015). These CPSs include products (key finders), work piece carriers, and assembly stations.

2.1.1.5.2 Virtualization

Virtualization allows CPSs to monitor physical processes. In virtualization, sensor data are linked to virtual plant models and simulation models, thus creating a virtual copy of the physical world. In the

TABLE 2.1
Design Principles of Industry 4.0 Components

	Cyber-Physical Systems	Internet of Things	Internet of Services	Smart Factory
Interoperability	X	X	X	X
Virtualization	X	-	-	X
Decentralization	X	-	-	X
Real-Time Capability	-	-	-	X
Service Orientation	-	-	X	-
Modularity	-	-	X	-

Source: Hermann et al., (2015).

SmartFactory^{KL} plant, the virtual model includes the condition of all CPSs and provides all necessary information, like next working steps or safety arrangements (Gorecky et al., 2014). Virtualization can help organizations handle the increasing technical complexity of production today (Hermann et al., 2015). Of course, if there is a failure, a human is notified.

2.1.1.5.3   Decentralization
As systems increase in complexity, it gets harder to control them centrally. However, embedded computers enable CPSs to make decisions on their own – except in cases of failure, when humans must get involved (Bauernhansl et al., 2014). Yet for reasons of quality assurance and traceability, it is a good idea to keep track of the whole system at any time. In the SmartFactory^{KL} plant, RFID tags inform machines which working steps are necessary, eliminating the need for central planning and control (Schlick et al., 2014).

2.1.1.5.4   Real-Time Capability
Ideally data should be collected and analyzed in real time. The SmartFactory^{KL} plant comes close to this capability because the status of the plant is constantly tracked and analyzed. What this means is that SmartFactory^{KL} can react to the failure of one machine and reroute products to another (Schlick et al., 2014).

2.1.1.5.5   Service Orientation
We have already mentioned the value of offering services online through IoS. The SmartFactory^{KL} plant features a service-oriented architecture. All CPSs offer their functionalities as an encapsulated web service (Hermann et al., 2015). Recall that the plant manufactures key finders, and each key finder has an RFID tag that provides relevant production data. To offer better service, SmartFactory^{KL} uses a product-specific process based on the customer-specific requirements provided by the RFID tag (Schlick et al., 2014).

2.1.1.5.6   Modularity
Modular systems can quickly adapt to changing requirements. They are quickly and easily replaced or expanded, depending on the situation, that is, seasonal fluctuations or new product characteristics. The SmartFactory^{KL} plant uses the Plug&Play principle to add new modules to the production process. Plug&Play refers to the use of standardized software and hardware interfaces (Schlick et al., 2014). In effect, a new piece of equipment can be "plugged in" and "played" right away. In the SmartFactory^{KL} plant, new modules are identified automatically and can be used immediately via the IoS (Hermann et al., 2015).

2.1.2   REFERENCE ARCHITECTURE MODEL INDUSTRY 4.0 (RAMI 4.0)

Several institutions and firms in Germany – the leader of Industry 4.0 activities and ideas – recently developed a new architecture to handle the complexities of Industry 4.0: Reference Architecture Model Industry 4.0, or RAMI 4.0.

2.1.2.1   RAMI 4.0
Three companies, BITCOM, VDMA, and ZWEI, decided to develop a three-dimensional (3D) architecture to better represent the various manually interconnected features of the technical/economic properties of Industry 4.0, including the digitization of networks, products, and services, as well as new market models.

RAMI 4.0 is actually a small modification of the Smart Grid Architecture Model, originally developed for communication in networks of renewable energy sources. The RAMI 4.0 model incorporates approximately 15 industrial branches. In Figure 2.2, the layers on the vertical axis

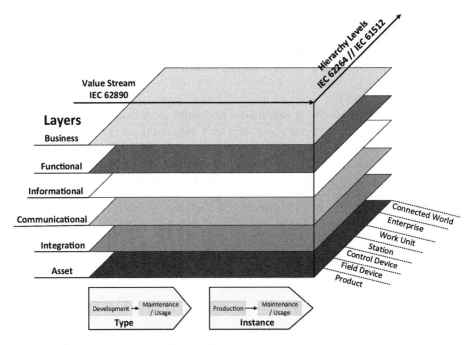

FIGURE 2.2 RAMI 4.0 model (VDI/VDE-GMA, 2016).

represent the different aspects, including functions, information, communication, the market aspect, and the integrative ability of components (Manzei et al., 2016; VDI/VDE-GMA, 2015, 2016).

The left-hand horizontal axis in Figure 2.2 displays the product lifecycle and its value stream, a very important criterion in modern engineering. Improved products, machines, and other layers of the Industry 4.0 architecture will be possible with the total digitization of the market chain. The right horizontal axis in the figure describes the components in Industry 4.0 and their function. Note that the highest level is the connected world (Zezulka et al., 2016).

2.1.3 INDUSTRY 4.0 COMPONENT MODEL

BITCOM, VDMA, and ZWEI also developed an Industry 4.0 component model to help producers and system integrators create hardware and software components. The model enables a better description of the cyber-physical features of virtual and cyber-physical objects and processes and the communication between them.

The hardware and software components of future production systems will be able to use the features specified in the Industry 4.0 component model to carry out requested tasks. The most important feature is the communication ability between virtual objects and virtual processes and real objects and real production processes. The model provides standardized, secure, and safe real-time communication among all components of production, and data are available to everything and everyone in the production chain (Zezulka et al., 2016).

The Industry 4.0 component model is shown in Figure 2.3 (VDI/VDE-GMA, 2015) and its construction is explained in the following section.

2.1.3.1 Specifications of Industry 4.0 Component Model

Different objects with different communication abilities are integrated as Industry 4.0 components on the Industry 4.0 platform. Figure 2.3 shows how an object becomes an Industry 4.0 component. The object is a standard technological object without the interactive features of an Industry 4.0

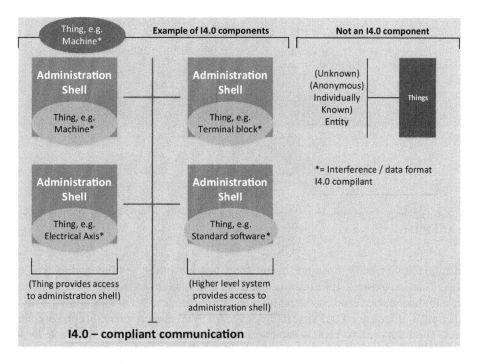

FIGURE 2.3 Industry 4.0 component model (VDI/VDE-GMA, 2015).

component. When it is surrounded by an administration shell (i.e., a data container), it becomes an Industry 4.0 component. The administration shell covers both the virtual representation of the object and its technical function.

Figure 2.3 shows four types of objects that can become Industry 4.0 components: a machine, a terminal block, an electrical axis, and software.

1. An entire machine can become an Industry 4.0 component via its control system, such as a programmable logic controller (PLC). This can be done by the producer and PLC integrator.
2. A strategically important system, such as an electrical axis, can become an Industry 4.0 component. This can be done by the component manufacturer.
3. A terminal block can become an Industry 4.0 component. This can be done by the electrical engineer.
4. The software can be an Industry 4.0 component. This could be standard software for large sets of machines. The supplier may also sell extended functions for products separately.

The Industry 4.0 component model has the following requirements (VDI/VDE-GMA, 2015):

1. The structure of the network of Industry 4.0 components must be such that any end points (i.e., Industry 4.0 components) can be connected. For this to happen, the components must have a common semantic model.
2. Industry 4.0 components must meet the requirements of different areas of interest, for example, the business office and the production area.
3. Industry 4.0 compliant communication must allow the data of a virtual representation of an Industry 4.0 component to be kept either in the object or in a higher level IT system.

Communication is extremely important for Industry 4.0 systems, so at least one information system must maintain a connection with the object. This means the object must have some passive

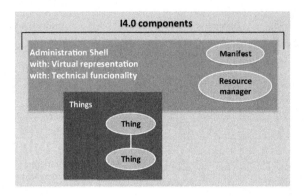

FIGURE 2.4 Industry 4.0 component (Umsetyungsstrategie Industrie 4.0: Ergebnisbericht der Platform Industrie 4.0., 2015).

communication ability but it does not mean it has to have Industry 4.0-compliant communication ability. Thus, an existing object can be extended to constitute an Industry 4.0 component. For example, a Profinet device can become an Industry 4.0 component (VDI/VDE-GMA, 2015).

Another pillar of Industry 4.0 is virtual representation. Virtual representation contains data on the real object. These data can be kept on the Industry 4.0 component and made available by Industry 4.0 compliant communication, or they can be stored in an IT system which makes them available via Industry 4.0 compliant communication. The manifest is a directory of the individual data contents of the virtual representation, including meta-information as well as basic data on the Industry 4.0 component. The virtual representation also includes data on individual objects' lifecycle phases; for example, computer-aided design (CAD) data, terminal diagrams, and manuals. In fact, the virtual representation of an object is based on data.

Figure 2.4 displays an Industry 4.0 component (Umsetyungsstrategie Industrie 4.0: Ergebnisbericht der Platform Industrie 4.0., 2015).

The administration shell of any Industry 4.0 component has to contain specific information expressed in a proper Industry 4.0 vocabulary and using an Industry 4.0 format. Depending on the purpose and context, the Industry 4.0 component can have more than one administration shell.

An Industry 4.0 component can have a technical function. The component can be software used for various tasks, such as local planning in connection with the object, or project planning, configuration, operator control, and services.

Deployment of Industry 4.0 components in a digital factory is shown in Figure 2.5. First, components are mapped in the repository by means of their administration shells and their mutual connections. Next, the physical factory is represented in digital form in the repository in administration shells and their connections using dynamic actualization (Manzei et al., 2016).

Industry 4.0 components can enter into and initiate all possible cross-connections within the Industry 4.0 factory.

2.1.4 SERVITIZATION

Industry 4.0 is changing the shape of the manufacturing sector. With the introduction of CPSs (Schaefer, 2015), real and virtual worlds are merging, with equipment, products, and people increasingly connected via the Internet. These connected systems interact to analyze data, predict failure modes, reconfigure themselves, and continuously adapt to changes in customer demand. Thus, Industry 4.0 presents many opportunities to firms who have servitized or are looking to do so (Huxtable & Schaefer, 2016).

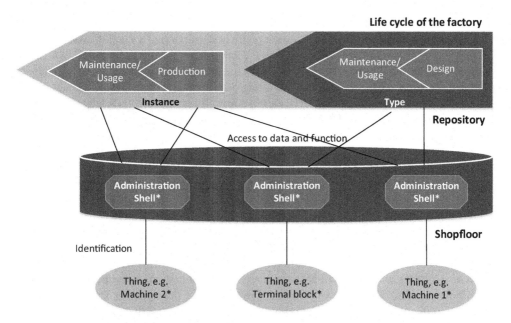

FIGURE 2.5 Repository of digital factory (Umsetyungsstrategie Industrie 4.0: Ergebnisbericht der Platform Industrie 4.0., 2015).

2.1.4.1 Concept of Servitization

During recent decades, the term "servitization" has been used by academia to capture the attitude of firms, previously known as pure manufacturers, who are increasingly bundling (or integrating) services with the goods they produce and sell (Schmenner, 2009). Although it is often considered an innovative trend, in his research on the history of US manufacturing, Schmenner (2009) suggested servitization has antecedents that go back at least 150 years. However, the term itself was coined by Vandermerwe and Rada in 1988.

The meaning of servitization goes beyond the mere adding of ancillary services to the offer of products to indicate the tendency of manufacturers to get closer to customers and their underlying needs (Schmenner, 2009). It can be useful in the business models of manufacturing firms, given their willingness to add complementary services to what is already provided to help customers. Essentially, companies offer the same products, but with different business models, these can be commercialized in different ways, yielding different returns. Servitization is often associated with activities for creating additional value and developing distinctive competences and capabilities, possibly leading to competitive advantages in economies previously just based on cost (Baines et al., 2009).

Simply stated, servitization is a strategy focusing on a service orientation to satisfy customers' needs, to achieve competitive advantage, and to enhance the performance of the firm (Ren & Gregory, 2007). Servitizing firms deliver capabilities to the customer in order to differentiate themselves and gain a competitive edge (Jacob & Ulaga., 2008). Their goal is not to sell "just" products but to offer solutions to customers (Baines et al., 2007), by integrating services with the products, creating a bundle (Vandermerwe & Rada, 1988).

One example of servitization is Rolls-Royce's "TotalCare" (Neely, 2009). Others are Xerox's "pay per click" scanning, printing, and copying (Baines, 2013) and Hilti's "Tool Fleet Managing Services" (Hilti, 2017). All illustrate attempts to deliver customers "value in use" (Baines et al., 2009). This means offering capabilities to solve users' problems, rather than selling products that can be used by them independently (Bureca, 2017).

2.1.4.2 Defining "Servitization"

Services are an "economic activity that does not result in ownership of a tangible asset".

The first use of the term servitization was by Vandemerwe and Rada in their 1988 European Management Journal article titled "Servitization of Business: Adding Value by Adding Services". They defined servitization as "the increased offering of fuller market packages or 'bundles' of customer focused combinations of goods, services, support, self-service and knowledge in order to add value to core product offerings". Under their definition, "services are performed and not produced and are essentially intangible".

Other definitions of servitization appear in Table 2.2.

The delivery of product-based services is central to all definitions, and they generally agree with Vadamerve and Rada (1988). One slight deviation is Lewis, Staudacher, and Slack (2004) who refer to the idea of a functional product. In the product-service systems (PSS) literature, this is considered a specific type of product-service offering. This highlights the many similarities between the servitization and PSS research communities. Although they emerged from differing perspectives on the world, they are converging toward a common conclusion that manufacturing companies should be focusing on selling integrated solutions or PSS. A link with servitization is also identified by Baines et al. (2007), who define a PSS as an integrated combination of products and services that deliver value in use. Although these two bodies of research have developed separately, it now seems appropriate to refine the servitization definition to encompass the PSS theme. Accordingly, we can say the following: servitization is the innovation of an organization's capabilities and processes to better create mutual value through a shift from selling products to selling PSS (Baines et al., 2009).

2.1.4.2.1 Drivers of Servitization

The servitization of manufacturing has risen exponentially over the last 20 years (Lightfoot et al., 2012).

- Economic: Traditional manufacturing has shifted production away from Western economies to emerging economies such as China and India. Lower labor rates in these nations mean Western firms cannot compete on cost alone and have transitioned to services.
- Environmental: Global populations are rising, and resources are being stretched. Western companies are looking to "do more with less". Services are considered to promote dematerialization, and servitization is seen as a viable strategy to meet these demands.
- Social and market: Evidence suggests service contributions to an economy have a direct link to wealth. Thus, the demand for services in Western economies is on the rise.
- Technology: ICT is a key enabler of servitization. Developments in ICT mean certain services, such as product monitoring and GPS position tracking, can be offered now (Huxtable & Schaefer, 2016).

TABLE 2.2
Servitization Definitions

Neely, 2009	"The innovation of organizations' capabilities and processes to better create mutual value through a shift from selling products to selling product service systems".
Baines et al., 2009	"Servitization is the concept of manufacturers offering services tightly coupled to their products".
Van Looy and Visnjic, 2013	"A trend in which manufacturing firms adopt more and more service components in their offerings".

Source: Huxtable and Schaefer (2016).

The main financial drivers often mentioned in the literature are higher profit margins and stability of income. For manufacturers with high installed product bases (e.g., aerospace, locomotive, and automotive), service revenues can be one or two orders of magnitude greater than new product sales (Wise & Baumgartner, 1999). There is often a higher potential revenue potential in these sectors (Slack, 2005). Companies that have enjoyed success with this approach include General Electric, IBM, Siemens, and Hewlett Packard; they have achieved stable revenues from services despite significant drops in sales (Sawhney et al., 2004). The increased lifecycle of many modern complex products, such as aircraft, is pushing the most significant revenues downstream toward in-service support. These product-service combinations tend to be less sensitive to price based competition and thus tend to provide higher levels of profitability than offering the physical product alone (Ward & Graves, 2005). Finally, product-service sales tend to be counter-cyclical or more resistant to the economic cycles that affect investment and goods purchase. This can help secure a regular income and balance the effects of mature markets and unfavorable economic cycles.

Competitive advantage is a key return on services. Competitive advantages achieved through services are often more sustainable; as they are less visible and more labor dependent, services are more difficult to imitate. Many authors note the increased commoditization of the markets, where differentiating strategies based on product innovation, technological superiority, or low prices are becoming incredibly difficult to maintain. Services can enhance the customer value to the point where homogeneous physical products are perceived as customized. These increase barriers to competitors (Frambach et al., 1997).

Marketing opportunities are generally understood as the use of services to sell more products. The service component is well known to influence the purchasing decision. This is especially true in business-to-business or industrial markets where customers are increasingly demanding services. Companies are pressured to create more flexible firms, narrower definitions of core competences, and higher technological complexity, and these often lead to increasing pressures to outsource services (Lewis et al., 2004). Services are also claimed to create customer loyalty to the point where the customer can become dependent on the supplier. Services tend to induce repeat-sale and, by intensifying contact opportunities with the customer, they can put the supplier in the right position to offer other products or services. Finally, by offering services, companies gain insight into their customers' needs and can develop more tailored offerings (Baines et al., 2009).

In a nutshell, servitization has financial drivers (e.g., revenue stream and profit margin), strategic drivers (e.g., competitive opportunities and advantage), and marketing drivers (e.g., customer relationships and product differentiation) (Baines et al., 2009).

2.1.4.3 Features of Servitization

Manufacturing companies have been selling services for some time. Traditionally, however, the tendency has been for managers to view services as a necessary evil in the context of marketing strategies. Here, the main part of total value creation was considered to stem from physical goods, and services were only an add-on to products. There has been a dramatic change in the way services are produced and marketed by manufacturing companies. The provision of services has become a conscious and explicit strategy, with services a main differentiating factor in a totally integrated product and service offering. Today, the value proposition often includes services as fundamental value-added activities, thus reducing the product to only a part of the offering. Some companies find this an effective way to open the door to future business.

A key feature of servitization is customer-centricity. Customers are not just provided with products but with tailored "solutions". These deliver desired outcomes for specific customers or types of customers even if this requires the incorporation of products from other vendors. The use of "multi-vendor" products to deliver customer centric solutions is exemplified by Alstom's maintenance, upgrade, and operation of trains and signaling systems, or Rockwell's on-site asset management for maintenance and repair of automation products.

This customer orientation consists of two separate elements. First, a shift of the service offering from product-oriented to "user's processes oriented", or in other words, a shift from a focus on ensuring the proper functioning and/or customer's use of the product to pursuing the efficiency and effectiveness of the end-user's processes related to the product. Second, a shift of the nature of customer interaction from transaction-based to relationship-based, or a shift from selling products to establishing and maintaining a relationship with the customer (Oliva & Kallenberg, 2003).

There are a variety of forms of servitization, each with different features. Potential applications can be situated along a "product-service continuum". This is a continuum from traditional manufacturers where companies merely offer services as add-on to their products, through to service providers where companies have services as the main part of their value creation process. Companies look at their unique opportunities and challenges at different levels of "service infusion" and deliberately define their position (Gebauer, 2008). This is a dynamic process, with companies redefining their position over time and moving toward increasing service dominance (Baines et al., 2009).

2.1.5 INDUSTRY 4.0 SERVICES

This section integrates the new service offerings identified by the servitization literature to form an Industry 4.0 servitization framework (Huxtable & Schaefer, 2016).

2.1.5.1 Industry 4.0 Servitization Framework

The servitization literature commonly describes three categories of services that manufacturing firms can offer. Base services are outcome focused, on product provision, for example, spare parts. Intermediate services are centered on enhancing a product's use and condition, for example, helpdesks or condition monitoring. Advanced services focus on the performance of the product (Lightfoot & Baines, 2013). For example, when Rolls-Royce sells "power by the hour", this is considered an advanced service because the company is selling a solution with contractual guarantees. Typically, firms begin selling base services, and develop these services through the intermediate phase with the aim of offering advanced services in the future. Advanced services have a range of complexities, and effective delivery can take substantial time, resources, and planning.

Figure 2.6 shows the Industry 4.0 servitization framework (Huxtable & Schaefer, 2016).

2.2 PERFORMANCE-BASED CONTRACTING (PBC)

PBC is about buying performance, not transactional goods and services, through an integrated acquisition and logistics process delivering improved capability to a range of products and services. PBC is a support strategy that places primary emphasis on optimizing system support to meet the user's needs. Performance-based contracts delineate outcome performance goals, ensure responsibilities are assigned, provide incentives for attaining these goals, and facilitate the overall lifecycle management of system reliability, supportability, and total ownership costs. In practice, PBC involves a contracting agency (contracting the work to an external provider) and a contractor (responsible for completing the work set out in the contract). Several other parties are often involved, such as subcontractors, a legal team, and consultants. These parties work for both the contracting agency and the contractor, completing various elements of work associated with contract development, contracted work completion, or performance management/measurement (Wikipedia, 2020).

Although the fundamental concept of buying performance outcomes is common to each PBC arrangement, the PBC strategy for any specific program or commodity must be tailored to the operational and support requirements of the end item. While similar in concept, the application of PBC for a tactical fighter aircraft may be very different from a PBC strategy for an army ground combat system. There is no one-size-fits-all approach to PBC. Similarly, there is no template for sources of support in PBC strategies. Finding the right mix of support sources is based on best value

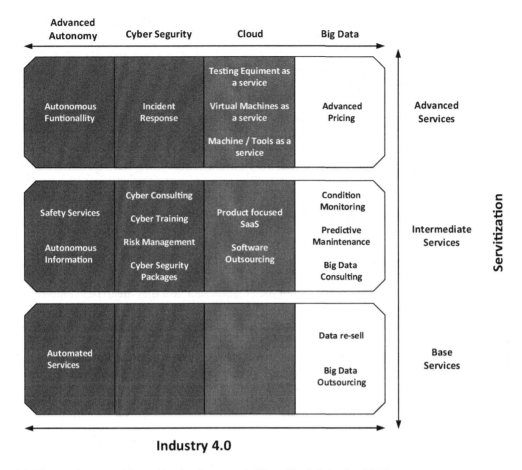

FIGURE 2.6 Industry 4.0 servitization framework (Huxtable & Schaefer, 2016).

determinations of inherent capabilities and compliance with statutes and policy. This process will determine the optimum PBC support strategy within the product support spectrum, and this can range from primarily organic support to a total system support package provided by a commercial original equipment manufacturer (Defense Acquisition University, 2005).

2.2.1 Performance-Based Contracting Metrics

A PBC approach focuses on developing strategic performance metrics and directly relating contracting payment to performance against these metrics. Common metrics include availability, reliability, maintainability, supportability, and total cost of ownership.

PCB requires incentivized, long-term contracts with specific and measurable levels of operational performance defined by the customer and agreed on by contracting parties. The incentivized performance measures aim to motivate the supplier to implement enhanced practices that offer improved performance and are cost effective. This stands in contrast to the conventional transaction-based approach, where payment is related to completion of milestones and project deliverables. In PBC, since a part or the whole payment is tied to the performance of the provider and the purchaser does not get involved in the details of the process, it becomes crucial to define a clear set of requirements for the provider. Occasionally governments fail to define the requirements clearly. This leaves room for providers to either intentionally or unintentionally misinterpret the requirements.

Recent studies highlight that the shift from transaction-based to outcome-based relationship requires a business model innovation (Wikipedia, 2020).

PBC is popular across a range of market sectors because it can reduce overall costs for buyers while providing profitable growth opportunities for sellers. PBC is deceptively simple. Performance-based contracts are different from conventional contracts, and a different approach is required to develop, implement, and manage them. Organizations must take care when implementing PBC, as failure can hurt long-term performance and cause financial, and reputational damage (Jacopino, 2020).

There are two other types of similar arrangements: performance-based lifecycle (PBL) product support and contractor logistics support. The former is much the same as PBC; the latter is different. In PBL product support, support providers are incentivized to reduce costs through innovation in contracts with either industry or government. PBL arrangements are tied to outcomes; they integrate the various product support activities of the supply chain with incentives and metrics. In contractor logistics support, support is provided by a contractor; it may or may not be incentivized.

PBC is focused on achieving specified performance quality levels, and payment is related to how well performance meets contract standards. To this end, performance-based contracts should do the following (PBL Guidebook, 2016):

- Describe the desired results or outcome to be obtained, not how they are to be achieved or the methods of performance.
- Use objective measures of performance (e.g., timeliness, quality, and quantity) that are both achievable and verifiable.
- Impose a financial penalty when services are not performed or do not meet contract requirements. Threat of a financial penalty will incentivize satisfactory contract performance.
- Include incentive fees that provide a financial reward (or some other reward) for superior performance.
- Where applicable, provide other rewards. For example, a performance-based contract might comprise five years with five additional one-year options to extend the contract, depending on performance. This kind of reward is often offered in government contracts (PBL Guidebook, 2016).

2.2.1.1   Performance-Based Contracting Implementation Process

A typical process for implementing PBC is as follows (Wikipedia, 2020):

1. Business case: Create a document reviewing potential risks, benefits, and other potential impacts of PBC, usually presented to senior managers to aid in their decision-making.
2. Outcomes: Create a short statement reflecting the desired result or final deliverable of the contract.
3. Measures: Define a set of performance measures that collectively measure the organization's performance against the outcome statement.
4. Levels: Set levels for the performance measures, that is, how well the contractor needs to perform.
5. Payment: Develop a set of payment curves setting out the pay for performance regime, that is, how much the contractor gets paid for performance level.
6. Incentives: Establish a group of incentives encouraging positive behaviors and discouraging negative behaviors.
7. Contract: Draft, review, workshop, and finalize a contract covering all aspects of the performance, payment, and terms and conditions of the relationship.
8. Review: Conduct an analysis of the outcomes of PBC, taking into account the differing definitions of success from the different groups involved in the contract.

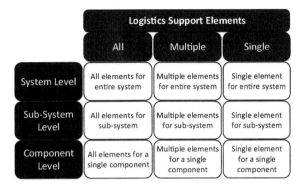

Logistics Support Elements			
	All	Multiple	Single
System Level	All elements for entire system	Multiple elements for entire system	Single element for entire system
Sub-System Level	All elements for sub-system	Multiple elements for sub-system	Single element for sub-system
Component Level	All elements for a single component	Multiple elements for a single component	Single element for a single component

FIGURE 2.7 PBC support integration.

This PBC implementation process is not intended to be rigid and inflexible. The steps can be applied to particular programs, businesses, and operational environments.

Contracts differ mainly in scale, covering a broad range from component level up to system platform level, as shown in Figure 2.7. An individual commodity is at one end of the spectrum and the entire system is at the other.

A commodity-type PBC is usually the easiest to implement, as it is easier to estimate the current baseline and level of support required and may often involve only a single commercial manufacturer. This contractor, having the most intimate knowledge of manufacturing processes, system reliability, and potential improvements, may be a prime candidate for entering into a public/private team relationship. Risk is one of the major cost drivers for contractors, and when the potential PBC contractor is also the original equipment manager, risks should be reduced.

2.2.2 Performance-Based Contracting in the Defense Industry

PBC is gaining ground in the areas of maintenance and spare parts procurement for armed forces around the world. Germany, the US, UK, and Australia are using PBC agreements to manage supplies of spare and replacement parts for their military planes. They have seen decreased costs and increased operational readiness, that is, more than 20% improvement in both areas. In what follows, we discuss PBC in the context of the military, but all comments are applicable in other industries.

In a PBC agreement, the contractor commits to achieving certain performance and availability targets. In the case of maintenance and spare parts procurement for the military, the contractor is responsible for the supply of spare parts or maintenance or logistic services to ensure availability. Depending on achieved availability and performance, the contractor's incentive increases or decreases, and with it, his profitability.

Traditional service contracts are based on work hours and material. Under PBC, this logic is reversed. Service contractors have more freedom in organizing and implementing the services.

The monetary incentive mechanism is the key element to ensure alignment of interest between the organization contracting the services (i.e., the military) and the service contractor. The military is looking for system availability and the contractor is seeking profitability. This exchange comprises the core of PBC.

For a PBC incentive model, the combined consideration of savings and achieved performance is important. Take the case of a contract for material supplies of some sort. If there was only a savings target in this type of contract, the contractor would save by procuring less, and availability would be adversely impacted. If only availability were measured, the contractor would be interested in putting as much material as possible on stock to ensure availability. Only a combination of savings and availability targets allows the interests of both parties to be met (Körner, 2020).

Organizations must address five action areas when shifting to PBC (Körner, 2020):

1. Improve quality of master data: Without good master data, it is impossible to derive demands or order material correctly. This often involves extensive preparatory work.
2. Refine predictive models: The forecasted demand is often based on historical data only. Demand planning can be enhanced by approaches such as PdM or the use of Big Data. Automating replenishments via modern enterprise resource planning (ERP) systems leads to further efficiency gains.
3. Introduce a differentiated material management: The procurement and inventory strategy should be based on both the quality of the material and the predictability of demand. For high-value material with high demand predictability, a just-in-time delivery by the supplier may be preferred; for high-value material with low predictability of demand, the use of a supplier reserve may suffice. For most small parts, replacement parts, standard parts, and low-priced materials, it may be more cost-effective to outsource the management to external service providers.
4. Manage supply chain risks: Because of budgetary constraints, there may be insufficient procurement of spare parts, and/or obsolescence management may be neglected. A gradual amendment of framework contracts along the entire supply chain may be necessary.
5. Measure and improve logistic processes: For long-term cost savings, it is necessary to continually measure and improve efficiency in the areas of procurement, warehousing, and logistics. Process mining is one solution; it renders process cycle times and process deviations transparent, and identifies approaches for process optimization.

PBC agreements provide an opportunity to expand the services business. However, to achieve cost savings and efficiency gains, it is necessary to develop a service and performance culture, besides introducing extensive changes to the organization and processes (Körner, 2020).

2.2.3 Challenges and Opportunities for Performance-Based Contracting

PBC can mean different things to different organizations. At times, performance reforms, including PBC, have been used as a political tool to increase the contracting-out of services. Simply stated, PBC involves the bidding for, monitoring, and evaluation of contracts. This requires the collection and use of service delivery performance data.

In PBC, there is often a shift away from the traditional low-bidder practice in requests for proposals toward best-value. This approach involves considering both the cost of the bid and the qualifications of the bidder, often based on knowledge of past projects. Performance indicators like citizen satisfaction, reliability of cost estimates, and reliability of time estimates based on previous work can inform decisions to accept or reject a bid. The primary goal of performance-information use in the bidding process is to reduce the problem of asymmetric information between agents. A private vendor may try to exploit an imbalance in available information; public managers can use tools like best-value bidding to even the scales.

The health services generally use high rates of contracting, but reports on how well PBC works in the health area are mixed. In at least one case, some metrics were worse after the implementation of PBC, and other operational metrics went unchanged (Ballard & Salorio, 2018). Arguably, the incentive to fully implement PBC and the sanctions for not implementing it may have been too small, suggesting the need for organizational buy-in. In any event, it is important to watch for unintended consequences in PBC. For example, in the case of public residential services, information on things like program discharges can be underreported or suppressed to hit a target occupancy.

For PBC to succeed, organizations need to devote significant time and effort to training all stakeholders. They need an ongoing monitoring system and a balanced series of metrics. PBC should incentivize both efficiency and quality and focus on real-world outcomes (Ballard & Salorio, 2018).

2.3 VIRTUAL ENGINEERING AND PROGNOSTICS FOR ADDED VALUE SERVICES

Industrial development is challenged by the growing complexity of product and process requirements, but drastically reducing time-to-market is key to maintaining a competitive edge. Strategies are being established to significantly improve the overall development process using less testing on physical builds in favor of various assembly checks, diagnosis, simulation, and risk analysis on digital models. Despite the successful practice of digital engineering, there is evidence that putting advanced IT systems into development processes cannot in itself lead to the ability to manage complexity or to achieve a new level of process performance. A new engineering methodology is required, one comprising significant technological and business improvements in products, processes, and services.

Virtual engineering is one such methodology. It refers to a range of scientific, technological, organizational, and business activities using advanced information and communication methods and tools with a focus on process and systems integration, immersive visualization, and "human-machine-human" interactions. In virtual engineering, design and validation activities occur collaboratively to test product designs, support decision-making, and enable continuous product optimization within interdisciplinary and cross-enterprise partnerships. The overall product development process must support the coordination, assessment, and concretion of engineering results of all involved partners with the support of virtual builds.

Virtual engineering integrates powerful information systems and tools within different tasks, from data generation and management to data sharing and communication (system view), and creates a network of processes and activities through the entire product lifecycle, thus enabling continuous product validation and optimization (process view).

Industry is already making use of virtual reality (VR) in systems engineering, product lifecycle management (PLM), VR technologies, and cross-enterprise and cross-cultural communication, hoping to realize benefits in the short term and to extend these benefits in the near future (Ovtcharova, 2010).

2.3.1 VIRTUAL ENGINEERING: A PARADIGM FOR THE 21ST CENTURY

Virtual engineering refers to the operationalization of virtualization in the field of engineering processes, a focused combination of engineering activities and a broad range of digital (online) data processing activities. It is trans-disciplinary and employs the same principles in several engineering fields to save costs, increase efficiency and cooperation, improve quality, and provide enhanced user experiences. It operates with very different technologies, methods, and tools in the various fields of engineering.

A product development process has three main components:

1. Developed artefact.
2. Development process.
3. Humans concerned with artefacts or processes, or both.

Within the product development process, there are three categories of computer support resources:

1. Procedural support resources: Information included in representations of artefacts, such as geometric modeling, and virtual and rapid prototyping.
2. Strategic support resources: Design processes, such as document management and workflow integration.
3. Societal support resources: Human and social aspects, such as customer research, user experiences, and human assets. (Horváth et al., 2010).

2.3.2 VIRTUAL ENGINEERING ENVIRONMENTS

2.3.2.1 General Aspects

The main aspects of success in the application of virtual engineering environments in industry are simulation and behavioral models used for dynamic visualization and interaction, collision detection, movement tracking, and 3D video projection of numerical models.

In such environments, real-time computing is a key issue. Technologies are interfaced and synchronized, for example, motion capture or man-machine interfaces (3D visualization, haptic devices, sound effects, etc.). New personal computer environments enable such computation, but algorithms and data formats have to be optimized.

The application of virtual engineering can be considered for the complete lifecycle of products, from the earliest design stages to their manufacturing, assembly, use, and maintenance phases. Production engineering is widely used to integrate multi-view and multi-technology models. Building ships, planes, or cars benefits from the use of industrial and production engineering in virtual engineering environments (Bernard, 2016).

2.3.2.2 Heterogeneous Data Formats

Mathematical models for graphical representations constitute a critical aspect of virtual engineering because reactivity and synchronization are needed in virtual engineering environments. In CAD applications, objects are mainly represented using geometric models and properties. The data formats correspond to discrete and continuous geometric entities, such as points, contours, voxels, curves, surfaces, volumes (boundary representation and constructive solid geometry), and parametric and variational models.

Information on material gradation has not yet become an integral part of CAD model data. Thus, down-line process planning assumes the presence of homogeneous material distribution throughout the interior of a solid. In discussing the requirements of rapid product development, however, other important aspects must be supported, such as color, material, and physical properties (like deformability properties). Information on exact geometry, materials, and tolerances needs to be represented in computer-compatible format. Tessellated representation is often used as a uniform and homogeneous format; such a use accelerates the computing and simulation algorithms for many virtual engineering applications (Bernard, 2016).

2.3.2.3 Virtual Reality Devices

VR and haptic devices constitute the most recent advances in man-machine interfaces for rapid product development and for virtual engineering more generally. Physical mock-ups are routinely used for applications such as assembly tests, accessibility, and space requirements. Virtual prototyping technology allows designers to test and improve their designs earlier and with more opportunities for multi-site collaborations.

Visual perception is considered the most important human sense. A high quality visual representation is thus essential for a good immersive impression of virtual products. In contrast to conventional screens that allow two-dimensional viewing, VR technology provides 3D stereoscopic viewing to support the depth perception of the eyes, thus providing better immersion. By using new hardware such as head-mounted-displays or multi-side projection walls (Figure 2.8), the view can be extended to 90–110 degrees, considered the necessary field-of-view for an immersive impression. The frequency of the image display is another concern in visual representation. Because VR allows dynamic scenes (flying-through, versatile objects), the image needs to be updated regularly. The update frequencies have to exceed 15 Hz to enable objects to appear to be in continuous motion (Bernard, 2016).

VR provides new possibilities to reduce the time needed from the first idea of a product to its mass production. It makes new simulation methods available and reduces costs without lowering product quality.

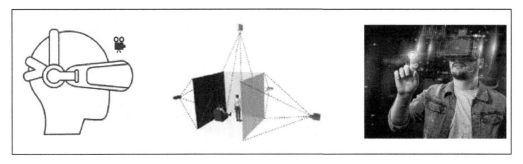

FIGURE 2.8 CAVE (multi-sided immersive environment) and two-sided projection screen with stereoscopic viewing (Bernard, 2016).

FIGURE 2.9 Virtual engineering devices with a space mouth, a 3DOF force-feedback device, and an exoskeleton (Bernard, 2016).

In contrast to the visual sense, the haptic sense is capable of sensing and interacting with the virtual environment. Haptic sensory information can be distinguished in kinetic and tactile information. Kinetic feedback is the presentation of all contact forces, including the weight of objects, to the human arm. Tactile feedback indicates the shape of objects to the touch receptors in the fingers. As the applications where force feedback is used are varied, no standards for haptic interfaces have been developed. Existing products are classified according to their different features (e.g., maximum forces or degrees of freedom (DOFs)) for the use of different applications (Figure 2.9) (Bernard, 2016).

With haptic and visual feedback, the geometrical data of a product received by 3D CAD systems can be displayed directly. In addition to displaying product assembly, simulations can be run to provide information on space requirements, necessary tools, and accessibility.

For human immersion and interaction in a virtual scene, it is necessary to be able to capture the related location of the human body (or part of it) in the virtual environment. Called "tracking", this functionality is needed for a realistic modeling of human movements (to reproduce them in the animation of a human mannequin) and for the real-time interaction management of a human body with a dynamic virtual environment if not using a direct man-machine interface (e.g., 3D mouse and force-feedback interface). Figure 2.10 shows two solutions for tracking, an electro-magnetic one and an optic one (Bernard, 2016).

2.3.3 Simple Definition of Value Added

In any business, adding value is the key to receiving more payment from customers than is needed to pay suppliers. Simply stated, "value-added work" is something that changes the product in a way that is visible to the final customer. In the making of a metal ruler, cutting the ruler off to exactly 1 m would be value-added work. Measuring the ruler to determine where the ruler should be cut does

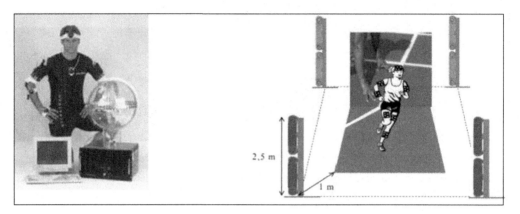

FIGURE 2.10 Tracking systems: electro-magnetic (Optotrak from Polemus) and optic (TR 16 cameras from ACTICM) (Bernard, 2016).

not change the product and is not considered value-added work. A ruler cannot be made without being cut, but perhaps the measurement task can be simplified by jamming the ruler stock against a calibrated stop so that the cutting blade will automatically cut at the proper place.

This definition is useful in the manufacturing environment because it helps identify cost items that could be reduced without affecting the customer product, such as inventory, inspection, transportation within the plant, or completion of any internal paperwork required by company policies. Studies of these non-value-added costs can help reengineer a business.

Many systems engineering tasks may not involve touching the product at all. Specially trained technicians must be called when a systems engineer wants to connect or disconnect a satellite cable from the system test setup to prevent damaging mistakes. Writing code is considered a programming job, not a systems engineering job. If systems engineering roles do not involve changing the product, how can they add value?

First, "product" has more than one meaning. Touching computer hardware may not be in the systems engineering domain. But most contracts include document deliverables (and more recently, electronic data deliverables), which systems engineers tend to create. While they are working on deliverables, they add value following the definition given above.

Second, the simple definition of value added given here represents a static and production-oriented view of the world. It ignores the dynamics of startup and the necessity of dealing with indeterminate behavior. Systems engineering adds value primarily in dynamic and chaotic processes, such as creating the first of a line of items, or when specifications change constantly, or in the interesting junction between the world of engineering and the world of people and politics. Many of the "first of a kind" systems that are built would fail miserably without adequate systems engineering (Sheard, 2017).

2.3.3.1 Four Types of Value-Added Work

The definition of value we gave at the start of the section corresponds to direct work, as shown in the first example below. But it is also important to recognize the value of indirect work. Value-added types of work are the following (Sheard, 2017):

1. Accomplishing work that directly changes the product received by the customer: This constitutes direct work. Often direct charge numbers are given to this type of value added, and others are considered "overhead".
2. Structuring the work effort (managing): Despite the current trend toward reducing management, every project must be well managed (planned, organized, tracked, and controlled).

TABLE 2.3
Systems Engineering Roles

Role	Abbreviations	Short Name
1	RO	Requirements Owner
2	SD	System Designer
3	SA	System Analyst
4	VV	Validation/Verification Engineering
5	LO	Logistics/Ops Engineer
6	G	Glue Among Subsystems
7	CI	Customer Interface
8	TM	Technical Manager
9	IM	Information Manager
10	PE	Process Engineer
11	CO	Coordinator
12	CA	Classified Ads SE

Source: Sheard (2017).

Managers do some of this, systems engineers or integrated product teams may do some, and, in the ideal state, everyone manages his or her own work. When this increases the efficiency and effectiveness of direct work, it adds value.

3. Defining the problem: Although "the problem" in the engineering arena is sometimes assumed to be defined by the product requirements, there is also a business problem of defining how to create the product. Setting a vision, guiding, and leading are included, because choices are continually made about what to approach first and how to pull together the contributions of different players. Leadership improves the efficiency and effectiveness of the managing and sometimes of the work itself.

4. Reducing risk: This is more important in some contexts than in others. Certainly NASA in the "man on the moon" decade valued risk aversion more highly than a software startup company building its initial public product might. When it is important to minimize risk, system analysts can be in high demand to model the system and its environment, before money is spent developing what might be the wrong system.

Table 2.3 shows the various systems engineering roles. Table 2.4 indicates the type of value added each provides (Sheard, 2017).

2.3.4 VIRTUAL ENGINEERING AND R&D

The virtual development process has the following advantages (Laschet Consulting, 2020):

- High-quality and faster development processes.
- Lower research and development (R&D) costs.
- Early detection of weak points.
- Significantly improved quality.
- Development of basic knowledge (i.e., prediction of technical characteristics).
- Systemized development.

TABLE 2.4
Type of Value of the Roles

Role	Role Name	Primary Type of Value
RO	Requirements Owner	C) Understanding Need
SD	System Designer	A) Accomplishing Work
SA	System Analyst	D) Reducing Risk
VV	Validation/Verification Engineering	D) Reducing Risk
LO	Logistics/Ops Engineer	C) Understanding Need
G	Glue Among Subsystems	A) Accomplishing Work D) Reducing Risk
CI	Customer Interface	C) Understanding Need
TM	Technical Manager	B) Managing
IM	Information Manager	B) Managing
PE	Process Engineer	B) Managing, and C) Understanding Need
CO	Coordinator	B) Managing
CA	Classified Ads SE	A) Accomplishing Work (assumed)

Source: Sheard (2017).

- Comparison of engineering solutions and configurations, thus enabling alternative development steps.
- Minimized test periods and commissioning procedure.
- Improved test quality.
- Efficient team work.
- Better understanding of measurements, field tests, calculations, observations, subjective evaluations, and so on.
- Easy access to results.
- Special evaluation functions to objectively assess system behavior (objective system evaluation).
- Application of powerful simulation tools while reducing model complexity.

All of these advantages provide a reasonable basis for making strategic decisions during the product development process Figure 2.11 shows virtual engineering as part of the virtual development process (Laschet Consulting, 2020).

2.3.5 Ubiquitous Computing Technologies for Next Generation Virtual Engineering

In the future, virtual engineering will address more complex innovation challenges, originating in the need for the ecological and social sustainability of products and services, the shortage of fossil-based energy resources, the limitation and scarcity of certain industrial materials, and the ever-growing need for socially contextualized innovation. The boundaries of virtual engineering will dissolve, and there will be a more robust integration of the policies, infrastructure, and approaches. The fusion will extend to hardware, software, and firmware platforms and systems. It can also be assumed that some currently emerging technologies will become technological and methodological enablers of virtual engineering processes.

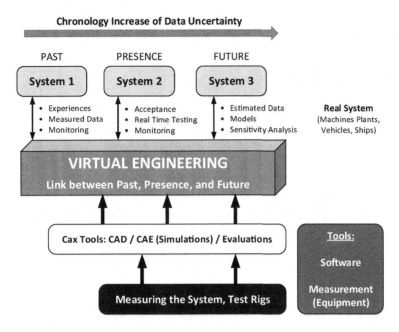

FIGURE 2.11 Conception of virtual engineering (Laschet Consulting, 2020).

Ubiquitous computing technologies will have an impact on all areas of future virtual engineering (Horváth et al., 2010). Ubiquitous computing technologies can operate in the domain of:

1. Digital information exploration and extraction.
2. Collaborative detection and elicitation of physical signals.
3. Transmission of digital information from short to long range.
4. Wired and wireless networking in a predefined and/or *ad hoc* manner.
5. Conversion of remote physical information to human sensations at a distance.

Facilitated by various application enablers, ubiquitous computing technologies will be applied in the realms of organs, artefacts, and environments. Figure 2.12 shows a comprehensive reasoning model with a detailed classification of the technologies and a taxonomical decomposition of the most important application domains (Horváth et al., 2010).

The ubiquitous computing technologies form a holistic pool of resources and facilitate a component-based innovation of new product and service combinations. As indicated by Figure 2.12, ubiquitous computing technologies consist of a wide spectrum of portable, mobile, wireless, wearable and embedded sensors, transmitters, and actuators, as well as controlling, networking, and computing (reasoning) appliances. Their application in products and environments will increase smartness (Horváth et al., 2010).

Because of their specific characteristics, ubiquitous computing technologies lend themselves not only to sustainable products, but also to socio-technical solutions. The six most important characteristics are (Horváth et al., 2010):

1. Small physical sizes.
2. Low energy consumption.
3. Decreasing production and use costs.
4. Reduced ecological and cognitive impacts.

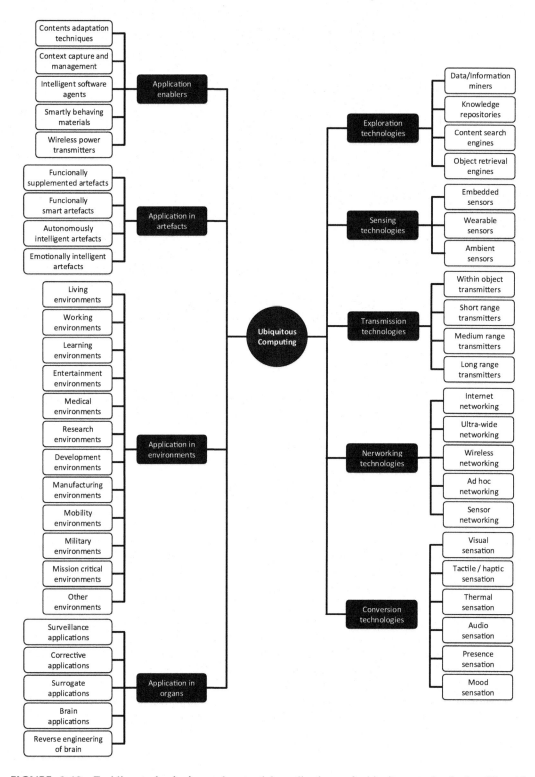

FIGURE 2.12 Enabling technologies and potential applications of ubiquitous technologies (Horváth et al., 2010).

5. High-level connectivity and flexibility.
6. Opportunity for a unique identification at a distance.
7. Proactive and adaptive (smart) behavior in applications.

These characteristics make it possible:

1. To use these technologies in large numbers (clusters) in various applications.
2. To embed them in artefacts, environments, and organs.
3. To realize a highly collaborative operation without user control.
4. To network without predefined servers and protocols.

Products involving current ubiquitous technologies are effective over a short distance:

1. Inside or on an artefact, or the human body.
2. Between carried and/or worn personal devices.
3. Among persons and personal devices and environment.

As a many-to-many communication medium, the Internet is becoming the ultimate repository of human knowledge in a publicly accessible form. In the future, a network of artefacts will communicate with human beings, other artefacts, and ambient environments – and eventually establish the IoT.

Research is already addressing the issue of establishing communication channels between the human brain and ubiquitous (embedded, wearable, and implanted) computing appliances by directly recording and transforming thought-modulated electroencephalogram (EEG) recordings into control signals. Current brain-computer interfaces are based on the detection and classification of motor-imagery-related EEG patterns and on the analysis of the dynamics of sensor-motor rhythms. A virtual world contact interface will give information to the muscles and nerves; these, in turn, will deliver the information to the brain to convert it into contact (tactile and haptic) sensation.

These radical innovations are forcing us to reconsider the ways of obtaining, processing, and communicating digital information for product design, engineering, and innovation. The main target field of applying smart technologies is providing support for inspiration, conceptualization, realization, and experiencing, the four main activity elements of creative work. They facilitate context awareness, content sensitiveness, knowledge intensiveness, and human centeredness. Omnipresent computing technologies will support socially based virtual innovation processes. These inclusive product innovation processes will feature a rapid transfer and synthesis of scientific and social knowledge, extensive information collection, widely based ideation and concept development, and concept testing and validation, all of which can be facilitated by dedicated computational resources (Horváth et al., 2010).

2.4 PRODUCT LIFECYCLE MANAGEMENT AND PREDICTIVE MAINTENANCE: CHANGING ROLE OF SUPPLIERS AND END USERS

PLM is a new strategic approach to efficiently manage product-related information over the whole lifecycle. Its concept appeared in the late 1990s, going beyond engineering aspects of products and providing a shared platform for creation, organization, and dissemination of product-related knowledge across extended enterprises.

PdM is very important in PLM. It can help manufacturers determine the condition of in-service products to predict when maintenance should be performed. It is an effective way to save cost and time, and to avoid unexpected equipment failures.

In general, the product lifecycle consists of three phases: beginning of life (BoL), including design and manufacturing; middle of life (MoL), including use, service and maintenance; and

EoL where products are disassembled, remanufactured, recycled, reused, or disposed of. During BoL, product data are quite complete and are supported by enterprise information systems, such as computer-aided manufacture design, ERP, supply chain management, and manufacturing execution systems. However, during MoL and EoL, because the customer owns the product, there is no corresponding information support system; the data almost break down after the delivery of the product to the customer. As a result, actors involved in each lifecycle stage make decisions based on incomplete and inaccurate product lifecycle information of other phases, leading to operational inefficiencies and hindering the implementation of true lifecycle management (Ren & Zhao, 2015).

Fortunately, information and communication technologies, especially wireless technologies such as RFID, sensors, and smart tags, have the potential to track and analyze product lifecycle data, allowing maintainers to make efficient decisions without spatial and temporal constraints.

In 2004, the Product Lifecycle Management and Information tracking using Smart Embedded systems (PROMISE) project proposed the concept of closed-loop PLM. The intent was to develop technology, software, and tools for decision-making based on data gathered through a product lifecycle. Lee and Suh (2009) proposed a ubiquitous product lifecycle conception called the UPLS system, which can obtain product lifecycle data and realize the information interaction at each stage. Georgiadis and Athanasiou (2010) studied PdM and remanufacturing applications based on closed-loop PLM. To solve the problem of separating the physical product from the related product data in late phases of product lifecycle (e.g., use, maintenance, and disposal), Erkayhan (2007) suggested the use of RFID to manage product lifecycle data, so generated product data could be read automatically and forwarded to different IT systems.

However, the application of emerging technologies brings new challenges. As companies begin to use advanced IT for real-time monitoring and tracking of business processes and products, a large amount of data on the product's lifecycle are produced, including in the areas of R&D, manufacturing, usage, maintenance, and recycling. These data are increasing at an unprecedented speed, creating what is called "Big Data" (Ren & Zhao, 2015).

Big Data technology is lagging behind, however, especially in PLM. Many applications focus on web analytics and customer behavior. Some researchers have considered how to use Big Data and advanced analysis in manufacturing (Auschitzky et al., 2014). Such research is important, as manufacturers taking advantage of advanced analytics can reduce process flaws, saving time and money.

Traditional PLM must be changed in the era of Big Data. The first challenge is establishing a Big Data capture and integration architecture to sense and exchange real-time data during the lifecycle. The second is discovering the previously unknown and potentially useful patterns and knowledge in Big Data (Ren & Zhao, 2015).

2.4.1 Product Lifecycle Management Approach

PLM is a strategic business approach for the effective management and use of corporate intellectual capital. Challenges faced by product development teams include globalization, outsourcing, mass customization, fast innovation, and product traceability. Meeting these challenges requires collaborating environments and knowledge management along the product lifecycle stages. PLM systems are gaining acceptance for their ability to manage all information about a product throughout its full lifecycle, from conceptualization to operations and disposal. But the PLM philosophy and systems aim at providing support to an even broader range of engineering and business activities.

PLM applies a consistent set of business solutions to support the collaborative creation, management, dissemination, and use of a product from concept to EoL – integrating people, processes, business systems, and information. It is an integrated, information-driven strategy that speeds up the innovation and launch of successful products, built on a common platform that serves as a single repository of all product-related knowledge, data, and processes.

As a business strategy, PLM lets distributed organizations innovate, produce, develop, support, and retire products, as if they were a single entity. It captures best practices and lessons learned, creating a storehouse of valuable intellectual capital for systematic and repeatable reuse.

As an IT strategy, PLM establishes a coherent data structure that enables real-time collaboration and data sharing among geographically distributed teams. PLM lets companies consolidate multiple application systems while leveraging existing legacy investments during their useful lives. Through adherence to industry standards, PLM minimizes data translation issues, while providing users with information access and process visibility at every stage of the product's life.

As shown in Figure 2.13, PLM systems support the management of a portfolio of products, processes, and services from initial concept, through design, launch, production, and use to final disposal, coordinating the various players, both internal and external, who must collaborate along the way (Gecevska et al., 2010).

The PLM concept permits the following (Gecevska et al., 2010):

- Updated product information shared within the organization among design, manufacturing, marketing, and procurement divisions.
- Internal team's collaboration with external users, suppliers, and customers for iterating new designs.
- Repository of product information retained for design reuse and part redundancy reduction.
- Systematic analysis of customer or market requirements.
- Sourcing team streamlined to identify a list of preferred suppliers for purchasing custom and standard parts.
- Resource management streamlined through analysis of the cost-benefits of allocating resources for specific projects.

Management and distribution of enterprise information using PLM are realized on various data levels (Gecevska et al., 2010):

- ICT: Compliance with existing legacy system; integration of PLM and ERP/CAD systems.
- Processes: Fragmented and unalterable; modeling, controlling, and improving.
- Data and objects: Different data formats; standard data representation (e.g., Initial Graphics Exchange Specification (IGES) and Standard for the Exchange of Product Model Data (STEP)); preserving data integrity over time; supporting data evolution.
- Methods and tools: Specific tools (CAD and computer-aided engineering (CAE)); new development methodologies (e.g., Six Sigma and Axiomatic design).
- People and organization: Functional organization promotes incommunicability; supply chain approach.

2.4.1.1 "Outside-In": New Approach to PLM

Traditionally, PLM solutions were toolkits that often required extensive customization to meet business requirements. They were implemented to support an internal, engineering-centric viewpoint. The cost and effort required to get basic capabilities in place within engineering groups meant initiatives stagnated in the product development department. Product innovation was constrained by the internal walls of the enterprise. This "inside-out" legacy approach focused on part and bill of material (BOM) management, CAD integration, and engineering change management. In the best-case scenario, ERP integration considered minimal analytics capabilities.

In today's fast-paced and customer-centric businesses, leading organizations use analytic, data-driven approaches to anticipate customer needs, leverage supply chains, and ensure customer satisfaction. This "outside-in" approach links the voice of the customer with the voice of the product into the digital twin delivering the visibility necessary for smarter innovation. As new technologies

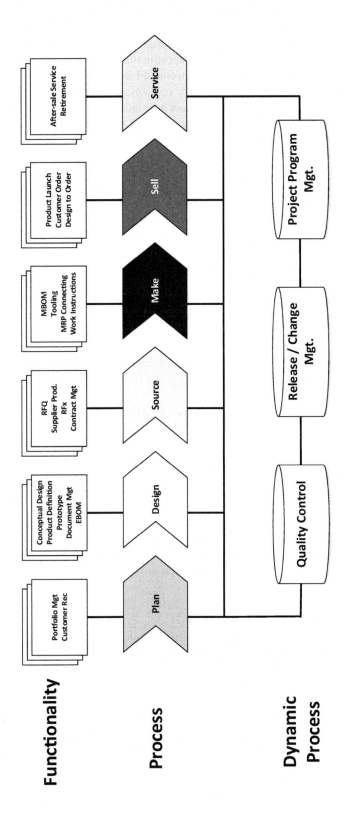

FIGURE 2.13   Structure of PLM system (Gecevska et al., 2010).

like Industrial IoT (IIoT) and PdM are incorporated into next generation PLM, we will see some interesting results, including improved profitability of service operations, the use of guided selling to improve sales efficiency, and even the creation of new business models for original equipment managers, such as products as a service (CIMdata, 2019).

2.4.2 EVOLUTION OF PLM

The creation of CAD software freed us from drafting tables and print rooms but introduced a bigger problem of how to manage, distribute, find, and reuse CAD files. Product data management (PDM) was introduced to tackle this problem and has evolved over the years to support ever-increasing business needs.

Before 1990, PDM solutions were CAD-centric and focused almost entirely on CAD file or document management. This quickly expanded into managing BOM and engineering change processes but remained focused on product development processes and designer and engineer productivity.

In the 1990s, globalization, outsourcing, and time to-market pressures forced companies to expand their PDM deployments. Early PLM solutions were introduced with robust security and collaboration features, as well as support for many processes and functions throughout the product lifecycle, including quality planning, manufacturing, product compliance (e.g., REACH, RoHS, and Conflict Minerals), product costing, and many others to address challenges beyond core product development. While PDM is still the core of any PLM solution, legacy tools were neither complete nor user friendly and required expensive and extensive customization.

After 2000, PLM emerged, incorporating more capabilities across the lifecycle, including innovation management, requirements management, and improved connections with downstream manufacturing, supply chain processes, and commercialization processes. In many cases, these capabilities were acquired and integrated with legacy tools. While many companies were able to leverage this new functionality, it was complex and required extensive customization.

Meanwhile, the pressures driving the initial development of PDM and early PLM solutions continued to grow. Businesses demanded better solutions to meet their product and process innovation requirements and wanted in on the latest digital transformations and Industry 4.0 initiatives. In the past few years, the convergence of the cloud, IoT, connected products, IIoT, Industry 4.0, machine learning, augmented reality (AR), and VR have helped companies transform how they design, manufacture, service, and sell their products. In effect, PLM solutions were re-architected as product innovation platforms able to support digital transformation and Industry 4.0 initiatives (CIMdata, 2019).

2.4.3 PRODUCT LIFECYCLE MANAGEMENT AND PREDICTIVE MAINTENANCE

When developing products and their associated services, it is important to recognize trends and adapt strategies to accommodate them. It is essential to innovate, because customers' expectations of individual products and efficient services are growing.

The key to success is the data gathered in every phase of the product lifecycle. With the increasing number of documents, however, production and customer service must have the same current data. Discrepancies between the various stages of the product lifecycle create critical obstacles. Integrated solutions are required to map all product-related processes of a company.

Ideally, a PLM solution enables end-to-end processes and flowing product data, from brainstorming, planning, development, product structure, to costing, change management, approval control, manufacturing, to service and operation. Important subsystems running side by side in different software worlds are combined in one central source of information, allowing the processes to work together optimally (Kruse, 2017).

2.4.3.1 Repair before Standstill

In the past, an asset had to fail before it was repaired, and production then came to a standstill. Today, intelligent systems recognize a malfunction before it occurs. PdM can save enormous costs and open up new business models for service departments via sensors that allow remote monitoring via software.

PdM enables permanent monitoring of components directly on the machine. For example, sensors measure vibration, temperature, or humidity. Maintenance software records these sensor data, evaluates them, and detects a possible failure of components at an early stage. Defective components are identified regardless of the usual maintenance intervals and can be replaced before damage occurs – precluding the necessity to stop production. Furthermore, the constant data analysis gives users of smart machines a more precise picture of their systems (Kruse, 2017).

2.4.4 Digital Transformation

Products generate revenue, so improving products and related services by improving product creation processes will generate the value customers expect and pay for. This is not a new concept. The pressures on companies to innovate include globalization, regulation, cost, quality expectations, and product complexity, and these all continue to increase.

Companies must be able to calculate and anticipate the changing demands and evolution of their customers. Increasing complexity means innovations more commonly involve a connected combination of products, software, and services. Meeting customer expectations in the digital age requires fast access to real-time global product data and nimble business operating models.

This fundamental shift in the paradigm gives companies an opportunity to evolve into a trusted source of business value while increasing collaboration with their customers to feed the innovation cycle. This continuous loop unifies products, services, customers, employees, and partners, while helping organizations deliver ongoing value, creating memorable experiences, enabling on-demand fulfilment, and providing personalized services.

Technology is increasingly able to support digital transformation. Product innovation platforms are able to create, capture, and manage digital twin data. Modern integration technologies enable connections to other platforms, including social platforms, so social monitoring can be used to define and refine requirements, and IoT can be used to drive PdM solutions that leverage exact product configuration data stored in the digital twin. Advanced analytics leveraging machine learning and deep learning are starting to be used to understand how products are used, allowing product planners to create new features and remove unused ones, closing the loop between the customer and the product developer. In essence, a product innovation platform is invaluable when customers demand speed of information, cost efficiencies, and exceptional services (CIMdata, 2019).

2.4.5 PLM Business Value

When the enterprise implements the PLM concept, it can move forward strategically while achieving short-term results and can establish a platform for innovation. As the enterprise addresses specific business issues and builds a solid foundation for future success through a PLM platform, it will realize measurable innovation benefits both immediately and over the long term, as shown in Figure 2.14 (Gecevska et al., 2010).

Traditionally, companies brought their products to market in time-consuming serial processes that delayed the participation of downstream contributors, such as suppliers, manufacturing experts, and service/maintenance providers. By allowing the enterprise to execute as many lifecycle tasks as possible in parallel processes, PLM enables the enterprise to streamline and collapse critical stages in the product lifecycle. PLM delivers aligned, accurate, and highly synchronized product knowledge to multiple disciplines early in the product lifecycle – thereby avoiding the cost and

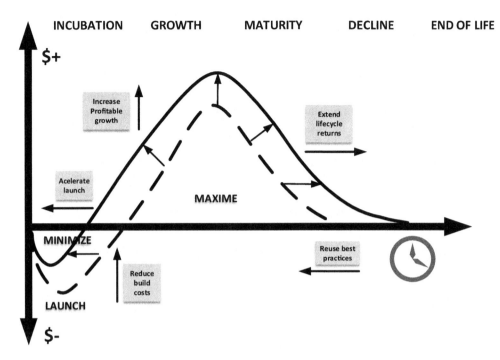

FIGURE 2.14 PLM business value (Gecevska et al., 2010).

scheduling impact that comes when late suggestions and unexpected concerns arise from downstream players. PLM allows the enterprise to beat the competition to market with innovative product content that carries first-to-market advantages and drives early product sales (Gecevska et al., 2010).

2.4.5.1 Increase Profitable Growth

PLM allows the enterprise to create, capture, and share the product-related requirements, expectations, and preferences of targeted customers and markets and align these with specific innovative content that customers want for a price they can afford at the time when it is needed. The PLM concept gives new product ideas in the context of quickly rising customer requirements while offering cost-effective manufacturability. Global cross-functional teams collaborate in real time on the development process, each contributing its own unique experience and perspective. Knowledge and "lessons learned" are captured for potential reuse in a process of continual innovation. PLM facilitates mass customization by enabling companies to rapidly and cost-effectively deliver customized products that satisfy the needs of individual customers and targeted market segments. It combines the advantages of configuration management with option and variant management. These state-of-the-market capabilities allow flexible and continuous portfolio planning (Gecevska et al., 2010).

2.4.5.2 Reduce Build Costs

PLM allows the enterprise to reduce cost across all stages of the product lifecycle. This, in turn, minimizes the cost of the product's planning, development, manufacture, and support.

For example, by leveraging PLM to understand the time and resource impacts of proposed design changes and requirement changes, the company's team can make decisions that minimize lifecycle and product costs. By using PLM to catch design flaws up front in the lifecycle, the team can avoid the cascading rework and cost associated with changing products during the manufacturing stages of the product lifecycle. The team can also use PLM to incorporate the concerns of the maintenance and service groups into the product designs and minimize warranty costs. By digitally creating and re-using the manufacturing plans, plant information, and manufacturing processes, the company can reduce the overall operational costs. It can also use PLM to implement virtual prototyping to reduce the validation costs associated with physical prototyping.

By implementing the PLM concept, an enterprise can cost-effectively deliver product enhancements, derivatives, niche offerings, and add-ons that extend the profitable duration of the product lifecycle. PLM facilitates this objective by enabling the enterprise to create product platforms that accelerate start-up processes, minimize take-to-market cost, and maximize the revenue generated by a product's initial release.

With PLM, the company can maximize the reuse of the best-practice processes, intellectual capital, human resources, product plans, production plans, production facilities, and value chains across a continuing set of take-to-market programs and a complete set of product and production management capabilities (Gecevska et al., 2010).

2.5 RUL ESTIMATION AS ENABLING TECHNOLOGY FOR CIRCULAR ECONOMICS

2.5.1 REMAINING USEFUL LIFE (RUL)

Remaining useful life (RUL) is the time remaining for a component to perform its functional capabilities before failure, or the duration from the current time to the end of a component's useful life (Xiongzi et al., 2011). The concept is shown graphically in Figure 2.15.

2.5.1.1 Classification of Techniques for RUL Prediction

Several prognostics prediction methods are used to determine the RUL of subsystems or components. They are presented in Figure 2.16 and discussed in the following sections (Okoh et al., 2014).

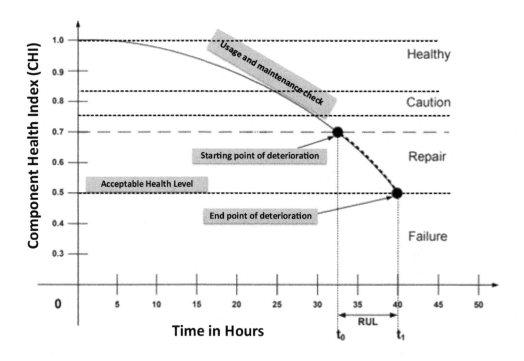

FIGURE 2.15 Component health index against time (hours) (Xiongzi et al., 2011).

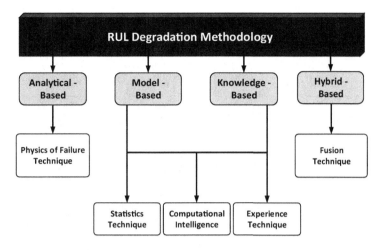

FIGURE 2.16 Classification of techniques for RUL predictions (Okoh et al., 2014).

2.5.1.2 Types of Prediction Techniques

Prediction can be based on statistical, experience-based, computational, physics-of-failure (PoF), or fusion techniques.

1. Statistics: This technique analyzes data using such methods as autoregressive moving average (ARMA) and exponential smoothing. These methods apply random variables to new data to improve the distribution of unknown parameters. Regression identifies the relations between variables and parameter values to predict RUL (Cheng & Pecht, 2007). In normal operating conditions, ARMA is used to recognize the dynamic behavior of components. Another statistical approach used in the medical and biomedical field is the proportional hazard model which, when applied to lifecycle issues, can make accurate and reliable RUL predictions (Okoh et al., 2014).

2. Experience: This approach rests on expert judgment. Knowledge is either explicit or tacit and gained from subject matter experts. It is used for degradation maintenance decision-making whereby processes and objects are under consistent observation. Understanding is obtained from data gathered from failure events and developmental test events. An analysis of the data enables the extraction of features based on degradation mechanisms to facilitate the construction of datasets. It also facilitates the introduction of rules for classification of the information to determine the RUL of an asset directly by setting a predefined threshold level.

3. Computational intelligence: This method, also known as soft computing, includes fuzzy logic and neural networks which are dependent upon parameters and input data to create the desired output. Artificial neural networks (ANNs) use data from continuous monitoring systems and require training samples. ANNs, usually called "black-boxes", provide little insight into the internal structures (Xiongzi et al., 2011), but data collected from sensors can be translated through ANN to predict an asset's RUL. Alternative approaches are Bayesian prediction and support vector machines which make statistical estimates of condition for limited samples to define a predictive learning base.

4. PoF: This technique requires parametric data and includes such methods as continuum damage mechanics, linear damage rules, nonlinear damage curves, and two-stage linearization. Life curve modification method of stress and load interaction, crack growth concept, and energy based damage models are also available.

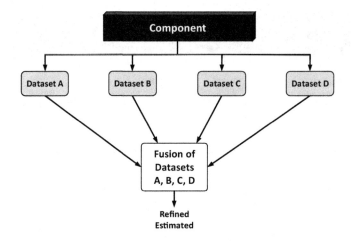

FIGURE 2.17   Sample fusion estimate (Okoh et al., 2014).

5. Fusion: This is the merging of multiple data into a refined state. The approach extracts, pre-processes, and fuses data for accurate and fast forecasting of the RUL of an asset, as illustrated in Figure 2.17. A better way to incorporate fusion is to classify data with the aid of the fuzzy method to improve the accuracy of the RUL estimate. In the context of uncertainty in RUL estimation, on-demand information collected from different sensors is fused by either centralized or decentralized means to accurately predict useful life using principal component analysis (Okoh et al., 2014).

2.5.2   WHAT IS THE CIRCULAR ECONOMY?

Broadly speaking, a circular economy is one in which there is no such thing as waste. Examples in the natural environment are the nutrient, carbon, and water cycles: minerals are released from rocks, taken up by plants and other organisms, released with the help of detritivores after death and taken up again, or laid back down as sediment which then returns to rock. Proponents of a circular economy seek to learn from and mimic this seamless, waste-free approach to resource management. The Ellen MacArthur Foundation provides a useful definition:

A circular economy is an industrial system that is restorative or regenerative by intention and design … it replaces the "end of life" concept with restoration, shifts towards the use of renewable energy, eliminates the use of toxic chemicals, which impair re-use, and aims for the elimination of waste through the superior design of materials, products, systems and within this, business models. Such an economy is based on a few simple principles. First, at its core, a circular economy aims to design out waste. Waste does not exist – products are designed and optimized for a cycle of disassembly and re-use. These tight component and product cycles define the circular economy and set it apart from disposal and even recycling where large amounts of embedded energy and labor are lost. Secondly, circularity introduces a strict differentiation between consumable and durable components of products. Unlike today, consumables in the circular economy are largely made of biological ingredients or nutrients that are at least non-toxic and possibly even beneficial and can be safely returned to the biosphere – directly or in a cascade of consecutive uses. Durables such as engines or computers, on the other hand, are made of technical nutrients unsuitable for the biosphere, like metals and most plastics. These are designed from the start for re-use. Thirdly the energy required to fuel this cycle should be renewable by nature, again to decrease resource dependence and increase system resilience.

Ellen MacArthur Foundation, 2012

Although a circular approach may not be a feasible solution in every circumstance, the conventional take, make, use, and dispose route cannot continue to be the dominant economic model – not just from a waste management point of view but also because of the issue of resource scarcity (techUK, 2015).

A circular economy is both restorative and regenerative; it aims to keep products, components, and materials at their highest utility and value at all times, while distinguishing between technical and biological cycles. It can be conceived as a continuous positive development cycle. It simultaneously preserves resources, optimizes yields, and minimizes system risks by managing finite stocks and renewable flows.

The concept of the circular economy is particularly relevant in the urban context, as it offers designers, planners, policymakers, and businesses a framework to rethink systems: how to design and operate them in a manner that will preserve, restore, and regenerate natural, social, and financial capital (Sukhdev et al., 2015).

2.5.3 DIGITAL TECHNOLOGY: ENABLING TRANSITION

Digital technology has enabled a fundamental shift in the way the economy functions, offering possibilities for radical virtualization, dematerialization, and greater transparency on product use and material flows, while creating new ways of operating and participating in the economy for producers and users. Through the collection and analysis of data on materials, people, and external conditions, digital technology has the potential to identify the challenges of material flows in cities, outline the key areas of structural waste, and inform more effective decision-making on how to address these challenges and provide systemic solutions. It will also be critical to ensure the security of data and systems.

The following technologies have been identified as enablers of circular economy activity in cities (Sukhdev et al., 2015):

- Asset tagging: Asset tagging technologies provide information on the condition and availability of products, components, or materials. In turn, this information can help extend the use of an asset, increase its utilization, loop or cascade it through additional use cycles, and help regenerate natural capital. For example, prolonging the lifecycle of a car by monitoring its usage patterns and condition through sensors can trigger alerts about problems as they appear to allow a fast and easy fix.
- Geo-spatial information: When combined with asset tagging, geo-spatial information can provide visibility on the flow of materials, components, products, and people across the city (including patterns of optimal mobility routes, energy demand peaks and valleys, congestion, and waste generation). For example, the ability to visualize traffic and pollution information on base maps, layered with valuable insights from other sources (i.e., census data and material information data) allows experts to predict and plan targeted strategies to address issues of congestion and pollution. It also allows citizens to understand what those data show in a format that is accessible and understandable.
- Big Data management: Current computation capability allows us to lay general patterns of human behavior over aggregated information received from asset tracking and geo-spatial mapping activities, for example, predicting energy consumption patterns at a local level, thus suggesting transport options that avoid peak hour traffic flows in real time. Leveraging advanced processing capacities, computers now can perform complex and agile analyses that

will help to determine and deploy the most effective resource solutions at an unprecedented speed and quality.

- Connectivity: The widespread and easy access to smartphone and application technology allows increased connection between people and between people and products, thus leveraging asset tracking capabilities. This enables circular business models such as leasing and sharing platforms, reverse logistics, take-back systems, and distributed remanufacturing. For example, business models such as Uber, Airbnb or BlaBlaCars would not be feasible without an accessible app that connects assets on offer with those who would like to use them.

2.5.4 CIRCULAR ECONOMY BUSINESS MODELS

Adopting circular business models enables companies not only to act sustainably, but also to create competitive advantage. Business models can be defined in the following way (Antikainena et al., 2018):

1. The business model is a unit of analysis distinct from the product, company, or network. It is centered on a focal company, but its boundaries are wider than those of the company.
2. Business models emphasize a system-level approach to explain how companies run their businesses.
3. Business models explain both value creation and value capture.

In addition to the extensive business model literature, there is a growing literature stream on sustainable business models and sustainable business model innovation suggesting that instead of concentrating purely on creating economic value for companies, the value created for stakeholders, including environmental and social value, needs to be taken into account. In this way, business models and business model innovation contribute to the sustainable development of the company and society.

A circular business model can be defined as "the rationale of how an organization creates, delivers, and captures value with and within closed material loops". Circular business models are networked by nature and thus require different actors in different value networks to work together toward common objectives. There is a need to consider the changes in value creation for a broad range of actors, as there are often game-changing alterations in business models.

Circular business models can be divided into three groups: slowing, closing, and narrowing the loop. Slowing the loop is based on the idea of extending the product lifecycle by design and maintenance. Closing the loop concentrates on efficient recycling of materials and can be implemented, for example, by industrial symbiosis. Narrowing the loop aims at using fewer resources per product and can be significantly boosted by intelligent technologies. Since these are closely supportive of each other, circular business models often include multiple groups or even all of them (Antikainena et al., 2018).

2.5.5 TECHNIQUES FOR RUL ESTIMATION AND MAINTENANCE INVESTMENT OUTCOMES

A variety of distribution models have been used to represent the probability of failure. For example, they are used in the railroad industry to predict defect formation in rails (Orringer, 1990). The presence of defects and the defect rate (defects/mile) are criteria for planning rail-defect testing and rail replacement. Because of public-safety concerns, the risk tolerance for defect-caused rail breaks resulting in derailments is very low (National Research Council, 2012).

2.5.5.1 Engineering Analysis

Traditional engineering analyses and models are used to predict remaining component life. Examples include fatigue analysis, wear-rate analysis, and corrosion effects on structural strength.

Components are typically analyzed on an as-needed basis. Computations of stress or strain, coupled with material properties (e.g., strength and dimensions) and operating conditions, are often used in a model to estimate an outcome (National Research Council, 2012).

2.5.5.2 Cost and Budget Models

Predicting outcomes of maintenance and repair investments requires estimates of costs or budget needs and consequences. The traditional approach involves using cost estimates developed for individual projects. The expected outcomes of a list of projects can be married to the costs of the projects, and the outcome of maintenance and repair investment can be predicted. However, this approach has several problems. First, when the overall program and cost are being developed with planned outcomes, the projects to support the program may or may not have been developed and their costs estimated. Second, the estimated costs may have been developed two or more years previously and may require updating. Third, developing project-level cost estimates can be expensive so it is not typically done unless there is a high certainty of project funding. Finally, the costing approach is not particularly sensitive to what-if consequence analyses. Any change requires the cost estimator to compute iteratively the cost, incorporating the changes.

Parametric cost or budget models have been developed to overcome the problems of cost estimating models. These models use costs correlated to particular measures to provide a reasonable estimate that is sufficiently accurate for planning purposes. They may be economic-based (involving, e.g., depreciation or average service life) or engineering-based (involving, e.g., actual and predicted condition or adjusted service life). Detailed project estimates are completed later in the project planning and execution process, when there is greater certainty about the availability of funding (National Research Council, 2012).

2.5.5.3 Operations Research Models

Operations research (OR) models have been applied to management of some types of infrastructure, such as bridges (Golabi & Shepard., 1997). Although OR models are well-suited to maintenance and repair investments, they are seldom applied. With OR techniques, an objective function (e.g., minimizing energy consumption) can be established subject to budget, labor, and other constraints. Multiple criteria can be considered, and an optimal mix of projects and a prediction of the outcomes can be identified.

Stochastic optimization models are used when decision-makers are faced with uncertainty and must determine whether to act now or to wait and see. There is uncertainty in future asset deterioration, budget levels, the effects of maintenance and repair actions, and so on. Stochastic optimization models evaluate managerial recourse, and this provides an opportunity to fix problems if worst-case scenarios occur.

Another OR model, the Markov decision process (MDP) model, links asset condition and optimal long-term maintenance strategy. MDP models are used to manage networks of assets and components. The application of these models in asset management is limited, in that, for example, an asset's condition may result from several different deterioration processes and, thus, require different remedial actions. Deterioration processes and required remedial actions are difficult to define in a model (National Research Council, 2012).

2.5.5.4 Simulation Models

Simulation models are used to analyze the results of what-if scenarios and can be used instead of or in conjunction with OR models. Simulation models require more data than OR models, generally cannot guarantee an optimum solution (OR solutions can guarantee this), and can be very cumbersome to run (potentially long simulation times) (National Research Council, 2012). However, simulation models are generally more detailed than OR models and are used in practice to obtain useful solutions to real problems, whereas the model simplifications required to obtain "convex" formulations in OR models may result in OR models being unrealistically simplified representations of real systems.

2.5.5.5 Proprietary Models

The private sector has long been active in asset management and has collected large amounts of facility data. Some companies have used the data to develop asset management models. Generally, the models purport to predict condition and budgets, and outcomes related to both. However, such models are proprietary, so little about how they work, or about their assumptions, robustness, and accuracy, is publically known or peer-reviewed (National Research Council, 2012).

REFERENCES

Acatech, 2013. Umsetzungsempfehlungen für das Zukunftsprojekt Industrie 4.0 – Abschlussbericht des Arbeitskreises Industrie 4.0.

Acatech, 2015. Umsetzungsstrategie Industrie 4.0 Ergebnisbericht der Plattform Industrie 4.0. acatech.

Antikainena M., Uusitaloa T., Kivikytö-Reponen P., 2018. Digitalization as an Enabler of Circular Economy. 10th CIRP Conference on Industrial Product-Service Systems, IPS2 2018, 29–31 May 2018, Linköping, Sweden.

Auschitzky E., Hammer M., Rajagopaul A., 2014. How big data can improve manufacturing. McKinsey Glob. Inst. July 1, 2014. www.mckinsey.com/insights/operations/how_big_data_can_improve_manufacturing. Viewed: May 23, 2020.

Baines T. S., 2013. Servitization impact study: how UK based manufacturing organisations are transforming themselves to compete through advanced services. Aston Centre for Servitization Research and Practice, Aston Business School, Birmingham, UK.

Baines T. S., Lightfoot H., Evans E., Neely A., Greenough R., Peppard J., Roy R., Shehab E., Braganza A., Tiwari A., Alcock J., Angus J., Bastl M., Cousens A., Irving P., Johnson M., Kingston J., Lockett H., Martinez V., Michele P., Tranfield D., Walton I., Wilson H., 2007. State-of-the-art in product-service systems. Proc. of the Institution of Mechanical Engineers, Part B: J. of Engineering Manufacture, 221(10), 1543–1552.

Baines T. S., Lightfoot H. W., Benedettini O., Kay J. M., 2009. The servitization of manufacturing. A review of literature and reflection on future challenges. Journal of Manufacturing Technology Management, 20(5).

Ballard A., Salorio A., 2018. Challenges and Opportunities for Performance-based Contracting. PATIMES American Society for Public Administration. https://patimes.org/challenges-opportunities-performance-based-contracting/. Viewed: May 21, 2020.

Bauernhansl T., Hompel M., Vogel-Heuser B., 2014. Industrie 4.0 in Produktion, Automatisierung und Logistik: Anwendung, Technologie, Migration.

Bernard A., 2016. Virtual engineering: methods and tools. IRCCyN (Institut de Recherche en Communications et Cybernétique de Nantes). Industrial Engineering Research Project, Ecole Centrale de Nantes, Nantes Cedex 3 – France. August 18, 2016.

Bureca A., 2017. The Role of the Internet of Things from a Servitization Perspective. Opportunities and challenges for the implementation of Product- Service Systems. A multiple case study. Double Degree Program: Master's degree in Management – LUISS MSc. in Innovation and Industrial Management – GU. Academic Year 2016–2017.

Buxmann P., Hess T., Ruggaber R., 2009. Internet of Services. Business & Information Systems Engineering 5, 341–342.

Cheng S., Pecht M., 2007. Multivariate state estimation technique for remaining useful life prediction of electronic products. AAAI Fall Symposium Artificial Intelligence. Prognostics, Arlington, VA; 2007.

CIMdata, 2019. Next Generation—Digitally Connected PLM. Digital transformation powered by a modern platform to enable the digital thread and digital twins supporting IoT, machine learning, and advanced analytics. Global Leaders in PLM Consulting. Sponsored by Oracle.

Defense Acquisition University, 2005. Performance Based Logistics: A Program Manager's Product Support Guide. DEFENSE ACQUISITION UNIVERSITY PRESS. Standard Form 298 (Rev. 8–98). Prescribed by ANSI Std Z39-18. March 2005.

Ellen MacArthur Foundation, 2012. Towards a circular economy: Economic and business rationale for an accelerated transition' (Report Volume 1). www.ellenmacarthurfoundation.org/business/reports/ce2012. Viewed: May 19, 2020.

Erkayhan S., 2007. The Use of RFID enables a holistic Information Management within Product Lifecycle Management (PLM). In RFID Eurasia, 2007 1st Annual (pp. 1–4). IEEE.

Frambach R. T., Wels-Lips I., Gündlach A., 1997. Proactive product service strategies: an application in the European health market. Industrial Marketing Management, 26, 341–52.

Fraunhofer-Institut für Materialfluss und Logistik (IML), 2014. Smart face – Smart Micro factory for Electric Vehicles with Lean Production Planning. Smart Face – Smart Micro Factory für Elektrofahrzeuge mit schlanker Produktionsplanung. Christoph Mertens, Supply Chain Engineering. www.iml.fraunhofer.de/content/dam/iml/de/documents/OE%20220/Referenzen/jahresbericht2016/Smart%20Face_Smart%20Micro%20Factory.pdf. Viewed: May 13, 2020.

Gausemeier J., Czaja A., Dülme C., 2015. Innovationspotentiale auf dem Weg zu Industrie 4.0. In: Wissenschafts- und Industrieforum Intelligente Technische Systeme 2015, Heinz Nixdorf Institut.

Gebauer H., 2008. Identifying service strategies in product manufacturing companies by exploring environment–strategy considerations. Industrial Marketing Management, 37, 278–291.

Gecevska V., Chiabert P., Anisic Z., Lombardi F., Cus F., 2010. Product lifecycle management through innovative and competitive business environment. Journal of Industrial Engineering and Management. 3(2): 323–336.

Georgiadis P., Athanasiou E., 2010. The impact of two-product joint lifecycles on capacity planning of remanufacturing networks. European Journal of Operational Research, 202 (2010), 420–433.

Giusto D., Iera A., Morabito, G., Atzori G., 2010. The Internet of Things.

Golabi K., Shepard R., 1997. Pontis: A system for maintenance optimization and improvement of U.S. bridge networks. Interfaces, 27(1), 71–88.

Gorecky D., Schmitt M., Loskyll M., 2014. Mensch-Maschine-Interaktion im Industrie 4.0-Zeitalter.

Gubán M., Kovás G., 2017. Industry 4.0 Conception. Acta Technica Corviniensis – Bulletin of Engineering Tome X [2017] Fascicule 1 [January – March] ISSN: 2067–3809.

Günthner W., Klenk E., Tenerowicz-Wirth P., 2014. Adaptive Logistiksysteme als Wegbereiter der Industrie 4.0.

Hermann M., Pentek T., Otto B., 2015. Design Principles for Industrie 4.0 Scenarios: A Literature Review. Technische Universität Dortmund. Fakultät Maschinenbau. Audi Stiftungslehrstuhl Supply Net Order Management. Working Paper No. 01/2015.

Hilti, 2017. www.hilti.com/content/hilti/W1/US/en/services/tool-services/fleet-management.html. Viewed: May 13, 2020.

Hirsch-Kreinsen H., Weyer J., 2014. Wandel von Produktionsarbeit – Industrie 4.0. Soziologisches Arbeitspapier 38, TU Dortmund.

Horváth I., Gerritsen B., Rusák Z., 2010. A new look at virtual engineering. Gépészet 2010, Budapest, 25–26. May 2010. G-2010-Section-No.

Huxtable J., Schaefer D., 2016. On servitization of the manufacturing industry in the UK. Changeable, agile, reconfigurable & virtual production. Procedia CIRP, 52 (2016), 46–51.

Industrie 4.0 Whitepaper FuE-Themen, 2015. Plattform Industrie 4.0, Stand: 7. April 2015. www.din.de/blob/67744/de1c706b159a6f1baceb95a6677ba497/whitepaper-fue-themen-data.pdf. Viewed: May 15, 2020.

Jacopino A., 2020. Guest Presentation on Managing a Performance Based Contract during Free Online Course from IACCM. Performance Based Contracting (PBC) Blog. May 2020. https://performancebasedcontracting.com/. Viewed: May 21, 2020.

Jacob F., Ulaga W., 2008. The transition from product to service in business markets: An agenda for academic inquiry. Industrial Marketing Management, 37(3), 247–253.

Kagermann H., 2014. Chancen von Industrie 4.0 nutzen.

Kagermann H., Lukas W., Wahlster W., 2011. Industry 4.0: Mit dem Internet der Dinge auf dem Weg zur 4. industriellen Revolution. VDI nachrichten, 13.

Kagermann H., Wahlster W., Helbig J., 2013. Recommendations for implementing the strategic initiative Industrie 4.0: Final report of the Industrie 4.0 Working Group.

Körner M., 2020. Performance- based contracting: An innovative contract and business model for the defense industry. ACTRANS. https://actrans.de/en/performance-based-contracting-an-innovative-contract-and-business-model-for-the-defence-industry/. Viewed: May 21, 2020.

Kruse R., 2017. Small service miracle: PLM and predictive. Head of Center of Excellence, Discrete Industries, itelligence AG – maintenance. June 14, 2017. https://itelligencegroup.com/de/local-blog/kleines-servic ewunder-plm-und-predictive-maintenance/. Viewed: May 23, 2020.

Laschet Consulting, 2020. Virtual engineering. Consulting & Virtual Engineering. Laschet Consulting GmbH. Friedrich-Ebert-Str. 75. 51429 Bergisch Gladbach, GERMANY. www.laschet.com/en/engineering/virtual-engineering/. Viewed: May 21, 2020.

Lee B. E., Suh S. H., 2009. An architecture for ubiquitous product life cycle support system and its extension to machine tools with product data model. The International Journal of Advanced Manufacturing Technology, 42(2009), 606–620.

Lewis M., Staudacher A. P., Slack N., 2004. Beyond products and services: opportunities and threats in servitization. Proceedings of the IMS International Forum. Cernobbio, 17–19, May: IMS International Forum Italy.

Lightfoot H., Baines T., Smart P., 2012. The servitization of manufacturing: A systematic literature review of interdependent trends. International Journal of Operations & Production Management, 33, 2012.

Lightfoot H., Baines T., 2013. Made to Serve, Wiley.

Lucke D., Constantinescu C., Westkämper E., 2008. Smart factory: A step towards the next generation of manufacturing. In: Mitsuishi, M., K. Ueda and F. Kimura, eds., Manufacturing Systems and Technologies for the New Frontier, the 41st CIRP conference on manufacturing systems, Tokyo, Japan, 115–118.

Manzei C., Schleupner L., Heinze R., 2016. Industrie 4.0 im internationalen Kontext.

National Research Council (NRC), 2012. Predicting Outcomes of Investments in Maintenance and Repair of Federal Facilities. Washington, DC: The National Academies Press.

Neely A., 2009. Exploring the financial consequences of the servitization of manufacturing. Operations Management Research, 1(2), 103–118.

Okoh C., Roy R., Mehnen J., Redding L., 2014. Overview of remaining useful life prediction techniques in through-life engineering services: Product services systems and value creation. Proceedings of the 6th CIRP Conference on Industrial Product-Service Systems. Procedia CIRP, 16 (2014), 158–163.

Oliva R., Kallenberg R., 2003. Managing the transition from products to services. International Journal of Service Industry Management, 14(2), 1–10.

Orringer O., 1990. Control of Rail Integrity by Self-Adaptive Scheduling of Rail Tests. DOT/FRA/ ORD-90/05. Washington, DC: U.S. Department of Transportation, Federal Railroad Administration, Office of Research and Development.

Ovtcharova J. G., 2010. Virtual engineering: Principles, methods and applications. International Design Conference – Design 2010. Dubrovnik, Croatia, May 17–20, 2010.

PBL Guidebook, 2016. A guide to developing performance-based arrangements. Release: 2016 | U.S. Department of Defence. http://acqnotes.com/wp-content/uploads/2017/07/Performance-Based-Logist ics-Guidebook-March-2016.pdf. Viewed: May 21, 2020.

Ren G., Gregory M. J., 2007. Servitization in manufacturing companies: A conceptualization, critical review, and research agenda. In: Frontiers in Service Conference 2007, 2007-10-4 to 2007-10-7, San Francisco, CA, US.

Ren S., Zhao X., 2015. A predictive maintenance method for products based on big data analysis. International Conference on Materials Engineering and Information Technology Applications (MEITA 2015).

Sheard S. A., 2017. The value of twelve systems engineering roles. Software Productivity Consortium. October 29, 2017.

Sawhney M., Balasubramanian S., Krishnan V., 2004. Creating growth with services. MIT Sloan Management Review, 34(4), 34–43.

Schaefer D., 2015. An investigation case study approach. ME40049 Innovation and Advanced Design, University of Bath, Bath.

Scheer A. W., 2013. Industrie 4.0: Wie sehen Produktionsprozesse im Jahr 2020 aus?

Schlick J., Stephan P., Loskyll M., Lappe D., 2014. Industrie 4.0 in der praktischen Anwendung.

Slack N., 2005. Operations strategy: will it ever realize its potential. Gestao & Producao, 12(3), 323–332.

Spath D., Ganschar O., Gerlach S., Hämmerle M., Krause T., Schlund S., 2013. Produktionsarbeit der Zukunft – Industrie 4.0. Fraunhofer IAO, Fraunhofer Verlag.

Stock T., Seliger G., 2016. Opportunities of Sustainable Manufacturing in Industry 4.0. 13th Global Conference on Sustainable Manufacturing – Decoupling Growth from Resource Use. Institute of Machine Tools and Factory Management, Technische Universität Berlin.

Sukhdev A., Vol J., Brandt K., Yeoman R., 2015. Cities in the circular economy: The role of digital technology. Ellen MacArthur Foundation. www.ellenmacarthurfoundation.org/assets/downloads/Cities-in-the-Circular-Economy-The-Role-of-Digital-Tech.pdf. Viewed: May 19, 2020.

techUK, 2015. The circular economy: A perspective from the technology sector. techUK representing the future. September 2015.

Umsetyungsstrategie Industrie 4.0: Ergebnisbericht der Platform Industrie 4.0., 2015. Platform Industrie 4.0. April 2015. www.its-owl.de/fileadmin/PDF/Industrie_4.0/2015-04-10_Umsetzungsstrategie_Industrie_4.0_Plattform_Industrie_4.0.pdf. Viewed: May 13, 2020.

Vandermerwe, Rada, 1988. Servitization of business: Adding value by adding services. European Management Journal. 6(4).

Van Looy B., Visnjic I., 2013. Servitization: Disentangling the impact of service business model innovation on manufacturing firm performance. Journal of Operations Management, 31(4): 169–180.

VDI/VDE-GMA (Gesellschaft Messund Au–omatisierungstechnik), 2015. Status Report: Reference Architecture Model Industrie 4.0 (RAMI4.0) (Vol. 0).

VDI/VDE-GMA (Gesellschaft Messund Automatisierungstechnik), 2015(a). Statusreport Referenz architektur modell Industrie 4.0. VDI/VDE-GMA.

VDI/VDE-GMA (Gesellschaft Messund Automatisierungstechnik), 2016. Statusreport: Industrie 4.0 – Technical Assets Basic terminology concepts, life cycles and administration models.

Ward Y., Graves A., 2005. The provision integrated customer solutions by aerospace manufacturers. University of Bath. School of Management, Working paper series.

Wikipedia, 2020. Performance-Based Contracting (PBC). https://en.wikipedia.org/wiki/Performance-based_contracting. Viewed: May 21, 2020.

Wise R., Baumgartner P., 1999. Go downstream: the new profit imperative in manufacturing. Harvard Business Review, September/October, 133–141.

Xiongzi C., Jinsong Y., Diyin T., Yingxun W., 2011. Remaining useful life prognostic estimation for aircraft subsystems or components: A review. 10th IEEE International Conference on Electronic Measurement & Instruments (ICEMI) 2011.

Zezulka F., Marcon P., Vesely I., Sajdl O., 2016. Industry 4.0 – An Introduction in the Phenomenon. ScienceDirect, IFAC-PapersOnLine 49–25 (2016) 008–012.

3 RUL Estimation Powered by Data-Driven Techniques

3.1 APPROACHES TO MAINTENANCE: PHYSICAL MODEL-BASED VS. DATA-DRIVEN

There are two possible approaches to data analysis for predictive maintenance (Yakovleva & Erofeev, 2015):

1. Physical model-based.
2. Data-driven.

The first approach requires an accurate physical model of system behavior in either a normal or faulty condition. When the data captured from sensors are compared to the model's predictions, the health of the system can be inferred. The second approach uses data on past behavior to determine present performance and predict remaining useful life (RUL) (Yakovleva & Erofeev, 2015).

Physical methods include physics-of-failure models. Another is a simple crack-growth-model to predict the RUL of a system affected by a fatigue failure mechanism. The model-based technique requires a combination of experimentation, observation, geometry, and condition monitoring data to estimate damage resulting from a particular failure mechanism.

Data-driven techniques are derived from the use of historical "run-to-failure" (RTF) data. These techniques are often used for estimation based on a predetermined failure threshold. A "wavelet packet" decomposition approach and/or hidden Markov models (HMMs) can be used, as time-frequency features allow more precise results than time variables only. However, the approaches using historical data to predict asset life require knowledge of the physical nature of the asset (Okoh et al., 2016).

The data-driven method of estimating RUL is the topic of this chapter.

3.1.1 WHERE RUN-TO-FAILURE DATA COME FROM

The simplest maintenance plan is to opt for RTF maintenance where equipment runs until it breaks down and is then fixed. Although this is entirely after the fact, it is most likely that an organization will have thought ahead and planned ways to fix a possibly inevitable failure to avoid production losses. For example, companies using RTF maintenance will keep spare parts on hand to replace the failed part as quickly as possible and thus maintain equipment availability.

This maintenance strategy should not be confused with reactive maintenance because there is a plan in place – simply stated, the company has planned for the failure. This strategy is especially useful for assets whose failure is not a safety risk and that have little effect on production. For example, think of the failure of an ordinary light bulb. Nothing major happens when it fails. A new light bulb is obtained from stores, and the old one replaced (Fiix, 2020).

3.1.1.1 Advantages

The advantages of RTF maintenance are pretty obvious. First, it requires little planning. Maintenance does not need to be scheduled in advance; it simply happens after failure. Second, it is easy to understand and implement (Fiix, 2020).

DOI: 10.1201/9781003097242-3

3.1.1.2 Disadvantages

There are some equally obvious downsides to RTF. Asset failures are unpredictable; an organization may expect a failure at some point but will not know precisely when it will happen, making the efficient planning of staff and resources (i.e., spare parts) difficult. Third, it is expensive. Costs include production costs, spare parts, and labor costs. The maintenance team needs to hold spare parts in inventory to accommodate intermittent failures (Fiix, 2020).

3.2 NEURAL NETWORKS (NNS), FUZZY LOGIC, DECISION TREES, SUPPORT VECTOR MACHINES (SVMS), ANOMALY DETECTION ALGORITHMS, REINFORCEMENT LEARNING, CLASSIFICATION, CLUSTERING AND BAYESIAN METHODS, AND DATA MINING ALGORITHMS

3.2.1 NEURAL NETWORKS

Neural networks (NNs) are based on models of the biological brain structure. Artificial NNs, first developed by McCulloch and Pitts in 1943, are mathematical models reproducing the learning process of the human brain. They are used to simulate and analyze complex systems starting from known input/output examples. Instead of the neurons in a human brain, in this case, an algorithm processes data through its interconnected network of processing units. The NN can be considered a "black box". For any particular set of inputs (a particular scheduling instance), the black box will give a set of outputs that are suggested actions to solve the problem, even though output cannot be generated by a known mathematical function.

An NN is an adaptive system comprising several artificial neurons interconnected to form a complex network; they change their structure depending on internal or external information. In other words, this model is not programmed to solve a problem, but it learns how to do so by performing a training (or learning) process using examples. This data record, called the training set, comprises inputs with their corresponding outputs. The process reproduces almost exactly the human brain's ability to learn from previous experience (Open Textbooks, 2016).

3.2.1.1 What Is a Neural Network?

The basic idea of an NN is to simulate densely interconnected brain cells inside a computer to get it to learn things, recognize patterns, and make decisions in a humanlike way. It does not need to be programed to learn; it learns by itself, just like a human brain.

However, it is important to remember that NNs are (generally) software simulations: they're made by programming very ordinary computers, working in a very traditional fashion with their ordinary transistors and serially connected logic gates, to behave as though they're built from billions of highly interconnected brain cells working in parallel. No one has yet attempted to build a computer by wiring up transistors in a densely parallel structure exactly like the human brain. In other words, an NN differs from a human brain in exactly the same way that a computer model of the weather differs from real clouds, snowflakes, or sunshine. Computer simulations are just collections of algebraic variables and mathematical equations linking them together (in other words, numbers stored in boxes whose values are constantly changing). They mean nothing whatsoever to the computers they run inside – only to the people who program them (Woodford, 2019).

3.2.1.2 What Does a Neural Network Consist of?

A typical NN comprises artificial neurons (i.e., units) arranged in a series of layers, each of which connects to the layers on either side. It can have anywhere from dozens of units to millions of units. Some, known as input units, are designed to receive various forms of information from the outside world that the network will attempt to learn about, recognize, or otherwise process. Other units sit on the opposite side of the network and signal how it responds to the information it has learned; those

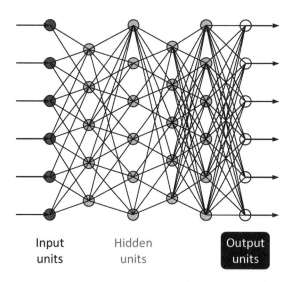

Input units   Hidden units   Output units

FIGURE 3.1 Fully connected neural network (Woodford, 2019).

are known as output units. In between the input units and output units are layers of hidden units, which, together, form the majority of the artificial brain.

Most NNs are fully connected, which means each hidden unit and each output unit is connected to every unit in the layers on either side. The connections between one unit and another are represented by a number called a weight, which can be either positive (if one unit excites another) or negative (if one unit suppresses or inhibits another). The higher the weight, the more influence one unit has on another. This corresponds to the way actual brain cells trigger one another across tiny gaps called synapses.

Figure 3.1 shows a fully connected NN is made up of input units (red), hidden units (blue), and output units (yellow), with all the units connected to all the units in the layers on either side. Inputs are fed in from the left, activating the hidden units in the middle, and making outputs feed out from the right. The strength (weight) of the connection between any two units is gradually adjusted as the network learns (Woodford, 2019).

3.2.1.3 How Does a Neural Network Learn Things?

Information flows through an NN in two ways. When it is learning (being trained) or operating normally (after being trained), patterns of information are fed into the network via the input units, which trigger the layers of hidden units, and these, in turn, arrive at the output units. This common design is called a feedforward network. Not all units "fire" all the time. Each unit receives inputs from the units to its left, and the inputs are multiplied by the weights of the connections they travel along. Every unit adds up all the inputs it receives in this way and (in the simplest type of network) if the sum is more than a certain threshold value, the unit "fires" and triggers the units it's connected to (those on its right).

For an NN to learn, there has to be an element of feedback involved – just as children learn by being told what they are doing right or wrong. In fact, we all use feedback to learn, all the time.

NNs learn things in exactly the same way, typically by a feedback process called backpropagation (sometimes abbreviated as "back prop"). This involves comparing the output a network produces with the output it was meant to produce, and using the difference between them to modify the weights of the connections between the units in the network, working from the output units through the hidden units to the input units – going backward, in other words. In time, backpropagation causes the network to learn, reducing the difference between actual and intended output to the point where the two exactly coincide, so the network figures things out exactly as it should (Woodford, 2019).

3.2.1.4 Neural Network Architecture

NNs consist of input layers, output layers, and hidden layers. The main job of an NN is to transform input into valuable output. NNs can be feedforward or feedback, depending on which way information flows (Sharma, 2018):

1. Feedforward networks: Signals travel toward the output layer. These networks have a single input layer and a single output layer but can have many or no hidden layers. There are two stages in the information flow: learning (when the network is "trained"); normal operation (after the network is "trained"). Feedforward networks are used in pattern recognition.
2. Feedback networks: Recurrent or interactive networks use memory (i.e., their internal state) to process input sequences. Signals can travel in both directions through network loops. Feedback networks are used for time series/sequential tasks.

3.2.1.5 Architectural Components

- Input layers, neurons, and weights: The basic unit in an NN, a neuron or node, receives input from an external source or other nodes. Weights are assigned to the neuron based on its relative importance compared to other neurons.
- Hidden layers and output layers: As the name suggests, the hidden layer is isolated from the external world. It can comprise several hidden nodes. It takes inputs from the input layer, performs calculations, and transforms the result to output nodes.

Let's assume the task is to make tea. The ingredients (tea, sugar, water, and milk) represent the input neurons. The amount of each ingredient is the weight. Then when the tea, sugar, water, and milk are mixed, the product is transformed into another state and color. This process of transformation is the activation function. The separate items within the mixture are the hidden layers. The warm drink is the output. (Sharma, 2018)

In NNs, almost every neuron influences and is connected to every other neuron, as seen in Figure 3.2 (Sharma, 2018).

3.2.1.6 Neural Network Algorithms

Many different algorithms are used to train NNs, and they have many different variants. Some of the most common are: feedforward, backpropagation, gradient descent, cost function, and sigmoid (Sharma, 2018).

The error of NNs is set according to a testing phase (to confirm the actual predictive power of the network while adjusting the weights of links). After building a training set of examples coming from historical data and choosing the kind of architecture to use (e.g., feedforward networks and recurrent networks), the most important step in implementing NNs is the learning process. Through the training, the network can infer the relations between input and output, defining the "strength" (weight) of connections between single neurons. From a very large number of extremely simple processing units (neurons), each performing a weighted sum of its inputs and firing a binary signal if the total input exceeds a certain level (activation threshold), the network manages to perform extremely complex tasks.

There are different categories of learning algorithms (Open Textbooks, 2016):

1. Supervised learning: The network learns the connection between input and output thanks to known examples coming from historical data.

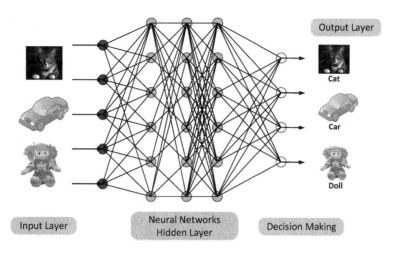

FIGURE 3.2   Diagram of the connections in neural networks (Sharma, 2018).

2. Unsupervised learning: Only input values are known, and similar stimulations activate close neurons; different stimulations activate distant neurons.
3. Reinforcement learning: A retro-activated algorithm is able to define new values of the connection weights starting from the observation of the changes in the environment.

Supervised learning by means of the back error propagation algorithm has become the most popular method of training NNs (Open Textbooks, 2016).

3.2.2   Fuzzy Logic

The role of inference in handling uncertainty is gaining importance in engineering applications. When engineers and scientists are confronted with problems which are impossible to solve numerically using traditional mathematical rules, the use of fuzzy logic is a good option. Fuzzy logic allows them to characterize and control a system whose model is not known or is ill-defined (Togai & Chiu, 1987). Fuzzy theory provides the capability to model human common-sense reasoning and decision-making (Calusdian, 1995).

Fuzzy logic is an extension of Boolean logic based on the mathematical theory of fuzzy sets, a generalization of classical set theory (Zadeh, 1965). By introducing the notion of degree into the verification of a condition, thus enabling a condition to be in a state other than true or false, fuzzy logic provides valuable flexibility for reasoning, making it possible to take inaccuracies and uncertainties into account.

One advantage of fuzzy logic is that the rules are set in natural language. For example, Table 3.1 gives some rules of conduct a driver follows, assuming he or she does not want to lose his/her driver's license (Dernoncourt, 2013).

Fuzzy logic has a weak connection to probability theory. Probabilistic methods that deal with imprecise knowledge are formulated in the Bayesian framework, but fuzzy logic does not need to be justified using a probabilistic approach. The common route is to generalize the findings of multivalued logic in such a way as to preserve part of the algebraic structure.

Fuzzy logic can be used to interpret the properties of NNs and to give a more precise description of their performance. Fuzzy operators can be conceived as generalized output functions of computing units. Fuzzy logic can also be used to specify networks directly without having to apply a learning algorithm. An expert in a certain field can sometimes produce a simple set of control rules

TABLE 3.1
Rules of Conduct That a Driver Follows

If the light is red…	If my speed is high…	and if the light is close…	then I brake hard.
If the light is red…	If my speed is low…	and if the light is far…	then I maintain my speed.
If the light is orange…	If my speed is average…	and if the light is far…	then I brake gently.
If the light is green…	If my speed is low…	and if the light is close…	then I accelerate.

Source: Dernoncourt (2013).

for a dynamic system with less effort than the work involved in training an NN. A classic example proposed by Zadeh (1965) is developing a system to park a car. It is straightforward to formulate a set of fuzzy rules for this task, but it is not immediately obvious how to build a network to do the same thing or how to train it. Fuzzy logic is now being used in many industrial and consumer electronic products for which a good control system is sufficient and where the question of optimal control does not necessarily arise (Rojas, 1996).

Basically, fuzzy logic is multivalue logic that allows intermediate values between conventional evaluations like true/false, yes/no, and high/low. Notions like rather tall or very fast can be formulated mathematically and processed by computers to apply a more human-like way of thinking in the programming of computers (Hellmann, 2001).

3.2.2.1 Fuzzy Logic Operators

Fuzzy logic operators are used to write logic combinations of fuzzy notions, that is, to perform computations on degrees of truth. Just as in classical logic, AND, OR, and NOT operators can be defined. For example:

Interesting Apartment = Reasonable Rent AND Sufficient Surface Area
(Chevrie & Guély, 1998).

3.2.2.1.1 Choice of Operators

There are many variants of operators; the most common are the "Zadeh" operators.

In what follows, the degree of truth of a proposal A is noted as $\mu(A)$ (Chevrie & Guély, 1998).

3.2.2.1.2 Intersection of Sets

The logic operator corresponding to the intersection of sets is AND. The degree of truth of the proposal "A AND B" is the minimum value of the degrees of truth of A and B:

$$\mu(A \text{ AND } B) = MIN(\mu(A), \mu(B))$$

For example:
- "Low temperature" is true at 0.7.
- "Low pressure" is true at 0.5.
- "Low temperature AND low pressure" is therefore true at 0.5 = MIN (0.7; 0.5).

Note that this fuzzy AND is compatible with classical logic: 0 AND 1 yield 0 (Chevrie & Guély, 1998).

3.2.2.1.3 Union of Sets

The logic operator corresponding to the union of sets is OR. The degree of truth of the proposal "A OR B" is the maximum value of the degrees of truth of A and B:

$$\mu(A \text{ OR } B) = \text{MAX}(\mu(A), \mu(B))$$

For example:
- "Low temperature" is true at 0.7.
- "Low pressure" is true at 0.5.
- "Low temperature OR low pressure" is therefore true at 0.7.

Note that this fuzzy OR is compatible with classical logic: 0 OR 1 yields 1 (Chevrie & Guély, 1998).

3.2.2.1.4 Complement of a Set
The logic operator corresponding to the complement of a set is the negation.

$$\mu(\text{NOT } A) = 1 - \mu(A)$$

For example:
- "Low temperature" is true at 0.7.
- "NOT low temperature" that we will normally write as "temperature NOT low" is therefore true at 0.3.

Note that the negation operator is compatible with classical logic: NOT(0) yields 1 and NOT(1) yields 0 (Chevrie & Guély, 1998).

3.2.2.1.5 Fuzzy Ladder
Ladder language or contact language is commonly used by automatic control engineers to write logic combinations, as it enables their graphic representation. The ladder can be used to describe fuzzy logic combinations.

Figure 3.3 gives an example of the comfort of ambient air: hot, damp air is uncomfortable; breathing is also difficult in air that is cold and too dry. The most comfortable thermal situations are those in which air is hot and dry, or cold and damp. This can be depicted by the fuzzy ladder in Figure 3.3 corresponding to the following combination (Chevrie & Guély, 1998):

Good comfort = (low temperature AND high humidity) OR (high temperature AND low humidity).

The figure represents a possible definition of the sensation of comfort felt by a person in a thermal environment in which air does not move.

3.2.2.1.6 Fuzzy Classification
Classification normally consists of two steps (Chevrie & Guély, 1998):

- Preparation: Determining the classes to be considered.
- On line: Assigning the elements to classes.

The notions of class and set are identical theoretically.
There are three types of assignment methods according to the result produced:

- Boolean: Elements either belong or do not belong to the classes.
- Probabilistic: Elements have a probability of belonging to Boolean classes, for example, the probability that a patient has measles given his or her symptoms (diagnosis).
- Gradual: Elements have a degree of membership to the sets, for example, a lettuce belongs to a varying degree to the class of "fresh lettuces".

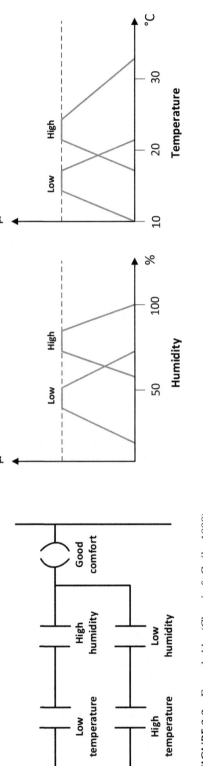

FIGURE 3.3   Fuzzy ladder (Chevrie & Guély, 1998).

Whether they produce a gradual, Boolean or probabilistic result, classification methods can be developed from:

- An experiment (e.g., "fuzzy ladder" mentioned above).
- Examples used for learning purposes (e.g., for neuron network classifiers).
- Mathematical or physical knowledge of a problem (e.g., the comfort of a thermal situation can be evaluated from thermal balance equations).

Gradual (or fuzzy) classification methods can be used in control loops (Chevrie & Guély, 1998).

3.2.2.2   Fuzzy Rules

3.2.2.2.1   Fuzzy Logic and Artificial Intelligence

The tool most commonly used in fuzzy logic applications is the fuzzy rule base. A fuzzy rule base is made of rules which are normally used in parallel but which can also be concatenated in some applications. The purpose of fuzzy rule bases is to formalize and implement a human being's method of reasoning. As such they can be classed in the field of artificial intelligence.

A rule is of the type: IF "predicate" THEN "conclusion". For example: IF "high temperature and high pressure" THEN "strong ventilation and wide open valve".

Fuzzy rule bases, just like conventional expert systems, rely on a knowledge base derived from human expertise. Nevertheless, there are major differences in the characteristics and processing of this knowledge (see Figure 3.4) (Chevrie & Guély, 1998).

A fuzzy rule comprises three functional parts summarized in Figure 3.5 (Chevrie & Guély, 1998).

3.2.2.2.2   Predicate

A predicate (also known as a premise or condition) is a combination of proposals by AND, OR, NOT operators. The "high temperature" and "high pressure" proposals in the previous example are combined by the AND operator to form the predicate of the rule (Chevrie & Guély, 1998).

3.2.2.2.3   Inference

The most commonly used inference mechanism is "Mamdani". It represents a simplification of the more general mechanism based on "fuzzy implication" and the "generalized modus ponens" (Chevrie & Guély, 1998).

3.2.2.2.4   Conclusion

The conclusion of a fuzzy rule is a combination of proposals linked by AND operators. In the previous example, "strong ventilation" and "wide open valve" are the conclusion of the rule.

Fuzzy rule base	Conventional rule base (expert system)
Few rules	Many rules
Gradual processing	Boolean processing
Concatenation possible but scarcely used	Concatenated rules A OR B → C, C → D, D AND A → E
Rules processed in parallel	Rules used one by one, sequentially
Interpolation between rules that may contradict one another	No interpolation, no contradiction

FIGURE 3.4   Fuzzy rule base and conventional rule base (Chevrie & Guély, 1998).

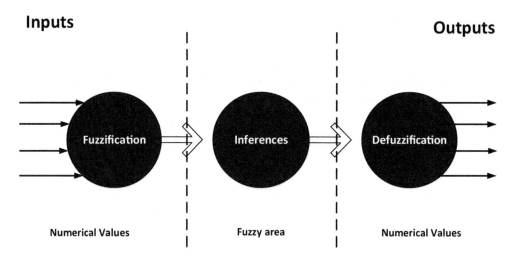

FIGURE 3.5   Fuzzy processing (Chevrie & Guély, 1998).

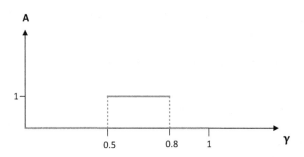

FIGURE 3.6   Characteristic function of a crisp set (Hellmann, 2001).

"OR" clauses are not used in conclusions as they would introduce uncertainty into the knowledge (the expertise would not make it possible to determine which decision should be made). This uncertainty is not taken into account by the Mamdani inference mechanism as it only manages imprecisions. Therefore, the Mamdani fuzzy rules are not, in theory, suitable for a diagnosis of the "medical" kind for which conclusions are uncertain. The theory of possibilities (Zadeh, 1965) offers an appropriate methodology in such cases.

Likewise, negation is not used in conclusions for Mamdani rules. If a rule were to have the conclusion "ventilation not average", it would be impossible to say whether this means "weak ventilation" or "strong ventilation". This would be another case of uncertainty (Chevrie & Guély, 1998).

3.2.2.3   Fuzzy Sets and Crisp Sets

The very basic notion of fuzzy systems is a fuzzy (sub)set. Classical mathematics uses what we call crisp sets. For example, the possible interferometric coherence values are the set X of all real numbers between 0 and 1. From this set X, a subset A can be defined (e.g., all values $0 < \gamma < 0.2$). The characteristic function of A (i.e., this function assigns a number 1 or 0 to each element in X, depending on whether the element is in the subset A or not) is shown in Figure 3.6 (Hellmann, 2001).

The elements assigned the number 1 can be interpreted as the elements that are in the set A, and the elements assigned the number 0 are the elements not in the set A. This concept is sufficient for many areas of applications, but it lacks flexibility for some applications, like the classification of remotely sensed data analysis.

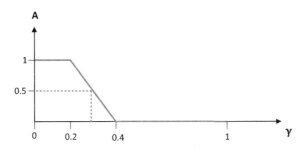

FIGURE 3.7 Characteristic function of a fuzzy set (Hellmann, 2001).

For example, it is well known that water shows low interferometric coherence γ in synthetic aperture radar (SAR) images. Since γ starts at 0, the lower range of this set ought to be clear. The upper range is harder to define. As a first attempt, we set the upper range to 0.2. Therefore, we get B as a crisp interval B = (0, 0.2). However, this means a γ value of 0.20 is low but a γ value of 0.21 is not. Obviously, this is a structural problem, for if we moved the upper boundary of the range from γ =0.20 to an arbitrary point, the same question can be asked. A more natural way to construct the set B would be to relax the strict separation between low and not low. This can be done by allowing not only the (crisp) decision Yes/No, but also more flexible rules like "fairly low". A fuzzy set allows us to define such a notion.

The aim is to use fuzzy sets to make computers more intelligent. Therefore, the idea above has to be coded more formally. In the example, all the elements are coded with 0 or 1. A straightforward way to generalize this concept is to allow more values between 0 and 1. In fact, an infinite number of alternatives can be allowed between 0 and 1, that is, the unit interval I = (0, 1).

The interpretation of the numbers, now assigned to all elements, is much more difficult. Of course, the number 1 assigned to an element means the element is in the set B, and 0 means the element is definitely not in the set B. All other values mean a gradual membership in set B. This is shown in Figure 3.7. The membership function is a graphical representation of the magnitude of participation of each input. It associates a weighting with each of the inputs that are processed, defines functional overlap between inputs, and ultimately determines an output response. The rules use the input membership values as weighting factors to determine their influence on the fuzzy output sets of the final output conclusion (Hellmann, 2001).

The membership function, operating in this case on the fuzzy set of interferometric coherence γ, returns a value between 0.0 and 1.0. For example, an interferometric coherence γ of 0.3 has a membership of 0.5 in the set low coherence (see Figure 3.7).

It is important to point out the distinction between fuzzy logic and probability. Both operate over the same numeric range and have similar values: 0.0 represents False (or non-membership), and 1.0 represents True (or full membership). However, the probabilistic approach yields the natural language statement, "There is a 50% chance that γ is low", while the fuzzy terminology corresponds to "γ's degree of membership within the set of low interferometric coherence is 0.50". The semantic difference is significant. The first view supposes γ is or is not low; it is just that we only have a 50% chance of knowing which set it is in. By contrast, the fuzzy terminology of the second view supposes γ is "more or less" low, or some other term corresponding to the value of 0.5 (Hellmann, 2001).

3.2.2.4 Fuzzy Logic Applications

The major applications of fuzzy logic are the following (Ansari, 1998):

1. Fuzzy control in industry: Recent applications include water quality control, automatic train operating systems, elevator control, control of smart locomotives, cement kiln control, and

power electronics, for example, speed control of a DC motor or induction motor efficiency optimization control.

2. Fuzzy logic-based products: Fuzzy logic has been introduced into consumer goods like washing machines and television sets. Recent work has been directed toward developing fuzzy logic-based handheld portable products (Deshpande, 1996).

3.2.3   Decision Trees

A decision tree is a decision support tool that uses a tree-like graph or model of decisions and their possible consequences, for example, to determine the outcome of chance events. It is one way to display an algorithm that only contains conditional control statements.

A decision tree is a flowchart-like structure in which each internal node represents a "test" of an attribute (e.g., whether a flipped coin comes up heads or tails), each branch represents the outcome of the test, and each leaf node represents a class label (i.e., the decision made after computing all attributes). The paths from root to leaf represent classification rules.

Tree-based learning algorithms are considered to be among the best and most commonly used supervised learning methods. They empower predictive models with high accuracy, stability, and ease of interpretation. They map nonlinear relationships quite well and can be adapted to solve any kind of problem at hand (classification or regression) (Brid, 2018).

3.2.3.1   Common Terms Used with Decision Trees

- Root node: The root node represents the entire sample or population; it is divided into two or more homogeneous sets.
- Splitting: Splitting refers to the process of dividing a node into sub-nodes.
- Decision node: A sub-node that splits into further sub-nodes is called a decision node.
- Leaf/ terminal node: Nodes that do not split are called leaf or terminal nodes.
- Pruning: When sub-nodes of a decision node are removed, the process is called pruning. It is the opposite of splitting.
- Branch/sub-tree: A subsection of the entire tree is called a branch or sub-tree.
- Parent and child node: A node divided into sub-nodes is a parent node; a sub-node is the child of a parent node (Brid, 2018).

3.2.3.2   How Decision Trees Work

A decision tree is a type of supervised learning algorithm (with a predefined target variable) used in classification problems. It works for both categorical and continuous input and output variables. The population or sample is split into two or more homogeneous sets (or sub-populations) based on the most significant splitter/differentiator in the input variables (Brid, 2018).

3.2.3.3   Types of Decision Trees

The type of decision tree depends on the type of target variable. It can be categorical or continuous (Brid, 2018).

3.2.3.4   Decision Tree Applications

Decision trees have a natural "if… then… else…" construction that makes them fit easily into a programmatic structure. They also are well suited to categorization problems where attributes or features are systematically checked to determine a final category. For example, a decision tree could be used to determine the species of an animal (Brid, 2018).

3.2.4 SUPPORT VECTOR MACHINES

Support vector machines (SVMs) are supervised learning methods used for classification, regression, and outlier detection (Scikit-Learn, 2019). However, SVMs are mostly used in classification problems.

3.2.4.1 How an SVM Works

Support vectors are simply the coordinates of individual observation. Each data item is plotted in the SVM algorithm as a point in n-dimensional space, where n is the number of features. The value of each feature is the value of a particular coordinate. As shown in Figure 3.8, the SVM classifier is the frontier which best segregates the classes, in this case, open and shaded circles (Sunil, 2017).

3.2.4.2 Advantages of Support Vector Machines

The advantages of SVMs are the following (Scikit-Learn, 2019):

- Effective in high-dimensional spaces, even when the number of dimensions is greater than the number of samples.
- Versatile, as different kernel functions can be specified for the decision function.
- Memory efficient, as they use a subset of training points in the decision function (i.e., support vectors).

3.2.4.3 Disadvantages of Support Vector Machines

The disadvantages of SVMs include the following (Scikit-Learn, 2019):

- If the number of features is much greater than the number of samples, overfitting can occur; the choice of kernel functions and regularization terms is crucial.
- SVMs do not directly provide probability estimates; these must be calculated using an expensive five-fold cross-validation.

3.2.4.4 SVM Applications

SVMs can be used to solve various real-world problems, including the following (Wikipedia, 2020):

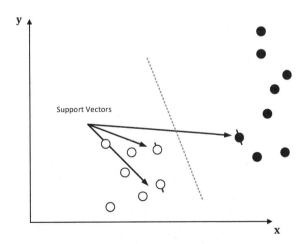

FIGURE 3.8 SVM algorithm (Sunil, 2017).

1. Text and hypertext categorization: SVMs can significantly reduce the need for labeled training instances in both inductive and transductive settings. In addition, some methods for shallow semantic parsing are based on SVMs.
2. Image classification: SVMs can achieve significantly higher search accuracy than traditional query refinement schemes after just three to four rounds of relevance feedback. This holds for image segmentation systems (Vapnik, 2014).
3. Satellite data classification: Supervised SVMs can classify SAR data.
4. Character recognition: SVMs can recognize handwritten characters.
5. Scientific classification: SVMs are used in the biological and other sciences for classification. For example, they can classify proteins with up to 90% of the compounds classified correctly.

Some new and emerging areas of usage have been suggested as well. For example, permutation tests based on SVM weights have been proposed as a possible mechanism to interpret SVM models. Post hoc interpretation of SVM models to identify features used by the model to make predictions is another relatively new area of research of interest to the biological sciences.

3.2.4.5   Selecting the Right Hyper-Plane

There may be many different hyper-planes. The following examples explain how to identify the correct one.

1. Scenario 1: In Figure 3.9, there are three hyper-planes (A, B, and C). The task is to identify the right hyper-plane to classify the stars and circles. The rule of thumb is to select the hyper-plane which best segregates the stars and circles (the two classes). In this scenario, hyper-plane B out-performs A and C.
2. Scenario 2: Figure 3.10 shows three hyper-planes (A, B, and C). All segregate the classes well. The task is to decide which is better.
   In this scenario, maximizing the distance between the nearest data point (either class) and the hyper-planes will determine the right hyper-plane. This distance is called the margin. Figure 3.11 solves the problem posed by Scenario 2. The margin for hyper-plane C is higher than for A and B. Hence, C is the best hyper-plane. A reason for selecting the hyper-plane with a higher margin is robustness. With a low margin, there is good chance of misclassification.

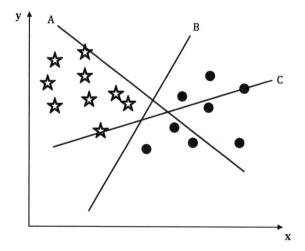

FIGURE 3.9   Identification of the right hyper-plane (Scenario 1) (Sunil, 2017).

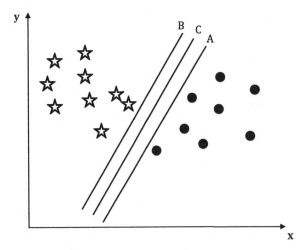

FIGURE 3.10   Identification of the right hyper-plane (Scenario 2) (Sunil, 2017).

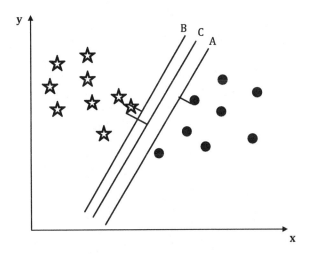

FIGURE 3.11   Identification of the right hyper-plane (Scenario 2) (Sunil, 2017).

3. Scenario 3: The third scenario appears in Figure 3.12. There are only two hyper-planes, A and B. Hyper-plane B has a higher margin than A, so it may seem a better choice. But there is a caveat: SVM selects the hyper-plane which classifies the classes accurately before maximizing the margin. In this case, hyper-plane B has a classification error. However, A has classified everything correctly, making A the right hyper-plane.
4. Scenario 4: It is not possible to segregate the two classes in Figure 3.13 using a straight line, as one of the stars lies in the territory of the circle, making it an outlier.

   However, the SVM algorithm has a feature allowing it to ignore outliers and simply find the hyper-plane with the maximum margin. The result is shown in Figure 3.14. Hence, it is possible to say SVM classification is robust to outliers.
5. Scenario 5: This scenario (Figure 3.15) will not permit a linear hyper-plane between the two classes. The preceding scenarios have been for a linear hyper-plane.

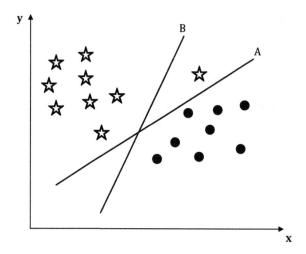

FIGURE 3.12   Identification of the right hyper-plane (Scenario 3) (Sunil, 2017).

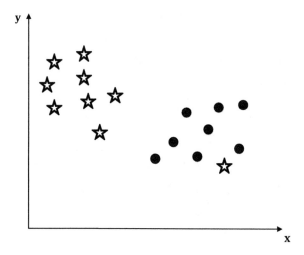

FIGURE 3.13   Classification in two classes (Scenario 4) (Sunil, 2017).

Fortunately, SVM can solve this problem by introducing an additional feature. For example, say we add a new feature $z = x^2 + y^2$. The data points can be plotted on the x and z axes (Figure 3.16). Note that:

- All values for z will always be positive because z is the squared sum of both x and y.

Also note that:

- In the original plot, circles appear close to the origin of the x and y axes, leading to lower values of z; stars are relatively further away, leading to higher values of z.

In the SVM classifier, it is easy to have a linear hyper-plane between the stars and circles. This feature does not need to be added manually. The SVM algorithm has a technique called the kernel

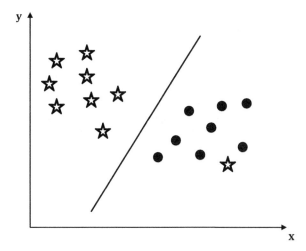

FIGURE 3.14 Classification in two classes (Scenario 4) (Sunil, 2017).

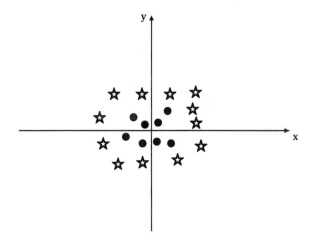

FIGURE 3.15 Finding the hyper-plane to segregate classes (Scenario 5) (Sunil, 2017).

trick that takes low-dimensional input space and transforms it to a higher-dimensional space. In this fashion, an inseparable problem is converted into a separable one. To explain in simple language, through some very complex data transformations, it discovers a way to separate the data based on the labels or defined outputs. It is commonly used in nonlinear separation problems.

Figure 3.17 shows the hyper-plane in the original input space.

3.2.5 ANOMALY DETECTION ALGORITHMS

Anomaly detection is a technique used to identify patterns that do not conform to expected behavior, called outliers. It has many applications in business, from intrusion detection (identifying strange patterns in network traffic that could signal a hack) to system health monitoring (spotting a malignant tumor in a magnetic resonance imaging scan), and from fraud detection in credit card transactions to fault detection in operating environments (Choudhary, 2017).

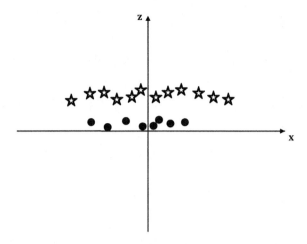

FIGURE 3.16   Finding the hyper-plane to segregate classes (Scenario 5) (Sunil, 2017).

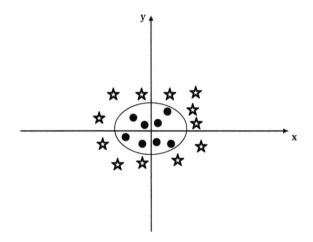

FIGURE 3.17   Hyper-plane in original input space (looks like a circle) (Sunil, 2017).

3.2.5.1   What Are Anomalies?

Anomalies can be broadly categorized as:

- Point anomalies: A single instance of data is anomalous if it is too far off from the rest. Business use case: Detecting credit card fraud based on "amount spent".
- Contextual anomalies: The abnormality is context specific. This type of anomaly is common in time-series data. Business use case: Spending $100 on food every day during the holiday season is normal, but may be odd otherwise.
- Collective anomalies: A set of data instances collectively helps in detecting anomalies. Business use case: Someone is trying to copy data from a remote machine to a local host unexpectedly, an anomaly that would be flagged as a potential cyber-attack.

Anomaly detection is similar to – but not entirely the same as – noise removal and novelty detection. Novelty detection is concerned with identifying an unobserved pattern in new observations not included in training data – like a sudden interest in a new channel on YouTube during Christmas, for instance. Noise removal is the process of immunizing analysis from the occurrence of unwanted observations; in other words, removing noise from an otherwise meaningful signal (Choudhary, 2017).

3.2.5.2 Anomaly Detection Algorithms

3.2.5.2.1 Density-Based Anomaly Detection

Density-based anomaly detection is based on the k-nearest neighbor (k-NN) algorithm. The assumption is that normal data points occur around a dense neighborhood and abnormalities are far away. The nearest set of data points is evaluated using a score, which could be Euclidean distance or a similar measure depending on the type of the data (categorical or numerical). K-NN is a simple, non-parametric lazy learning technique used to classify data based on similarities in distance metrics such as Euclidean, Manhattan, Minkowski, or Hamming distance.

Data can also be classified based on relative density. This is better known as the local outlier factor. This concept is based on a distance metric called reachability distance (Choudhary, 2017).

3.2.5.2.2 Clustering-Based Anomaly Detection

Clustering is one of the most popular concepts in the domain of unsupervised learning. The basic assumption is that similar data points tend to belong to similar groups or clusters, as determined by their distance from local centroids.

K-means is a widely used clustering algorithm. It creates k similar clusters of data points. Data instances that fall outside these groups could potentially be marked as anomalies (Choudhary, 2017).

3.2.5.2.3 Support Vector Machine-Based Anomaly Detection

An SVM is another effective technique for detecting anomalies. An SVM is typically associated with supervised learning, but there are extensions (e.g., OneClassCVM) that can be used to identify anomalies as an unsupervised problem (i.e., training data are not labeled). The algorithm learns a soft boundary in order to cluster the normal data instances using the training set; then, using the testing instance, it tunes itself to identify the abnormalities that fall outside the learned region.

Depending on the use case, the output of an anomaly detector could be numeric scalar values for filtering on domain-specific thresholds or textual labels (such as binary/multi labels) (Choudhary, 2017).

3.2.6 Reinforcement Learning

Reinforcement learning is simultaneously a problem, a class of solution methods that work well on the class of problems, and the field that studies these problems and their solution methods. Reinforcement learning problems involve learning what to do to maximize a numerical reward signal. In effect, they map situations to actions. They are closed-loop problems because the learning system's actions influence its later inputs. Moreover, the learner is not told which actions to take, as in many forms of machine learning, but instead must discover which actions yield the most reward by trying them out. In the most interesting and challenging cases, actions may affect more than the immediate reward. They may also affect the next situation and all subsequent rewards.

These three characteristics – being closed-loop, not having direct instructions on what actions to take, and having the consequences of actions, including reward signals, play out over extended time periods – are the three most important distinguishing features of reinforcement learning problems.

One of the challenges in reinforcement learning, and not in other kinds of learning, is the trade-off between exploration and exploitation. To obtain a lot of reward, a reinforcement learning agent must prefer actions it has tried in the past and found to be effective in producing reward. But the agent must go beyond what it knows and try actions it has not selected before. The agent has to exploit what it already knows in order to obtain reward, but it also has to explore in order to make better action selections in the future. The dilemma is that neither exploration nor exploitation can be pursued exclusively without failing at the task. The agent must try a variety of actions and progressively favor those that appear to be best. On a stochastic task, each action must be tried many times to gain a reliable estimate of its expected reward. The exploration–exploitation dilemma has been intensively studied by mathematicians for many decades.

Another key feature of reinforcement learning is that it explicitly considers the whole problem of a goal-directed agent interacting with an uncertain environment. This contrasts with many approaches that consider sub-problems without addressing how they might fit into a larger picture. For example, researchers have developed theories of planning with general goals, but without considering planning's role in real-time decision-making, or the question of where the predictive models necessary for planning would come from. Although these approaches have yielded many useful results, their focus on isolated sub-problems is a significant limitation.

Reinforcement learning takes the opposite tack, starting with a complete, interactive, goal-seeking agent. All reinforcement learning agents have explicit goals, can sense aspects of their environments, and can choose actions to influence their environments. Moreover, it is usually assumed from the beginning that the agent has to operate despite significant uncertainty about the environment it faces. When reinforcement learning involves planning, it has to address the interplay between planning and real-time action selection, as well as the question of how environment models are acquired and improved. When reinforcement learning involves supervised learning, it does so for specific reasons that determine which capabilities are critical and which are not. For learning research to make progress, important sub-problems have to be isolated and studied, but they should be sub-problems that play clear roles in complete, interactive, goal-seeking agents, even if all the details of the complete agent cannot yet be filled in.

One of the most exciting aspects of modern reinforcement learning is its substantive and fruitful interaction with other engineering and scientific disciplines. Reinforcement learning is part of a decades-long trend within artificial intelligence and machine learning toward greater integration with statistics, optimization, and other mathematical subjects. For example, the ability of some reinforcement learning methods to learn with parameterized approximators addresses the classical "curse of dimensionality" in operations research and control theory. Reinforcement learning has also interacted strongly with psychology and neuroscience, with substantial benefits going both ways. Of all the forms of machine learning, reinforcement learning is the closest to the kind of learning that humans and other animals do, and many of the core algorithms of reinforcement learning were originally inspired by biological learning systems. And reinforcement learning has given back, both through a psychological model of animal learning that better matches some of the empirical data, and through an influential model of parts of the brain's reward system (Sutton & Barto, 2015).

3.2.6.1 Terminologies Used in the Field of Reinforcement Learning

The following are some basic terminologies used in reinforcement learning (Moni, 2019):

1. Agent: The learner and the decision-maker.
2. Environment: Where the agent learns and decides what actions to perform.
3. Action: A set of actions which the agent can perform.
4. State: The state of the agent in the environment.
5. Reward: For each action selected by the agent, the environment provides a reward, usually a scalar value.
6. Policy: The decision-making function (control strategy) of the agent, representing a mapping from situations to actions.
7. Value function: Mapping from states to real numbers, where the value of a state represents the long-term reward achieved starting from that state and executing a particular policy.

8. Function approximator: Inducing a function from training examples. Standard approximators include decision trees, NNs, and nearest-neighbor methods.

9. Markov decision process: A probabilistic model of a sequential decision problem, where states can be perceived exactly, and the current state and action selected determine a probability distribution for future states. Essentially, the outcome of applying an action to a state depends only on the current action and state (and not on preceding actions or states).

10. Dynamic programming: A class of solution methods for solving sequential decision problems with a compositional cost structure.

11. Monte Carlo methods: A class of methods for the learning of value functions, estimating the value of a state by running many trials starting at that state, then averaging the total rewards received on those trials.

12. Temporal difference algorithms: A class of learning methods, based on the idea of comparing temporally successive predictions. It is possibly the most fundamental idea in reinforcement learning.

13. Model: The agent's view of the environment, mapping state-action pairs to probability distributions over states. Note that not every reinforcement learning agent uses a model of its environment.

3.2.6.2 Applications

Reinforcement learning has been applied to various fields, including electric power systems, healthcare, finance, robotics, marketing, natural language processing, transportation systems, and gaming. Gaming offers an excellent environment for reinforcement learning; the agent can explore different trials in a virtual world because the cost of exploration is affordable. The following subsections discuss some applications (Hammoudeh, 2018).

1. Hyper-parameter selection for NNs

The determination of an NN architecture and the selection of hyper-parameters require an iterative process of trial and evaluation. Hence, this can be formulated as a reinforcement learning problem.

Zoph and Le (2017) designed a recurrent NN (RNN) generating the hyper-parameters for NNs. The RNN was trained with reinforcement learning by searching in varying hyper-parameter spaces to maximize accuracy; the accuracy of the generated model on a validation set was considered a reward signal. This approach achieved competitive results when compared with state-of-the-art methods (Zoph & Le, 2017). The concept was later extended to find optimization methods for deep NNs and achieved better results than standard optimization methods such as stochastic gradient descent (SGD), SGD with momentum, root mean square propagation, and adaptive moment estimation (Hammoudeh, 2018).

2. Intelligent transportation systems

Intelligent transportation systems take advantage of the latest information technologies for handling traffic and facilitating transport networks.

Adaptive traffic signal control (ATSC) can lessen traffic congestion by dynamically adjusting signal timing plans in response to traffic fluctuations. A multi-agent reinforcement learning approach was proposed by El-Tantawy (2013) to solve the ATSC problem where each controller (agent) is responsible for the control of traffic lights around a single traffic junction. Multi-agent reinforcement learning combines game theory with single agent reinforcement learning.

Some of the challenges of multi-agent reinforcement learning approach are the following: the exploration-exploitation tradeoff, curse of dimensionality, stability, and non-stationarity.

The non-stationarity challenge of multi-agent learning occurs because multiple agents are learning simultaneously; every agent has a different learning problem but the agent's optimal policy changes as the policies of other agents change (Hammoudeh, 2018).

Van der Pol and Oliehoek (2016) developed the previous work on the traffic light control problem by eliminating the simplifying assumptions and introducing a new reward function.

3. Natural language processing

Reinforcement learning has been used in various natural processing tasks, such as text generation, machine translation, and conversation systems (Hammoudeh, 2018). Dialogue systems are also an important part of natural language processing.

Dialogue systems are programs that interact with natural language. They are generally classified into two categories: chatbots and task-oriented dialogue agents. The task-oriented dialogue agents interact through short conversations in a specific domain to help complete specific tasks. The chatbots are designed to handle generic conversations and imitate human to human interactions.

3.2.7 CLASSIFICATION, CLUSTERING, AND BAYESIAN METHODS

3.2.7.1 Classification

The classification problem is one of the main sub-domains of discriminant analysis and closely related to many fields in statistics. Classification means to assign an element to the appropriate population in a set of known populations based on certain observed variables. It is an important direction of the development of multivariate statistics and has applications in many different fields (Tai, 2017). This problem interests statisticians in both theoretical and applied areas.

According to Tai (2017), there are four main methods to solve the classification problem:

1. Fisher method.
2. Logistic regression method.
3. SVM method.
4. Bayesian method.

Because the Bayesian method does not require normal data conditions and can classify two and more populations, it has many advantages and is widely used by scientists (Tai, 2017).

The Bayesian approach to unsupervised learning provides a probabilistic method to inductive inference. In Bayesian classification, class membership is expressed probabilistically; that is, an item is not assigned to a unique class; instead it has a probability of belonging to each of the possible classes. The classes provide probabilities for all attribute values of each item. Class membership probabilities are then determined by combining all these probabilities. Class membership probabilities of each item must sum to 1. Thus, there are no precise boundaries for classes: every item must be a member of some class, even though we do not know which one. When every item has a probability of no more than 0.5 in any class, the classification is not well defined because it means that classes are abundantly overlapped. When the probability of each instance is about 0.99 in its most probable class, the classes are well separated.

Let $D = \{X_1,..., X_m\}$ denote the observed data objects, where instances or items X_i are represented as ordered vectors of attribute values $X_i = \{X_{i1},..., X_{ik}\}$. Unsupervised classification aims at determining the best class description (hypothesis) h from some space H that predicts data D. The term "best" can be interpreted as the most probable hypothesis given the observed data D and some prior knowledge on the hypotheses of H in the absence of D, that is, the prior probabilities of the various hypotheses in H when no data have been observed. Bayes' theorem provides a way to compute the probabilities of the best hypothesis, given the prior probabilities, the probabilities of observing the data given the various hypotheses, and the observed data (Talia et al., 2015).

Let $P(h)$ denote the prior probability that the hypothesis h holds before the data have been observed. Analogously, let $P(D)$ denote the prior probability that the data will be observed, that is, the probability of D with no knowledge of which hypothesis holds. $P(D|h)$ denotes the probability of observing D in some world where the hypothesis h is valid. In unsupervised classification, the main problem is to find the probability $P(h|D)$, that is, the probability hypothesis h is valid, given the observed data D. $P(h|D)$ is called the posterior probability of h and expresses the degree of belief in h after the data have been seen. Thus, the set of data items biases the posterior probability, while the prior probability is independent of D. Bayes' theorem provides a method to compute the posterior probability (Talia et al., 2015):

$$P(h|D) = \frac{P(D|h)P(h)}{P(D)} \tag{1}$$

which is equivalent to

$$P(h|D) = \frac{P(D|h)P(h)}{\sum_h P(D|h)P(h)} \tag{2}$$

because of the theorem of total probability which asserts that if events $h_1, \ldots, h_n$ are mutually exclusive with $\sum_{i=1}^{n} P(h_i) = 1$, then $P(D) = \sum_{i=1}^{n} P(D | h_i) P(h_i) = 1$.

When the set of possible h is continuous, the prior becomes a differential, and the sums over h are integrals. Thus, this computation becomes difficult to realize.

The expectation–maximization (EM) algorithm is a Bayesian solution used to find the local maximum likelihood or maximum a posterior (MAP) estimates of parameters when the model depends on unobserved latent variables. Instead of considering all the possible states of the world and, consequently all the possible hypotheses h specifying that the world is in some particular state, in EM, only a small space of models is considered, and the assumption S that one of such models describes the world is believed true. A model consists of two sets of parameters: a set of discrete parameters T which describes the functional form of the model, such as number of classes and whether attributes are correlated, and a set of continuous parameters V that specifies values for the variables appearing in T, needed to complete the general form of the model.

Given a set of data D, EM searches for the most probable pair V, T which classifies D, $P(DVT | S) = P(D | VTS) P(VT | S)$. This is done in two steps (Talia et al., 2015):

1. For a given T, EM seeks the MAP parameter values V.
2. Regardless of any V, EM searches for the most probable T, from a set of possible Ts with different attribute dependencies and class structure.

There are two levels of search: parameter level search and model level search. When the number J of classes and their class model are fixed, the space of allowed parameter values is searched to find the most probable V. This space is real valued and contains many local maxima; thus, finding the global maximum is not an easy task. The algorithm randomly generates a starting point and converges on the maximal values nearest to the starting point. Model level search chooses the best model structure. All instances in D are assumed independent; thus,

$$P(D | VTS) = \prod_i P(X_i | VTS) \tag{3}$$

and it is necessary to find the probability $P(X_i \mid VTS)$ of each instance X_i.

The joint probability $P(DVT \mid S)$ generally has many local maxima. To find them, NASA designed a variant of the EM algorithm, AutoClass, based on the fact that at a maximum, the class parameter V_j can be estimated from sufficient weighted statistics, summarizing all the information relevant to a model. The weights w_{ij} give the probability that an instance X_i is a member of class J. These weights must satisfy $\Sigma_j \, w_{ij} = 1$. This approximate information can be used to re-estimate the parameters V, and this new set of parameters permits re-estimation of the class probabilities. Repeating these two steps yields a local optimum. To find as many local maxima as possible, AutoClass generates pseudo-random points in the parameter space, converges to a local optimum, records the results, and repeats the same steps (https://ti.arc.nasa.gov/tech/rse/synthesis-projects-applications/autocl ass/autoclass-c/).

3.2.7.2   Clustering

Clustering is an important unsupervised learning problem. Commonly used clustering methods include the following (Heller, 2007):

1. *K*-means clustering and mixture modeling: These are the most common methods for canonical and flat clustering.
2. Hierarchical clustering: In hierarchical clustering, the goal is not to find a single partitioning of the data, but a hierarchy (generally represented by a tree) of partitioning to reveal interesting structures in the data at multiple levels of granularity. Hierarchical clustering algorithms may be agglomerative or divisive. The former look at ways of merging data points together to form a hierarchy; the latter separate the data repeatedly into finer groups.
3. Spectral clustering: A similarity matrix is computed for all pairs of data points. An eigenvalue decomposition is performed, data points are projected into a space spanned by a subset of the eigenvectors, and one of the previously mentioned clustering algorithms (typically *K*-means or hierarchical clustering) is used to cluster the data.

3.2.7.3   Bayesian Methods

Bayesian methods provide a way to reason coherently about the world around us, in the face of uncertainty (see Section 3.2.7.1). Bayesian approaches are based on a mathematical handling of uncertainty, initially proposed by Bayes and Laplace in the 18th century and further developed by statisticians and philosophers in the 20th century. Bayesian methods have recently emerged as models of human cognitive phenomena in areas such as multi-sensory integration, motor learning, visual illusions, and neural computation and as the basis of machine learning systems (Heller, 2007).

Bayes' rule states that:

$$P(\theta \mid x) = \frac{P(x \mid \theta) P(\theta)}{P(x)} \tag{4}$$

and can be derived from basic probability theory. Here x might be a data point and θ some model parameters. $P(\theta)$ is the probability of θ and is referred to as the prior; it represents the prior probability of θ before observing any information about x. $P(x \mid \theta)$ is the probability of x conditioned on θ and is also referred to as the likelihood. $P(\theta \mid x)$ is the posterior probability of θ after observing x, and $P(x)$ is the normalizing constant.

Letting $P(x, \theta)$ be the joint probability of x and θ and marginalizing θ, we have:

$$P(x) = \int P(x, \theta) \, d\theta \tag{5}$$

Thus, $P(x)$ can be considered the marginal probability of x.

For a dataset of N data points, where $D = \{x_1, x_2, \ldots, x_N\}$, model m with model parameters θ can be expressed as:

$$P(m|D) = \frac{P(D|m)*P(m)}{P(D)} \tag{6}$$

This quantity can be computed for many different models m. We can select the one with the highest posterior probability as the best model for our data:

$$P(D|m) = \int P(D \mid \theta, m) P(\theta \mid m) d\theta \tag{7}$$

Equation (7) gives the marginal likelihood, and this is necessary to compute Equation (6).

The probability of new data points, x^*, which have not yet been observed can also be predicted:

$$P(x^*|D, m) = \int P(x^* \mid \theta) P(\theta \mid D, m) d\theta \tag{8}$$

where

$$P(\theta|D, m) = \frac{P(D|\theta, m)*P(\theta \mid m)}{P(D \mid m)} \tag{9}$$

is the posterior probability of model parameters θ conditioned on data D, and computed using Bayes' rule.

Bayesian probability theory can be used to represent degrees of belief in uncertain propositions. In fact, if we represent beliefs numerically, with only a few basic assumptions, this numerical representation of beliefs results in the derivation of basic probability theory (Cox, 1946; Jaynes, 2003). A game theory result, called the Dutch Book Theorem, states that if we are willing to place bets in accordance with our beliefs, unless those beliefs are consistent with probability theory (including Bayes rule), there is a Dutch book of bets that we will be willing to accept that is guaranteed to lose money regardless of the outcome of the bets (Heller, 2007).

Bayesian methods inherently invoke Occam's razor, the principle that, of two explanations accounting for all the facts, the simpler one is more likely to be correct. It is applied to a wide range of disciplines, including religion, physics, and medicine. Consider two models m_1 and m_2, where m_2 contains m_1 as a special case (e.g., linear functions m_1 are special cases of higher-order polynomials m_2). The marginal likelihood (Equation (7)) for m_2 will be lower than for m_1 if the data are already being modeled well by m_1 (e.g., a linear function). However, m_2 will model some datasets (e.g., nonlinear functions) better than m_1. Therefore, Bayesian methods do not typically suffer from the overfitting problems commonly encountered in other methods (Heller, 2007).

3.2.8 DATA MINING ALGORITHMS

In simple terms, a data mining (or machine learning) algorithm is a set of calculations and heuristics used to create a model from data. In the first step, the algorithm analyzes the data provided, looking for specific types of patterns or trends. Next, it uses the results of this analysis over many iterations to find the optimal parameters to create the data mining model. Finally, these parameters are applied to the entire dataset to extract actionable patterns and detailed statistics (Microsoft, 2020).

A data mining model can take many different forms and be used for many different purposes. A mining model can be:

- A set of clusters describing how cases in a dataset are related.
- A decision tree predicting an outcome and describing how different criteria affect that outcome.
- A mathematical model forecasting sales.
- A set of rules describing how products are grouped together in a transaction, and the probabilities of products being purchased together.

3.2.8.1 Types of Data Mining Algorithms

There are many different types of data mining algorithms.

- Classification algorithms: This type of algorithm is used to predict discrete variables, based on the other attributes in the dataset.
- Regression algorithms: Regression algorithms are used to predict continuous numeric variables, such as profit or loss, based on other attributes in the dataset.
- Segmentation algorithms: As the name suggests, these algorithms divide data into groups or clusters of similar items.
- Association algorithms: Association algorithms are used to find correlations (i.e., associations) between attributes in a dataset. A common application is to find relations between items people buy and thus target sales strategies.
- Sequence analysis algorithms: This type of algorithm is used to summarize frequently occurring sequences, such as a series of clicks in a website, or a series of log events preceding machine maintenance.

These are not necessarily used alone. Analysts may use one algorithm to determine the most effective inputs (i.e., variables), and then apply a different one to predict a specific outcome based on those data. Multiple algorithms can be used within a single solution to perform separate tasks: for example, regression analysis can be used to obtain financial forecasts, or an NN algorithm can be used to analyze factors that influence forecasts.

The SQL Server Data Mining from Microsoft is one example of a data mining solution. It permits multiple models on a single mining structure. Thus, a single data mining solution may include a decision tree model, a clustering algorithm, and a naive Bayes model, so the analyst can get different views on the data (Microsoft, 2020).

3.2.8.2 Top Data Mining Algorithms

Some of the main data mining algorithms are the following (Bakshi, 2017):

1. The C4.5 algorithm generates a classifier in the form of a decision tree after it is given a set of data representing things that have already been classified. The decision trees generated can be used for further classification. C4.5 is often called a statistical classifier.
2. K-means clustering, also known as nearest centroid classifier or the Rocchio algorithm, is a method of vector quantization used for cluster analysis. It creates k groups from a set of objects, where the members of a group are similar. It is often used to explore a dataset.
3. SVMs are supervised learning models with associated learning algorithms which analyze data. An SVM model is created to represent points in space; these are mapped so that the examples of the separate categories are divided by a clear gap. SVMs are used for regression analysis and classification.
4. A *priori* algorithms are used for frequent item set mining and association rule learning in transactional databases. The individual items that appear frequently in the database are identified and extended to larger item sets, as long as those item sets appear often enough in the database. The frequent item sets determined by *a priori* algorithms can be used to determine association rules which then highlight general trends.

5. EM is an iterative method used to find MAP or maximum likelihood estimates of parameters in statistical models, generally for unobserved latent variables.

6. PageRank (PR) is named after Larry Page, one of the founders of Google. PR is an algorithm used by Google Search to rank the websites in search engine results.

7. Adaptive boosting or AdaBoost is a machine learning meta-algorithm. It can be used in combination with many other types of learning algorithms to improve performance. A disadvantage of this algorithm is that it is sensitive to noisy data and outliers.

8. The k-NN algorithm is a type of lazy or instance-based learning. The input consists of the k closest training examples in the feature space, and the output depends on whether the algorithm is being used for classification or regression. The algorithm is among the simplest of all machine learning algorithms. This non-parametric method is used for both classification and regression.

9. Naive Bayes classifiers are simple probabilistic classifiers based on the application of Bayes' theorem and strong independent assumptions between the features. They are considered to be highly scalable.

10. Classification and regression tree (CART) is a decision tree learning technique that outputs either classification or regression trees. Like C4.5, CART is a classifier. Many reasons for using C4.5 apply to CART. Both are decision tree learning techniques, and both are easy to interpret and explain.

3.3 CONVENTIONAL NUMERICAL TECHNIQUES: WAVELETS, KALMAN FILTERS, PARTICLE FILTERS, REGRESSION, DEMODULATION, AND STATISTICAL METHODS

3.3.1 Wavelets

Wavelet analysis is a "numerical microscope" used in signal and image processing. It has the advantage of multi-resolution properties and various basis functions, with an enormous potential to solve partial differential equations (PDEs). It has attracted attention for its ability to analyze rapidly changing transient signals.

Methods include wavelet weighted residual method, wavelet finite element method, wavelet boundary method, wavelet meshless method, and wavelet-optimized finite difference method.

Wavelets are a set of self-similar mathematical functions used to approximate more complex functions via super positioning principles. Wavelets also allow a signal to be divided into different frequency and time components. They have been applied extensively in the fields of biology, chemistry, physics, engineering, and mathematics, and are a popular choice in data analysis as they consistently outperform conventional methods such as principal component analysis (PCA) and Fourier analysis. Any application using the Fourier transform can be formulated using wavelets to provide more accurately localized temporal and frequency (Li & Chen, 2014).

Wavelets are based on the fundamental theory of expressing a complicated function by a set of self-similar functions by super positioning, a principle introduced by Joseph Fourier in the 1800s. Wavelet analysis is closely related to Fourier analysis as both enable a given function to be expressed in terms of summation of basic functions. In Fourier analysis, these basis functions are the sine trigonometric functions. Wavelet analysis, however, uses a specially generated set of self-similar, orthonormal basis functions localized in space and time. This enables wavelet analysis to be performed at different scales or resolutions.

The major advantage of wavelets over Fourier basis functions is that the Fourier basis function, the sine function, is non-local; that is, the sine function ranges from negative to positive infinity, and therefore does poorly in approximating sharp, localized irregularities in data. Fourier analysis provides good results if the function to be approximated is relatively smooth and periodic. Windowed Fourier analysis was introduced to overcome this problem. However, windowed Fourier

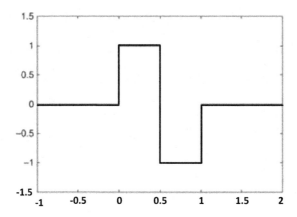

FIGURE 3.18   Haar wavelet (Donald et al., 2009).

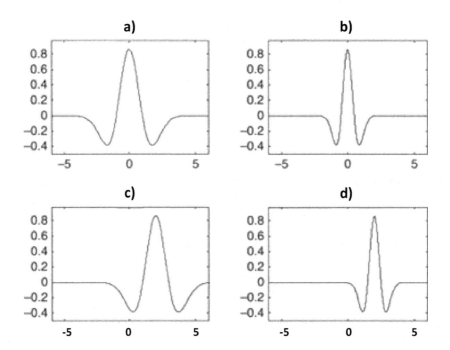

FIGURE 3.19   Mexican Hat wavelet. (a) Mother wavelet, (b) dilation by a factor of 2, (c) translation of 4 to the right, and (d) combination of (b) and (c) (Donald et al., 2009).

analysis still uses the sine function as its basis, and this function has an infinite domain. Wavelets use basis functions localized in finite domains (compact support), making them well suited to detect both sharp irregularities and smooth oscillations in data (Donald et al., 2009).

Basis functions used by the wavelet transform are all generated from one initial function called the mother wavelet. The simplest mother wavelet is the Haar wavelet (Figure 3.18).

Another example is the Mexican hat wavelet (Figure 3.19).

The wavelet basis is formed by translating and dilating the mother wavelet, as illustrated in Figure 3.19. Wavelet analysis of a signal measured over time is performed with a contracted, higher-frequency version of the mother wavelet, whereas frequency analysis is performed by translating the same wavelet. Any function can be represented by linear combinations of these wavelet functions on the mother wavelet. As a result of this wavelet decomposition, data are encoded by the coefficients of the wavelet expansion. Data operations performed using just the corresponding wavelet coefficients can lead to insightful discoveries. Because each coefficient corresponds to a particular level of detail, the coefficients may be truncated to represent the function to a particular tolerance, allowing the data to be represented by a relatively small number of coefficients (Donald et al., 2009).

3.3.1.1 Solving Partial Differential Equations (PDEs) Using Wavelets

The application of wavelets to the solution of difficult PDEs in various areas of physics and engineering has been very limited. Although the application shows promise, wavelet-based numerical algorithms are in their infancy. A large communication gap must be diminished before researchers can take advantage of this new mathematical tool.

Common numerical techniques for the solution of physical problems fall into three classes: finite difference and finite volume methods, finite element methods, and spectral methods. Sometimes the latter two are considered subsets of the method of weighted residuals.

Briefly, the finite difference method consists of defining the different unknowns by their values on a discrete (finite) grid and replacing differential operators by difference operators using neighboring points. In the method of weighted residuals, the unknown solution is approximated by a linear combination of a set of linearly independent trial or basis functions. In the finite element methods, the trial functions are only piecewise continuous and non-vanishing on certain elements of the domain, whereas the spectral method uses basis functions that are infinitely differentiable and non-vanishing on the whole domain.

The method of weighted residuals proceeds by substituting the approximation into a differential equation and imposing the condition that the integral over the residual, weighted by some weighting function, is zero. Different choices of the weighting function give rise to different methods, which collectively are known as methods of weighted residuals. Two well-known examples are the collocation method and the Galerkin method. In the collocation method, the weighting functions are taken to be Dirac delta functions. In the Galerkin method, the weighting functions are chosen from the same family as the test functions.

If the solution of a physical problem has regular features, any of these numerical techniques can be applied. However, in many physical problems, the solution has a multiplicity of very different spatial and temporal scales. This situation occurs in such cases as strongly time-dependent non-Newtonian convection, the formation of shock waves in compressible gas flow, pattern formation in hydrodynamic systems, turbulent flow around bluff bodies, and dendritic crystal growth.

Multiple spatial scales can change over time, putting great strain on conventional numerical methods. Spectral methods have problems capturing large irregularities in solutions. The main difficulty of existing adaptive finite difference or finite element methods is the development of a computationally efficient and robust procedure that dynamically adapts the computational grid to local structures of the solution.

The basic idea of wavelet decomposition is to represent a function in terms of building blocks, that is, wavelets, localized in both position and scale. The good wavelet localization properties in physical and wavenumber spaces can be contrasted with the spectral approach, which employs infinitely differentiable functions but with global support and small discrete changes in the resolution. Finite difference, finite volume, and finite element methods use bases with small compact support but poor continuity properties. Wavelets appear to combine the advantages of spectral and finite difference bases. It can be expected that numerical methods based on wavelets will attain both good spatial and spectral resolution (Vasilyev et al., 1997).

3.3.2 KALMAN FILTERS

Since its formulation (Kalman, 1960), the Kalman filter (KF) has been applied to many practical problems, especially in aeronautics and aerospace. As applications become more numerous, however, some problems have been discovered, such as the problem of divergence because of the lack of reliability of the numerical algorithm or inaccurate modeling of the system under consideration (Jazwinski, 1970). Several modified implementations of the KF have been suggested to avoid these numerical problems. Many of these modifications are based on heuristics (e.g., stabilized KF, or conventional KF with lower bounding) and, as such, often require more experience to implement them effectively.

A more reliable KF implementation is the square root filter (SRF) (Potter &. Stem, 1963). The reliability of the filter estimates is expected to be better because of the use of numerically stable orthogonal transformations for each recursion step. SRF implementation requires more computations than the conventional KF, however. As a result, modified versions of the SRF have been developed, such as UDU-algorithms and the Chandrasekhar form. These implementations can be made as efficient as the conventional KF, or for the Chandrasekhar SRF even more efficient for some special experimental conditions (Verhaegen & Van Dooren, 1986).

The KF's generalized model-based approach to optimal estimation seems ideal for accelerating the transition from a conceptual definition of an estimation problem to its final algorithm implementation – bypassing the selection and testing of alternative suboptimal designs. But this has not been the case for engineering disciplines such as communications and speech processing. For one thing, KF robustness issues remain even after over 30 years of refining the details of implementation. For another, the processing speed for KF solutions cannot approach the many-MHz update cycle times demanded for modern signal processing algorithms (Grewal, 2010).

Issues of KF numerical stability were well known from the early days of KF applications. In fact, the optimality of the estimation process suggests sensitivity to various errors. Widely implemented solutions for these stability issues include the following (Grewal, 2010):

1. Increase arithmetic precision.
2. Implement some form of square root filtering.
3. Symmetrize the covariance matrix at each step.
4. Initialize the covariance appropriately to avoid large changes.
5. Use a fading memory filter.
6. Use fictitious process noise.

Work on KFs discusses these issues but stops short of suggesting universal solutions. For example, many authors show the injection of fictitious process noise prevents Kalman gains from approaching zero, thereby stabilizing the otherwise divergent solution. But they do not discuss strategies for selecting the fictitious noise characteristics. Other authors cover numerical issues, as well as a suite of candidate solutions, but stop short of offering a generally applicable implementation.

KF execution speed improvement has received less scrutiny. To enhance robustness and allow mechanization into fixed point arithmetic, thereby offering methods for transitioning a design to high-speed digital signal processors, the covariance matrix must be symmetric and positive definite; otherwise, it cannot represent valid statistics for state vector components. It was recognized during the early years of KF applications that factored-form KFs (SRFs) are the preferred implementation for applications demanding high operational reliability. Factored terms of the covariance matrix are propagated forward between measurements and updated at each measurement. The covariance matrix, reformed by multiplying its factors together, is positive-semi-definite.

One widely used factored-form KF is the UDU^T filter. The covariance matrix P is defined in terms of the matrix factors U and D as

$$P = UDU^T \qquad (10)$$

The UDU^T KF form has worked well, and specialized digital implementations have been developed so that processing times for factored covariance filters are nearly the same as for traditional covariance propagation methods.

The individual terms of the covariance matrix can be interpreted as

$$P_{ij} = \sigma_i \sigma_j \rho_{ij}, \tag{11}$$

where P_{ij} is the ij^{th} entry of the covariance matrix, σ_i is the standard deviation of the estimate of the ith state component, and ρ_{ij} is the correlation coefficient between the ith and j^{th} state component. Both σ_i and ρ_{ij} contain important physical information defining the progress of KF estimation, in terms of the current success of estimation and the likelihood of future numerical issues. However, the individual terms within matrices propagated for factored-form filters have no useful physical interpretation unless the covariance matrix and, in turn, the statistical parameters in Equation (11), are computed.

The KF is optimal under assumptions that the model is correct and thus can exhibit intolerance to model error. Initial (pre-estimation) state values may be completely unknown, leading to assigning crude estimates for initial states with associated large initial standard deviations. Moreover, initial state correlations are often unknown and assumed to be zero. Such expedient initial conditions often lead to extreme initial transient behavior and early filter failure.

A class of adaptive KF methods has been used to address impacts of modeling uncertainty. Filter real-time performance is evaluated using residual tests with the process noise and/or measurement noise increased if the residual variance is observed to be outside the expected range. However, because the closed adaptation loop must be slower than the estimator response time, the filter may become unrecoverable after the point where the divergence is noticed. A method that can anticipate future anomalous performance is preferable.

One such approach is to recognize that instability often results from extracting too much information too quickly from the estimation process. A slower rate of convergence and/or limiting the lower threshold of the estimated standard deviation is preferable to potential divergence. One way to achieve this result is to compute the KF updated covariance with the physics-based model parameters and test the resulting covariance matrix. If too much estimate improvement is predicted, then either the measurement noise or the process noise levels can be increased and the covariance update recomputed before processing the data at this step. Such an iterative procedure leads to a limited information filter where the maximum standard deviation improvement (per measurement) for any state is limited. In a similar way, we could limit the lower extent of a state standard deviation to an absolute value or to a percentage of its original uncertainty level. The difficulties of such adaptive approaches are primarily computational. A filter form could readily be devised to cycle through the covariance update computations, iteratively adjusting noise models to limit the information extraction.

The computer word length issues that plagued early KF implementations have been mitigated by the power of today's computers. However, KF solutions rarely receive consideration for extreme speed embedded applications, such as communications, speech and video data processing, where fixed point solutions are preferred (Grewal, 2010).

3.3.3 Particle Filters

Particle filters (PFs), also known as the sequential Monte Carlo method, can deal with problems with strong nonlinearity, without requiring linearization. The basic idea of the PF is the following. Suppose the mathematical model is a nonlinear stochastic dynamic system, and our goal is to estimate the hidden states of the system by combining model predictions and noisy partial observations of the system. We can do this with the Bayes filter (or the optimal filter), where the posterior probability density function of the hidden states is estimated by Bayes' rule recursively. A difficulty is that, in general, the posterior distribution does not admit an analytical form. Many approximation approaches have been proposed to address this problem, such as the extended KF. This kind of approach can be problematic when the system is highly nonlinear or the posterior distribution is strongly non-Gaussian. The PF method approximates the posterior distribution with Monte Carlo sampling without making assumptions of linearity on the dynamic model or of Gaussianity on the noise. Specifically, PF employs a number of independent random realizations called particles, sampled directly from the state space, to represent the posterior probability and update the posterior by involving the new observations. The particle system is properly located, weighted, and propagated recursively by the Bayesian formula.

Since its introduction, PF has been applied in many areas, such as signal processing, economics, robotics, and geophysics, to name a few.

Considerable effort has been devoted to analyzing the statistical error of PF and its convergence properties. Weak convergence of the state estimates of the PF to the estimates of the optimal filter has been established with a convergence rate of $O(1/\sqrt{M})$, where M is the number of particles. The convergence is uniform if the number of particles increases over time or the kernel of the PF is weakly dependent on the past. Such conditions require all particles to lie in a compact support subset of the space, but this cannot always be satisfied in practice. In such cases, the error due to the inaccuracy of the particle-based approximation may grow quickly.

Little attention has been paid to analyzing the errors of PF caused by its numerical implementation. In fact, the state equations in almost all systems must be solved numerically in practice, and this inevitably introduces errors into the filtering process (Han et al., 2015).

3.3.4 REGRESSION

Regression analysis can be used to approximate an overdetermined system which has more equations than unknowns. This technique is useful when finding the exact solution is too expensive because of measurement errors or random noise in the data (Wikibooks, 2018). It has numerous applications in various disciplines, including finance.

In regression analysis, a set of statistical methods is used to estimate relations between a dependent variable and one or more independent variables. It can be used to assess the strength of the relations between the variables and to model their future relations.

Regression analysis can be linear, multiple linear (multilinear), or nonlinear (see Figure 3.20). The most common models are linear and multilinear. Nonlinear regression models are commonly

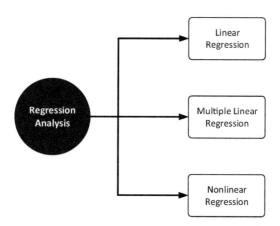

FIGURE 3.20 Regression analysis variations (CFI, 2020).

used for more complicated datasets where the dependent and independent variables have nonlinear relations (CFI, 2020).

3.3.4.1 Linear Regression

Linear regression finds a linear function that most closely passes through the given data points – in other words, it finds the regression (function) line that best fits the data. The metric must be defined to measure the goodness of fit. If all data points lie on the function, it is a perfect fit; otherwise, there are errors in the function representation of the data. The deviations of the data points from the function must also be measured.

Linear regression is the most popular regression model. In this model, we want to predict the response to n data points (x_1, y_1), (x_2, y_2),..., (x_n, y_n) using a regression model given by

$$y = a_0 + a_1 x, \tag{12}$$

where a_0 and a_1 are the constants of the regression model.

A measure of goodness of fit, that is, how well $a_0 + a_1 x$ predicts the response variable y, is the magnitude of the residual ε_i at each of the n data points:

$$E_i = y_i - (a_0 + a_1 x). \tag{13}$$

Ideally, if all the residuals ε_i are zero, we may have found an equation in which all the points lie on the model. Thus, minimization of the residual is an objective of finding the regression coefficients.

The most popular method to minimize the residual is the least squares methods; the estimates of the constants of the models are chosen such that the sum of the squared residuals is minimized; that is, minimize $\sum_{i=1}^{n} E_i^2$ (Wikibooks, 2018).

3.3.4.2 Multilinear Regression

Multilinear regression models the relations between several independent variables (predictors) and a response variable. Once the model is built, it can be used for prediction. In almost all real-world regression analyses, multiple predictors are used to model multiple factors that affect the system's response (dependent variable) (Wikibooks, 2018).

3.3.4.3 Nonlinear Regression

Nonlinear regression models the relationships in observational data using a nonlinear combination of the model parameters and one or more independent variables. Some nonlinear regression problems can be transformed into the linear domain (Wikibooks, 2018).

3.3.5 DEMODULATION

The process of demodulation always requires a nonlinear operation on a signal in order to estimate a baseband signal proportional to the modulation of the carrier. Based on this nonlinearity, demodulation methods can be broadly classified as methods using rectification (non-synchronous detection) and methods using mixing with a reference oscillator signal (synchronous detection). For demodulators of the latter class, the reference signal can be either a square wave, most commonly used for analog implementations, or a sinusoid, most commonly used for digital implementations. Within the class of demodulators using mixing, further classification can be made based on how the $2f_c$ component from the mixing process is filtered out. While open-loop methods rely on either general or numerically precise low-pass filters, closed-loop methods employ feedback of the parameterized signal states to eliminate this component. An overview of the demodulator classification is shown

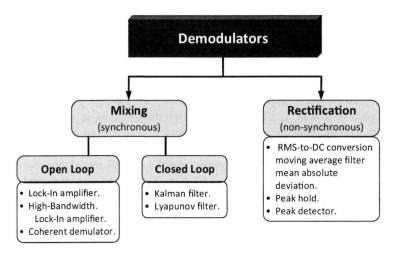

FIGURE 3.21   Classification of demodulation methods (Ruppert et al., 2017).

in Figure 3.21. Each class has distinct properties in tracking bandwidth, implementation complexity and sensitivity to other frequency components (Ruppert et al., 2017).

The linear parameterization used by the demodulation techniques based on mixing is derived from a sine wave with known carrier frequency ω_c, unknown amplitude $A = A(t)$ and unknown phase $\phi = \phi(t)$ of the form (Ruppert et al., 2017)

$$y(t) = A(t)\sin\left[\omega_c t + \phi(t)\right]. \qquad (14)$$

The signal can be rewritten as a sum of its quadrature and in-phase components by applying trigonometric identities to obtain a linear parameterization (note: the time dependency for slowly changing parameters is left out for the sake of readability):

$$y(t) = A\cos\left(\omega_c t\right)\sin\left(\phi\right) + A\sin\left(\omega_c t\right)\cos\left(\phi\right)$$
$$\text{quadrature component} \quad \text{in-phase component}$$
$$= \left[\cos\left(\omega_c t\right)\sin\left(\omega_c t\right)\right] + \left[A\sin\left(\phi\right)A\cos\left(\phi\right)\right]^T \qquad (15)$$
$$= c(t)x$$

The entries of the vector $c(t)$ are termed the quadrature and in-phase sinusoids and the entries of the state vector

$$x = \begin{bmatrix} x_1 & x_2 \end{bmatrix}^T$$

are termed the quadrature and in-phase states. In this form, amplitude and phase can be directly calculated as (Ruppert et al., 2017):

$$A = \sqrt{x_1^2 + x_2^2} \qquad (16)$$

$$\phi = \arctan\left(\frac{x_1}{x_2}\right). \qquad (17)$$

3.4   STATISTICAL APPROACHES

3.4.1   GAMMA PROCESS

A gamma process is a random process with independent gamma distributed increments. Often written as $\Gamma(t, \gamma, \lambda)$, it is a pure-jump increasing Lévy process with intensity measure $v(x) = \gamma x^{-1} exp(-\lambda x)$ for positive x. Jumps whose size lies in the interval $[x, x + dx]$ occur as a Poisson process with intensity $v(x) dx$. Parameter γ controls the rate of jump arrivals, and scaling parameter γ inversely controls the jump size. The process is assumed to start from a value 0 at $t = 0$.

The gamma process can be parameterized in terms of the mean (μ) and variance (v) of the increase per unit time, equivalent to $\gamma = \mu^2 / v$ and $\lambda = \mu / v$ (Wikipedia, 2020a).

3.4.1.1   Gamma Process Model

To deal with data scarcity and uncertainties, a stochastic process model for the time-dependent reliability analysis of corrosion-affected structures and infrastructures can be considered. To model the monotonic progression of a deterioration process, the stochastic gamma process model can be used for the corrosion progress. The gamma process is a stochastic process with independent, non-negative increments having a gamma distribution with an identical scale parameter and a time-dependent shape parameter.

A stochastic process model, such as the gamma process, incorporates the temporal uncertainty associated with the evolution of deterioration.

The gamma process is suitable to model gradual damage monotonically accumulating over time, such as wear, fatigue, corrosion, crack growth, erosion, consumption, creep, swell and a degrading health index. An advantage of modeling deterioration processes using gamma processes is that the required mathematical calculations are relatively straightforward (Mahmoodian & Li, 2016).

3.4.2   HIDDEN MARKOV MODEL

A Markov chain is useful when we need to compute a probability for a sequence of observable events. In many cases, however, the events we are interested in are hidden: we don't observe them directly. For example, normally we don't observe part-of-speech tags in a text. Rather, words are seen, and we must infer the tags from the word sequence. These are called "hidden tags" because they are not observed (Jurafsky & Martin, 2019).

An HMM allows us to talk about both observed events (like words that we see in the input) and hidden events (like part-of-speech tags) that we think of as causal factors in our probabilistic model. An HMM is specified by the components shown in Table 3.2 (Jurafsky & Martin, 2019).

A first-order HMM instantiates two simplifying assumptions. First, as with a first-order Markov chain, the probability of a particular state depends only on the previous state (Jurafsky & Martin, 2019):

1. Markov assumption:

$$P(q_i \mid q_1 \ldots q_{i-1}) = P(q_i \mid q_{i-1}) \tag{18}$$

Second, the probability of an output observation o_i depends only on the state that produced the observation q_i and not on any other states or any other observations:

2. Output independence:

$$P(o_i \mid q_1 \ldots q_i, \ldots, q_T, \; o_1 \ldots o_i, \ldots, o_T) = P(o_i \mid q_i). \tag{19}$$

TABLE 3.2
HMM Components

$Q = q_1 q_2 \ldots q_N$	**Set of N States**
$A = a_{11} \ldots a_{ij} \ldots a_{NN}$	Transition probability matrix A, each a_{ij} representing the probability of moving from state i to state j, s.t. $\sum_{j=1}^{N} a_{ij} = 1 \; \forall_i$
$O = o_1 o_2 \ldots o_T$	Sequence of T observations, each one drawn from a vocabulary $V = v_1, v_2, \ldots, v_V$
$B = b_i(o_t)$	Sequence of observation likelihoods, also called emission probabilities, each expressing the probability of an observation o_t being generated from a state i
$\pi = \pi_1, \pi_2, \ldots, \pi_N$	Initial probability distribution over states. π_i is the probability that the Markov chain will start in state i. Some states j may have $\pi_j = 0$, meaning they cannot be initial states. Also, $\sum_{j=1}^{N} \pi_i = 1$.

Source: Jurafsky and Martin (2019).

To exemplify these models, consider a task invented by Jason Eisner (2002). Imagine that you are a climatologist in the year 2799 studying the history of global warming. You cannot find any records of the weather in Baltimore, Maryland, for the summer of 2020, but you do find Jason Eisner's diary, which lists how many ice creams Jason ate every day that summer. The goal is to use these observations to estimate the temperature every day. We will simplify this weather task by assuming there are only two kinds of days: cold (C) and hot (H). So the Eisner task is as follows:

Given a sequence of observations O (each an integer representing the number of ice creams eaten on a given day) find the "hidden" sequence Q of weather states (H or C) which caused Jason to eat the ice cream.

Figure 3.22 shows a sample HMM for the ice cream task. The two hidden states (H and C) correspond to hot and cold weather, and the observations (drawn from the alphabet $O = \{1,2,3\}$) correspond to the number of ice creams eaten by Jason on a given day (Jurafsky & Martin, 2019).

Rabiner (1989) introduced the idea that HMMs should be characterized by three fundamental problems shown in Table 3.3 (Jurafsky & Martin, 2019).

3.4.3 REGRESSION-BASED MODEL

In practice, researchers start by selecting a model to estimate and then use their chosen method (e.g., ordinary least squares) to estimate its parameters. Regression models include the following components (Wikipedia, 2020b):

- Unknown parameters, often denoted as a scalar or vector β.
- Independent variables observed in data, often denoted as a vector X_i (where i denotes a row of data).
- Dependent variable observed in data, often denoted using the scalar Y_i.
- Error terms not directly observed in data, often denoted using the scalar e_i.

Note that, depending on the field of application, other terms are used instead of dependent and independent variables.

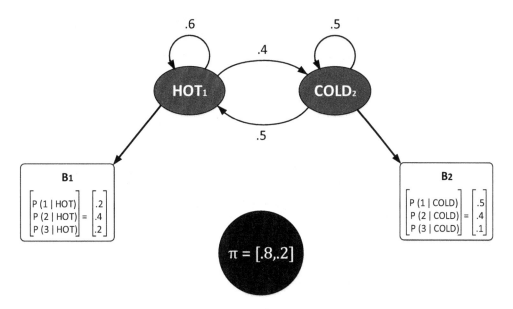

FIGURE 3.22 Hidden Markov model relating numbers of ice creams eaten by Jason (the observations) to the weather (*H* or *C*, the hidden variables) (Jurafsky & Martin, 2019).

TABLE 3.3
Fundamental Problems of Hidden Markov Models (HMM)

Problem 1 (Likelihood)	Given an HMM $\lambda = (A, B)$ and an observation sequence O, determine the likelihood $P(O|\lambda)$.
Problem 2 (Decoding)	Given an observation sequence O and an HMM $\lambda = (A, B)$, discover the best hidden state sequence Q.
Problem 3 (Learning)	Given an observation sequence O and the set of states in the HMM, learn the HMM parameters A and B.

Source: Jurafsky and Martin (2019).

Most regression models propose Y_i is a function of X_i and β. Thus:

$$Y_i = f\left(X_i, \beta\right) + e_i \tag{20}$$

where e_i represents an additive error term denoting un-modeled determinants of Y_i or random statistical noise.

The goal is to estimate the function $f\left(X_i, \beta\right)$ that most closely fits the data. The form of this function can be based on knowledge of the relations between Y_i and X_i that does not rely on the data. If no such knowledge is available, the analyst selects a flexible or convenient form. For example, a simple univariate regression, $f\left(X_i, \beta\right) = \beta_0 + \beta_1 X_i$, suggests the analyst thinks $Y_i = \beta_0 + \beta_1 X_i + e_i$ reasonably approximates the statistical process generating the data.

Once the analyst decides on a statistical model, various tools are available to estimate the parameters β. For example, the least squares (and ordinary least squares) method finds the value of β that minimizes the sum of squared errors $\sum_i \left(Y_i - f\left(X_i, \beta\right)\right)^2$. The selected regression method will ultimately provide an estimate of β, usually denoted $\hat{\beta}$ to distinguish the estimate from the true (unknown) parameter value that generated the data. With this estimate, the analyst can use the fitted value $\hat{Y}_i = f\left(X_i, \hat{\beta}\right)$ to make predictions or to assess the model's ability to explain the data. Whether the analyst is more interested in the estimate $\hat{\beta}$ or in the predicted value $\hat{Y}_i$ will depend on the research context and goals.

Least squares are widely used because the estimated function $f\left(X_i, \hat{\beta}\right)$ approximates the conditional expectation $E(X_i \mid Y_i)$. That said, alternative variants (e.g., least absolute deviations or quantile regression) are useful to model other functions $f\left(X_i, \beta\right)$.

We need sufficient data to estimate a regression model. Say we have access to N rows of data with one dependent and two independent variables $\left(Y_i, X_{1i}, X_{2i}\right)$. Also suppose that we want to estimate a bivariate linear model via least squares: $Y_i = \beta_0 + \beta_1 X_{1i} + \beta_2 X_{2i} + e_i$. If we have access to $N = 2$ data points, infinitely many combinations $\left(\hat{\beta}, \hat{\beta}_1, \hat{\beta}_2\right)$ explain the data equally well. We can choose any combination that satisfies $\hat{Y}_i = \hat{\beta}_0 + \hat{\beta}_1 X_{1i} + \hat{\beta}_2 X_{2i}$. These all lead to $\sum_i \hat{e}_i^2 = \sum_i (\hat{Y}_i - (\hat{\beta}_0 + \hat{\beta}_1 X_{1i} + \hat{\beta}_2 X_{2i}))^2 = 0$ and, thus, are valid solutions minimizing the sum of squared residuals.

There are infinitely many options, because the system of $N = 2$ equations is solved for three unknowns, making the system underdetermined. Alternatively, we can visualize infinitely many three-dimensional (3D) planes that go through $N = 2$ fixed points.

To estimate a least squares model with k distinct parameters, we must have $N \geq k$ distinct data points. If $N > k$, we usually do not have a set of parameters that will perfectly fit the data. The quantity $N - k$ appears frequently in regression analysis and is called the degrees of freedom in the model.

Moreover, the independent variables $\left(X_{1i}, X_{2i}, \ldots, X_{ki}\right)$ must be linearly independent: we must not be able to reconstruct any of them by adding and multiplying the remaining ones. This condition ensures $X^T X$ is an invertible matrix and, therefore, a solution $\hat{\beta}$ exists (Wikipedia, 2020b).

3.4.4 Relevance Vector Machine (RVM)

The relevance vector machine (RVM) method assumes knowledge of probability in the areas of Bayes' theorem and Gaussian distributions, including marginal and conditional Gaussian distributions. It also assumes familiarity with matrix differentiation, the vector representation of regression, and kernel (basis) functions (Fletcher, 2010).

RVM is a machine learning technique that uses Bayesian inference to obtain parsimonious solutions for regression and probabilistic classification. The functional form of RVM is identical to the SVM, but it provides probabilistic classification (Wikipedia, 2020c).

It is actually equivalent to a Gaussian process model with a covariance function (Wikipedia, 2020c):

$$k\left(x, x'\right) = \sum_{j=1}^{N} \frac{1}{\alpha_j} \varphi\left(x, x_j\right) \varphi\left(x', x_j\right), \tag{21}$$

where φ is the kernel function (usually Gaussian), α_j are the variances of the prior on the weight vector $\omega \sim N\left(0, \alpha^{-1} I\right)$, and $X_1, \ldots, X_N$ are the input vectors of the training set.

The Bayesian formulation of RVM avoids the set of free parameters of SVM (the latter usually require cross-validation-based post-optimizations). However, RVMs use an EM-like learning method and are therefore at risk of local minima. By contrast, the standard sequential minimal optimization-based algorithms employed by SVMs are guaranteed to find a global optimum (of the convex problem) (Wikipedia, 2020c).

3.4.5 Autoregressive (AR) Model

Autoregressive (AR) models (also called conditional models, Markov models, or transition models) predict future behavior based on past behavior. They are used for forecasting when there is some correlation between values in a time series and the values preceding them and following them. The process is essentially a linear regression of the data in the current series against one or more past values in the same series.

In an AR model, the value of the outcome variable Y at some point t in time is directly related to the predictor variable X. This is the same as in simple linear regression models. Where simple linear regression and AR models differ is that in AR, Y is dependent on X and previous Y values.

Autoregression is a stochastic process, with degrees of uncertainty or randomness built in. Because of the randomness, the prediction of future trends may be adequate but will never be 100% accurate. Usually, it's close enough to be useful in most scenarios (Glen, 2015).

In statistics, econometrics, and signal processing, an AR model is a representation of a type of random process; as such, it is used to describe certain time-varying processes in nature, economics, and so on. The AR model specifies that the output variable depends linearly on its own previous values and on a stochastic term (an imperfectly predictable term); thus, the model is in the form of a stochastic difference equation (or recurrence relation which should not be confused with differential equation). Together with the moving-average (MA) model, it is a special case and key component of the more general ARMA and AR integrated MA (ARIMA) models of time series, which have more complicated stochastic structures. It is also a special case of the vector AR model, which consists of a system of more than one interlocking stochastic difference equation in more than one evolving random variable.

Unlike the MA model, the AR model is not always stationary, as it may contain a unit root (Wikipedia, 2020d).

3.4.5.1 AR(p) Models

An AR(p) model is an AR model in which specific lagged values of y_t are used as predictor variables. Lags occur when results from one time period affect those in following periods.

The value given for p denotes the order. For example, an AR(1) is a "first order AR process". The outcome variable in a first order AR process at some point in time t is related to time periods that are one period apart, in other words, the value of the variable at $t - 1$. The outcome variable in a second or third order AR process is related to data two or three periods apart, respectively (Glen, 2015).

The AR(p) model is defined as (Glen, 2015)

$$y_t = \delta + \varphi_1 y_{t-1} + \varphi_2 y_{t-2} + ... + \varphi_p y_{t-p} + A_t, \tag{22}$$

where

$y_{t-1}, y_{t-2}, ..., y_{t-p}$ are the past series values (lags),
A_t is white noise (randomness), and
δ is defined by

$$\delta = \left(1 - \sum_{i=1}^{p} \emptyset_i\right)\mu, \tag{23}$$

where μ is the process mean (Glen, 2015).

3.4.6 Threshold Autoregressive (TAR) Model

The threshold AR (TAR) model proposed and explained by Tong (1983) is contained within the state-dependent (regime-switching) model family, along with the bilinear and exponential AR models (Gibson & Nur, 2011).

The simplest class of TAR models is the self-exciting TAR (SETAR) model of order p introduced by Tong (1983) and specified by the following equation (Gibson & Nur, 2011):

$$
Y_t = \begin{cases}
a_0 + \displaystyle\sum_{j=1}^{p} a_j Y_{t-j} + \epsilon & \text{if } Y_{t-d} \le r \\[2em]
\left(a_0 + b_0\right) + \displaystyle\sum_{j=1}^{p} (a_j + b_j) Y_{t-j} + \epsilon_t & \text{if } Y_{t-d} > r
\end{cases}
\tag{24}
$$

TAR models are piecewise linear. The threshold process divides one-dimensional (1D) Euclidean space into k regimes, with a linear AR model in each regime. Such a process makes the model non-linear for at least two regimes, but remains locally linear (Tsay, 1989). One of the simplest TAR models equates the state determining variable with the lagged response, producing what is known as a SETAR model (Gibson & Nur, 2011).

A comparatively recent development is the smooth transition AR (STAR) model, developed by Terasvirta and Anderson (1992). The STAR model of order p model is defined by

$$
Y_t = a_0 + a_1 Y_{t-1} + \ldots + a_p Y_{t-p} + \left(b_0 + b_1 Y_{t-1} + \ldots + b_p Y_{t-p}\right) G\left(\frac{Y_{t-p} - r}{z}\right) + \varepsilon_t
\tag{25}
$$

where d, p, r, and $\{\varepsilon_t\}$ are defined as above, z is a smoothing parameter $z \in \Re^+$, and G is a known distribution function assumed to be continuous. Transitions are now possible along a continuous scale, making the regime-switching process "smooth". This helps overcome the abrupt switch in parameter values characteristic of simpler TAR models (Gibson & Nur, 2011).

3.4.7 Bilinear Model

A bilinear model uses two sets of weights to do the reconstruction. For example, say we have two sets of scalars z_i and y_i which possibly have a different number of dimensions each and a larger set of basis vectors $m_{i,j}$. The reconstruction is the weighted sum $x = \sum_{i,j} m_{i,j} z_i y_i$. This is called a bilinear model because summing over i or j results in a simple linear model similar to PCA.

While PCA is a constrained problem, and the optimal solution can be found using an eigendecomposition or SVD, the bilinear model is under constrained. So constraints must be imposed to solve the parameter estimation problem.

A bilinear model is a function of two (or more) variables and is independently linear in both variables (Quora, 2014). An example is the following:

$$
\forall u, v, w, x
$$

$$
f(u + v, w) = f(u, w) + f(v, w)
$$

$$
f(v, w + x) = f(u, w) + f(u, x).
$$

A simple example is the dot product of two vectors in normal Euclidean space. Normal and matrix multiplication is also bilinear, so various translations of the plane are bilinear (Quora, 2014).

3.4.8 PROJECTION PURSUIT

The term "projection pursuit" was first used by Friedman and Tukey (1974) to name a technique for the exploratory analysis of reasonably large and reasonably multivariate datasets. Projection pursuit reveals structure in the original data by offering selected low-dimensional orthogonal projections of it for inspection. For projection from, say, two dimensions to one dimension, it is possible to examine essentially all such projections to select those of interest: the appearance of the projected dataset does not change abruptly as the projection direction changes, and the space of projection directions, although forming a continuum, is of low dimensionality. For projection from higher dimensions, the appearance of the projected data still changes smoothly, but it becomes increasingly impractical to explore possible projections exhaustively because of the high dimension of the space of projection directions. Friedman and Stuetzle (1981) describe interactive procedures for such exploration, but this does not extend the applicability of the approach as far as is needed. An automatic procedure for selecting potentially interesting projections is required. Projection pursuit is the process of making such selections by the local optimization over projection directions of some index of "interestingness" (Jones & Sibson, 1987).

Pattern variation in more than three dimensions cannot be directly appreciated. Even 3D point clouds are not easy to display without the use of computer graphics hardware of the kind employed in molecular modeling. In practice, then, projection will be onto 1D or two-dimensional space. It will considerably limit the impact of the approach if only 1D projections can be inspected, so it is important that both the theory and the computational practicability of the method extend at least to projection onto a plane.

Exploratory multivariate analysis is usually based on the hope that much of the data are redundant, and the main features can be described in terms of a tendency for the point cloud to concentrate into clusters, or about a curve or a generally non-flat surface. Even when subtler levels of structure are sought, it is usually necessary, as a first step, to be aware of these systematic variations in the density of the point cloud. PCA is a familiar exploratory technique of this kind; it is, in fact, a projection pursuit method in which the index of interestingness is the proportion of total variance accounted for by the projected data (Jones & Sibson, 1987).

In a nutshell, the projection pursuit class of exploratory projection techniques contains statistical methods to analyze high-dimensional data using low-dimensional projections. The goal is to discover possible nonlinear and thus interesting structures hidden in the data. Analysts use an index to measure how interesting the structures are (Wiwi, 2020).

3.4.9 MULTIVARIATE ADAPTIVE REGRESSION SPLINES (MARS)

Multivariate adaptive regression splines (MARS), a non-parametric modeling method, extends the linear mode by incorporating nonlinearities and interactions between variables. It is a flexible tool that automates the construction of predictive models: selecting relevant variables, transforming the predictor variables, processing missing values, and preventing overshooting using a self-test. It is also able to predict, taking into account structural factors that might influence the outcome variable, thereby generating hypothetical models. The end result could identify relevant cut-off points in data series.

Developing a good regression model takes time and considerable modeling experience. However, with MARS, regression models can be developed in a systematic and automatic way without limiting the assumptions that traditional regression models must fulfill.

MARS is a flexible tool that automates the construction of prediction models, allowing the selection of relevant variables, the transformation of predictor variables, establishing the interactions of predictor variables, the treatment of missing values and a self-test to protect against overfitting.

Finally, it can reveal patterns and relationships that are difficult, if not impossible, for other methods to reveal (Vanegas & Vásquez, 2017).

3.4.9.1   Explanation of MARS Method

MARS is a generalization of recursive partitioning regression, which divides the space of the predictor variables into different sub-regions.

The model can be written as

$$y_t = f(x_t) = \beta_0 + \sum_{i=1}^{k} \beta_i \beta(x_{it}),$$ (26)

where y_t is the response variable at time t, and β_i are the model parameters for the respective variables x_{it}, ranging from $i = 1,\ldots, k$. The value β_0 represents the intercept, the base functions $B(x_{it})$ are functions that depend on the respective variables x_{it}, where each $B(x_{it})$ can be written as $B(x_{it}) = \max(0, x_{it} - c)$ or $B(x_{it}) = \max(0, c - x_{it})$, c is a threshold value, and k represents the number of explanatory, including interactions of the predictor variables. The space partition points and the model parameters are obtained from the analyzed data. The number of resulting base functions indicates the complexity of the model.

MARS generates cut points for the different variables. Points are identified by basal functions, which indicate the beginning and end of a region. In each region in which the space is divided, a base function of a variable is adjusted to be linear. The final model is constituted as a combination of the generated base functions.

To determine these cut points, use a stepwise forward/backward stepwise algorithm. First, the forward stepwise algorithm generates an overestimated model with a large number of base functions; then, using the backward stepwise algorithm, the nodes that least contribute to the global adjustment are eliminated. The algorithm stops when the constructed approximation includes a maximum number of functions set by the researcher (Vanegas & Vásquez, 2017).

The following can be used to identify the best model (Vanegas & Vásquez, 2017):

- Cross-validation criterion: Measure of fit to data and penalty. Required because of the complexity of the model and the increase in variance. According to this criterion, a simpler model may be preferred over a more complex one (Friedman 1991).
- Determination coefficient (adjusted R^2): Coefficient of the observed and the predicted value, to ensure model adequacy.
- Mean absolute error ratio: Determined by the observed values and the predicted value. The best model is the one with the lowest error. The mean absolute ratio is expressed as

$$\sum_{i=1}^{n} \frac{\left[\left[\frac{|Observed\_value - Predicted\_value|}{Observed\_value}\right]\right]}{n}?$$ (27)

The statistical software available to apply MARS includes the statistical package R, MATLAB, Python, Salford Predictive Modeler (SPM 8), Statistica Data Miner-StatSoft, and Adaptivreg for SAS (Vanegas & Vásquez, 2017).

3.4.10   Volterra Series Expansion

Volterra series expansion is widely employed to represent the input-output relationship of nonlinear dynamical systems. It is based on the Volterra frequency-response functions (VFRFs), which can either be estimated from observed data or by means of a nonlinear governing equation, when the Volterra series is used to approximate an analytical model. In the latter case, the VFRFs are usually evaluated by the harmonic probing method. This operation is straightforward for simple systems but may reach a level of such complexity, especially when dealing with high-order nonlinear systems or calculating high-order VFRFs, that it may lose its attractiveness (Carassale & Kareem, 2010).

3.4.10.1 Volterra Series: Background and Definitions

The nonlinear system is represented by the following equation (Carassale & Kareem, 2010):

$$x(t) = H[u(t)], \tag{28}$$

where $u(t)$ and $x(t)$ = input and output, respectively, and t = time. If the operator $H[\cdot]$ is time-invariant and has finite memory, its output $x(t)$ can be expressed far enough from the initial conditions through Volterra series expansion

$$x(t) = \sum_{j=0}^{\infty} x_j(t) = \sum_{j=0}^{\infty} H_j[u(t)] \tag{29}$$

in which each term x_j is the output of an operator H_j, homogeneous of degree j, referred to as the jth order Volterra operator. The zeroth-order term $x_0 = H_0$ is a constant output independent of the input, while the generic j^{th}-order term ($j \geq 1$) is given by the expression

$$H_j[u(t)] = \int_{\tau_j \in R^j}^{0} h_j(\tau_j) \prod_{r=1}^{j} u(t - \tau_r) d\tau_j \qquad (j = 1, 2, \ldots) \tag{30}$$

where $\tau_j = [\tau_1, \ldots, \tau_j]$ is a vector containing the j integration variables and the functions h_j = Volterra kernels. The first-order term is the convolution integral typical of linear dynamical systems, with h_1 being the impulse-response function. The higher-order terms are multiple convolutions, involving products of the input values for different delay times.

The series defined in Equation (29) is, in principle, composed of infinite terms and, for practical applications, needs to be conveniently truncated retaining the terms up to some order n. Within this framework, the operator H is represented by the parallel assemblage of a finite number ($n + 1$) of Volterra operators $H_0, \ldots, H_n$ (Figure 3.23) and is referred to as an nth-order Volterra system.

The expression of the Volterra operators provided by Equation (30) can be slightly generalized defining the multilinear operators

$$H_j\{u_1(t), \ldots, u_j(t)\} = \int_{\tau_j \in R^j}^{0} h_j(\tau_j) \prod_{r=1}^{j} u(t - \tau_r) d\tau_j \quad (j = 2, \ldots) \tag{31}$$

where $u_1, \ldots, u_j = j$, in general, different scalar inputs. From Equation (31), it obviously results that $H_j\{u, \ldots, u\} = H_j[u]$:

$$H_j(_j) = \int_{\tau_j \in R^j}^{0} e^{-i_j^T \tau_j} h_j(\tau_j) d\tau_j \quad (j = 1, \ldots), \tag{32}$$

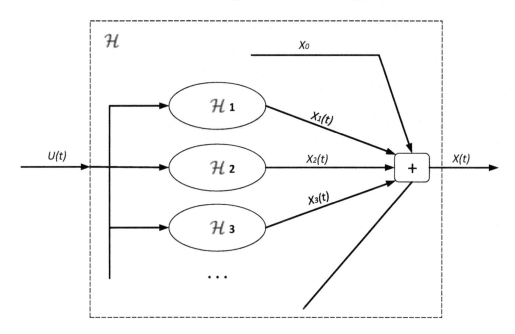

FIGURE 3.23 Block diagram for a Volterra system (Carassale & Kareem, 2010).

where $_j = \begin{bmatrix} \omega_1 \ldots \omega_j \end{bmatrix}^T$ is a vector containing the j circular frequency values corresponding to τ_1, ..., τ_j in the Fourier transform pair. For completeness of the notation, the zeroth-order VFRF is defined as the zeroth-order output; that is, $H_0 = x_0$.

The Volterra operators defined by Equation (30) are not unique and can always be chosen in such a way that the corresponding multilinear operators $H_j \{u_1, \ldots, u_j\}$ are symmetric (i.e., independent of the ordering of the j inputs). This assumption works as a consequence of the symmetry of the Volterra kernels and VFRFs, respectively, as noted in Equations (31) and (32). It follows that any non-symmetric Volterra operator, defined by the VFRF $\tilde{H}_j$, can be replaced by its equivalent symmetric one, whose VFRF is given by (Carassale & Kareem, 2010):

$$H_j\left(_j\right) = sym\left[\tilde{H}_j\left(_j\right)\right] = \frac{1}{j!} \sum_{\substack{all\ jth-order \\ permuting \\ matrices\ E}} \tilde{H}_j\left(E_j\right) \tag{33}$$

REFERENCES

Ansari A. Q., 1998. The basics of fuzzy logic: A tutorial review. Department of Elect. Engg., Faculty of Engg. and Technology, Jamia Millia Islamia (A Central University). New Delhi-110025, India. Computer Education - Stafford - Computer Education Group, U.K., February 1998.

Bakshi K., 2017. A list of top data mining algorithms. December 16, 2017. TechLeer Media Pvt Ltd. www.techleer.com/articles/438-a-list-of-top-data-mining-algorithms/. Viewed: May 29, 2020.

Brid R. S., 2018. Decision Trees – A simple way to visualize a decision. GreyAtom. October 26, 2018. https://medium.com/greyatom/decision-trees-a-simple-way-to-visualize-a-decision-dc506a403aeb. Viewed: May 29, 2020.

Calusdian R. F., 1995. Fuzzy reality. IEEE Potentials, April/May 1995.

CFI, 2020. Regression analysis. Corporate Finance Institute (CFI). https://corporatefinanceinstitute.com/resources/knowledge/finance/regression-analysis/. Viewed: May 30, 2020.

Carassale L., Kareem A., 2010. Modeling nonlinear systems by Volterra series. Journal of Engineering Mechanics. 801–818.

Chevrie F., Guély F., 1998. Fuzzy logic. Cahier technique No.191. ECT 191 first issued, December 1998. Groupe Schneider's.

Choudhary P., 2017. Introduction to anomaly detection. Learn Data Science. February 15, 2017. https://blogs.oracle.com/datascience/introduction-to-anomaly-detection. Viewed: May 28, 2020.

Cox R. T., 1946. Probability, frequency and reasonable expectation. American Journal of Physics, 14, 1–13.

Dernoncourt F., 2013. Introduction to fuzzy logic. MIT, January 2013.

Deshpande S., Patki A. B., Raghunathan G. V., 1996. Towards rapid prototyping CAD environment for fuzzy logic based portable products. Paper accepted for International Discource FLAMOC' 96 Sydney Australia, January 15–18, 1996.

Donald D. A., Everingham Y. L., McKinna L. W., Coomans D., 2009. Feature selection in the wavelet domain: adaptive wavelets. In: Brown, Steven, Tauler, Roma, and Walczak, Beata, (eds.) Comprehensive Chemometrics: chemical and biochemical data analysis. Elsevier, Oxford, UK, pp. 647–679.

Eisner J., 2002. An interactive spreadsheet for teaching the forward-backward algorithm. In Proceedings of the ACL Workshop on Effective Tools and Methodologies for Teaching NLP and CL, 10–18.

Fletcher T., 2010. Relevance vector machines explained. October 19, 2010.

Friedman J. H., 1991. Multivariate adaptive regression splines. Annals of Statistics, 19, 1–141.

Friedman J. H., Stuetzle W., 1981. Projection pursuit classification. Unpublished manuscript.

Friedman J. H., Tukey J. W., 1974. A projection pursuit algorithm for exploratory data analysis, IEEE Transactions on Computers C. 23, 881–890.

Gibson D., Nur D., 2011. Threshold autoregressive models in finance: a comparative approach. Proceedings of the Fourth Annual ASEARC Conference, 17–18 February 2011, University of Western Sydney, Paramatta, Australia.

Glen S., 2015. Autoregressive model: Definition & the AR process. Statistics How To. Statistics for the rest of us! August 19, 2015. www.statisticshowto.com/autoregressive-model. May 30, 2020.

Grewal M., Kain J., 2010. Kalman filter implementation with improved numerical properties. Number FP-09-062.1. IEEE Transactions on Automatic Control, October 2010.

Hammoudeh A., 2018. A concise introduction to reinforcement learning. February 2018.

Han X., Li J., Xiu D., 2015. Error analysis for numerical formulation of particle filter. discrete and continuous. Dynamical Systems Series B. 20(5), July 2015, 1337–1354.

Heller K. A., 2007. Efficient Bayesian methods for clustering. Thesis. University College London, United Kingdom.

Hellmann M., 2001. Fuzzy logic introduction. Laboratoire Antennes Radar Telecom, F.R.E CNRS 2272, Equipe Radar Polarimetrie. Universite de Rennes 1. March 2001.

Jaynes E. T., 1988. How does the brain do plausible reasoning? In G. J. Erickson & C. R. Smith (Eds.), Maximum Entropy and Bayesian Methods in Science and Engineering (pp. 1–24). Dordrecht: Kluwer.

Jazwinski A. H., 1970. Stochastic processes and filtering theory. Academic.

Jones M. C., Sibson R., 1987. What is projection pursuit? Journal of the Royal Statistical Society: Series A. 150, Part 1, 1–36.

Jurafsky D., Martin J. H., 2019. Speech and language processing. Chapter a: Hidden Markov models. Draft.

Kalman R. R., 1960. A new approach to linear filtering and prediction problems. Journal of Basic Engineering, 82(1), 35–45.

Li B., Chen X., 2014. Wavelet-based numerical analysis: A review and classification. Finite Elements in Analysis and Design. 81, April 2014, 14–31.

Mahmoodian M., Li C. Q., 2016. Handbook of materials failure analysis with case studies from the oil and gas industry. Royal Melbourne Institute of Technology, Melbourne, Australia. 235–255.

Moni R., 2019. Reinforcement learning algorithms: An intuitive overview. SmartLab AI, February 18, 2019. https://medium.com/@SmartLabAI/reinforcement-learning-algorithms-an-intuitive-overview-904e2dff5bbc. Viewed: May 28, 2020.

Okoh C., Roy R., Mehnen J., 2016. Predictive maintenance modelling for through-life engineering services. The 5th International Conference on Through-life Engineering Services (TESConf 2016). EPSRC Centre for Innovative Manufacturing in Through-Life Engineering Services. Procedia CIRP 59 (2017) 196–201.

Open Textbooks, 2016. Neural networks (NNs). The Open University of Hong Kong. January 19, 2016. www.opentextbooks.org.hk/ditatopic/27154. Viewed: May 23, 2020.

Potter J. E., Stem R. G., 1963. Statistical filtering of space navigation measurements. In Proc. 1963 AZAA Guidance Contr. Conference, 1963.

Rabiner L. R., 1989. A tutorial on hidden Markov models and selected applications in speech recognition. Proceedings of the IEEE, 77(2), 257–286.

Rojas R., 1996. Neural networks. Chapter 11: Fuzzy logic. Springer-Verlag.

Ruppert M. G., Harcombe D. M., Ragazzon M. R. P., Moheimani R. S. O., Fleming A. J., 2017. A review of demodulation techniques for amplitude-modulation atomic force microscopy. Guest Editor: T. Glatzel. Beilstein Journal of Nanotechnology 8, 1407–1426.

Scikit-Learn, 2019. Support vector machines. Scikit-learn developers (BSD License). https://scikit-learn.org/stable/modules/svm.html. Viewed: May 28, 2020.

Sharma V., 2018. How neural network algorithms works: An overview. December 31, 2018. Vinod Sharma's Blog: Quantum Technology, AI, ML, Data Science, Big Data Analytics, Blockchain and Fintech. https://vinodsblog.com/2018/12/31/how-neural-network-algorithms-works-an-overview. Viewed: May 23, 2020.

Sunil Ray S., 2017. Understanding support vector machine (SVM) algorithm from examples (along with code). September 13, 2017. www.analyticsvidhya.com/blog/2017/09/understaing-support-vector-machine-example-code. Viewed: May 27, 2020.

Sutton R. S., Barto A. G., 2015. Reinforcement learning: An introduction. The MIT Press.

Tai V., 2017. Classifying by Bayesian method and some applications. Published: November 2, 2017. doi: 10.5772/intechopen.70052. www.intechopen.com/books/bayesian-inference/classifying-by-bayesian-method-and-some-applications. Viewed: May 29, 2020.

Talia D., Trunfio P., Marozzo F., 2015. Data analysis in the cloud. models, techniques and applications. Computer Science Reviews and Trends. ISBN 978-0-12-802881-0.

Terasvirta T., Anderson H M., 1992. Characterizing nonlinearities in business cycles using smooth transition autoregressive models, John Wiley & Sons, Vol. 7.

Togai M., Chiu S., 1987. A fuzzy logic chip and a fuzzy inference accelerator for real-time approximate reasoning. Proceedings del decimoséptimo IEEE International Symposium on Multiple-Valued, 1987.

Tong H., 1983. Threshold models in non-linear time series analysis. Springer-Verlag New York Inc.

Tsay R. S., 1989. Testing and modeling threshold autoregressive processes. Journal of the American Statistical Association, 84, 231–240.

Van Der Pol E., Oliehoek F. A., 2016. Coordinated deep reinforcement learners for traffic light control. NIPS'16 Work. Learn. Inference Control Multi-Agent System, No. Nips, 2016.

Vapnik V. N., 2014. Invited speaker. IPMU Information Processing and Management 2014.

Vasilyev O. V., Yuen D. A., Paolucci S., 1997. Solving PDEs using wavelets. Algorithms and applications. Department Editors: Ralf Gruber and Jacques Rappaz. Computers in Physics, 11, 429.

Verhaegen M., Van Dooren P., 1986. Numerical aspects of different implementations. IEEE Transactions On Automatic Control, Vol. AC-31. No. 10.

Wikipedia, 2020a. Gamma process. https://en.wikipedia.org/wiki/Gamma_process#:~:text=A%20gamma%20process%20is%20a,a%20Poisson%20process%20with%20intensity. Viewed: May 30, 2020.

Wikipedia, 2020b. Regression analysis. https://en.wikipedia.org/wiki/Regression_analysis#Regression_model. Viewed: May 30, 2020.

Wikipedia, 2020c. Relevance vector machine. https://en.wikipedia.org/wiki/Relevance_vector_machine. Viewed: May 30, 2020.

Woodford C., 2019. Neural networks. ExplainThatStuff. April 4, 2019. www.explainthatstuff.com/introduction-to-neural-networks.html. Viewed: May 23, 2020.

Yakovleva E., Erofeev P., 2015. Data-driven models for run-to-failure time prediction for aircraft engines. Institute for Information Transmission Problems RAS, Bolshoy Karetny per. 19, Moscow, 127994, Russia.

Zadeh L. A., 1965. Fuzzy sets. Information and Control. 8, 338–353.

Zoph B., Le Q. V., 2017. Neural architecture search with reinforcement learning. In the International Conference on Learning Representations (ICLR).

4 Context Awareness and Situation Awareness in Prognostics

4.1 IT VS. OT

In the industrial sector, information technology (IT) and operational technology (OT) are usually understood separately. Although they are different concepts, they can be combined. However, it is important to know the differences to ensure their convergence.

Although the technologies can work together to improve their functionalities, their uses are very different, and the environments in which they must be maintained are different. Figure 4.1 shows what can happen in OT when using IT-based cybersecurity (Oasys, 2019a).

4.1.1 IT and OT – What's the Difference?

IT, broadly speaking, concerns the use of computers or telecommunications systems to gather, store, retrieve, and send information.

OT is a relatively newer term. It comprises hardware and software that detect or cause change by directly monitoring and/or controlling physical systems and processes (Desai, 2016).

4.1.2 When Worlds Collide – Industrial Internet

Traditionally, IT and OT have had fairly separate roles within an organization. With the emergence of the Industrial Internet and the integration of complex physical machinery with networked sensors and software, the lines are blurring.

One of the main reasons industrial systems are being brought online is to deliver smart analytics – using data generated from the machines to modify and optimize the manufacturing process, thus entering traditional IT territory.

With the Industrial Internet, the historically closed systems that relied on physical security to ensure integrity are being replaced by more general Internet connectivity. The shift from closed to open systems brings an even greater interdependence and overlap, along with new security concerns (Desai, 2016).

4.1.3 New Concerns for Both Sides

Greater connectivity and integration are obviously beneficial for smart analytics and control, but more connections and networked devices mean more opportunities for security breaches. While security has always been a priority for both IT and OT in traditional systems, these networked systems are presenting new scenarios and risk profiles to both. IT needs to start thinking like OT and vice versa (Desai, 2016).

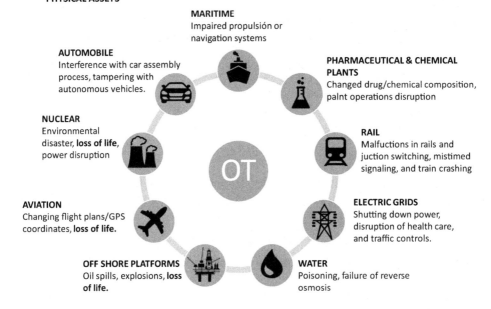

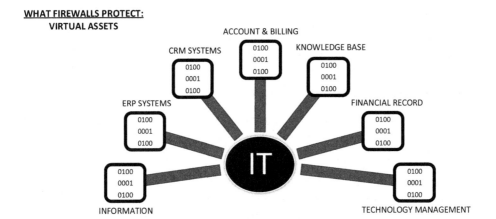

FIGURE 4.1   Infographic: OT vs. IT cybersecurity (Waterfall, 2016).

4.1.3.1   New Concerns for IT

- Greater scope of impact: There are obvious detrimental results of a security incident in a traditional enterprise environment, but the effects of an incident on an interlinked industrial system are on a completely different scale. Consider the repercussions if an electricity grid went offline, or if a car's engine control system was hacked, and drivers no longer had complete control.
- Physical risks and safety: Unlike more traditional enterprise systems, networked industrial systems bring an element of physical risk that IT teams have not had to think about. An interruption in service or machine malfunction can result in injury to plant floor employees or the production of faulty goods, potentially harming end users.

- Outdated or custom systems: IT is used for frequent and consistent software patches and upgrades, but linked industrial environments tend to be more systemic, as one small change can trigger a domino effect. Many legacy plant control systems may be running outdated operating systems that cannot easily be swapped out or a custom configuration that isn't compatible with IT's standard security packages (Desai, 2016).

4.1.3.2 New Concerns for OT
- Physical risks and safety: OT has been implementing safety measures in industrial systems for decades, but some threats are now potentially outside of its control. Taking machines and control systems out of a closed system brings the threat of hacked machines, potentially injuring employees (e.g., overheating and emergency shut-offs overridden).
- Productivity and quality control: Losing control of the manufacturing process or any related devices is an obvious problem. Consider a scenario where a malicious party is able to shut down a plant, halting production entirely, or reprogram an assembly process to skip a few steps, resulting in a faulty product that could potentially injure end users down the road.
- Data leaks: While data breaches have long been a top concern for traditional IT, they are new territory to OT which is used to working with closed systems. Given the nature of the types of industrial systems that are coming online, such as utilities, aviation, and automobile manufacturing, ensuring the privacy of transmitted data is critical.
- Working with IT: IT teams generally have little experience with industrial systems and their traditional security solutions are generally not compatible with legacy control systems (Desai, 2016).

4.1.4 FINDING COMMON GROUND

While OT and IT may have different backgrounds framing their concerns about the transformation brought about by the Industrial Internet of Things (IIoT), the main underlying concerns for both are retaining control of systems and machines and ensuring the safety of employees and customers. Key components of any potential security solutions should include (Desai, 2016):

- Identifying and authenticating all devices and machines within the system, both within manufacturing plants and in the field, to ensure only approved devices and systems are communicating with each other. This would mitigate the risk of a hacker inserting a rogue, untrusted device into the network and taking control of any systems or machines.
- Encrypting all communications between these devices to ensure privacy of the data being transmitted.
- Ensuring the integrity of the data generated from these systems. Smart analytics are a major driver in the adoption of the Industrial Internet, but those analytics are worthless if the data are inaccurate.
- Enabling the ability to perform remote upgrades down the road and ensuring the integrity of those updates, assuming the manufactured goods contain software or firmware themselves.

If things continue as they are today, the separation between OT and IT will gradually disappear. In the meantime, it is essential that both sides consider each other's experience and point of view and work together to achieve a secure and productive Industrial Internet (Desai, 2016).

4.1.5 DIFFERENCES BETWEEN IT AND OT

Although IT and OT can be expected to converge in the future, there are still many differences that must be resolved (Oasys, 2019a).

4.1.5.1 Technological Needs

The first factor to take into account is the needs of both technologies. In IT, the number of technological components is usually similar to the number of professionals active in an office, for example. OT deploys a large number of devices widely distributed among themselves, and the number of people is usually lower. Broadly understood, IT devices always require a professional to control them, while OT devices are more autonomous.

4.1.5.2 Conditions of Conservation

OT environments tend to be much harsher. In many cases, the technology must be able to withstand various weather conditions (e.g., winter/summer). IT devices are usually located in controlled environments where there are few to no changes.

4.1.5.3 Security

OT works with machines and devices and is part of IIoT. Not surprisingly, it is much more common to find risks when working with this technology. IT systems prioritize the confidentiality of data over other elements. However, OT and industrial control systems must be available because the industrial sector depends on the functionality of machinery. Thus, integrity is prioritized over confidentiality (Oasys, 2019a).

4.1.6 REGULATIONS AND PROTOCOLS

In the OT field, regulations are generally tailored to each industry sector. However, IT can be used in various sectors, regardless of the scope. Therefore, regulations tend to be much less comprehensive and much more open. In this sense, this type of technology is generally regulated by more globalized international organizations, while OT follows specific procedures agreed upon by independent regulators who depend on the sector (Oasys, 2019a).

4.1.7 DATA VS. PROCESSES

In IT systems, communication routes are often congested because of the large amounts of information sent and received. In OT, the information infrastructure is simpler. In fact, OT organizations often deploy a small set of control applications to manage and maintain systems. In addition, this environment remains relatively static. The priorities are different, in that IT seeks to analyze data to make optimal decisions and OT seeks to ensure the quality of physical processes (Oasys, 2019a).

4.1.8 UPDATE FREQUENCY

IT technology is more vulnerable and therefore needs constant updates. As these are more dynamic environments, however, it is easy to find errors and solve them. However, OT systems must remain running for long periods of time; they cannot be patched often, as this would require a reboot. If these systems are deactivated, all production processes are stopped with all the economic losses this entails. Unfortunately, this often results in obsolete OT systems being used (Oasys, 2019a).

Industry 4.0, that is, connected industry, is arriving in all sectors. It represents the opportunity to unite OT and IT to improve productivity and business competitiveness, through intelligent devices that allow the operation of the IIoT (Oasys, 2019b).

4.1.9 IIoT DEVICES IN INDUSTRY 4.0

Although the industrial sector is already highly automated, IIoT has great potential for further improvement (Oasys, 2019b):

- IIoT allows the direct management of the productive processes when all devices are connected to each other and to a control center.
- IIoT devices are not simple sensors but have autonomy within their configuration or previous programming; they can keep information if there is a problem in the communications and send it later, so data are never lost.
- A management center collects all data, leading to improved decision-making (Oasys, 2019b).

4.1.10 What Industrial Processes Will Improve IT and OT Integration?

4.1.10.1 Energy
Smart devices will increase their energy consumption in hours when energy is cheaper and reduce it when energy is more expensive.

4.1.10.2 Environment
Smart devices will measure, for example, meteorological and energy data so the enterprise can meet environmental pollution levels established by law and adapt its production.

4.1.10.3 Production
Smart devices will make it possible to adapt the customer's demand to production, to optimize production, and save costs.

4.1.10.4 Quality Control
Smart devices will allow companies to manufacture products with both precision and speed.

4.1.10.5 Maintenance
Smart devices carry the potential for inaugurating predictive maintenance that reduces breakdowns and allows companies to design better maintenance and reduce losses (Oasys, 2019b).

4.2 CONTEXT DEFINITIONS AND CONTEXT CATEGORIZATION

4.2.1 Definition of Context

The environment or setting in which something exists is that thing's context. Thus, when something is contextualized, it is placed in an appropriate setting (Merriam-Webster, 2020).

Context helps readers understand what they otherwise wouldn't be able to comprehend. It is a much-needed assistant, helping readers define unknown words and make sense of outside information.

In writing, it is often necessary to provide new words, concepts, and information to help develop a thought. For example, maybe you need to include a fact to support your claim or a quotation to illustrate your analysis of a literary work. Whenever you use a fact or quotation from another source, it is important that you tell the reader a bit about that information first. This is what we mean by context. You need to surround that piece of information with text that illuminates its meaning and relevance. That is why context, when broken down, literally means "with text". It helps readers understand that which otherwise, they wouldn't be able to comprehend (Carlisle, 2020).

Some common definitions of context in the domain of context-aware computing are the following:

- Definition by example: The context is described by providing examples. While these definitions might be helpful, matching new situations can be problematic as it is not possible to give examples (Rosenberger & Gerhard, 2018).

- Definition by synonyms: As with examples, this can be helpful for new users but problematic in unforeseen situations because there is no other equivalent word to use (Rosenberger & Gerhard, 2018).
- Definition by aspects: These definitions describe context by its aspects like who the user is, where, and with whom. The advantage of this approach is that new situations can be included or excluded depending on whether they satisfy the stated aspects. It is generally recognized that aspect definitions are most useful for industrial applications (Rosenberger & Gerhard, 2018).

4.2.2 OPERATING CONTEXT

In the world of operations, maintenance, and reliability engineering, context is frequently used. It can be defined as the current condition, environment, and culture in which a piece of equipment is operated. This would include but not be limited to the following (Plucknette, 2013):

- Temperature: Hot, cold, or severe swings/stable.
- Dirty or dusty atmosphere/clean atmosphere.
- Wet/dry area.
- Corrosive, erosive, or abrasive environment/non-corrosive, and so on.
- Dark or dimly lit/brightly lit.
- Noisy/acceptable levels of noise.
- Culture's goals and expectations clearly defined/not clearly defined, high/low level of demand.
- Operating outside/inside design expectations or performance standards.
- Asset condition working/faulty, that is, loose, improperly supported, improperly installed, improper design, and damaged.
- Proper/improper operation, that is, start up, shutdown, product change, setting, speed, flow, and pressure.
- Human error, that is, no check lists or procedures (Plucknette, 2013).

4.2.3 CONTEXT AWARENESS FOR ASSET MAINTENANCE DECISIONS

A context-aware system actively and autonomously provides appropriate services or information to users, taking advantage of those users' contextual information while requiring little interaction with them.

These are complicated systems. They can perform a number of different jobs, including data representation, management, reasoning, and analysis, and they have many different components. Moreover, there are various types of context-aware systems, making it hard to generalize a context-aware system process. That being said, a context-aware system process can generally be divided into four steps (Galar, 2014), as shown in Figure 4.2.

In the first step, the system acquires real-world contextual information from physical and virtual sensors. The system stores these data in its repository. Note that the kind of data model used to represent the context information is very important; context models are diverse and have unique

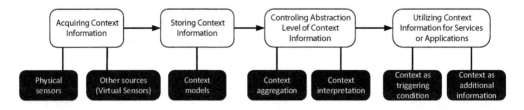

FIGURE 4.2   General process in context-aware systems (Galar, 2014).

characteristics. To more easily use the stored context data, the system controls their abstraction level by interpreting or aggregating them. In the final step, the system applies the abstracted context data to context-aware applications (Galar, 2014).

4.2.4 CONTEXT-DRIVEN MAINTENANCE

Sense making is the challenge for today's Big Data sets. To this end, concepts from electronic maintenance must migrate to intelligent maintenance. Today's maintainers must deal with many different data sources. Therefore, a system framework should support the integration of various data sources with different formats and natures. It should provide facilities for data wrapping and be able to mediate different data formats; it should also provide interfaces for external data wrappers and mediators.

In addition, the system should be able to add new sources and mediation procedures and handle the necessary data validation and consistency checking. For efficient operations, different data spaces must be managed at different system levels.

At the data management level, the following must be managed and merged (Galar, 2014):

- Database containing database baseline.
- Synthetic database containing derived calculations.
- External sources not included in the database.
- Information on managing databases.
- Information on managing wrappers and mediators.
- Archived data.

Figure 4.3 shows typical data related to transportation facilities that must be managed and merged (Galar, 2014).

4.2.5 CLASSIFICATION OF CONTEXT TYPES

Researchers tend to identify and classify the different context types based on their perspective. One of the most popular classifications separates contexts into primary and secondary types. Primary contexts include time, identity, location, and activity. With these, it is possible to fully capture any given situation. All other contexts are secondary and can be derived from the primary ones (Dey, 2001).

There are some contradictions to this general understanding. A contradictory example is machine condition, such as the temperature of a machine, which would be considered secondary even though it cannot be derived from any of the established primary contexts. It is not always possible to distinguish between more and less important contexts, as it depends largely on the use case. Considering the provision of information in the workshop, it can easily be shown how the importance of a context depends on the present situation. A worker inspects a product based on his smart device. In this case, the device and the size of its screen are important for the user, as the use of a screen below a certain threshold greatly complicates the input of information. If the same worker has to navigate to a certain location, the size of the screen is not so important, since the smartest devices can provide navigation in a suitable format. And when the worker records working hours with a chip card, the device will not even be considered as it will not be part of the interaction.

Chen and Kotz (2000) solved this problem by dividing contexts into active and passive groups. Active contexts are all those necessary to identify the current entity and its conditions. Passive contexts are all other contexts. As indicated, only the current situation is relevant to the classification, so that in the same use case, a context can be active and passive, depending on the situation. This concept can be demonstrated by considering a contextual information system. If the system

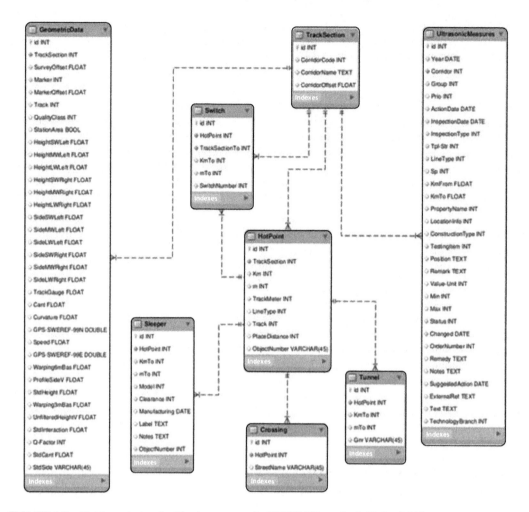

FIGURE 4.3   Database design for Sweden case study (OPTIRAIL project) (Galar, 2014).

tries to provide the maintenance team only with an error notification, the roles of the users are important and represent an active context. If the system provides all users with general information, their roles are not of interest and therefore represent a passive context.

4.2.6   Categorization by Context

4.2.6.1   Introduction

The traditional approach to document categorization is categorization by content, as information for categorizing a document is extracted from the document itself.

In a hypertext environment like the Web, the structure of documents and the link topology can be exploited to perform what we call categorization by context: the context surrounding a link in an HTML document is used to categorize the document referred by the link.

Categorization by context is capable of dealing with multimedia material, as it does not rely on the ability to analyze the content of documents. Categorization by context leverages the categorization activity implicitly performed when someone places or refers to a document on the Web. By focusing the analysis on the documents used by a group of people, we can build a catalogue tuned to the needs of that group.

Categorization by context is based on the following assumptions (Attardi et al., 1998):

1. A Web page which refers to a document must contain enough hints about its content to suggest reading it.
2. Such hints are sufficient to classify the document.

The classification task must be capable of identifying such hints. One obvious hint is the anchor text for the link, but additional hints may be present elsewhere in a page: page title, section headers, list descriptions, and so on. All these hints make up the context for the link.

Categorization by context exploits the structure of Web documents and Web link topology to determine the context of a link. Such context is then used to classify the document referred to by the link (Attardi et al., 1998).

4.2.6.2 Categories of Context

A categorization of context types helps application designers uncover the most likely pieces of context that will be useful in their applications. Some suggest context types include location, environment, identity, and time (Ryan et al. 1997). Others say context includes where people are, who they are with, and what resources are close at hand (Shilit et al., 1994).

Essentially, context-aware applications look at the who, where, when, and what of entities and use this information to determine why a particular situation is happening. An application doesn't actually determine why a situation is occurring – the designer of the application does this. The designer uses incoming context to determine why a situation is occurring and uses this to encode some action in the application.

For example, in a context-aware tour, a user carrying a handheld computer approaches some interesting site resulting in information relevant to the site being displayed on the computer. In this situation, the designer has encoded the understanding that when a user approaches a particular site (the "incoming context"), it means the user is interested in the site (the "why"), and the application should display some relevant information (the "action") (Dey & Abowd, 1999).

Certain types of context are, in practice, more important than others. These are location, identity, activity, and time. The only difference between this list and the definition of context provided above by Ryan, Pascoe, and Morse (1997) is the use of "activity" rather than "environment". Environment is a synonym for context and does not add to the actual investigation of context. Activity answers a fundamental question – what is occurring in the situation? The second definition given above (where you are, who you are with, and what objects are around you) only includes location and identity information. To characterize a situation, activity and time information is also needed.

Location, identity, time, and activity are the main types of context to characterize the situation of a particular entity. These types of context not only answer the questions of who, what, when and where; they also act as indexes of other sources of contextual information.

For example, given the identity of a person, many related data can be acquired, such as phone numbers, addresses, email addresses, date of birth, list of friends, and relationships with other people in the environment. With the location of an entity, we can determine what other objects or people are near the entity and what activity is taking place near it. The primary context pieces for an entity can be used as indexes to find the secondary context (e.g., email address) for that same entity, as well as the primary context for other entities and related entities (e.g., other people in the same location).

This initial categorization is a simple two-tiered system. The four primary pieces of context already identified are on the first level. All the other types of context are on the second level. The secondary pieces of context share a common characteristic: they can be indexed by primary context because they are attributes of the entity with primary context. For example, a user's phone number is a piece of secondary context, and it can be obtained by using the user's identity as an index into an information space like a phone directory. There are some situations in which multiple pieces of primary context are required to index into an information space. For example, the forecasted weather is context in an outdoor tour guide that uses the information to schedule a tour for users. To obtain the forecasted weather, both the location for the forecast and the date of the desired forecast are required.

This characterization helps designers choose context to use in their applications, structure the context they use, and search out other relevant context. The four primary pieces of context indicate the types of information necessary for characterizing a situation and their use as indexes provide a way for the context to be used and organized (Dey & Abowd, 1999).

4.2.7 CATEGORIZATION OF CHARACTERISTICS OF CONTEXT-AWARE APPLICATIONS

Schilit, Adams, and Want (1994) proposed a categorization of context sensitive applications with two orthogonal dimensions: if the task is to obtain information or execute a command and if the task is executed manually or automatically. Applications that retrieve information for the user manually based on the available context are classified as immediate-select applications. A list of objects (printers) or places (offices) is presented, and the elements relevant to the user's context are emphasized or made easier to choose. Applications that retrieve information for the user automatically based on available context are classified as automatic context reconfiguration. This system level technique creates an automatic link to an available resource based on the current context. Applications that execute user commands manually based on the available context are classified as contextual command applications. They are executable services available because of the context of the user or whose execution is modified depending on the context of the user. Finally, applications that execute user commands automatically based on the available context use context-triggered actions. These services run automatically when there is a correct combination of context and are based on simple if-then rules (Dey & Abowd, 1999).

Pascoe (1998) proposed a similar taxonomy of context-aware features. There is considerable overlap between the two taxonomies but some crucial differences as well. His taxonomy aimed at identifying the core features of context awareness, as opposed to the previous taxonomy, which identified classes of context-aware applications. In reality, the following features of context-awareness map well to the classes of applications in the Schilit, Adams, and Want taxonomy (1994).

The first feature is contextual sensing, the ability to detect contextual information and present it to the user, augmenting the user's sensory system. This is similar to proximate selection, except in this case, the user does not necessarily need to select one of the context items for more information (i.e., the context may be the information required). The next feature is contextual adaptation, the ability to execute or modify a service automatically based on the current context. This maps directly to Schilit's context triggered actions. The third feature, contextual resource discovery, allows context-aware applications to locate and use services and resources relevant to the user's context. This maps directly to automatic contextual reconfiguration. The final feature, contextual augmentation, is the ability to associate digital data with the user's context. A user can view the data when he or she is in that associated context. For example, a user can create a virtual note providing details about a broken television and attach the note to the television. When another user is close to the television or attempts to use it, s/he will see the virtual note (Dey & Abowd, 1999).

Pascoe (1998) and Schilit and colleagues (1994) both list the ability to exploit resources relevant to the user's context, the ability to execute a command automatically based on the user's context, and the ability to display relevant information to the user. Pascoe goes further in terms of displaying relevant information to the user by including the display of context, not just information that requires further selection (e.g., showing the user's location vs. showing a list of printers and allowing the user to choose one). Pascoe's taxonomy has a category not found in Schilit and colleagues' taxonomy: contextual augmentation, or the ability to associate digital data with the user's context. Finally, Pascoe's taxonomy does not support the presentation of commands relevant to a user's context. This presentation is called contextual commands in Schilit and colleagues' taxonomy (Dey & Abowd, 1999).

Dey and Abowd (1999) proposed a categorization combining the ideas from these two taxonomies and taking the three major differences into account. Similar to Pascoe's taxonomy, it is a list of the context-aware features that context-aware applications may support. There are three categories:

1. Presentation of information and services to a user.
2. Automatic execution of a service.
3. Tagging of context to information for later retrieval.

4.2.8   CONTEXT CATEGORIZATION, ACQUISITION, AND MODELING

This section discusses contextual database support. The term context here refers to the situation in which access to the user's database occurs. Context can be user-centered or environmental, as shown in Figure 4.4 (see Section 4.2).

Context is viewed as an n-dimensional space, constructed of n contextual attributes. Each context dimension is represented by a contextual attribute, which describes a context perspective. The domain of a contextual attribute can be a scalar value, a string value, a set of scalar/string values, or an empty value, that is, a null value, depending on the semantics of the application.

For example, the domain of the traffic jam contextual attribute may be a set of strings, specifying the names of the sites where a traffic jam occurs. When there is no traffic jam, traffic jam = $\emptyset$ (Feng et al., 2014).

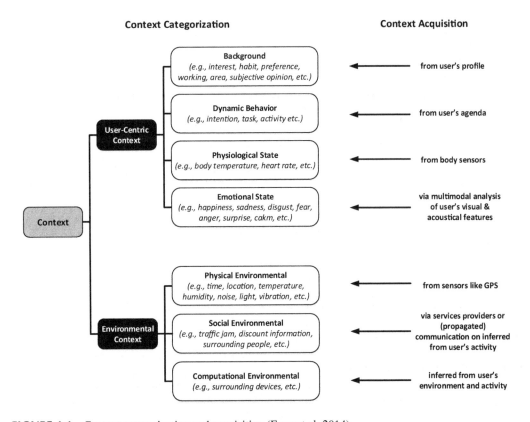

FIGURE 4.4   Context categorization and acquisition (Feng et al, 2014).

Formally, we leave $\{a_1, a_2,..., a_n\}$ as a set of contextual attributes, whose domains are represented as Dom (a_1), Dom (a_2),... Sun (a_n). An n-dimensional context, denoted JContextK, is the Cartesian product JContextK: $a_1 \times a_2 \times.... \times a_n$ of these contextual attributes. Based on application requirements, abstract operators can be defined by supported contextual attributes. To support operations performed on different types of operands, we can define functions, the parameters of which can also involve contextual attributes. For example, given the address (e.g., street and area code) of a restaurant, its distance from a current location, measured in a geographic pair of latitude and longitude, can be implemented using the floating distance function of form (address: string, location: (float, float)). Geographic Information System systems allow translation from an implicit reference (e.g., an address) to an explicit geographic reference (e.g., latitude and longitude). These privately defined abstract operators/functions are called uniformly contextual operators/functions (Feng et al., 2014).

4.2.9 Categorization of Context in Mobile Map Services

The European project "Geospatial info-mobility service by real-time data-integration and generalization" (GiMoDig) aims to develop methods for real-time integration and generalization of national topographic databases so that they can be accessed and used by mobile users through a unified service.

To consider usability issues, pilot user tests were arranged early in the GiMoDig project. There were no real prototypes available for testing, so the purpose was to obtain some basic information on the usability of topographic maps in mobile devices and to gain experience on how to plan and carry out user tests during the later stages of the project. Based on the field tests, the contexts were reclassified.

Table 4.1 summarizes the context categorization used by GiMoDig. Categorization is divided into general contexts considered the most important for mobile map applications. A set of characteristics of each group is listed in the third column (Nivala & Sarjakoski, 2003).

4.3 CONTINUOUS CHANGE, TEMPORALITY, AND SPATIALITY

4.3.1 What Happens When Transformation Becomes the Rule

Today's reality is one of continuous change: digitization, reorganization, and constantly changing work environments are a simple matter of fact. It can be hard enough just to keep up. But in an ideal world, a company could go beyond simply keeping up to being able to control changes according to its culture (Teambay, 2014).

Humans are creatures of habit and are generally skeptical when it comes to change. The introduction of a new value culture, even the introduction of new software, can be problematic. Three things are essential for targeted change processes to be sustainable (Teambay, 2014):

1. Allocating time for change.
2. Adopting the right perspective.
3. Undertaking continuous engagement.

These are explained in more detail in the following sections.

4.3.1.1 Real Change Takes Time

Change is a continuous process that takes time. This means budgeting time to try things out and to be able to make mistakes. Even the changes will need to be changed. Transparent communications with all stakeholders are essential to avoid frustration.

TABLE 4.1
Categorization of Contexts and Their Features for Mobile Map Services

General Context Categories	Context Categories for Mobile Maps	Features
• Computing	• System	• Size of a display. • Type of the display (black – color screen). • Input method (touch panels, buttons, etc.). • Network connectivity. • Communication costs and bandwidth. • Nearby resources (printers, displays).
• User	• Purpose of use • User • Social • Cultural	• User's profile (experience, disabilities, etc.). • People nearby. • Social situation.
• Physical	• Location • Physical surroundings • Orientation	• Lighting. • Temperature. • Surrounding landscape. • Weather conditions. • Noise levels.
• Time	• Time	• Time of day. • Week. • Month. • Season of the year.
• History	• Navigation history	• Previous locations. • Former requirements and points of interest.

Source: Nivala and Sarjakoski (2003).

4.3.1.2 Find the Right Perspective

Not all changes apply everywhere and are right for everyone. For example, applicable changes will differ for retail stores, call centers, and mines. Regional, national, and international companies are vastly different, and even a particular company will have various subcultures contained within it. Different areas and departments operate in different circumstances: the IT department needs a different focus than the department for customer support (Teambay, 2014).

4.3.1.3 Continuous Engagement

One of the most important elements of successful change is the engagement and inclusion of everyone, from the company interns to employees and managers. From top executives to interns, everyone needs to be included and get the chance to impact the process. Engaging everyone involved helps the process receive the necessary support; it counters skepticism and fear of change and is a great way to discover problem areas needing change (Teambay, 2014).

4.3.2 CYCLE OF CONTINUOUS CHANGE

Continuous change can be divided into four phases: influence (selling the idea of change), authority (making the decision to change), technology (updating technology to accommodate the change), and culture (changing the work culture). This is shown in Figure 4.5 and explained in more detail in the following sections (Lawrence et al., 2006).

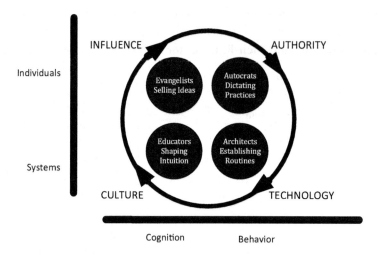

FIGURE 4.5   Cycle of continuous change (Lawrence et al., 2006).

4.3.2.1   Phase 1: Using Influence to Sell Ideas

Change begins with an idea – an insight, intuition, or belief that motivates someone to question the way things are done. But for an idea to initiate processes of change, it needs to be articulated and presented in ways that influence people. In some cases, the idea may simply need to be taken to the right people and presented in a straightforward manner, but many ideas require more adept handling. An idea may need to be reframed in dramatically different ways for different audiences: attaching the idea to other ideas or plans that are already accepted; enrolling well-respected organizational members to act as spokespersons for the idea; demonstrating the concrete economic benefits of adopting the idea; showing others how the idea might help advance their own careers. This step involves persuasion. The figure labels the person who is able to do this an "evangelist" (Lawrence et al., 2006).

4.3.2.2   Phase 2: Using Authority to Change Practices

A good plan may fail because its advocates do not understand what is needed to translate it into action or do not have the resources necessary to do so. To move organizational members to action typically requires more than persuasion. Effecting collective action usually requires someone with the formal, legitimate power to tell people what to do, how to do it, and when. The prudent use of authority is necessary for three reasons.

First, although key organizational members may have accepted a new idea, they might be uncertain or anxious about putting it into practice. What if the idea flops, or what if others don't follow? At this point, someone in charge needs to approve (or reject) the new direction, not just in theory but also in everyday practice.

Second, the new idea could generate a wide range of interpretations. This is especially true with ideas that are genuinely novel or involve intangible concepts like innovation or collaboration. Detailing in practical terms the new behaviors or practices that are required (and those that need to stop) will certainly help. But such prescriptions often need to be backstopped with authority.

Third, authority may be needed to overcome any resistance. New ideas can be threatening, and their implementation can generate tremendous anxiety, especially when they put employees' existing skills and relationships in jeopardy. The result is often resistance, both active and passive. That is not to say continuous change processes incur continuous resistance; however, they almost inevitably lead to some resistance, often at the point when ideas need to be translated into action (Lawrence et al., 2006).

4.3.2.3 Phase 3: Embedding Change in Technology

Making new routines stick requires more than the force of individuals. It typically requires technology. Thus, to institutionalize change, organizations need an architect to design the systems necessary to embed the change in corporate routines, ensuring its maintenance independent of shifting responsibilities. The goal is to entrench the desired behaviors and practices so deeply that they become not only routine but taken for granted. That can be accomplished with information, manufacturing, and financial systems, as well as with the physical work space – the floors, walls, and areas that bring people together, keep them apart, and channel their movement through the organization.

In each case, technology is critical because it can rapidly and effectively eliminate undesirable options and facilitate collective behavior. That said, phase 3 of the change cycle is a precarious one that needs to happen quickly, efficiently, and visibly; if it doesn't, enthusiasm could dissipate, frustrating those invested in the process (Lawrence et al., 2006).

4.3.2.4 Phase 4: Managing Culture to Fuel the Cycle of Change

The final phase is perhaps the most critical but often the most neglected – ensuring that the change process leaves a legacy that surpasses its original objectives. Doing so requires a culture for fostering innovation that extends and elaborates the initial ideas and practices.

Although traditional change efforts often include a cultural component, the emphasis is typically on instilling a new set of values, such as quality, that relate to the program's initial aims. By contrast, the cultural component of continuous change processes is forward-looking with a focus on helping employees gain the expertise and motivation not only to enact the change but also to extend and elaborate it. The goal is an environment that fosters innovation and strategic thinking, setting the stage for the birth of new ideas that can then be picked up and integrated. Practically speaking, this involves implementing routines and practices that help establish a cycle of improvement, learning, and strategic adaptation that is self-perpetuating (Lawrence et al., 2006).

4.3.3 TEMPORAL DATA AND DISCOVERY

Temporal data are often represented as a sequence, sorted in a temporal order. A time series is a time-ordered sequence of numerical observations taken over time. An example is the series of numbers <1, 3.5, 2, 1.7,...>. In a univariate time series, each observation consists of a value for a single variable, and in a multivariate time series, each observation consists of values for several variables (Karimi, 2019).

Some approaches to time series processing assume the presence of a variable representing time, with numeric values for all other variables. Attempts have been made to fit constant or time-varying mathematical functions to time series data. A time series can be regular or irregular. In a regular time series, data are collected at predefined intervals. An irregular time series does not have this property, and data can arrive at any time, with arbitrary temporal gaps in between.

A deterministic time series can be predicted exactly, while the future values in a stochastic time series can only be determined probabilistically. The former is a characteristic of artificial and controlled systems, while the latter applies to many natural systems. Simple operations like determining the minimum or maximum values of certain variables, finding trends (e.g., increases or decreases in the value of stocks), cyclic patterns (e.g., seasonal changes in the price of commodities), and forecasting are common applications of time series data (Karimi, 2019).

Many approaches to the discovery of rules from time series data involve pre-processing the input by extracting global or local features from the data. Global features include the average value or the maximum value, while local features include an upward or downward change or a local maximum value. Another example of discovering temporal traits by pre-processing time series data is the discovery of increasing or decreasing trends before rule extraction. While the study of time series is pursued widely, in some cases, the results may not be useful or even meaningful.

Some researchers have used multiple streams of data to describe simultaneous observations of a set of variables. The streams of data may come from different sensors of a robot or the monitors in an intensive care unit, for example. The values in the streams are recorded at the same time and form a time series (Oates & Cohen, 1996). Other researchers have proposed an algorithm that can find rules, called "structures", relating the previous observations to future observations (McCarthy & Hayes, 1969). Such temporal data appear in many application areas (Roddick and Spiliopoulou, 1999).

An event sequence is a series of temporally ordered events, with either an ordinal time variable, which gives the order but not a real-valued time, or no time variable. The main difference between an event sequence and a time series is that a time series is a sequence of real numbers, while an event sequence can contain variables with symbolic domains. Each event specifies the values for a set of variables. A recurring pattern in an event sequence is called a frequent episode. Recent research has emphasized finding frequent episodes with varying numbers of events between key events that identify the event sequence. Algorithms such as dynamic time warping and its variants measure the similarity of patterns stretched differently over time (Karimi, 2019).

Temporal sequences are often considered passive indicators of the presence of temporal structure in data, and we do not know whether they represent causal relationships. Despite having different terminology, all the domains listed so far in this section have the common characteristic of recording the values of some variables and placing them together in a record. Time series, event sequences, and streams of data all try to find temporal rules, called patterns, episodes, and structures, respectively, from the input data (for a review, see Höppner, 2003; Karimi, 2019).

4.4 MODELING CONTEXT AND REPRESENTATION METHODS

4.4.1 CONTEXT MODEL

A context model (or context modeling) defines how context data are structured and maintained. It plays a key role in supporting efficient context management. It produces a formal or semi-formal description of the context information present in a context-aware system. In other words, the context is the surrounding element for the system, and a model provides the mathematical interface and a behavioral description of the surrounding environment.

A key role of a context model is to simplify context-aware applications and introduce greater structure into their development (Wikipedia 2019).

4.4.2 EVOLUTION OF CONTEXT MODELING AND REASONING

A number of context modeling and reasoning approaches have been developed over the past decade – from very simple early models in the early years to today's next-generation models. At the same time, researchers have worked on developing context management systems able to collect, manage, evaluate, and disseminate context information.

Over the years, a large number of context-sensitive applications based on various context models have been applied in a number of domains. This, in turn, has influenced the set of requirements defined for context modeling and reasoning and therefore also influenced research on context information models with high expressive power and good computational reasoning, that is, able to support reasoning about context.

This section describes and evaluates state-of-the-art context models that are generic (i.e., suitable for any type of application) and meet most of the requirements of context modeling, management, and reasoning (Bettini et al., 2009).

4.4.2.1 Requirements

4.4.2.1.1 Heterogeneity and Mobility

Contextual information models deal with a wide variety of contextual information sources that differ in their update rate and semantic level. Certain contextual information is perceived. Sensors can observe certain states of the physical world and provide fast, near-real-time access, providing raw data (e.g., a Global Positioning System (GPS) or camera stream) that must be interpreted before applications can use them. Data provided by the user, such as user profiles, are updated less frequently but generally require no further interpretation. Context data can also be derived from existing context information. Context data obtained from databases or digital libraries, such as geographic map data, are often static.

A context model should be able to express different types of context information, while the context management system should be able to manage information according to its type.

Many contextual applications are mobile (i.e., they run on a mobile device) or rely on sources of mobile context information (e.g., mobile sensors). This adds to the heterogeneity of the data. Thus, the provisioning of context information must be adaptable to the changing environment. The location and spatial design of context information play important roles because of this requirement (Bettini et al., 2009).

4.4.2.1.2 Relationships and Dependencies

Various relationships between different types of context information must be captured to ensure the viability of an application. One such relationship is dependency, whereby certain context information entities may depend on other context information entities: to give an example, a change in the value of one entity (e.g., network bandwidth) may affect the value of another entity (e.g., remaining battery power).

Context-aware applications may need access to both past and future states (i.e., prognosis). Therefore, the context history is another feature of context information that must be captured by context models and managed by the context management system. The management of context histories is difficult if the number of updates is very high. In this case, it may not be feasible to store every value. This calls for the use of summarization techniques: position updates may be aggregated to a movement function using interpolation techniques, for example, or historical synopses of data may be created (Bettini et al., 2009).

4.4.2.1.3 Imperfection

Because of its dynamic and heterogeneous nature, context information has variable quality. These data may even be incorrect. Most sensors have an inherent inaccuracy (e.g., a few meters for GPS positions), and as the physical world changes over time, this inaccuracy will increase. In addition, the context information may be incomplete or conflict with other context information. Inherent data imperfections mean that a good context modeling approach must be able to model context information quality to support reasoning about context (Bettini et al., 2009).

4.4.2.1.4 Reasoning

Context-aware applications use context information to evaluate whether there is a change to the user's context or to that of the computing environment, but deciding whether any adaptation is necessary requires reasoning capabilities. Consequently, context modeling techniques must support context reasoning. They must be able to derive new context facts from existing ones and reason about high-level context abstractions that model real-world situations. These reasoning techniques should be computationally efficient (Bettini et al., 2009).

4.4.2.1.5 Usability of Modeling Formalisms

Ease of use is extremely important. Designers must be able to translate real world concepts into the modeling constructs, and applications must be able to easily manipulate context information (Bettini et al., 2009).

4.4.2.1.6 Efficient Context Provisioning

Efficient access to context information is difficult to achieve for large models and numerous data objects. To select relevant objects from a large dataset, context modeling must define the attributes for suitable access paths, that is, dimensions along which applications select context information, typically supported by indexes. These dimensions are often called the primary context. The secondary context is accessed using the primary context. Commonly used primary context attributes include identity, location, object type, time, or activity of user. The choice of primary context attributes is application-dependent; thus, given a certain application domain, a certain set of primary context attributes is used to build efficient access paths; for example, spatial indexes are used if location is a primary context (Bettini et al., 2009).

4.4.2.2   Early Approaches: Key-Value and Markup Models

Key-value context information models use simple key-value pairs to define the list of attributes and their values, while markup-based context information models use a variety of markup languages (e.g., XML). The composite capabilities/preference profile (CC/PP) for mobile devices is an early context modeling approach to use a resource description framework (RDF) and to include elementary constraints and relationships between context types. CC/PP represents both key-value and markup models, as it uses RDF syntax to store key-value pairs under appropriate tags.

CC/PP and other key-value and markup-based context information models are well described in the literature (Indulska et al. 2003; Lum & Lau 2002). The main criticisms of such models are the following:

1. Limited ability to capture a variety of context types.
2. Limited ability to capture relationships, dependencies, timeliness, and quality of context information.
3. Limited ability to perform consistency checking,
4. Limited ability to support reasoning on context, context uncertainty, or higher context abstractions (Bettini et al., 2009).

4.4.2.3   Domain-Focused Modeling

Certain types of context information can significantly enhance the functions of domain-specific context-aware applications. For example, the W4 context model, developed for context-aware browsing, supports the representation of context as Linda-like tuples (i.e., parallel and distributed models) and provides an interface to store and query these tuples (Bettini et al., 2009).

4.4.2.4   Toward More Expressive Modeling Tools

Early approaches to context modeling, for example, CC/PP, do not meet many of the requirements listed in Section 4.4.2.1. Other approaches, characterized by more expressive context modeling tools, provide better solutions.

For example, the object-role based modeling approach originated from information systems modeling. It provides an easy mapping from real-world context concepts to modeling constructs, and it uses a novel form of predicate logic to reason about high-level context abstractions in order to satisfy the heterogeneity, timeliness, reasoning, and usability requirements mentioned above (Bettini et al., 2009).

4.4.3   Modeling Approaches

This section surveys the most commonly used context modeling techniques. These techniques are classified by the schema of data structures used to exchange contextual information in the system (Priya & Kalpana, 2016).

4.4.3.1   Key Value Models

The model of key-value pairs (see Section 4.4.2.2) is the simplest and most frequently used data structure to model contextual information.

4.4.3.2 Markup Scheme Models

The hierarchical data structure of a markup scheme model consists of markup tags with attributes and content (see Section 4.4.2.2). Typical examples are the CC/PP model discussed previously and user agent profile model (Butler, 2001).

4.4.3.3 Graphical Models

A very well-known general purpose modeling instrument is unified modeling language (UML). It has a strong graphical component (UML diagrams). Because of its generic structure, UML can model context. Another example is the graphics oriented extension of the object role modeling approach discussed next.

4.4.3.4 Object-Oriented Models

Object-oriented context modeling approaches employ the main benefits of the object-oriented approach, namely encapsulation and reusability. All details are encapsulated within active objects and hidden in other components of the system.

4.4.3.5 Logic-Based Models

Logic-based models define the context using facts, predictions, or roles. A goal is to form new expressions or facts from previous ones. A logic defines the conditions from which a concluding expression or fact may be derived. An example of the use of such models is McCarthy's introduction of contexts as abstract mathematical entities in artificial intelligence (AI) (McCarthy, 1993; McCarthy & Buva, 1997).

4.4.3.6 Ontology-Based Models

This model represents a concept group in a given domain and the relationship between the different concepts. It depicts a domain using a graph of concepts; contextual relationships may be hierarchical or semantic. Ontology-based approaches represent knowledge, concepts, and relationships about a domain and describe specific situations in a domain. For example, given two atomic classes, Female and Person, the class Male can be defined as: Male $\equiv$ Person-Female. This model is discussed more fully in the next section.

4.4.3.7 Spatial Context Model

Space is a most important context in many context-aware applications. Most spatial context models are fact-based models. These models organize their context information by physical location (Priya & Kalpana, 2016).

4.5 ONTOLOGIES AND CONTEXT FOR REMAINING USEFUL LIFE ESTIMATION

4.5.1 DEFINITION OF ONTOLOGY

An ontology is a formal description of knowledge as a set of concepts within a domain, as well as the relationships among these concepts (see Section 4.4.3.6). To enable such a description, we need to formally specify the components: individuals (i.e., instances of objects or facts), classes, attributes, and relations. We also need to specify restrictions, rules, and axioms. Ontologies introduce a sharable and reusable knowledge representation and at the same time add new knowledge about the domain.

The ontology data model can be applied to a set of individual facts to create a knowledge graph – a collection of entities, where the types of entities and the relationships between them are expressed by nodes and edges between these nodes, respectively. By describing the structure of the knowledge in a domain, the ontology data model permits the knowledge graph to capture the data in it (Hub, 2020).

Some methods use formal specifications for knowledge representation, such as vocabularies and logical models. However, unlike these methods, ontologies express relationships and enable users to link multiple concepts to other concepts in a variety of ways.

4.5.2 ONTOLOGY USE CASES

Ontologies define the terms used to describe and represent an area of knowledge, making them extremely useful to capture relationships and foster knowledge management in many different applications. For example, the use of ontologies enables early hypothesis testing in the pharmaceutical industry by categorizing identified explicit relationships within a causality relation ontology. In other examples, ontologies are useful in semantic web mining, fraud detection, mining health records for insights, and semantic publishing.

Simply stated, ontologies are handy frameworks for representing shareable and reusable knowledge across a domain. Their good ability to describe relationships and their high interconnectedness make them attractive for modeling high-quality, linked, and coherent data (Hub, 2020).

4.5.3 BENEFITS OF USING ONTOLOGIES

By having the essential relationships between concepts built into them, ontologies enable automated reasoning on the data. Such reasoning is easy to implement in semantic graph databases that use ontologies as their semantic schemata. In many ways, ontologies function like a human brain, working and reasoning with concepts and relationships in ways that are evocative of how humans perceive interlinked concepts.

In addition to the reasoning feature, ontologies provide coherent and easy navigation as users move from one concept to another. Moreover, ontologies are easy to extend; relationships and concept matching are easy to add to existing ontologies. The model evolves with the growth of data without impacting the dependent processes and systems if something goes wrong or requires change.

Last but by no means least, ontologies can represent any data format, including unstructured, semi-structured, or structured data. This enables smoother data integration, fosters easier concept and text mining, and boosts data-driven analytics (Hub, 2020).

4.5.4 LIMITATIONS OF ONTOLOGIES

As the above discussion makes clear, ontologies are valuable tools for modeling data. However, they have some limitations.

One limitation is the availability of property constructs. To give one example, a recent version of the Web Ontology Language, OWL2, provides powerful class constructs but has a limited set of property constructs. A related limitation is the way OWL employs constraints. The constraints specify how data should be structured and prevent the addition of data inconsistent with these specifications. This is not always beneficial, as data from a new source can be structurally inconsistent with the constraints set using OWL. Consequently, these new data have to be modified before they can be integrated with existing data (Hub, 2020).

4.5.5 CONTEXT ONTOLOGY

Certain contexts are common and fundamental like user, location, computational entity, and activity for any current situation. These entities are the skeleton of every context and can point to any associated information (Malik & Jain, 2019).

TABLE 4.2
Comparison of Existing Context Ontologies

Comparison Parameters	CONON	CoBra	SOUPA	CoDA MOS	Smart Space	CAC ont
Location	✓	✓	✓	✓	✓	✓
User	✓	✓	✓	✓	✓	✓
Activity	✓	✓	✓	✓	✓	✓
Time			✓	✓	✓	
Security		✓	✓	✓	✓	✓
Space			✓	✓	✓	✓
Environment				✓	✓	✓
Representational Framework	OWL	OWL	OWL	OWL	OWL	OWL

Source: Malik and Jain (2019).

4.5.5.1 Existing Upper Ontologies
Existing context ontologies include CONON, CoBrA, SOUPA, CoDAMoS, SmartSpace, and CACOnt. They are compared in Table 4.2 (Malik & Jain, 2019).

4.5.5.2 Scope of Logical Content
The ontologies mentioned previously are very useful in their respective fields, but their scope and applicability may be diminishing as user requirements grow. We may need something more from each of them.

For example, CONON is not able to define context in terms of time, and this is a major drawback. Nor does it consider the security of user-sensitive information. While CoBrA solves this by providing security to user-sensitive information, it does not define time. CoBrA sometimes fails to give the required description of context; at other times, it is too specific. Therefore, the desired results may not be obtained. SOUPA, also known as the comprehensive model, is able to define the time of context but cannot define details of the environment or the entity's surroundings. Neither SOUPA nor CoBrA-Ont provides provenance, quality of context, or multiple representations. CoDAMoS adds mobile services, code mobility, and resource awareness to the SOUPA ontology. Yet CoDAMoS requires the use of complex contexts, and this is not user friendly. SMARTSPACE is a separate ontology. It is a good ontology for defining a smart home, but its scope is limited to smaller areas, and it cannot handle a larger area, like a city or town. CACOnt makes use of several models, including environment, user, service, space, and device models, but like CONON and CoBrA, it cannot define time of context (Malik & Jain, 2019).

4.5.5.3 Scope of Representational Framework
All the existing context ontologies are implemented in OWL. OWL is accepted as a knowledge representation standard and is the leading ontology language in academia and industry. It has been used by a significant number of development tools and applications. Despite its popularity, there are several limitations, including handling exceptions, no certainty factor, and no default assumption. To overcome these limitations, a new knowledge representation framework, the Extended Hierarchical Censored Production Rule framework, was recently developed (Malik & Jain, 2019).

4.5.6 Ontology Classifications

Ontologies can be classified according to the expressivity and formality of the languages used: natural language, formal language, and so on. Alternatively, they can be classified according to the scope of the objects described by the ontology (Roussey et al., 2011).

4.5.6.1 Classification Based on Language Expressivity and Formality

Depending on the expressivity of an ontology (i.e., a knowledge representation language), different kinds of ontology components can be defined: concepts, properties, instances, axioms, and so on. Figure 4.6 presents a set of components. Concepts are main components of ontologies, and they can be defined in different (and complementary) ways (Roussey et al., 2011):

- By their textual definitions: For example, the concept "person" is defined by the phrase "an individual human being".
- By a set of properties: For example, the concept "person" has the property "name", "birth date", and "address"; note that a property can be reused for several concepts.
- By a logical definition composed of several formulae: For example, the concept "person" is defined by the formula "Living Entity ∩ Moving Entity".

4.5.6.2 Classification Based on the Scope of the Ontology or the Domain Granularity

Figure 4.7 shows classification based on the scope of the objects described by the ontology. For instance, the scope of a local ontology is narrower than the scope of a domain ontology; domain

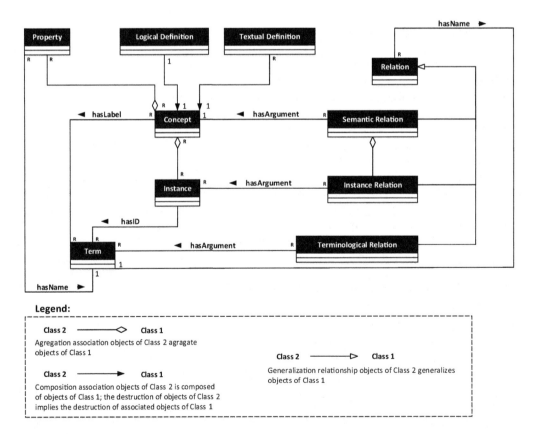

FIGURE 4.6 Unified modeling language class diagram representing ontology components and their relationships (Roussey et al., 2011).

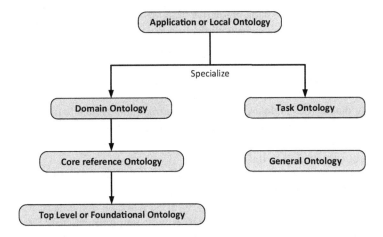

FIGURE 4.7 Ontology classification based on domain scope (Roussey et al., 2011).

ontologies have more specific concepts than core reference ontologies, which contain the fundamental concept of a domain. Foundational ontologies can be viewed as meta-ontologies that describe the top level concepts or primitives used to define other ontologies. Finally, general ontologies are not dedicated to a specific domain; thus, their concepts can be as general as those of core reference ontologies (Roussey et al., 2011).

4.5.7 CONTEXT DRIVEN REMAINING USEFUL LIFE (RUL) ESTIMATION

4.5.7.1 Introduction

Asset owners are interested in knowing the present condition of their assets and predicting the future condition. During the last 40 years, various diagnostic techniques have been developed for this purpose, many of which are based on signal analysis and statistical methods. However, the methods often require costly installation of signal analysis equipment handled by special skilled personnel. Methods using available operational data without the need of costly equipment and special skilled personnel would obviously be preferable.

The use of condition-based maintenance (CBM), based on fingerprint and operational data, gives information about operational conditions without increasing complexity (Johansson et al., 2014). Context-driven prognostics can be used to estimate remaining useful life (RUL) based on available fingerprint and operational data, considered as context data.

The aim is to improve the overall business effectiveness, under a triple perspective:

- Maintenance optimization: Optimize maintenance strategies based on the prediction of potential failures and guide the planning of maintenance operations to schedule these operations in convenient periods and avoid unexpected equipment failures.
- Energy optimization: Manage energy as a production resource and reduce its consumption and cost.
- Asset reliability: Provide the asset builder with real data about the behavior of the asset and its critical components (Johansson et al., 2014).

By integrating all the information from machines, fleets of machines, and even between different companies, the CBM platform can act as a hub of technology methods and analysis tools to provide the different user profiles (machine tool users, maintenance service providers, and machine tool manufacturers) with a unified framework delivering business processes targeting:

- Maintenance through a fleet-wide predictive maintenance strategy including:
    1. Supporting diagnostic processes with fleet-wide comparison.
    2. Associating monitored data with component operation condition on a larger scale.
    3. Associating monitored data with component health for a regular update of the prediction models used locally.
    4. Providing data context for predictive diagnostics.
    5. Providing diagnostics and past solutions for similar abnormal situations.
    6. Optimizing maintenance strategy under cost-effective parameters considering all the cost factors for planned maintenance operations, predictive inspections, machine breakdowns time, cost to repair, and so on.
- Fleet-wide performance assessment allowing:
    1. Fleet-scale key performance indicator (KPI) aggregation.
    2. Machine fleet relative performance (e.g., weakest machine in the fleet).
    3. Fleet energy efficiency assessment, based on consumption patterns. For this the context will be relevant, as the operational conditions have a decisive influence.
    4. Fleet energy optimization, showing cause-effect relationships for abnormal consumption.
- Product reliability through a closed loop with machine engineering and design by providing continuous feedback on:
    1. Machine operation health condition.
    2. Estimation of reliability, availability, maintainability, and safety (RAMS) parameters at the system level and at the component level.
    3. Failure root cause discovery.
    4. Data mining tools to support consolidation, aggregation, and synthesis (Johansson et al., 2014).

4.5.8 Methods for Prognostics and Remaining Useful Life Estimation

The science of prognostics is predicated on four fundamental notions:

- All electromechanical systems age as a function of use, passage of time, and environmental conditions.
- Component aging is a monotonic process that manifests itself in the physical and chemical composition of the component.
- Signs of aging (either direct or indirect) are detectable prior to overt failure of the component (i.e., loss of function).
- It is possible to correlate signs of aging with a model of component aging and thereby estimate RUL of individual components (Uckun et al. 2008).

Figure 4.8 illustrates the states of a component's lifecycle as a finite state machine. Note that there is no general agreement in the research community with respect to the terminology used for component aging (Uckun et al., 2008).

4.5.9 Data-Driven Methods for Remaining Life Estimation

Most RUL estimation efforts follow a data-driven approach. Unlike traditional reliability engineering approaches primarily interested in the initiation of faults, prognostics and health management requires a much more intensive data collection process to characterize damage accumulation and progression. In data-driven prognostics, the challenge is to capture and analyze a multidimensional and noisy data stream from a large number of channels (use conditions, environmental conditions, and direct and indirect measurements) from a population of similar components. In many cases,

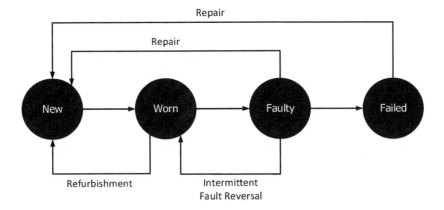

FIGURE 4.8 Component lifecycle represented as a finite state machine (Uckun et al., 2008).

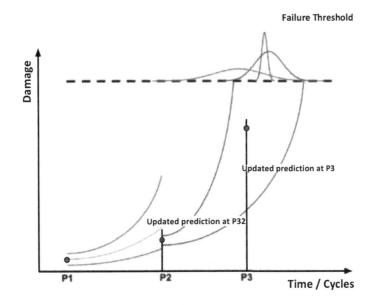

FIGURE 4.9 Uncertainty associated with prognostic RUL estimation as time progresses (Uckun et al. 2008).

data collection, storage, and analysis requirements are onerous. For example, vibration or acoustic-emissions data analysis requires sampling rates in the 10 kHz range and above.

The challenge of data-driven RUL estimation does not end with the complexity of the datasets. Unlike diagnosis which aims to isolate the root cause of a fault event that has already occurred, prognosis is the prediction of a future failure condition. As the future health of the component depends on future environmental and operational conditions, uncertainty management is an inherent element of RUL estimation. Figure 4.9 illustrates the uncertainty bands associated with prognostic RUL estimation. When the component is relatively new and accumulated damage is relatively minor, the uncertainty about the exact time of failure is high (the wide probability distribution illustrated in red). As the component accumulates more damage and the failure point approaches, there is usually much less uncertainty as to when the component fault may occur (the narrow probability distribution illustrated in blue) (Uckun et al., 2008).

4.6 CONTEXT UNCERTAINTY MANAGEMENT

4.6.1 UNCERTAINTY MANAGEMENT THEORY

Uncertainty management theory (UMT) addresses the concept of uncertainty management. Several theories have been developed to define uncertainty, identify its effects, and establish strategies for managing it.

UMT does not see uncertainty as negative. Rather, uncertainty is neutral – neither positive nor negative. Nonetheless, researchers of uncertainty management propose uncertainty can be used strategically for beneficial purposes while also acknowledging that the effects of uncertainty can be harmful, espousing an approach that requires examination of each situation, the parties involved, the issues at stake, and the desired objectives for determining the best method for managing uncertainty (Wikipedia 2020).

4.6.2 ASPECTS OF UNCERTAINTY

Uncertainty is an unavoidable aspect of everyday life. The degree to which it is felt in a given situation varies among individuals. Uncertainty depends on perspective, and different people have different tolerances for uncertainty. For some, the existence of uncertainty is stimulating, whereas others can be highly motivated to reduce even the slightest degree of uncertainty. Personal tolerance for uncertainty will determine how willing a person is to invest when the likelihood of the desired outcome is unclear, whether this investment is monetary, relational, or otherwise (Wikipedia 2020).

4.6.3 UNCERTAINTY MANAGEMENT IN CONTEXT-AWARE APPLICATIONS

Uncertainty arises when there is no clear knowledge of something, when there is a fear of error, or when it is unclear what is stated. Imperfect information has been seen as a synonym of uncertainty, in that something is not well defined, imprecise, or incomplete. Error is also used to define uncertainty. Inconsistency is another concept associated with uncertainty (Ranganathan et al., 2004). It refers to the existence of information that contradicts or presents situations that violate established operating rules (Reyes et al., 2009).

4.6.3.1 Uncertainty in Context-Aware Computing

In general, uncertainty in context may refer to three different notions.

4.6.3.1.1 Uncertain Context

An uncertain context can be created when:

1. The contextual information is generated from unreliable sources because of ignorance or lack of control of the mechanism or device used to acquire it.
2. Application malfunctions create doubts in users, in terms of the validity of the information or the quality of the service rendered (Benford et al., 2006).

4.6.3.1.2 Ambiguous Context

This type of uncertainty appears in such situations as the following:

1. Definitions of context using natural language: For example, the representation of information can be very abstract and difficult to relate to the real world, and this hampers users' interpretation (Beeharee & Steed, 2007). In other instances, the non-specificity expressed in the concepts that define objects can generate uncertainty (Poole & Smyth, 2005). Finally, it can be

difficult to put contextual information, such as geographic coordinates, into words or concepts that are meaningful to users (Hightower et al., 2005).

2. Information from different sources: The use of heterogeneous technologies increases the complexity of the acquisition and processing of context, while causing contradictions between the information provided. For instance, the complexity of handling the signals emitted by sensors to infer context can generate different or contradictory contexts in a given situation (Mantyjarvi & Seppanen, 2003). The use of different technologies also increases complexity in the acquisition task and in context processing (Sheik et al., 2008; Ranganathan & Campbell, 2003).

3. Contradictions, violation of rules, or inconsistencies: Examples include when users carry out different activities at the same time, when two or more services execute contrary operations at the same time and on the same object, when the generated context activates different rules at the same time and each leads to different actions, and when contradictions are generated by the occurrence of disjoint events (Reyes et al., 2009).

4.6.3.1.3 Wrong Context
The generation of incorrect contextual information can come from:

1. Accuracy and precision of instruments, devices, or technologies used to obtain context: For example, the GPS can be inaccurate for a number of reasons, but it is one of the most widely used location estimation technologies. Other location estimation algorithms use the intensity of the signal strength of a wireless network, a technique that tends to be more inaccurate.

2. Lack or absence of information: There are situations when there is insufficient information to derive context because data are incorrect or there are flaws in the devices being used. At times, the behavior of an application may be erratic because location information is incomplete (Benford et al., 2006). However, it is necessary to estimate context even in the absence of information.

3. Lapsed information when a context is not valid at a particular moment: Having an up-to-date context is important when it is necessary to obtain a context of adequate quality (Sheikh et al., 2008; Reyes et al., 2009).

4.6.4 LOCATING AND MODELING UNCERTAINTY

As also shown in Figure 4.10, uncertainty can be found in the following generic locations (Walker et al., 2003).

4.6.4.1 Context Uncertainty
Context refers to the conditions and circumstances (including stakeholder values and interests) that underlie the choice of the boundaries of the system and the framing of the issues and formulation of the problems to be addressed within the confines of those boundaries. Context uncertainty includes uncertainty about the external economic, environmental, political, social, and technological situation that forms the context for the problem being examined. The context could fall within the past, the present, or the future.

Uncertainties are often introduced in framing a decision situation because the context of the decision support is unclear. Actors in a decision situation often have different perceptions of reality, related to their different frames of reference or views of the world. That is why it is important to involve all stakeholders from the very beginning of the process of defining what the issue is. In recent years, expert groups have been accused of framing problems such that the context fits the tacit values of the experts and /or fits the tools the experts use to provide a "solution" to the problem. Such "decision support" is biased and manipulative.

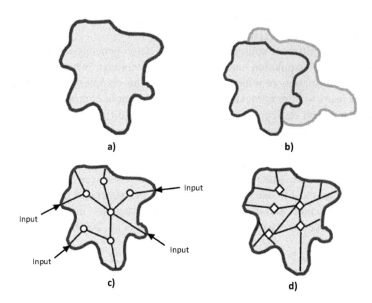

FIGURE 4.10 Location of uncertainty: (a) and (b) context uncertainty; (c) and (d) model structure uncertainty (Walker et al., 2003).

Deciding on a proper framing of context is a significant part of the problem, and reasonable alternative framings should be incorporated in the analysis. The concept and methodology of context validation proposed by Dunn (2001) can help to avoid problems arising from incorrect problem framing (Walker et al., 2003).

4.6.4.2 Model Uncertainty
There are two major categories of uncertainty within model uncertainty.

4.6.4.2.1 Uncertainty of Model Structure
The uncertainty of the model structure arises from a lack of sufficient understanding of the system (past, present, or future) that is the subject of policy analysis, including the behavior of the system and the interrelationships of its elements. The uncertainty about the structure of the system to be modeled implies that any one of many model formulations could be a plausible representation of the system, or that none of the proposed system models is an adequate representation of the real system. There may be insecurity about the current behavior of a system, the future evolution of the system, or both.

Model structure uncertainty involves uncertainty associated with the relationships between inputs and variables, between variables, and between variables and outputs, and refers to system limits, functional forms, definitions of variables and parameters, equations, assumptions, and mathematical algorithms (Walker et al., 2003).

4.6.4.2.2 Model Technical Uncertainty
Model technical uncertainty is the uncertainty generated by software or hardware errors, that is, hidden flaws in the technical equipment. Software errors arise from bugs in software, design errors in algorithms, and typing errors in model source code. Hardware errors arise from bugs. For example, the bug in the early version of the Pentium processor gave rise to numerical errors in a broad range of floating-point calculations performed on the processor (Walker et al., 2003).

4.6.4.3 Input Uncertainty

Input is associated primarily with data describing the reference (base case) system and the external driving forces with an influence on the system and its performance (Walker et al., 2003).

4.6.4.4 Parameter Uncertainty

Parameters are constants in the model, supposedly invariant within the chosen context and scenario (Walker et al., 2003).

4.6.4.5 Model Outcome Uncertainty

This is the accumulated uncertainty caused by the uncertainties in all of the above locations (context, model, inputs, and parameters) that are propagated through the model and reflected in the resulting estimates of the outcomes of interest. This is sometimes called prediction error, as it represents the discrepancy between the true value of an outcome and the model's predicted value. If the true values are known (this is rare, even for scientific models), a formal validation exercise can be carried out to compare the true and predicted values and establish the prediction error. However, practically all policy analysis models are used to extrapolate beyond known situations to estimate outcomes for situations that do not yet exist (Walker et al., 2003).

4.7 PROGNOSIS IN PRESCRIPTIVE ANALYTICS POWERED BY CONTEXT

4.7.1 What Is Prescriptive Analytics?

Simply stated, prescriptive analytics suggests various courses of action and outlines what the potential implications will be for each. It uses historical data to forecast what will happen in the future and what actions you can take to affect those outcomes (Brinkmann, 2019).

Prescriptive analytics is actually the most sophisticated type of business analytics and can bring the greatest intelligence and value to businesses. It aims at suggesting (prescribing) the best decision options to take advantage of the predicted future using large amounts of data. To do this, it incorporates the predictive analytics output and uses AI, optimization algorithms, and expert systems in a probabilistic context to provide adaptive, automated, constrained, time-dependent and optimal decisions (Lepenioti et al., 2020).

Prescriptive analytics has two levels of human intervention: decision support, for example, providing recommendations, and decision automation, for example, implementing the prescribed action. The effectiveness of the prescriptions depends on how well these models incorporate a combination of structured and unstructured data, represent the domain under study, and capture impacts of decisions being analyzed.

Figure 4.11 shows how prescriptive analytics work. As shown in the figure, the first step is descriptive analytics. Descriptive analytics aims to determine what is happening now by gathering and analyzing parameters related to the root causes of the event to be eliminated or mitigated. Descriptive analytics is able to detect patterns that indicate a potential problem or a future opportunity for the business.

On this basis, predictive analytics is able to predict whether an event will happen, when it is about to happen, and why it will happen. Predictive analytics can contribute significantly to the business value. However, this is closely related to the decisions made and the actions taken. In case of human decisions, it depends on knowledge and experience (Lepenioti et al., 2020).

Prescriptive analytics generates proactive decisions on the basis of the predictive analytics outcomes. Between the decision and the implementation of the action, there is a time interval for the preparation of the action. Moreover, an action may be better implemented at a specific time before the predicted event occurrence when the expected utility/loss is optimized.

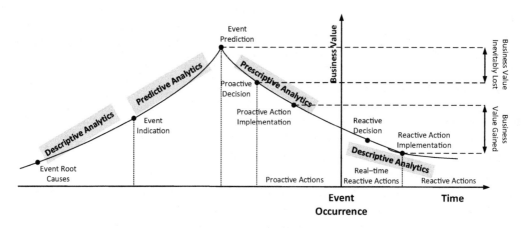

FIGURE 4.11 Business value of analytics with respect to time (Lepenioti et al., 2020).

When the event actually occurs, descriptive analytics may again be applied to derive insights about what happened and why it happened. In this case, descriptive analytics may deal with near real-time reactive actions or with more long-term actions (Lepenioti et al., 2020).

4.7.2 PRESCRIPTIVE MAINTENANCE: BUILDING ALTERNATIVE PLANS FOR SMART OPERATIONS

4.7.2.1 Prescriptive Maintenance Framework

This section discusses the Smart Prescriptive Maintenance Framework (SPMF). It describes the framework structure, the key enablers' elements and technologies, the key indicators used to measure the efficiency of the framework, and its adaptiveness and generalizability (Marques & Giacotto, 2019).

4.7.2.1.1 SPMF Framework

The SPMF framework is built on three domains of interest: the system's RAMS factors, the operating environment of the system, and the maintenance environment of the system (see Figure 4.12). Time and cost are treated as constraints in all domains (Marques & Giacotto, 2019).

Each domain has essential information leading to the development of a maintenance plan. The SPMF fusion flow diagram (Figure 4.13) for the prescriptive maintenance of a fleet of commercial aircraft shows how the framework works, the inputs needed, and the outputs expected in each phase.

From another perspective, the framework comprises five building blocks: input, fusion algorithm, output, the supported system, and the efficiency check process, as shown in Figure 4.14.

4.7.2.1.2 Inputs

The inputs building block shown in Figure 4.14 comprises the information needed to feed the fusion algorithm (Marques & Giacotto, 2019).

4.7.2.1.3 Fusion Algorithm and Its Output

Continuing the example of the fleet of aircraft shown in Figure 4.13, the fusion algorithm in Figure 4.14 is an AI problem-solving agent. In this case, once inputting is complete, the algorithm starts to search for the most probable effective action or sequence of actions to solve the maintenance problem. When the actions are selected, the algorithm recommends the best scheduling according to the minimum fleet availability requirement and the constraint of the direct (DMC) and indirect maintenance cost (IMC), thus providing the output of the fusion step (Marques & Giacotto, 2019).

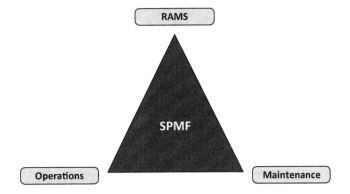

FIGURE 4.12   Framework domains of interest (Marques & Giacotto, 2019).

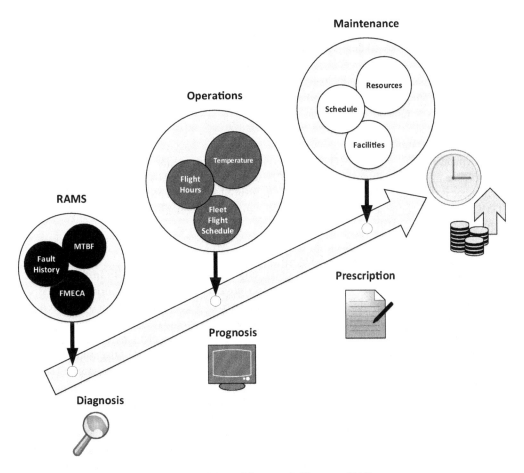

FIGURE 4.13   Framework fusion methodology (Marques & Giacotto, 2019).

4.7.2.1.4   Efficiency Check
After the implementation of the tasks, the algorithm itself verifies during the efficiency check if the maintenance actions are practical, the aircraft fleet availability meets the customer's requirements, and the maintenance costs are minimized. If all these are satisfied, the selection of actions is

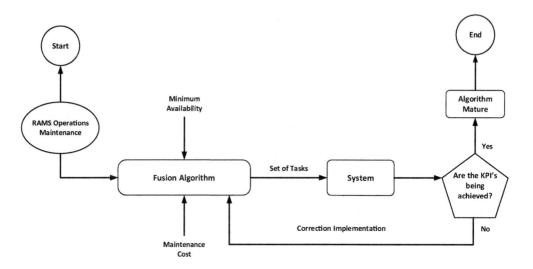

FIGURE 4.14   Framework workflow (Marques & Giacotto, 2019).

reinforced, and the algorithm is considered mature. If at least one of the conditions is not satisfied, the probability attached to each task is redistributed, and a new set of actions is proposed in the next iteration (Marques & Giacotto, 2019).

4.7.2.1.5   Key Indicators
The performance of the system is checked against two requirements: fleet availability and minimization of the system's maintenance cost. Both parameters are also key indicators of the algorithm performance: is the fleet availability above the threshold aligned with the operator`s intent? Are the DMC and IMC contained in comparison to historical maintenance cost or operator estimates (Marques & Giacotto, 2019)?

4.7.2.1.6   Adaptiveness and Context Awareness
Adaptiveness and context awareness are essential characteristics of the SPMF. In the example of the fleet of aircraft, these characteristics allow prescriptive maintenance generation specifically for each "tail number". Some inputs, for example, the number of flight hours or flight cycles, temperature and humidity, RUL, and fault history, are a function of each operational environment; that is, one aircraft presents different values from another, and the same happens to maintenance teams and maintenance, repair, and overhauls. The prescriptive approach favors the use of the ability to identify the individual characteristics of each system, instead of using their average performance (Marques & Giacotto, 2019).

4.7.3   Methods and Techniques for Prescriptive Analytics

Many different optimization methods and techniques can be used for prescriptive analytics: linear optimization, including mixed-integer, binary integer and fractional programming, nonlinear optimization methods, like binary quadratic and mixed integer nonlinear programming, stochastic optimization for handling uncertainty in the decision-making process, distributionally robust optimization, and statistical bootstrapping. Various simulation methods and approaches have been developed as well. Since data may be non-numeric, prescriptive solutions may rely on qualitative analysis, logic, reasoning, collaboration, and negotiation.

This encourages the use of decision rules and decision trees in the decision-making process. These include: decision rules to continuously improve business processes using real-time predictions and recommendations; business rules in combination with a simulation and optimization prescription mechanism; an information system for prescriptive maintenance in which the decision is derived according to rules in combination with mathematical functions; an architecture with the use of proactive event processing rules by combining complex event processing engines with predictive analytics (Lepenioti et al., 2019).

Machine learning techniques are commonly used in predictive analysis. Although their use is less studied for prescriptive analysis, the following methods have been tested: decision trees, real-time random forests (RF), *k*-nearest neighbors, kernel methods, Bayesian belief network, and auto-regressive integrated moving average in combination with stochastic simulation to identify significant KPIs. More sophisticated solutions consist of combinations of optimization, simulation, custom ratings and measures, search policies, and other heuristic techniques (Lepenioti et al., 2019).

4.7.4 CATEGORIES OF METHODS FOR PREDICTIVE AND PRESCRIPTIVE ANALYTICS

Figure 4.15 shows the classification of predictive analysis methods into probability models, machine learning/data mining, and statistical analysis. Figure 4.16 shows the classification of prescriptive analysis methods. However, we should note that the boundaries between probabilistic models,

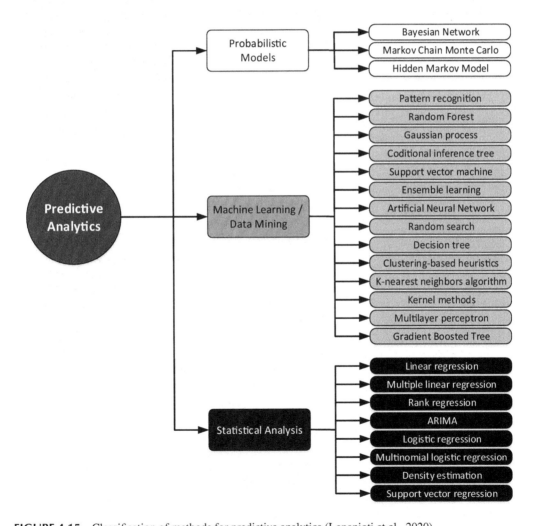

FIGURE 4.15 Classification of methods for predictive analytics (Lepenioti et al., 2020).

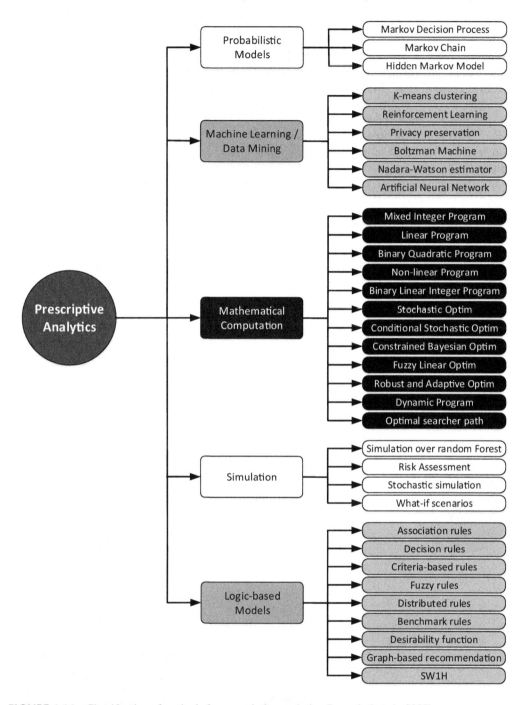

FIGURE 4.16 Classification of methods for prescriptive analytics (Lepenioti et al., 2020).

machine learning/data mining, mathematical programming, evolutionary computation, simulation, and logic-based models are not always clear (Lepenioti et al., 2020).

REFERENCES

Attardi G., Gullì A., Sebastiani F., 1998. Theseus: Categorization by context, Dipartimento di Informatica Università di Pisa, Italy.

Beeharee A., Steed A., 2007. Exploiting real world knowledge in ubiquitous applications. Personal and Ubiquitous Computing, 11, 429–437.

Benford S., Crabtree A., Flintham M., Drozd A., Anastasi R., Paxton M., Tandavanitj N., Adams M., Row-Farr J., 2006. Can you see me now? ACM Transactions on Computer-Human Interaction, 13(1), 100–133.

Bettini C., Brdiczka O., Henricksen K., Indulska J., Nicklas D., Ranganathan A., Riboni D., 2009. A survey of context modelling and reasoning techniques. Pervasive and Mobile Computing.

Brinkmann B., 2019. Comparing descriptive, predictive, prescriptive, and diagnostic analytics. Logi Analytics. www.logianalytics.com/predictive-analytics/comparingdescriptive-predictive-prescriptive-and-diagnostic-analytics. Viewed: June 15, 2020.

Butler, M. H., 2001. DELI: A delivery context library for CC/PP and UAProf. Hp Laboratories Technical Report Hpl, (260).

Carlisle A., 2020. What is Context? Definition & Application, Study.com 2020. https://study.com/academy/lesson/what-is-context-definition-application quiz.html. Viewed: June 11, 2020.

Chen G., Kotz, D., 2000. A survey of context-aware mobile computing research. Technical Report TR2000-381, Dept. of Computer Science, Dartmouth College.

Desai N., 2016. IT vs. OT for the Industrial Internet: Two sides of the same coin? Globalsing, April 27, 2016. www.globalsign.com/en/blog/it-vs-ot-industrial-internet.Viewed: June 08, 2020.

Dey A. K., Abowd G. D., 1999. Towards a better understanding of context and context-awareness, graphics, visualization and usability. Center and College of Computing.

Dey A. K., 2001. Understanding and using context. Personal and ubiquitous computing, 5(1), 4–7.

Dunn, W. N., 2001. Using the method of context validation to mitigate type III errors in environmental policy analysis. In: M. Hisschemoller, R. Hoppe, W.N. Dunn, J. Ravetz (eds.): Knowledge, Power and Participation in Environmental Policy. Policy Studies Review Annual. 12.

Feng L., Apers P., Jonker W., 2014. Towards context-aware data management for ambient intelligence, University of Twente, The Netherlands.

Galar D., 2014. Context driven maintenance: an e maintenance approach, management systems in production engineering Luleå University of Technology, DOI: 10.12914/MSPE-05-03-2014.

Hightower J., Consolvo S., LaMarca A., Smith I., Hughes J., 2005. Learning and recognizing the places we go. In Ubiquitous Computing: 7th International Conference on Ubiquitous Computing, UBICOMP 2005 (pp. 159–176). Springer.

Höppner F., 2003. Knowledge discovery from sequential data, PhD dissertation, Fachbereich für Mathematik und Informatik der Technischen Universität Braunschweig.

Hub K., 2020. What are ontologies? Ontotext, Europe Twins Centre. www.ontotext.com/knowledgehub/fundamentals/what-are-ontologies. Viewed: June 14, 2020.

Indulska J., Robinson R., Rakotonirainy A., Henricksen K., 2003. Experiences in using CC/PP in context-aware systems. In: M.-S. Chen, P.K. Chrysanthis, M. Sloman, A.B. Zaslavsky (Eds.), Mobile Data Management, in: Lecture Notes in Computer Science, 2574, Springer.

Johansson C., Simon V., Galar D., 2014. Context driven remaining useful life estimation, Luleå University of Technology, Luleå, 971 87, Sweden, Procedia CIRP 22 (2014) 181–185.

Karimi K., 2019. A brief introduction to temporality and causality. D-Wave Systems Inc. Burnaby, BC, Canada. https://arxiv.org/ftp/arxiv/papers/1007/1007.2449.pdf.Viewed: June 11, 2020.

Lawrence T. B., Dyck B., Maitlis S., Mauws M. K., 2006. The underlying structure of continuous change. MIT Sloan Management Review.

Lepenioti K., Bousdekis A., Apostolou D., Mentzas G., 2019. Prescriptive analytics: A Survey of Approaches and Methods, DOI: 10.1007/978-3-030-04849-5_39.

Lepenioti K., Bousdekis A., Apostolou D., Mentzas G., 2020. Prescriptive analytics: Literature review and research challenges, Elsevier, International Journal of Information Management, 50, 57–70.

Lum W. Y., Lau F.C.M., 2002. A context-aware decision engine for content adaptation. IEEE Pervasive Computing, 1(3), 41–49.

Malik S., Jain S., 2019. Ontology based context aware model, National Institute of Technology Kurukshetra, India.

Mantyjarvi J., Seppanen T., 2003. Adapting applications in handheld devices using fuzzy context information. Interacting with Computers, 15(4), 521–538.

Marques H., Giacotto A., 2019. Prescriptive maintenance: building alternative plans for smart operations, AerologLab-ITA, Aeronautics Institute of Technology, São José dos Campos, São Paulo/Brazil DOI 10.3384/ecp19162027, 2019.

McCarthy J., 1993. Notes on formalizing contexts, In Proceedings of the Thirteenth International Joint Conference on Artificial Intelligence (San Mateo, California, 1993), R. Bajcsy, Ed., Morgan Kaufmann (pp. 555–556).

McCarthy J., Buva., 1997. Formalizing context. In Working Papers of the AAAI Fall Symposium on Context in Knowledge Representation and Natural Language (Menlo Park, California, 1997), American Association for Artificial Intelligence (pp. 99–135).

McCarthy J., Hayes P. C., 1969. Some philosophical problems from the standpoint of artificial intelligence. Machine Intelligence, 4.

Merriam-Webster, 2020. Context, Dictionary, Merriam-Webster. www.merriam-webster.com/dictionary/context. Viewed: June 11, 2020.

Nivala A., M., Sarjakoski L., T., 2003. Need for context-aware topographic maps in mobile devices. Department of Geoinformatics and Cartography, Finnish Geodetic Institute, Finland.

Pascoe J., 1998. Adding generic contextual capabilities to wearable computers. 2nd International Symposium on Wearable Computers, 92–99.

Priya K., Kalpana Y., 2016. A review on context modelling techniques in context aware computing. International Journal of Engineering and Technology (IJET), e-ISSN: 0975-4024.

Oasys, 2019a. Differences between IT and OT, Zemsania Global Group, 12 July 2019. https://oasys-sw.com/diferencias-entre-it-y-ot. Viewed: June 08, 2020.

Oasys, 2019b. OT e IT Integration for an Industry 4.0, 2019. https://oasys-sw.com/integracion-ot-e-it-para-una-industria-4-0/. Viewed: June 08, 2020.

Oates T., Cohen P. R., 1996. Searching for structure in multiple streams of data. Proceedings of the Thirteenth International Conference on Machine Learning. (pp. 346–354).

Plucknette D., 2013. Operating context: What's included? Allied Reliability. www.alliedreliability.com/rcm-blitz-blog/2013/11/20/operating-context-whats-included.Viewed: June 11, 2020.

Poole D., Smyth C., 2005. Type uncertainty in ontologically grounded qualitative probabilistic matching. In L. Godo (Ed.), 8th European Conference, ECSQARU 2005, LNCS. 3571, 763. Springer Verlag.

Ranganathan A., Al-Muhtadi J., Campbell R. H., 2004. Reasoning about uncertain contexts in pervasive computing environments. IEEE Pervasive Computing Journal, 3(2), 62–70.

Ranganathan A., Campbell R. H., 2003. An infrastructure for context-awareness based on first order logic. Personal and Ubiquitous Computing, 7(6), 353–364.

Reyes P. D., Favela J., Castillo J. C., 2009. Uncertainty management in context-aware applications: Increasing usability and user trust, Springer Science+Business Media, LLC.

Roddick J.F., Spiliopoulou M., 1999. Temporal data mining: Survey and issues, Research Report ACRC-99-007. School of Computer and Information Science, University of South Australia.

Rosenberger P., Gerhard D., 2018. Context-awareness in industrial applications: Definition, classification and use case. ScienceDirect, Procedia CIRP, 72, 1172–1177.

Roussey C., Pinet F., Kang M., Corcho O., 2011. An introduction to ontologies and ontology engineering, G. Falquet et al., Ontologies in Urban Development Projects, Advanced Information 9 and Knowledge Processing 1, DOI 10.1007/978-0-85729-724-2_2, Springer-Verlag.

Ryan N., Pascoe J., Morse D., 1997. Enhanced reality fieldwork: the context-aware archaeological assistant. Gaffney V., Van Leusen M., Exxon S. (eds.) Computer Applications in Archaeology.

Sheikh K., Wegdam M., Sinderen M. V., 2008. Quality-of-context and its use for protecting privacy in context aware systems. Journal of Software, 3(3), 83–93.

Schilit B. N., Adams N. L., Want R., 1994. Context-aware computing applications. In IEEE Workshop on Mobile Computing Systems and Applications (Santa Cruz, CA, US, 1994).

Teambay, 2014. Continuous change: When transformation becomes the Rule 2014. https://teambay.com/continuous-change-when-transformation-becomes-the-rule/#/top. Viewed: June 10, 2020.

Uckun S., Goebel K., Lucas P. J. F., 2008. Standardizing research methods for prognostics. International Conference on Prognostics and Health Management.

Walker W. E., Harremoees J., Rotmans J. P., Sluijs D. V., Asselt V., Janssen P., Krauss K. V., 2003. Defining uncertainty a conceptual basis for uncertainty management in model-based decision support. 4(1), 5–17.

Waterfall 2016. Infographic: Cybersecurity, Waterfall Security Solutions Ltd. 2016. https://waterfall-security.com/ot-vs-it-cybersecurity-infographic.Viewed: June 08, 2020.

Wikipedia, 2019. Context model, https://en.wikipedia.org/wiki/Context_model. Viewed: June 13, 2020.

Wikipedia, 2020. Uncertainty management theory. https://en.wikipedia.org/wiki/Uncertainty_management_theory. Viewed: June 14, 2020.

5 Black Swans and Physics of Failure

5.1 PROGNOSIS PERFORMANCE OF DATA-DRIVEN ESTIMATORS

5.1.1 INTRODUCTION

The large amounts of data gathered continuously from a variety of different systems make it difficult to interpret the data to anticipate breakdowns. Most large industries have specialized engineers skilled in the use of high technology maintenance equipment, with special certification in the field of maintenance. Even so, it is hard to make immediate decisions and predict system failure. Computer systems that constantly record and analyze data to predict the remaining useful life (RUL) of critical components are particularly important in maintenance (Saha & Goebel, 2008; Pal et al., 2011).

In general, maintenance involves performing routine actions to obtain optimal availability of industrial systems (Montgomery et al., 2012). Maintenance routines can be broadly categorized into corrective and preventive maintenance (Kothamasu et al., 2006). In corrective maintenance, interventions are performed only when failure occurs. Preventive maintenance can be further divided into two main approaches: time-based maintenance and condition-based maintenance (CBM). In time-based maintenance, interventions are scheduled at periodic intervals regardless of the asset's health condition. Thus, the service life of critical components is not fully utilized (Soh et al., 2012). CBM uses machine run-time data to assess the critical component's state and schedule required maintenance actions before breakdown (Peng et al., 2010).

In Industry 4.0, predictive maintenance (PdM) will become increasingly viable. PdM uses the current health status of a given critical component to predict its future condition and plan maintenance actions.

Prognostics and health management (PHM) (Jardine et al., 2006) links degradation modeling research to PdM policies (Saha & Goebel, 2008; Pal et al., 2011). PHM consists of four main modules: fault detection, fault diagnostics, fault prognostics, and decision-making (Medjaher et al., 2012):

- Fault detection can be defined as the process of recognizing that a problem has occurred regardless of the root cause (Dong et al., 2012).
- Fault diagnostics is the process of identifying the faults and their causes (Choi et al., 2009).
- Fault prognostics is the prediction of when a failure might take place (Tobon-Mejia et al., 2012).
- Decision-making uses all the information gathered about the monitored system status to choose the optimal maintenance actions (Iyer et al., 2006).

Prognostics have recently attracted significant research interest because of the need for models to make accurate RUL predictions (Saha & Goebel, 2008; Pal et al., 2011).

RUL prediction of critical components is a non-trivial task for many reasons. Sensor signals, for instance, are usually obscured by noise; thus, it is challenging to process the signals and to extract information relevant to the RUL (Javed et al., 2013). Another problem is the prediction uncertainty because the end of life (EoL) may differ for two components made by the same

DOI: 10.1201/9781003097242-5

manufacturer and operating under the same conditions. Therefore, proposed models should include such uncertainties and represent them in a probabilistic form (Saha & Goebel, 2008; Pal et al., 2011).

5.1.2 MOTIVATION

Performance metrics have many components, as shown in Figure 5.1.

Managers of critical systems/applications have struggled to define concrete performance specifications. In most cases, performance requirements are either derived from previous diagnostics experience or are very loosely specified. Prognostics metrics depend on various parameters that must be specified by the customer as requirements that an algorithm should attempt to meet as specifications (Saxena et al., 2009).

The process of coming up with reasonable parameters must consider a complex interplay between several factors. These must be set within a systematic framework to ensure the practical implementation of prognostics. Providing feedback to algorithm developers and helping them improve their algorithms while trying to meet such specifications is another role of these metrics (Saxena et al., 2009).

New prognostics metrics require a change in thinking about what constitutes good performance. The time varying aspect of performance differentiates these metrics from other related domains. These metrics offer visual and quantitative assessment of performance as it evolves over time. The visual representation allows observations of performance. These metrics also permit the incorporation of available uncertainty estimates in the form of RUL distributions.

There is a dilemma between creating a complete but complicated metric and a simple but less generic metric. The latter offers ease of use, interpretability, and comprehensiveness, but the former has less chance of being adopted (Saxena et al., 2009).

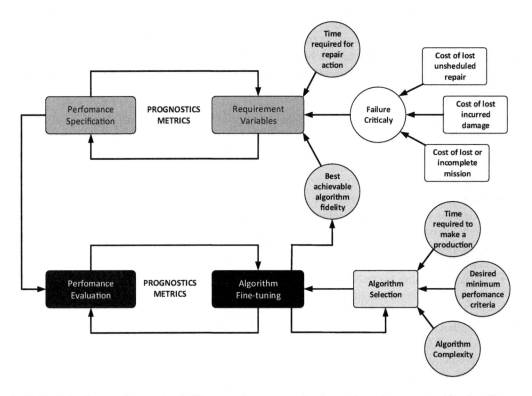

FIGURE 5.1   Prognostics metrics facilitate performance evaluation and requirement specification (Saxena et al., 2009).

5.1.3 Prognostic Performance Metrics

Four forecast metrics, prognostic horizon (PH), yield α-λ, relative accuracy (RA), and convergence, can be used to evaluate the offline performance of forecast performance (Saxena et al., 2009).

5.1.3.1 Offline vs. Online Performance Metrics

There is some confusion about the concepts of online and offline performance evaluation for prognostics. This confusion arises because prognostic performance evaluation is a causal problem that requires inputs from events expected to take place in the future. Specifically, we need to know the true EoL of the system to evaluate prediction accuracy. Another aspect that makes this evaluation complicated is the paradox of prognostics in real applications: if something is sensed to break in the future, it is immediately attended to prevent downtime. This alters the original system and leaves no way to confirm whether the prediction about failure was accurate. Therefore, it has been tricky to assess long-term prognostic results.

Given this situation, a range of performance metrics has been proposed for offline performance evaluation. This would assist in the development of the forecasting algorithm by providing a way to measure performance in cases when we know the true EoL and thus have adequate feedback.

Once these metrics have been refined and adjusted, they will continue to be developed to extend the concept to online performance evaluation. The online assessment will have to incorporate methods for dealing with uncertainties associated with particular future operating conditions. This will require significant advances in uncertainty representation, quantification, and management methods, making the discussion of online performance evaluation appropriate for future work (Saxena et al., 2009).

5.1.3.2 Offline Performance Evaluation

The four prognostic performance metrics, prognostic horizon, yield α-λ, RA, and convergence, follow a systematic progression in terms of the information they seek (Figure 5.2).

First, the prognostic horizon (PH) identifies whether an algorithm predicts the actual EoL within a specified error margin (specified by the parameter α) and if it does how much time it allows for any corrective action to be taken. In other words, it assesses whether an algorithm yields a sufficient PH.

If an algorithm passes the PH test, the α-λ performance identifies whether the algorithm performs within desired error margins (specified by the parameter α) of the actual RUL at any given time instant (specified by the parameter λ) that may be of interest to a particular application. This presents a more stringent requirement of staying within a converging cone of error margin as a system nears EoL.

If this criterion is met, the next step is to quantify the accuracy levels relative to actual RUL (Saxena et al., 2009).

These notions assume prognostics performance improves as more information becomes available with time, and, hence, by design, an algorithm will satisfy these metrics criteria if it converges to true RUL. The fourth metric, convergence, quantifies how fast the algorithm converges when it satisfies all previous metrics.

The group of metrics can be considered a hierarchical test with several levels for comparison of different algorithms in addition to the specific information these metrics provide individually on algorithm performance (Saxena et al., 2009).

5.1.3.3 Prognostic Horizon

The PH is the difference between the time index for EoL and the time index i when predictions meet specified performance criteria based on data accumulated up to time index i (Saxena et al., 2009):

$$PH = EoL - i, \qquad (1)$$

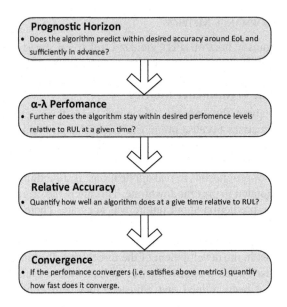

FIGURE 5.2   Hierarchical design of prognostics metrics (Saxena et al., 2009).

where

$$i = minj|\left(j \in \ell\right) \wedge \left(r_* \left(1 - \alpha\right) \le r^l \left(j\right) \le r_* \left(1 + \alpha\right)\right) \tag{2}$$

and where

  i is the first time index when predictions satisfy α- bounds;
  ℓ is the set of all time indexes when a prediction is made.
  l is the index for l^{th} unit under test (UUT);
  r_* is the ground truth RUL; and
  α is the allowable error bound on true EoL.

The PH produces a score that depends on length of life of an ailing system and the time scales. PH ranges between $t_{EoL} - t_P$ and $max\left[0, t_{EoL} - t_{EoP}\right]$. It obtains the best score when an algorithm always predicts EoL within a desired accuracy zone and the worst score when it never predicts within the accuracy zone (Saxena et al., 2009).

5.1.4   DATA-DRIVEN TECHNIQUES FOR PROGNOSTICS

Data-driven approaches commonly model the desired system output (but not necessarily the mechanics of the system) using historical data. Such approaches encompass "conventional" numerical algorithms, like linear regression or Kalman filters, as well algorithms found in machine learning and data mining. The latter algorithms include neural networks, decision trees, and support vector machines. Listed below are the most popular methods for data-based techniques used for forecasting. A review by Schwabacher and Goebel (2007) provides an extensive overview of data-based methods in the context of computational intelligence.

 One of the most popular data-driven approaches to prognostics is the artificial neural network (ANN). An ANN is a type of (typically nonlinear) model that establishes a set of interconnected functional relationships between input stimuli and desired output where the parameters of the functional relationship need to be adjusted for optimal performance.

Some of the conventional numerical techniques used for data-driven prognostics are wavelets, Kalman filters, particle filters, regression, demodulation, and statistical methods. Another popular technique used for prognostics is fuzzy logic. Fuzzy logic provides a language (with syntax and local semantics) into which we can translate qualitative knowledge about the problem to be solved. The fuzzy reasoning mechanism has powerful interpolation properties that, in turn, give fuzzy logic a remarkable robustness to variations in the system's parameters, disturbances, and so on (Goebel et al., 2008).

A core issue in making a meaningful prediction is to account for and subsequently bound various kinds of uncertainties arising from different sources, like process noise, measurement noise, and inaccurate process models. Long-term prediction of the time to failure (TTF) entails large-grain uncertainty that must be represented effectively and managed efficiently. For example, as more information about past damage propagation and future use becomes available, means must be devised to narrow the uncertainty bounds. Prognostic performance metrics should take the width of the uncertainty bounds into account. It is critical to choose methods that can take care of these issues in addition to providing damage trajectories.

Khawaja, Vachtsevanos, and Wu (2005) introduced a confidence prediction neural network that employs confidence distribution nodes based on Parzen estimates to represent uncertainty. The learning algorithm is implemented as a lazy or Q-learning routine that improves uncertainty of online prognostics estimates over time. Not all data-driven techniques can be expected to inherently handle these issues and thus must be combined with other methods suited for uncertainty management. Some such techniques used for dealing with uncertainty include Dempster–Shafer theory (Goebel et al., 2006) or a Bayesian framework with relevance vector machines combined with particle filters. The concept of prognostic fusion has also been suggested as a way to reduce uncertainty. Similar to multiple classifier fusion, the output from several different prognostic algorithms is fused such that the resulting output is more accurate and has tighter uncertainty bounds, on average, than the output of any individual algorithm alone (Goebel et al., 2008).

5.2 BLACK SWANS IN RISK ESTIMATION

5.2.1 Black Swan

The black swan metaphor is historically attributed to the complications that arise when deriving general rules from observed facts. How many white swans must be observed before inferring that all swans are white, and there are no black swans? Hundreds? Thousands? The problem is that we do not know where to start, and we do not have an analysis framework to determine if our estimation is appropriate (Taleb, 2004).

5.2.2 What Is a Black Swan Event?

A black swan event is an "outlier" or an exception that may have a large impact (Taleb, 2004).

It is described by Wikipedia as an "event that comes as a surprise, has a major effect and is often inappropriately rationalized after the fact with the benefit of hindsight".

Taleb (2007) uses black swans as a vehicle for discussing the human propensity to underestimate the degree to which randomness factors into everyday life. He applies three criteria to "black swan events":

First, it is an outlier, as it lies outside the realm of regular expectations, because nothing in the past can convincingly point to its possibility. Second, it carries an extreme "impact". Third, in spite of its outlier status, human nature makes us concoct explanations for its occurrence after the fact, making it explainable and predictable.

Taleb, 2007

5.2.3   BASIC APPROACHES TO MANAGING RISK AND BLACK SWANS

The easiest risk to handle is when the uncertainties are small – the knowledge base is strong, and accurate predictions are possible. For this type of problem, standard risk analysis using statistical methods can be used to make rational decisions, as seen, for instance, in the traffic and health areas. Yet surprises and black swans may also occur, and the issue is how to confront this type of risk (Aven, 2015).

It is obviously not straightforward to assess and manage a black swan type of risk, and in cases where the knowledge base is not strong, judgments need to be based on hypotheses and assumptions (Cornell, 2012).

Two approaches for dealing with black swan events are adaptive risk analysis and robust analysis.

Adaptive risk analysis is based on the acknowledgement that one optimal decision cannot be made; rather, a set of competing alternatives should be dynamically tracked to gain information and knowledge about the system and the effects of different courses of action. On an overarching level, the basic process is straightforward: we choose an action based on broad considerations of risk and other aspects, monitor the effect, and adjust the action based on the monitored results. This way, a black swan type of event may be avoided (Aven, 2015).

Adaptive risk analysis uses abductive thinking. Abduction can be seen as the process of noticing an anomaly and getting an explanatory hunch about it (Pettersen, 2013). By means of abduction, a new idea (or hypothesis) is created. This can be explained as a three-step process:

1. We notice a surprising fact.
2. We explore its qualities.
3. We make a guess to explain the surprising fact (Chiasson, 2014).

In a process plant, abductive thinking could occur in a scenario where we notice the pressure is increasing, we explore the phenomenon, and we provide a hypothesis to explain it. At that point, we can carry out testing to prove or disprove the hypothesis. After all, knowledge is built on theory. Rational prediction and analysis require theory and build knowledge through systematic revision and extension of theory based on a comparison of prediction with observation. Without theory, experience has no meaning, and without theory, there is no learning (Deming, 2000).

Bayesian decision analysis provides a strong theoretical framework for choosing optimal decisions when information is in the form of signals and warnings, but, in many cases, it is difficult to use in practice. Instead, based on a crude assessment of risk and other relevant concerns, we may look for procedures that prescribe what to do for a given signal/warning level. The idea is simply to make an assessment of various signals/warning levels and establish adequate decision rules on how to act in the different cases. This would give us a level of preparedness for specific signals/warnings but would not necessarily provide much support in the case of surprising events. It would be impossible to pre-scribe what to do in all cases; hence, the approach needs to be supplemented with other methods.

Following Cox (2012), the robustness problem is written in general terms as (C, P, u, a), where C is the consequences of the actions, P is the probability of C given the actions, u is the utility function of C, and a is the actions. However, it may be difficult to assign some of these values, for example, P, when the uncertainties are large. A robust approach is required. The key is to make decisions that are good for a set of values of, for example, C and P, and in this way, the approach can also with-stand some types of surprises. However, the set-up used for robustness analysis often excludes the possibility of many forms of black swans, as the framework reflects current knowledge and beliefs. The protective measures could, for example, be based on a probability model reflecting variation due to a set of key risk sources but fail to include an important one. Robust analysis is not easily conducted in practice (Aven, 2013). There are many ways of looking at robustness, and it is difficult to find arguments for why some are better than others (Aven, 2015).

It is helpful to distinguish management strategies for handling a risk agent (e.g., a chemical or a technology) from those needed for the risk-absorbing system (e.g., a building, an organism, or an ecosystem). With respect to the risk agent, assessments can be useful, but equally important are cautionary and precautionary strategies, including such principles as containment, substitution, safety factors, ALARP (as low as reasonably practicable), redundancy and diversity in designing safety devices, and best available technology. For risk-absorbing systems, robustness and resilience are two main categories of strategies or principles when studying risk problems characterized by moderate or large uncertainties. Measures to improve robustness include inserting conservatisms or safety factors as an assurance against variation, introducing redundant and diverse safety systems to meet multiple stress situations, avoiding high vulnerabilities, and establishing building codes and zoning laws to protect against specific hazards (Joshi & Lambert, 2011).

Risk-absorbing systems should be resilient so they can withstand surprises. Resilience is a protective strategy against unforeseen or unthinkable events (Aven, 2015). Resilience engineering has become an important field for the management of safety in socio-technical systems (Hollnagel et al., 2006). To be resilient, a system or an organization must be able to:

1. Respond to regular and irregular threats in a robust, yet flexible manner.
2. Monitor what is going on, including its own performance.
3. Anticipate risk events and opportunities.
4. Learn from experience.

Risk analysis also highlights signals and precursors of serious events. It is a common feature of most approaches that intend to meet surprises. If we look at basic insights from organizational theory and learning, we see that both this feature and resilience are main building blocks. A good example is the concept of collective mindfulness, linked to high reliability organizations (HROs), with its five principles: preoccupation with failure, reluctance to simplify, sensitivity to operations, commitment to resilience, and deference to expertise. There is a vast amount of literature providing arguments for organizations to organize their efforts in line with these principles to obtain high performance (high reliability) and effectively manage risks, the unforeseen, and potential surprises (Aven, 2015).

In addition, a continuous focus on learning and improvements should be mentioned as an approach for dealing with surprises and black swans, as noted in the previous section. The focus on improvements leads us to Taleb's concept of antifragility (2012). According to Taleb, the antifragile is a blueprint for living in a "black swan world", the key being to love randomness, variation, and uncertainty to some degree, and thus also errors.

5.2.4　What Is a Black Swan in a Risk Context?

A black swan in a risk context is understood as an event (or a combination of events and conditions) which is unforeseen and/or represents a surprise in relation to our knowledge and beliefs. When speaking about such events, the implication is always that their consequences are serious (Norwegian Oil & Gas, 2017).

5.2.4.1　Examples of Black Swans

Think of a container for liquid, which is normally filled with water and which people drink from every day. On one occasion, John drinks a liquid from the container which proves to be poisonous. This incident is a black swan for John, a surprise in relation to his knowledge and beliefs (and has serious consequences). Afterwards, an explanation is found for why the liquid was toxic. That is also characteristic of a black swan – it can be explained with hindsight.

A specific example of a black swan of this kind, based on Sande (2013), relates to open-air explosions. Until around 1960, it was accepted knowledge that natural gas cannot explode in the open air. This knowledge was supported by experiments, but the experiments used inadequate gas volumes. Later trials with larger quantities of various gas types showed all hydrocarbon gases can explode in the open air. However, different gases vary in terms of their "critical diameter" – the smallest diameter which must be present for a continuous detonation to occur (Norwegian Oil & Gas, 2017).

5.2.4.2   Three Types of Black Swans

There are three types of black swans (see Figure 5.3). The first is the extreme – the completely unthinkable – an event unknown to the whole scientific community, to everyone. A new type of virus provides an example. This type of event is often termed an unknown unknown.

The second type of black swan is an event that has not been included in the risk assessment, either because it is not known or because it has not been consciously considered. The poisonous liquid in the example above can be seen as a black swan on this basis. John's risk analysis was rather superficial, if he did one. In any case, the poison took him by surprise because he had not imagined that such a scenario was a possibility. In other words, it was unforeseen.

The third type of black swan is an incident that occurs even though its probability has been evaluated as insignificant. Continuing the liquid example, suppose a risk analysis has identified various types of poisonous liquids that could fill the container in special circumstances, but excludes a dangerous form due to a set of physical arguments. But this scenario occurs nonetheless. It occurs despite being considered impossible by the analyst. Actual conditions are not the same as those applied in the risk analysis, and the incident surprises even analysts. However, it is easy to explain in retrospect.

Alternatively, limited knowledge of the subject may have caused risk analysts to overlook a certain type of liquid (known to other specialists) in their analysis. Its appearance consequently becomes a black swan for analysts. This can be called a "known unknown", known to some, unknown to others. An important form of black swan is the combination of events and conditions that lead to disaster. Such combinations are precisely what characterizes serious accidents. Risk analyses cannot detect them, and even if they are detected, and would normally ignore them because of their negligible probability (Norwegian Oil & Gas, 2017).

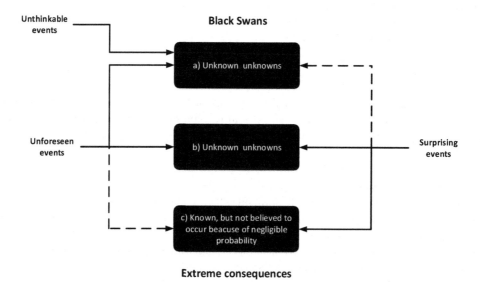

FIGURE 5.3   Hierarchical design of prognostics metrics (Norwegian Oil & Gas, 2017).

Look at the 2010 Deepwater Horizon disaster in the Gulf of Mexico. A blowout is not a surprising event in itself – it is an obvious event for risk analyses – but this event involved an unforeseen combination of events and conditions (PSA 2013):

- Erroneous assessments of pressure test results.
- Failure to identify formation fluid penetrating the well despite log data showing this was happening.
- Inability of the diverter system to divert gas.
- Failure of the cutting valve (blind shear ram) in the blowout precentor to seal the well.

This was probably not a relevant scenario ahead of the accident. But it happened (Norwegian Oil & Gas, 2017).

5.3   FAILURE MODES AND CAUSES MISSING IN DATA: BLACK SWANS

5.3.1   FAILURE MODE AND EFFECTS ANALYSIS

Failure mode and effects analysis (FMEA) is the process of reviewing as many components, assemblies, and subsystems as possible to identify potential failure modes in a system and their causes and effects. For each component, the failure modes and their resulting effects on the rest of the system are recorded in a specific FMEA worksheet. There are numerous variations of such worksheets. An FMEA can be a qualitative analysis, but may be quantitative when mathematical failure rate models are combined with a statistical failure mode ratio database.

FMEA was one of the first highly structured, systematic techniques for failure analysis. It was developed by reliability engineers in the late 1950s to study problems that might arise from malfunctions of military systems. An FMEA is often the first step of a system reliability study (Wikipedia, 2020(a)).

Types of FMEA analysis include the following:

- Functional.
- Design.
- Process.

5.3.2   FUNCTIONAL FAILURE MODE AND EFFECTS ANALYSIS

FMEA analysis should always be started by listing the functions that the design needs to fulfill. Functions are the starting point of a well done FMEA, and using functions as a baseline provides the best yield of an FMEA. An FMEA can be done on concept designs as well as detail designs, on hardware as well as software, no matter how complex the design.

When performing FMEA, interfacing hardware (or software) is first considered to be operating within specifications. After that, it can be extended by using one of the five possible failure modes of one function of the interfacing hardware as a cause of failure for the design element under review. This makes the design robust for function failure elsewhere in the system.

In addition, each part failure postulated is considered to be the only failure in the system (i.e., it is a single failure analysis). In addition to the FMEAs done on systems to evaluate the impact of lower level failures on system operation, several other FMEAs are done, with special attention to interfaces between systems, in fact, at all functional interfaces. The purpose of these FMEAs is to ensure irreversible physical and/or functional damage is not propagated across the interface as a result of failures in one of the interfacing units. These analyses are done at the piece part level for the circuits directly interfacing with the other units. The FMEA can be accomplished without a criticality analysis (CA), but a CA requires that the FMEA has previously identified system level critical failures. When both steps are done, the total process is called an FMECA (Wikipedia, 2020(a)).

5.3.3 Black Swans, Cognition, and the Power of Learning from Failure

5.3.3.1 Definitions of Failure

Failure can be defined simply as not achieving a stated objective or as a "deviation from expected and desired results". In this regard, the term failure is imprecise and may encompass multiple outcomes across various scales and project phases muddied by different perspectives on what comprises failure (Catalano et al., 2018).

Disciplines that learn to manage failure effectively determine what definitions best suit their organizational goals. Professionals in business, commercial and military aviation, and medicine have spent decades gathering data to refine definitions and conceptualize processes for managing and learning from failure, opting for utility rather than prescription to guide their definitions. For example, the business sector recognizes that fault and failure can be conflated (e.g., a purported human error may result, in fact, from a system failure). Therefore, conceptualizing failures across a spectrum from those resulting from the potentially useful exploratory testing of new approaches (i.e., praiseworthy) to those caused by conscious deviance from acceptable practice (i.e., blame-worthy) allows organizations to better understand and manage deviations from desired outcomes while maintaining high performance standards (see Figure 5.4).

In systems incorporating complex technology, such as commercial and military aviation, analysis of failure has evolved beyond reductionist binary definitions and simple blame-assigning exercises. Aviation-safety investigators have operationalized the "Swiss cheese" accident-causation model (see Figure 5.5) in recognition of the fact that failure is inevitable in complex systems, failure rarely has a single cause, human error is implicated in the majority of accidents involving technological systems, and attributions of blame drive errors underground and limit learning opportunities. For example, accepting that human error may compromise all systems, commercial airline flight-crew training intentionally does not set goals for zero errors, but for zero accidents, an achievable goal through the active process of mitigating human error when it occurs (United Airlines, 2016) and one that shifts blame away from individuals. The training program teaches pilots specific, evidence-based actions to reduce the number of errors they commit and to respond effectively when errors inevitably emerge. Errors (as distinct from violations) are not deemed blameworthy but are a rich source of data to be mined to further refine error-mitigation systems and the operational models used to develop crew training (Figure 5.5).

In some cases, failure may be most effectively defined as something we know when we see, as, for example, with a catastrophic plane crash or a species going extinct. In other cases, conditional or multipart definitions of failure most ensure the achievement of goals and learning. For example, military aviation may use conditional definitions of failure during warfare. The goal of complete destruction of an enemy's facility (defined as a target building razed to the ground by aerial attack) may not be defined as a failure if parts of the building are left standing. Upon review, a pragmatic assessment may be made of the outcome that combines evidence with expertise and experience. If the facility is deemed to no longer be functional, then the initial definition of failure is reassessed. Such pragmatic definitions result from acknowledging that failure is inevitable; the long-term outcome is ultimately of greatest importance, and the likelihood that we have enough data to make a 100% accurate assessment is inevitably zero. This is essentially a course correction based on new data, a critical step in learning from failure, and ideally an integral part of assessments.

It is necessary to accept the inevitable nature of failure to reduce the stigma of failure and allow effective pre-implementation risk assessment and post-implementation reflection and learning to occur. The discomfort with failure may also be evidenced by a strong aversion to uncertainty and a tendency to reframe poor project outcomes in a more positive light when reporting to stakeholders

FIGURE 5.4 Spectrum of reasons for failure (Catalano et al., 2018).

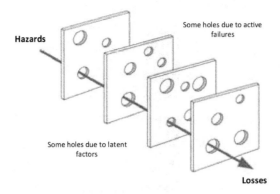

FIGURE 5.5 Swiss cheese model of human error accident causation in complex systems (Catalano et al., 2018).

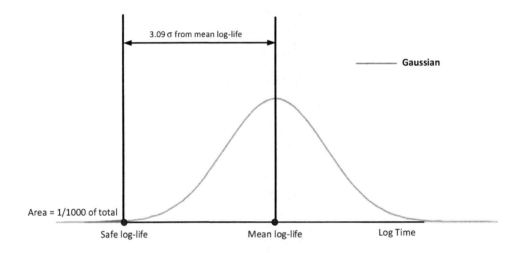

FIGURE 5.6 Probability distribution function (PDF) for Gaussian distribution, safe life = 1/1000 probability of failure (PoF) (Fraes, 2020).

(if reported at all). The aviation community is a good model to follow; it has highly developed non-attribution reporting systems to publish and disseminate error information to drive continuous improvement (Catalano et al., 2018).

5.3.4 BLACK SWANS AND FATIGUE FAILURES

5.3.4.1 The Question

Both the 2008 economic crash and the A380 aircraft engine failure can be considered black swan events, something coming as a surprise, having a major effect, and rationalized after the fact with the benefit of hindsight.

Black swans are examples of "fat tails" in the distribution of catastrophic events (see Figures 5.6 and 5.7). These fat tails occur where contributing factors combine to exacerbate each other. Taleb calls this a "scalable" property (2004). An example of a non-scalable property might be the distribution of people's heights, because nature physically limits how tall we can get. However, wealth is scalable – the wealthier we become, the more wealth we can gather, without limit (Fraes, 2020).

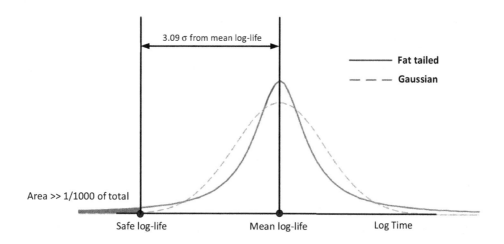

FIGURE 5.7 PDF with fat tails, safe life represents >> 1/1000 probability of failure (PoF) (Fraes, 2020).

In the case of the 2008 financial crash, the fat tails were due to the financial services' use of bad risk models and poor forecasting methods. In the aviation case, the problem was the result of a poorly made oil feed stub pipe in Engine #2, an issue known to the engine manufacturer. The pipe cracked, leading to a fire, explosion, and then very nearly the loss of the aircraft with 469 passengers and crew on board. This event fits the definition of a black swan because it was unanticipated, the impact on the aircraft was significant, and there was considerable after-the-fact rationalizing of the events that led to the engine failure.

5.3.4.2 Fatigue Failure and Fat Tails

In aviation, fatigue test results are intended to provide a minimum acceptable level of safety; however, thick tails may exist, and as a result, the actual risks are not quantified. This is important because those who make decisions in the field of structural integrity need to quantify their risks (Fraes, 2020).

Fatigue failure modeling generally uses a normal logarithmic distribution (Figure 5.6), which is a thin-tailed (Gaussian) distribution.

Poor construction, misuse, and poor maintenance can fatten queues. They are all "scalable" and thus can all exacerbate each other. A minor part of an engine can crack and uncontrollable engine failure occurs. A poorly maintained part can start to corrode or fail, leading to a cascade of other failures. Similarly, if the use of an aircraft (or any other system) changes, this can significantly alter the accumulation of fatigue damage and/or create new unexpected failure modes. All of these things involve small unforeseen problems that can lead to big changes in if or when a fatigue failure is likely to occur. For this reason, thick tails and black swan events can be expected in real-world aircraft use (Fraes, 2020).

5.4 PROBABILISTIC PHYSICS OF FAILURE APPROACH TO RELIABILITY

5.4.1 PHYSICS OF FAILURE: AN INTRODUCTION

Physics of failure (PoF) can be used to design reliable products to avoid failure, based on the knowledge of root cause failure mechanisms and using failure reliability technology that studies the failure regularities from the failure reasons and failure mechanisms of products.

PoF employs the knowledge of the processes and mechanisms that induce failure to predict reliability and thus improve performance (Wikipedia, 2020b). This science-based approach uses modeling and simulation to design-in reliability. It sheds light on system performance and helps to reduce decision risk both during design and after the equipment is put into the field by modeling the root causes of failure, for example, wear, fatigue, fracture, and corrosion. A clear understanding of PoF is necessary for applications that are difficult to test (Sadiku et al., 2016).

The PoF concept considers the relations between the requirements of the product and its physical characteristics, variations in different manufacturing processes, the reactions of product elements and materials to loads or stressors, their interactions under loads/stressors, and their influence on fitness for use in certain use and time conditions (Wikipedia, 2020b).

5.4.2 Physics of Failure Models

5.4.2.1 Introduction to Physics of Failure Models

Simply stated, PoF models deal with the relationships between stresses and materials. PoF modeling uses knowledge of lifecycle stresses, loading profiles, and failure mechanisms to build mathematical models that are useful for the following:

- Modeling TTF.
- Selecting product materials.
- Designing tradeoff analysis.
- Determine mitigation strategies.
- Minimizing demonstration time or accelerated testing.
- Improving prognostics during product use.

By drawing on scientific theory and research, PoF approaches create rigorous mathematical models able to estimate the impact of different environmental or use conditions on reliability. To that end, the models use experimental data, computer simulations, and probabilistic factors (Schenkelberg, 2020).

The traditional approach to reliability assessment is based on empirical models fitted to field data. This approach uses modeling and simulation to identify failure mechanisms prior to physical testing. The three traditional approaches are:

1. Using standard handbooks.
2. Statistically analyzing data.
3. Performing life testing experiments.

A traditional approach may be inappropriate for new products or a product with new technologies.

Deterministic and statistical models play a critical role in the prediction of reliability. Researchers have developed a wide variety of models to predict the life span of materials and components. There are three types of degradation models:

1. Physics-statistics based models.
2. Parametric statistics-based models.
3. Non-parametric statistics-based models.

Physics-based models provide a more accurate estimate of reliability than statistics-based models (Sadiku et al., 2016).

5.4.3 Deterministic vs. Empirical Models

5.4.3.1 Deterministic

Deterministic PoF models attempt to model the path to failure using a mathematical model, starting with the identification of potential failure mechanisms. The method requires stress information for each failure site; relevant information includes structural geometry, loading conditions, and material properties. The models are able to describe various phenomena leading to either sudden or gradual failure: corrosion, degradation, diffusion, erosion, and so on. Failures may occur due to accumulated damage or deterioration of the system under conditions of applied stress (Schenkelberg, 2020).

5.4.3.2 Empirical

Empirical models use test or field data to estimate the TTF of a system under specific stress conditions. However, they do not attempt to model the detailed interactions between stress and failure mechanisms. Peck's temperature-humidity relationship is an example of an empirical model.

An empirical model may simply be a curve fitting experimental data, but the design of the testing and model frequently rely on a detailed understanding of the failure mechanisms involved. The Arrhenius model is a good example. The model describes thermally activated failure mechanisms. It may be in the form of a deterministic model based on the activation energy of the molecular rate reaction; alternatively, it may be in the form of an equation used for empirical modeling (Schenkelberg, 2020).

5.4.4 Reasons to Use PoF-Based Modeling in Reliability

- When we want to avoid repeating long and costly tests:
  - Reduce the development time.
  - Reduce cost to build cheaper products.
- When it is impractical to build many identical units for testing:
  - Large systems like off-shore platforms or space vehicles.
  - One-of-a-kind or very expensive systems.
  - Products that must work properly the first time.
- When there is no prototype to test during the design.
- When highly reliable products and systems that don't fail are required:
  - Lifetime is long and/or system is non-repairable.
  - Internal control or safety related devices limit the stress.
- When dynamic predictions of reliability optimization are required.
- When we want to predict the occurrence of rare or extreme events (Modarres, 2019).

5.4.5 PoF Model Development Steps

5.4.5.1 Strengths of PoF Modeling

- Based on sound science and experimental data.
- Offers a well-defined path to modeling aging and degradation.
- Integrates well with modern machine learning.
- Provides unit-specific reliability predictions.

5.4.5.2 Weaknesses of PoF Modeling

- More expensive to build.
- Hard to specialize applications involving multiple, interactive failure mechanisms.
- Difficult to extend lab test data to field applications involving complex stresses (Modarres, 2019).

5.4.5.3 PoF Model Development Steps

1. Specify component's operating limits, pertinent characteristics, and operating requirements.
2. Define operating environment and profile.
3. Use the profile to assess the applied static and dynamic mechanical, thermal, electrical, and chemical stresses.
4. Identify hot spots exposed to the highest stress.
5. Identify failure mechanisms that become activated and their interactions.
6. Determine materials characteristics and their vulnerabilities to the applicable failure mechanisms.
7. Propose a mathematical model that correlates load (stresses) applied to amount or rate of degradation.
8. Use generic data or accelerated reliability test data to estimate the PoF model parameters, uncertainties, and model error.
9. Validate and revise the model, considering adequacy of the PoF mathematical model fit to the data.
10. Determine a level of degradation beyond which the component fails to operate or endure more damage.
11. Using the PoF model and the endurance limit, estimate the time- or cycle-to-failure, including uncertainties associated with such estimation.
12. Perform computer-based simulation to estimate expected life or remaining life of an item (Modarres, 2019).

5.4.6 Physics of Failure Procedure

The application of the PoF approach comprises the four steps illustrated in Figure 5.8 (Mati & Sruk, 2014).

The first step is to determine environmental factors by specifying or measuring them.

The next step is to isolate potential failures triads (site, mode, and mechanism). The failure-triad determining sequence can differ depending on the available data. This step unifies the identification of failure sites and site corresponding failure modes and the determination of mechanisms

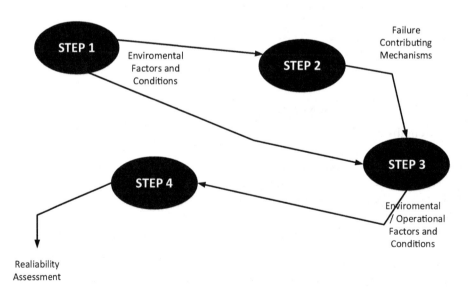

FIGURE 5.8   Generic PoF procedure (Mati & Sruk, 2014).

contributing to a potential failure mode. A comparative analysis of good components and failed components, those failed either in field environment or in laboratory environment, may be helpful.

The third step is to filter out contributing environmental and/or operational factors.

The last step is to find proper functional dependencies of all stresses and to identify applicable models, that is, the best fitting models for the specified operational/environmental conditions. It is also important to determine the model's validity bounds. Then, when the proper equations are known, the life of components for any given operational/environmental condition inside its validity can be determined.

5.4.7 PROBABILISTIC PHYSICS OF FAILURE

Because of the inevitable variations of variables involved in processes contributing to failures, the PoF approach has to be probabilistic. New probabilistic PoF (PPoF) methodologies should be developed to assess the reliability of components by involving variations caused by the following:

- Environmental factors: The variations as a result of environmental factors are specific not only for the field environment but also for accelerated qualification or demonstration testing; for example, during temperature cycling in an environmental chamber, there is always a discrepancy between the air temperature in the chamber and the temperature of the components due to their thermal inertia.
- Mission profiles: The uncertainty of mission profiles includes both nominal values vs. extreme values and dynamics in the domain of environmental factors.
- Manufacturing processes: Manufacturing processes combine variations in tooling and in material properties (Mati & Sruk, 2014).

A logical, straightforward way of implementing PPoF is to extend the existing PoF framework by applying methods for including variability of either key variables or all variables. Another method proposes building engineering parameters into prior beliefs. Predictions using this method often require expert knowledge that itself introduces huge uncertainty. Moreover, by utilizing the Bayesian approach, this method requires failure evidence to change the prior beliefs. These requirements imply that this PPoF method is not so effective during the development process – consequently, the method has disadvantages similar to those of the classical reliability methodology (Mati & Sruk, 2014).

Haggag et al. (2000) used some elements of PPoF principles to determine the failure-time distribution of deep-submicron MOSFET transistors and optical interconnects through a common defect activation distribution.

However, the PPoF approach is in its infancy. An effective PPoF methodology represents a target for reliability scientists. PPoF methodologies will bridge the last gap in reliability engineering by establishing a relationship among probability, time, and environment (Mati & Sruk, 2014).

5.4.8 PROGNOSTICS AND HEALTH MANAGEMENT USING PHYSICS OF FAILURE

PoF uses the knowledge of a product's lifecycle loading and failure mechanisms to perform reliability modeling, design, and assessment. The approach is based on the identification of potential failure modes, failure mechanisms, and failure sites at a particular lifecycle loading condition. The stress at each failure site is obtained as a function of both the loading conditions and the product geometry and material properties. The application of PoF modeling to electronic components and devices provides powerful support for prognostic capabilities.

PHM assesses the extent of deviation or degradation from the expected normal operating condition. For electronics, this provides data to meet the following critical goals (Gu & Pecht, 2008):

1. Giving advance warning of failures.
2. Minimizing unscheduled maintenance, extending maintenance cycles, and conducting timely repairs.
3. Reducing equipment lifecycle costs by decreasing inspections, downtime, and inventory.
4. Improving qualification.
5. Assisting in the design and logistical support of present and future systems.

5.4.8.1 PoF-Based PHM Implementation Approach

The PoF methodology is founded on the premise that failures result from fundamental mechanical, chemical, electrical, thermal, and radiation processes. The objective of the PoF methodology in the PHM process is to calculate the cumulative damage due to various failure mechanisms for a product in a given environment. The approach to implementing PoF in PHM can be based on a failure mode, mechanism, and effect analysis (FMMEA), as shown in Figure 5.9. This approach includes design capture, identification of potential failure, and reliability assessment.

Design capture refers to the collection of structural and material information to generate a model. This step involves characterizing the item of interest at all levels – that is, parts, systems, and physical interfaces.

The potential failure identification step involves using the geometry and material properties of the product, together with the measured lifecycle loads acting on the product, to identify the potential failure modes, mechanisms, and failure sites in the product. This task is best performed through virtual qualification, a simulation-based methodology used to identify and rank the potential failure mechanisms.

The reliability assessment step involves identifying appropriate PoF models for certain specified failure mechanisms. A load-stress analysis considers material properties, product geometry, and lifecycle loads. Then, the cycles to failure are determined using the computed stresses and the failure models, and the accumulated damage is estimated using a damage model.

The PoF methodology can provide a systematic approach to reliability assessment early in the design process (Gu & Pecht, 2008).

5.4.9 Application of PoF for PHM

A PoF-based prognostics approach is applicable in quite different areas, ranging from new products to legacy systems.

When a product has not yet been manufactured, it is obviously impossible to use a data-driven method, as there will be no data on failure to train the algorithm. With the PoF method, however, we simply change the material properties or geometries to model the product. After all, most new products have similarities with older products; thus, similar products can be referenced using FMMEA.

For legacy systems, training data are difficult to obtain. With no understanding of the failure mechanisms and how they influence collected parameters, it is difficult to assess RUL. The use of the PoF approach in PHM is based on an understanding of the structure and lifecycle conditions of the legacy system and its failure modes and mechanisms. First, available information (e.g., previous loading conditions, maintenance records etc.) is used to assess the legacy system's health status. Second, the system's health status is calibrated using individual unit data to derive an assessment of individual legacy systems' health. Third, sensors and prognostic algorithms are used to continually update health status to provide up-to-date system prognosis (Pecht & Gu, 2009).

The PHM methodology is a type of prognosis. Prognostics is the process of predicting the future reliability of a product by assessing the extent of deviation or degradation from its expected normal operating conditions. Similarly, health monitoring is a process of measuring and recording the extent of deviation and degradation from a normal operating condition.

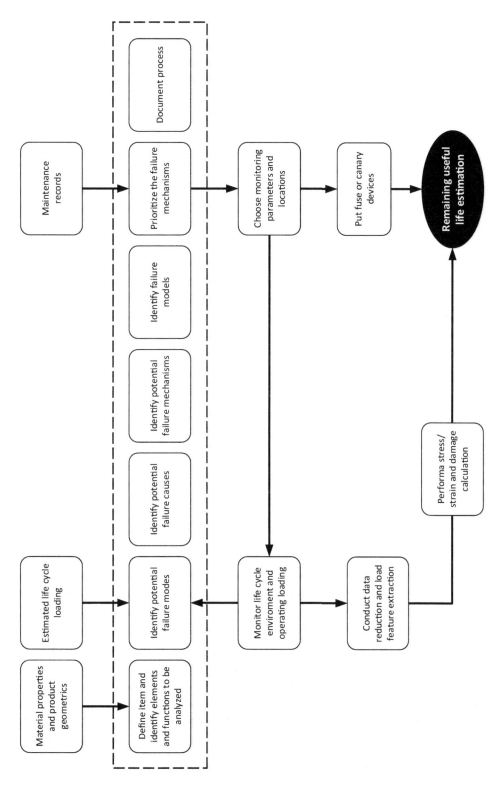

FIGURE 5.9 PoF-based PHM approach (Gu & Pecht, 2008).

Traditional reliability predictions are frequently inaccurate. PHM is more suitable for reliability prediction and RUL assessment, as it considers actual operational and environmental loading conditions. Researchers are currently attempting to build physics-based damage models for electronics, to obtain product lifecycle data and to assess the uncertainty in RUL prediction to make the use of PHM more realistic. Researchers are also considering advanced sensor technologies, communication technologies, decision-making methods, and return on investment methods. Given the increasing amount of electronics in the world and the competitive drive to create more reliable products, PHM may soon become a cost-effective way to predict the reliability of electronic products and systems (Schenkelberg, 2020).

5.4.10 Probabilistic Physics of Failure Degradation Models

PPoF degradation models are usually developed for specific applications and conditions. For example, there are several PPoF models for electromigration of semiconductors describing the hot carrier injection, time-dependent dielectric breakdown, and negative bias temperature instability. More generalized performance degradation models include Arrhenius dependence, Eyring dependence, power dependence, and exponential models (Yanga et al., 2013).

5.4.11 Why Physics of Failure is Preferred to Mean Time between Failures (MTBF) for Reliability Testing

The importance of data in a design for reliability (DfR) practice cannot be understated. Success revolves around reliable design, which can be approached from one of two methodologies: mean time between failures (MTBF) or PoF.

The merits of the two approaches are widely debated. MTBF is reactive, meaning data are gathered based on physical testing during production, but there's no practical application to design improvements. You gain insights into failure, but not failure avoidance. Conversely, PoF mathematical modeling is proactive. It is used to analyze and improve the design during initial concepts and the design phase. With PoF, failure risk is minimized or eliminated well before production (Hillman, 2016).

5.4.11.1 Mean Time between Failures

MTBF is critical to the classic DfR model. Its calculations are the foundation for all reliability predictions compiled in the Military Handbook for Reliability Prediction of Electronic Equipment (MIL-HDBK-217), the original reliability prediction resource. The handbook's failure rate models of various electronic system components are based on the best field data available for parts and systems, the measurement of which ultimately requires physical failure of a system during operation. This forced downtime is just one of the shortcomings of MTBF. Other significant limitations include (Hillman, 2016):

- Reliability predictions based on MIL-HDBK-217 are known to be widely inaccurate, with one study reporting deviations in excess of 500%.
- Failures are assumed to be random in nature.
- Data can be manipulated to reach the desired MTBF, such as modifying the quality factors for each component.
- MTBF results are often misinterpreted as reporting that no failures occurred within a given timeframe, instead of correctly read as the amount of time that elapses between one failure and the next.

The nature of MTBF provides no motivation for failure avoidance and is better suited for logistics and procurement. MTBF reflects the number of components that fail instead of the rate at which they fail (Hillman, 2016).

5.4.11.2 Physics of Failure

PoF is focused on unplanned failure avoidance. It uses the principles of physics and chemistry to enable modeling and simulation of the root causes of failure, for example, thermal cycling, fracture, wear, vibration, shock, fatigue, and corrosion, to design-in reliability. PoF leverages this knowledge to predict reliability, the probability of failure over the product's life, and improve product performance.

PoF employs scientific principles to create complex algorithms that are substituted for the physical testing required by MTBF. There is no waiting for physical failure to calculate reliability.

PoF algorithms are limited in that they assume a "perfect design", and they can be influenced by environmental factors like temperature, vibration, humidity, and power cycling. But these drawbacks are outweighed by the advantages of using science to capture an understanding of failure mechanisms and evaluate useful life under actual operating conditions (Hillman, 2016).

PoF uses root cause analysis and total learning to increase reliability by:

- Providing an understanding of system performance.
- Reducing decision risk during design and in the field.
- Recognizing potential failure mechanisms, sites, and modes.
- Identifying the appropriate failure models and input parameters for material characteristics, damage properties, failure site geometry, manufacturing defects, and operating loads.

Altogether, PoF provides a better alternative to testing than MTBF in that it allows more informed and quantifiable decisions to be made in the design stage, eliminating the need for physical failure or reliance on data known to be inaccurate. Finally, PoF more readily provides for the development and expansion of new tools that improve reliability testing (Hillman, 2016).

5.5 MECHANISMS OF FAILURE AND ASSOCIATED POF MODELS

5.5.1 Degradation Mechanisms

Component failure is never desirable. In the field of industrial systems, a timely awareness of failure mechanisms can lead to better maintenance decisions based on threshold levels, confidence intervals, and RUL estimates (Okoh et al., 2014).

Complicated manufacturing systems cost more to implement, operate, support, and maintain. Complexity-based failure modeling could enhance preventive maintenance. Current research in this area includes work on complexity-based failure modeling for planned maintenance (Meselhy et al., 2009). In their study of operational complexity, ElMaraghy and Urbanic (2004) considered human characteristics relative to system performance. In fact, man, machine, and software interactions are crucial in manufacturing systems with operational complexity. Further detail is needed when connectivity affects the degree of complication (ElMaraghy et al., 2005).

In general, future research will continue to develop PdM strategies to reduce failure, increase productivity, enhance reliability and availability, and reduce costs and downtime by estimating the RUL of assets or equipment in-service (Gouriveau & Ramasso, 2010).

In the healthcare field specifically, research on RUL can be applied to point-of-care devices used in remote environments to tell users when to take medications (Ajai et al., 2009). The adaptation and implementation of RUL will help these devices estimate the time for medication (Okoh et al., 2014).

5.5.1.1 Types of Degradation Mechanisms

Common degradation mechanisms include wear, corrosion, fracture, and deformation.

Wear is the loss of material over time resulting from component use. Wear (or resistance to wear) can be estimated using a weighting method, that is, by weighting the component before and after use. To give one example of its measurement, variables to consider in the wear of metallic engine

FIGURE 5.10   Metal discs showing corrosion on the surface (Okoh et al., 2014).

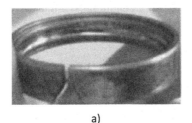   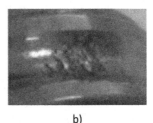

a)   b)

FIGURE 5.11   Bearing (a) external ring failure, and (b) inner ring failure with fracture (Okoh et al., 2014).

components include speed, surface finish/texture, surface hardness, friction coefficient, number of cycles, load, and time (Okoh et al., 2014).

Corrosion refers to the loss of material through chemical deterioration resulting from electrical or biological reactions, including oxidation and sulphidation (see Figure 5.10). To give an example of its measurement, an electro-chemical technique showing the speed at which reinforcing steels are corroding and identifying degraded areas can be used to measure the rate of corrosion.

Fracture is the separation of material by means of cracking or disintegration, occurring as a result of chemical effects, shock, or stress and rendering a component incapable of performing its designed function. This failure mechanism occurs via loading which is independent of time. A slow change in structure can lead to fracture (see Figure 5.11). Strain ratio is critical, as are stress rate and temperature (Johnson & Cook, 1985; Okoh et al., 2014).

Deformation refers to a change in the geometry or shape of a component, including bending, shrinking, stretching, and twisting. Deformation has a cumulative effect on strain on a component because of the applied force, but it can be time-dependent or time independent (see Figure 5.12). Types of deformation include creep, plastic, and elastic deformation.

- Creep deformation: In this type of deformation, the component gradually deforms over time in the presence of high temperature and thermal cycle stress until failure.
- Plastic deformation: In plastic deformation, the material exceeds its elastic limit. This results in a permanent change to the physical structure of the material even with load removal.
- Elastic deformation: This type of deformation results from applied loads; when the load is removed, the component returns to its original condition.

Measurement techniques for deformation include Monte Carlo-based uncertainty, optical measurement systems, digital image correlation, the phase shift method, and the intensity method (Okoh et al, 2014).

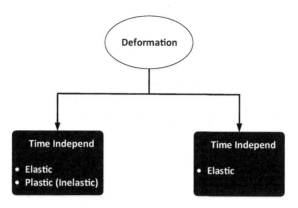

FIGURE 5.12   Types of deformation (Okoh et al., 2014).

5.5.2   Review of Prognostics and Health Management

Prognostic approaches predict the RUL of an asset by assessing the degradation or deviation of a system from a benchmarked healthy system state. Traditional methods of RUL prediction include data-driven or modeling methods, for example, PoF, as discussed previously. As prognostics are applied to increasingly complex systems and the demands for accurate predictions increase, there is growing adoption of fusion prognostics to integrate the benefits of both methods and overcome their limitations through an integrated RUL prediction. The successful application of PHM methods is dependent on tailoring the prognostic model to the given application.

PHM methods have been used in numerous industrial applications such as machinery, electronic products, avionics and vehicles (Roman et al., 2018).

5.5.2.1   Physics of Failure Approach

PoF modeling allows the assessment of a system under its actual operating conditions. In contrast to a data-driven approach, the PoF method exploits prior knowledge of system geometry, material properties, loading conditions, failure mechanisms, and environmental factors to estimate RUL. There are several critical assessment stages prior to the application of the PoF based prognostics model: FMMEA, data reduction and feature extraction, lifecycle load data procurement, damage accumulation, and assessment of uncertainty and prediction. PoF approaches are often divided into two phases. The first is the collection of data, determining the current system state. This is known as damage accumulation. The second phase is the mathematical modeling of the failure degradation.

The foundation of the PoF method is the identification of the critical areas in systems, that is, stress concentration regions, contact surfaces, or similar regions of concern. As an example, in a study on the application of such models on a gull-wing power supply chip, the first stage was to perform a finite element analysis on the solder joint to quantify the strain/stress values caused by mechanical and thermal cyclic loading (Simons & Shockey, 2006). Superimposing the values with the exact locations leads to the creation of boundary conditions for the PoF model. It is crucial at this stage to include data from all sources available, such as environment data, sensor data, and maintenance and inspection records, so that a comprehensive understanding of the system is produced and other locations where sensing is required are identified.

There are other methods of determining failure modes. Methods such as numerical stress analysis and accelerated aging tests have been used in numerous case studies (Roman et al., 2018).

Once the design capture phase has been completed, the mathematical model can be generated based on the preliminary information collected. Built around the previous identification stage of potential failure modes and mechanisms, the physical model aims at predicting the future damage of the system. The extent and degradation of the asset depends on the magnitude and duration of the load acting upon it. Therefore, the degradation can be tracked and life can be estimated by correlating the damage to the load. The damage is computed by inputting the extracted loading features, such as cyclic mean or ramp rate, from raw data, such as vibration amplitude or temperature, and loading these into the mathematical failure model. For instance, damage caused by temperature can be calculated in the time domain using the Coffin-Manson model, while damage caused by vibration can be determined in the time domain using Basquin's model and in the frequency domain using the first-order Steinberg model (Pecht & Gu, 2009). Even though the mathematical damage accumulation model approach can be described using precise first/second-order nonlinear differential equations, many assumptions and approximations are employed. Thus, model validation is cumbersome despite recent efforts to develop model validation techniques using statistical Bayesian approaches.

An alternative solution to statistical validation is to authenticate the model against benchmarked data obtained in accelerated aging tests. Unfortunately, accelerated aging tests on a complex system require time and expensive equipment that few companies/institutes can afford. As a rule of thumb, the more complex the system, the more model parameters are required, thus exposing one of the weaknesses of the approach. Model parameter identification becomes a delicate issue and can be tackled using estimation algorithms such as Kalman filters, particle filters, or Bayesian methods to update or estimate unknown parameters based on measured data. Correlation issues between model parameters, coupled with uncertainty from data, can have a major impact on the RUL, and the predicted RUL can be significantly different from reality.

There are three major challenges to the implementation of a physics-based prognostics approach:

1. Lack of sufficient knowledge in FMMEA stage.
2. Limited mathematical failure models databases.
3. Inability to obtain and validate parameters in the mathematical model.

It is clear that an approach taking advantage of the historical data pattern analysis and understanding the failure model and degradation stages is required (Roman et al., 2018).

5.5.3 FAILURE MODES, CAUSES, MECHANISMS, AND MODELS

A failure mode is a way in which a failure is observed to occur. It can also be defined as the way a component, subsystem, or system could fail to meet or deliver the intended function. All possible failure modes should be considered in analysis. Methods to identify failure modes include numerical stress analysis, accelerated tests to failure (e.g., highly accelerated life test), past experience, and expert judgment. The failure mode needs to be directly observable through visual inspection, electrical measurement, or other tests and measurements.

A failure cause is a specific process, design, or environmental condition that initiates the failure, and whose removal will eliminate the failure. Knowledge of the causes of potential failure can shed light on the failure mechanisms driving the failure modes of a given component or system. One way to look for causes is to review the lifecycle loads item by item to evaluate whether any could cause failure (Pecht, 2009).

Failure mechanisms are physical, chemical, thermodynamic, or other processes that result in failure. They are either overstress or wear-out mechanisms. Overstress failure occurs when a single load (or stress) condition exceeds a fundamental material strength. Wear-out failure occurs when loads (or stresses) applied over an extended period or number of cycles cause cumulative damage. With the existing technology, PHM can only be applied to wear-out failure. Table 5.1 summarizes some typical wear-out failure mechanisms for electronics (Pecht, 2009).

TABLE 5.1
Failure Mechanisms, Relevant Loads, and Models in Electronics

Failure Mechanisms	Failure Sites	Relevant Loads	Failure Models
Fatigue	Die attach, wire bond/tape-automated bonding, solder leads, bond pads, traces, vias/plated through-holes, interfaces	ΔT, T_{mean}, dT/dt, dwell time, ΔH, ΔV	Nonlinear power law (Coffin–Manson)
Corrosion	Metallizations	M, ΔV, T	Eyring (Howard)
Electromigration	Metallization	T, J	Eyring (Black)
Conductive filament formation	Between metallization	M, ∇V	Power law (Rudra)
Stress driven diffusion voiding	Metal traces	S, T	Eyring (Okabayashi)
Time-dependent dielectric breakdown	Dielectrics layers	V, T	Arrhenius (Fowler-Nordheim)

Source: Pecht (2009).
Δ: cyclic range, V: voltage, T: Temperature, S: Stress, ∇: Gradient, M: Moisture, J: Current density, H: Humidity.

Failure models quantify failure by evaluating TTF or the likelihood of a failure for a given geometry, material construction, environmental condition, or operational condition. In the case of wear-out mechanisms, failure models use both stress and damage analysis to quantify the cumulative damage.

Canary devices can be used to provide advance warning of failure caused by specific wear-out failure mechanisms. "Canary" is derived from one of coal mining's earliest systems for warning of the presence of hazardous gas. A canary is more sensitive to hazardous gases than humans, so the death or sickening of the canary indicated the need to get out of the mine. The same approach has been employed in PHM. Canary devices are integrated into a specific component, device, or system design and incorporate failure mechanisms that occur first in the embedded device. These embedded canary devices (also called prognostics cell) are non-critical elements of the overall design providing early incipient failure warnings before actual system or component failure (Pecht, 2009). When using the canary device PHM approach, the geometries or material properties of the prognostics cell can be scaled to accelerate the failure under user conditions, on the basis of potential failure mechanisms. When using the modeling of stress and damage approach, environmental and usage load profiles are captured using sensors. Sensor data are then converted into a format that can be used in the failure models.

In a component or a system's lifecycle, a number of failure mechanisms may be activated by environmental and operational parameters at various stress levels, but for the most part, only a few operational and environmental parameters and failure mechanisms are responsible for failures. High-priority mechanisms have high combinations of occurrence and severity. Prioritizing these mechanisms leads to more effective use of resources (Pecht, 2009).

5.5.4 FAILURE MODES, MECHANISMS, AND EFFECTS ANALYSIS

FMMEA is a method based on assessing the root cause of failure mechanisms of a given item. To reiterate the previous discussion, a potential failure mode is the way in which a failure manifests itself, and failure mechanisms are the processes by which electrical, physical, chemical, and mechanical stresses induce failures either individually or in combination (see Section 5.5.3).

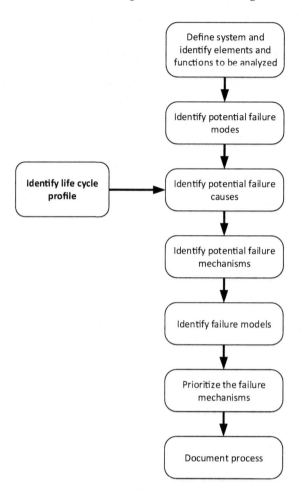

FIGURE 5.13 FMMEA methodology (Mathew et al., 2012).

FMMEA is built on an understanding of the relations between component requirements and the physical characteristics of the component and their variation in the production process, the interactions of component materials with loads (stresses), and their influence on the component's susceptibility to failure depending on the use conditions. Figure 5.13 contains a schematic diagram of the FMMEA steps (Mathew et al., 2012). Also see Ganesan et al. (2005) for a detailed description of the FMMEA methodology.

The purpose of FMMEA is to identify potential failure mechanisms for all the potential failure modes and prioritize these mechanisms. To this end, it uses lifecycle environmental and operating conditions and the duration of the intended application with knowledge of the active stresses and potential failure mechanisms. To ascertain the criticality of the identified failure mechanisms, the method includes the calculation of a risk priority number (RPN) for each mechanism. A higher RPN means a higher ranking among the failure mechanisms. Figure 5.14 shows the prioritization of failure mechanisms based on the calculated RPN for each mechanism (Mathew et al., 2012).

The RPN is the product of the probability of detection, occurrence, and severity of each failure mechanism:

- Detection: The probability of detecting the failure modes associated with the failure mechanism.

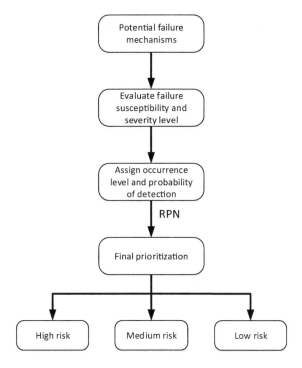

FIGURE 5.14   Prioritization of failure mechanisms (Mathew et al., 2012).

- Occurrence: How frequently a failure mechanism is expected to result in failure.
- Severity: The seriousness of the effect of the failure caused by a failure mechanism.

Figure 5.15 shows the axis of a three-dimensional risk matrix. From the estimation of the critical/ dominant failure mechanisms that affect an item, the appropriate environmental and operational loads and performance parameters can be selected to monitor its health. FMMEA represents a major improvement over traditional reliability methods, as it internalizes the concept of failure mechanisms at every step of decision-making. The use of failure mechanisms as the basis of reliability assessment has been accepted by such major organizations as IEEE, EIA/JEDEC, and SEMATECH (Mathew et al., 2012).

5.6   TIME-DEPENDENCE OF MATERIALS AND DEVICE DEGRADATION

5.6.1   Condition-Based Prediction of Time-Dependent Reliability in Composites

5.6.1.1   Introduction

In general, damage prognosis is challenging not only because of its complexity and multidisciplinary nature, but also because of its direct impact on safety and cost. While structural health monitoring (SHM) technology has considerably developed over the past two decades, little effort has gone into integrating SHM with prognostics science for lifecycle reassessment and condition-based maintenance. The latter is especially significant for composite materials because of their increasing use in high-performance areas, such as aeronautics or space. Composites are well-known for their high strength-to-weight ratios, but they are also susceptible to damage from the beginning of the lifespan. This damage can be hard to detect and usually becomes a critical issue in the reliability and competitiveness of composite structures. Continuous assessment of the health state using state-of-the-art SHM technology, and based on that, the prediction of the remaining time the asset or system is

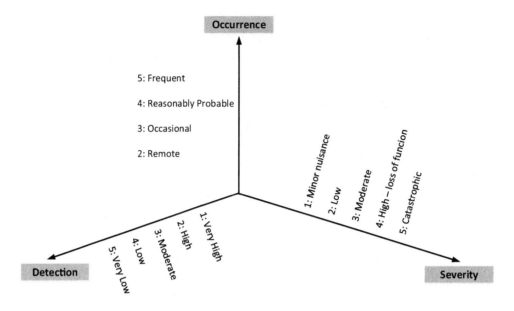

FIGURE 5.15   Risk matrix (Mathew et al., 2012).

expected to continue performing the required function is important for the efficient and reliable use of composite materials (Chiachíoa et al., 2014).

5.6.1.2   Fatigue Damage Modeling

The progression of fatigue damage in composites involves either a progressive or a sudden change in the macroscale mechanical properties, such as stiffness or strength, as a consequence of different fracture modes that evolve at the microscale along the lifespan of the structure. Longitudinal stiffness loss can be a macroscale damage variable. In contrast to the strength variable, it can be measured using in-situ non-destructive methods, something important to a filtering-based reliability approach. At the microscale level, matrix micro-cracking is selected as the dominant fracture mode for the early stage of damage accumulation.

To accurately represent the relationship between the internal damage and its manifestation through macroscale properties, several families of damage mechanics models have been proposed. These models can be roughly classified into:

1. Analytical models.
2. Semi-analytical models.
3. Computational models.

The second and third seem promising, but they are computationally prohibitive in a filtering-based prognostics approach, where a large number of model evaluations is required.

Depending on the level of assumptions adopted to model the stress field in the presence of damage, analytical models can be classified (from simpler to more complex) as shear-lag models, variational models, and crack opening displacement-based models. The shear-lag models have received the most attention, and there are many extensions and variations of the models. Shear-lag models use one-dimensional approximation of the equilibrium stress field after cracking to derive expressions for the stiffness properties of the cracked material. The main modeling assumption of

TABLE 5.2
Mapping of Degradation Mechanisms vs. Prediction Techniques

Degradation Mechanism	Statistics	PoF	CI	Experience	Fusion
Fracture	X	X	X	-	X
Wear	X	X	X	-	X
Deformation	-	X	-	X	-
Corrosion	X	X	X	-	X

Source: Okoh et al. (2014).

TABLE 5.3
Techniques vs. Types of Data

Techniques	Large Dataset	Few Dataset	Numerical Dataset	Categorical Dataset	Ordinal Dataset
Fusion	X	-	X	X	X
Statistics	-	X	X	X	-
CI	X	-	X	X	X
Experience	X	X	-	-	X
PoF	X	-	X	-	-

Source: Okoh et al. (2014).

shear-lag models is that the axial load is transferred to uncracked plies by the axial shear stresses at the interfaces (Chiachíoa et al., 2014).

5.6.2 MAPPING DEGRADATION MECHANISM AND TECHNIQUES

Table 5.2 shows the techniques mapped against failure/degradation mechanisms. When data are mapped against techniques in this way, it is easier to decide which technique to use for a degradation mechanism. When planned maintenance at six-month intervals is required to service a component, the decision can be based on the use of past information to predict its life span. When through-life engineering service (TES) strategies are used for maintenance, repair, and overhaul (MRO), a computational intelligence technique (e.g., fuzzy logic) is employed when damage, such as a fracture or crack, is identified. The threshold level will be a specified crack length; once this threshold is reached, the component will be replaced. If a fracture is within the tolerance margin, the component will be repaired.

In Table 5.3, datasets are mapped against failure prediction techniques. A model to predict RUL will include pre-processing, fusion, and post-processing steps. The confidence interval used to justify the maintenance decision is very important when a statistical method of prediction is selected. PoF requires large amounts of parametric data for better and more confident prediction, and the data must be regularly updated to successfully use TES strategies for better MRO decision-making. In ordinal datasets, low, medium, and high classifications can be used in fusion, computational intelligence, and experience techniques to estimate RUL during MRO. These techniques can be embedded in information system applications for, for example, product lifecycle management (PLM) (Okoh et al., 2014).

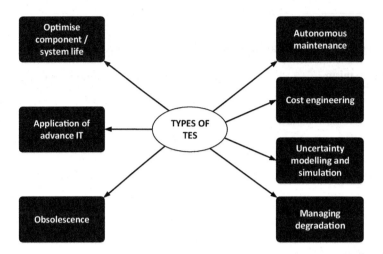

FIGURE 5.16   Types of through-life engineering services (Okoh et al., 2014).

5.6.3 THROUGH-LIFE ENGINEERING SERVICES, DEGRADATION MECHANISMS, AND TECHNIQUES TO PREDICT RUL

Figure 5.16 shows the scope of TES. It is important for an organization to align its MRO function with its operations strategy. This requires the correct application of technology, as well as a reliance on expert knowledge and experience. Accurate RUL prediction and improved MRO decision-making will yield significant benefits to the organization. The following tools and methodologies, when supported by obsolescence management, capability assessment, and cost estimations, have the potential to improve the design function, leading to an improvement in quality, reliability, availability, and safety, while yielding feedback to manufacturers (Okoh et al., 2014):

1. Simulation tools, adaptability procedures, modular maintenance systems, and informed disposal decisions can facilitate a reliable prediction of RUL.
2. Increased application of advanced information technology, such as PLM for distribution and collaboration, condition monitoring, and prognosis, can reduce downtime and increase availability.
3. Degradation management is a key aspect in TES and the maintenance of autonomous systems. The development of these capabilities in a collaborative environment can increase a component's lifetime.
4. Cost engineering provides a performance-based service approach and a whole-life cost model for whole system maintenance.
5. Uncertainty modeling and simulation techniques based on technological and business uncertainties can improve component design.

5.7   UNCERTAINTIES AND MODEL VALIDATION

5.7.1 MODEL VALIDATION AND PREDICTION

From a mathematical perspective, validation is the process of assessing whether or not the quantity of interest (QOI) for a physical system is within some tolerance level, determined by the intended use of the model, of the model prediction.

In simple settings, validation can be accomplished by directly comparing model results to physical measurements for the QOI and computing a confidence interval for the difference or carrying out a hypothesis test of whether or not the difference is greater than the tolerance. In other settings, a more complicated statistical modeling formulation may be required to combine simulation output, various kinds of physical observations, and expert judgment to produce a prediction with accompanying prediction uncertainty, which can then be used for the assessment. This more complicated formulation can also produce predictions for system behavior in new domains where no physical observations are available (National Academy of Science, 2012).

5.7.2 Model Validation Statement

Validation is a process involving measurements, computational modeling, and subject-matter expertise for assessing how well a model represents reality for a specified QOI and domain of applicability. Although it is often possible to demonstrate that a model does not adequately reproduce reality, the generic term validated model does not make sense. There is, at most, a body of evidence suggesting the model will produce results consistent with reality (with a given uncertainty). In other words, a simple declaration that a model is "validated" cannot be justified. Rather, a validation statement should specify the QOIs, accuracy, and domain of applicability.

The body of knowledge that supports the appropriateness of a given model and its ability to predict the QOI in question, as well as the key assumptions used to make the prediction, are important pieces of information to include in the reporting of model results. Such information will allow decision-makers to better understand the adequacy of the model, as well as the key assumptions and data sources on which the reported prediction and uncertainty rely (National Academy of Science, 2012).

5.7.3 Uncertainties in Physical Measurements

The overall task of assessing uncertainty in a computational estimate of a physical QOI involves many smaller uncertainty quantification (UQ) tasks, illustrated hierarchically in Figure 5.17 (National Academy of Science, 2012).

The first UQ task is to quantify uncertainties in model inputs, often by specifying ranges or probability distributions. Model inputs include those that do not vary from problem to problem

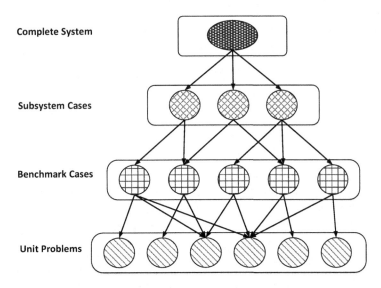

FIGURE 5.17 Validation phases based on hierarchically decomposing a physical system and the models representing it (National Academy of Science, 2012).

(acceleration of gravity, thermal conductivity of a given material, etc.), as well as those that are problem-dependent (e.g., boundary and initial conditions).

A key UQ task is to propagate input uncertainties through the calculation to quantify the effects of those uncertainties on the computed QOIs. Whether or not the computational model is an adequate representation of reality, understanding the mapping of inputs to outputs in the model is a key ingredient in the assessment of prediction uncertainty and in gaining an understanding of model behavior. It is conceptually possible to generate a large set of Monte Carlo samples of inputs, run these random input settings through the model, and collect the resulting model outputs to accomplish the forward propagation of input uncertainty. However, the computational demands of the model often preclude the possibility of carrying out a large ensemble of model runs, and the number of points required to sample a high-dimensional input-parameter space densely is prohibitively large. Understanding low-probability high-consequence events is difficult using standard Monte Carlo schemes because such events are rarely generated (National Academy of Science, 2012).

Another UQ task is quantification of variability in the true physical QOI, which can arise from random processes or from "hidden" variables that are absent from the model. Appropriate methods for quantification depend on the source and nature of the variability.

An important UQ task is the aggregation of uncertainties that arise from different sources. The uncertainties in the QOI that are due to uncertain inputs, true, physical variability, numerical error, and model error must be combined into a quantitative characterization of the overall uncertainty in the computational prediction of a given physical QOI (National Academy of Science, 2012).

5.7.4　TYPES OF UNCERTAINTY

While there are many different ways to classify uncertainty, the taxonomy prevalent in the risk assessment community categorizes uncertainties according to their fundamental essence. Thus, uncertainty is classified as one of the following:

- Aleatory: The inherent variation in a quantity that, given sufficient samples of the stochastic process, can be characterized via a probability density distribution.
- Epistemic: Uncertainty due to lack of knowledge by the modelers, analysts conducting the analysis, or experimentalists involved in validation.

The lack of knowledge can pertain to, for example, the modeling of the system of interest or its surroundings, simulation aspects such as numerical solution error and computer round-off error, or lack of experimental data. In scientific computing, there are many sources of uncertainty, including the model inputs, the form of the model, and poorly characterized numerical approximation errors. All of these sources of uncertainty can be classified as purely aleatory, purely epistemic, or a mixture of the two (Roy & Oberkampf, 2011).

5.7.4.1　Aleatory Uncertainty

Aleatory uncertainty (also called irreducible uncertainty, stochastic uncertainty, or variability) is uncertainty due to inherent variation or randomness and can occur among members of a population. Alternatively, it may be due to spatial or temporal variations.

Aleatory uncertainty is generally characterized by either a probability density function (PDF) or a cumulative distribution function (CDF). A CDF is simply the integral of the PDF from minus infinity up to the value of interest. An example of an aleatory uncertainty is a manufacturing process which produces parts that are nominally 0.5 m long. Measurement of these parts will reveal that the actual length for any given part will be different from 0.5 m. With a sufficiently large number of samples (i.e., information), both the form of the CDF and the parameters describing the distribution of the population can be determined. The aleatory uncertainty in the manufactured part can only be changed by modifying the manufacturing or quality control processes; however, for a given set of processes, the uncertainty due to manufacturing is considered irreducible (Roy & Oberkampf, 2011).

5.7.4.2   Epistemic Uncertainty

Epistemic uncertainty (also called reducible uncertainty or ignorance uncertainty) is uncertainty that arises because of a lack of knowledge on the part of the analyst or team of analysts conducting the modeling and simulation. If knowledge is added (through experiments, improved numerical approximations, expert opinion, higher fidelity physics modeling, etc.), the uncertainty can be reduced. If sufficient knowledge (which costs time and resources) is added, the epistemic uncertainty can, in principle, be eliminated.

Epistemic uncertainty is traditionally represented as either an interval with no associated PDF or a PDF which represents degree of belief of the analyst (as opposed to frequency of occurrence of an event in aleatory uncertainty). Epistemic uncertainty can be represented as an inter-valued quantity, meaning that the true (but unknown) value can be any value over the range of the interval, with no likelihood or belief that any value is more true than any other value. The Bayesian approach to UQ characterizes epistemic uncertainty as a PDF that represents the degree of belief of the true value on the part of the analyst.

5.7.4.3   Epistemic vs. Aleatory Uncertainty

The distinction between aleatory and epistemic uncertainty is not always easily determined during characterization of input quantities or the analysis of a system. For example, consider the manufacturing process mentioned above, where the length of the part is described by a PDF; that is, it is an aleatory uncertainty. If we are only able to measure a small number of samples (e.g., three) from the population, we will not be able to accurately characterize the PDF representing the random variable. In this case, the uncertainty in the length of the parts could be characterized as a combination of aleatory and epistemic uncertainty. By adding information, that is, by measuring more samples of manufactured parts, the PDF (both its form and its parameters) could be more accurately determined. When we obtain a sufficiently large number of samples, we can characterize the uncertainty in length as a purely aleatory uncertainty given by a precise PDF, that is, a PDF with scalar values for all the parameters that define the chosen distribution.

In addition, the classification of uncertainties as either aleatory or epistemic depends on the question being asked. In the manufacturing example given above, if we ask "What is the length of a specific part produced by the manufacturing process?" the correct answer is a single true value that is not known, unless the specific part is accurately measured. If we ask "What is the length of any part produced by the manufacturing process?" the correct answer is that the length is a random variable given by the PDF determined using the measurement information from a large number of sampled parts (Roy & Oberkampf, 2011).

5.7.5   Quantitative Validation of Model Prediction

Suppose a computational model is constructed to predict an unknown physical quantity. Quantitative model validation methods involve the comparison of model prediction and experimental observation. We can use the following notations:

- Y represents the "true value" of the system response.
- Y_m is the model prediction of this true response Y.
- Y_D is the experimental observation of Y (Ling & Mahadevan, 2011).

The development of validation metrics is usually based on assumptions of Y, Y_m, and Y_D, and these assumptions relate to the various sources of uncertainty and the types of available validation data.

To select appropriate validation methods, the first step is to identify the sources of uncertainty and the type of validation data.

The available validation data can be from fully characterized or partially characterized experiments. In the case of fully characterized experiments, the model/experimental inputs x are measured and reported as point values. The true value of the physical quantity (Y) and the output of model Y_m corresponding to these measured values of x will be deterministic if there are no other uncertainty sources in the physical system and the model. Note that Y and Y_m can still be stochastic because of other uncertainty sources other than the input uncertainty. For example, Young's modulus of a certain material can be stochastic due to variations in the material micro-structure, and the output of a regression model for given inputs is stochastic because of the random residual term. If the experiment is partially characterized, some of the inputs x are not measured or are reported as intervals, and the uncertainty in x should be considered. In the Bayesian approach, the lack of knowledge (epistemic uncertainty) about x is represented by a probability distribution (subjective probability). Then, since both Y and Y_m are considered as functions of x, they are treated using probability distributions.

Note that Y_D results from the addition of measurement uncertainty to the true value of the physical quantity Y; that is, $Y_D = Y + \varepsilon_D$ where ε_D represents measurement uncertainty. Hence, the uncertainty in the experimental observation (Y_D) can be split into two parts, the uncertainty in the physical system response (Y) and the measurement uncertainty in experiments (ε_D). It should be noted that experimental data with poor quality cannot provide useful information on the validity of a model (Ling & Mahadevan, 2011).

Table 5.4 summarizes the applicability of the validation methods discussed above in different scenarios.

Methods considered:

1. Classical hypothesis testing.
2. Bayesian interval hypothesis testing.
3. Bayesian equality hypothesis testing.
4. Reliability-based method.
5. Area metric-based method.

Note: Y_D is always treated as a random variable due to measurement uncertainty

After selecting a validation method and computing the corresponding metric, an important aspect of model validation is to decide if we should accept or reject the model prediction based on the computed metric and the selected threshold. The flowchart in Figure 5.18 describes a systematic procedure for quantitative model validation (Ling & Mahadevan, 2011).

TABLE 5.4
Scenarios of Validation and the Corresponding Methods

Experimental Data	Quantity Y (to be predicted)	Prediction Y_m (from model)	Applicable Methods
Fully characterized	Stochastic	Deterministic	1, 2, 4
	Deterministic	Stochastic	1, 2, 4, 5
	Stochastic	Stochastic	1, 2, 3, 4, 5
Partial characterized	Stochastic	Stochastic	1, 2, 3, 4, 5

Source: Ling and Mahadevan (2011).

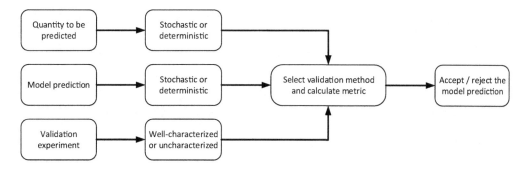

FIGURE 5.18 Decision process in quantitative model validation (Ling & Mahadevan, 2011).

5.7.6 Uncertainty in Prognostics

The presence of uncertainty has a significant impact on RUL prediction. When the state estimates, future loading conditions, and operating conditions are uncertain, the future states and the RUL will also be uncertain. While non-probabilistic methods, such as fuzzy logic, possibility theory, Dempster–Shafer theory, evidence theory, are often used for the treatment of uncertainty, probabilistic methods are more commonly used in prognostics. The choice makes sense: probabilistic approaches are contextually meaningful for uncertainty representation and quantification, as they are consistent with decision-theory analysis (Sankararaman, 2020).

Although the mathematical axioms and theorems of probability are well-established, researchers still disagree on interpretations of probability. There are two major interpretations, one based on physical probabilities and the other on subjective probabilities.

Physical probabilities (or objective or frequentist probabilities) are related to random physical systems, such as rolling dice or tossing coins. Each trial of an experiment leads to an event (i.e., a subset of the sample space). Over repeated trials, each event tends to occur at a persistent rate, known as the relative frequency. Relative frequency is expressed in terms of physical probabilities. Thus, physical probabilities are defined only in the context of randomness.

Subjective probabilities can be assigned to any statement. The statement does not need to be about an event which is a possible outcome of a random experiment. In fact, subjective probabilities can be assigned even in the absence of random experiments. The Bayesian methodology is based on subjective probabilities, considered the degrees of belief on the extent to which the statement is supported by existing knowledge and available evidence. The terms "subjectivist" and "Bayesian" are essentially synonymous (see Chapter 8). Randomness in the context of physical probabilities is equivalent to a lack of information in the context of subjective probabilities (Calvetti & Somersalo, 2007). Even deterministic quantities can be represented using probability distributions which reflect the analyst's subjective beliefs about such quantities (Sankararaman, 2012).

Which interpretation is more suitable for prognostics? In the context of condition-based monitoring or online health monitoring, only one system is being monitored, so at any time instant, no physical randomness is associated with the system. Therefore, following the frequentist interpretation of probability, any quantity associated with it, even an uncertain quantity, cannot be represented using a probability distribution.

Yet the system state is commonly estimated in health monitoring by means of particle filters and Kalman filters. Both approaches compute probability distributions for the state variables. This means the subjective (i.e., Bayesian) approach is being used to quantify uncertainty. Such filtering approaches are known as Bayesian tracking methods partly because they use Bayes' theorem and partly because they fall into the realm of subjective probability. To take this a step further, this implies the uncertainty estimated via filtering algorithms reflects the analyst's degree of belief and is not related to actual physical probabilities.

Common sources of uncertainty in prognostics are measurement errors, model uncertainty errors, the methods used for present state estimation, and the estimation of future loading and operating conditions:

- Measurement errors: In health monitoring or condition-based monitoring, measurements from the system are continuously available, but they may be uncertain because of sensor bias or sensor noise.
- Model uncertainty errors: Models built to represent system behavior have different types of uncertainty. For one thing, the model parameters may be uncertain; for another, the model form may not capture the true underlying behavior. This is sometimes expressed as process noise, but this representation is not accurate.
- Present state estimate: State estimation is commonly performed using Bayesian tracking methods (e.g., Kalman filter, particle filter). These approaches estimate the state of the system as a random variable, thus adding additional uncertainty to prognostics.
- Future loading and operating conditions: It is difficult if not impossible to be certain when predicting these conditions, so there is always an element of subjectivity while assessing the uncertainty of such variables (Sankararaman, 2020).

Several PHM approaches quantify risk based on UQ in an algorithm's output and incorporate it into a corresponding cost-benefit equation through monetary concepts. However, it is very difficult to accurately quantify the various sources of uncertainty in practical applications. In fact, the uncertainty may be undesirably high. Although some uncertainties can be reduced by improving techniques of measurement or modeling, their elimination is practically impossible. Representing them in prognostics is therefore extremely important, as it directly affects decision-making. (Sankararaman, 2020).

REFERENCES

Ajai O., Tiwari A., Alcock J. R., 2009. Informatics- based product- service-systems for point-of-care devices. Proceedings of the First CIRP Conference on Industrial Product-Service Systems.

Aven T., 2013. On how to deal with deep uncertainties in a risk assessment and management context. Risk Analysis, 33(12), 2082–2091.

Aven T., 2015. Reliability engineering and system safety, University of Stavanger, Stavanger, Norway, 134, 83–91.

Calvetti D., Somersalo E., 2007. Introduction to Bayesian scientific computing: Ten lectures on subjective computing, Vol. 2, Springer.

Catalano A., S., Redford K., Margoluis R., Knight A. T., 2018. Black swans, cognition, and the power of learning from failure. Conservation Biology, 32(3).

Chiachíoa J., Chiachíoa M., Sankararaman S., Saxena A., Goebel K., 2014. Condition-based prediction of time-dependent reliability in composites, Dept. Structural Mechanics and Hydraulic Engineering, University of Granada, Spain, 2014.

Chiasson P., 2014. Abduction as an aspect of retroduction. In The Commens Encyclopedia: The Digital Encyclopedia of Peirce Studies. www.commens.org/encyclopedia/article/chiasson-phyllis-abduction-aspect-retroduction

Choi K., Singh S., Kodali A, Pattipati K. R., Sheppard J. W., Namburu S. M., Chigusa S, Prokhorov D. V., Qiao L, 2009. Novel classifier fusion approaches for fault diagnosis in automotive systems, IEEE Transactions on Instrumentation and Measurement, 58(3), 602–611.

Cornell P., 2012. On black swans and perfect storms: risk analysis and management when statistics are not enough. Risk Analysis, 32(11), 1823–1833.

Cox L., 2012. Confronting deep uncertainties in risk analysis. Risk Analysis, 32, 1607–1629.

Deming W. E., 2000. The New Economics. 2nd ed. MIT CAES.

Dong J., Verhaegen M., Gustafsson F., 2012. Robust fault detection with statistical uncertainty in identified parameters. IEEE Transactions on Signal Processing, 60(10), 5064–5076.

ElMaraghy H. A., Kuzgunkaya O., Urbanic R. J., 2005. Manufacturing systems configuration complexity. CIRP Annals Manufacturing Technology, 54(1), 445–450.

ElMaraghy W. H., Urbanic R. J., 2004. Assessment of manufacturing operational complexity. CIRP Annals Manufacturing Technology, 53(1), 401–406.

Fraes S. D., 2020. Black swans and fatigue failures, linkedin, Posted on January 7, 2020. www.linkedin.com/pulse/black-swans-fatigue-failures-stephen-dosman-fraes

Ganesan S., Eveloy V., Das D., Pecht M., 2005. Identification and utilization of failure mechanisms to enhance FMEA and FMECA. In: Proceedings of the IEEE Workshop on Accelerated Stress Testing & Reliability (ASTR), Austin, Texas, October 2–5, 2005.

Goebel K., Eklund N., Bonanni P., 2006. Fusing Competing Prediction Algorithms for Prognostics, Proceedings of 2006 IEEE Aerospace Conference.

Goebel K., Saha B., Saxena A., 2008. A Comparison of Three Data-Driven Techniques for Prognostics, NASA Ames Research Center, January 2008.

Gouriveau R., Ramasso E., 2010. From real data to remaining useful life estimation: An approach combining neuro-fuzzy predictions and evidential Markovian classifications. 38th ESReDA Seminar Advanced Maintenance Modelling.

Gu J., Pecht M., 2008. Prognostics and health management using physics-of-failure, 54th Annual Reliability and Maintainability Symposium (RAMS), Las Vegas, Nevada.

Haggag A., McMahon W., Hess K., Cheng K., Lee J., Lyding J., 2000. A probabilistic-physics-of-failure/short-time-test approach to reliability assurance for high-performance chips: Models for deep-submicron transistors and optical interconnects. IEEE International Integrated Reliability Workshop, 2000 Oct 23–26; S. Lake Tahoe, California, USA (pp. 179–182).

Hillman C., 2016. Why physics of failure is preferred to MTBF for reliability testing. DfR Solutions, Mar 24, 2016, www.dfrsolutions.com/blog/why-physics-of-failure-is-preferred-to-mtbf-for-reliability-testing. Viewed: June 03, 2020.

Hollnagel E., Woods D., Leveson N., 2006. Resilience engineering: Concepts and precepts. Ashgate.

Iyer, N., Goebel, K., Bonissone, P., 2006. Framework for post-prognostic decision support. IEEE Aerospace Conference, 9(1), 3962–3971.

Jardine A. K. S., Lin D., Banjevic D., 2006. A review on machinery diagnostics and prognostics implementing condition based maintenance. Mechanical Systems and Signal Processing, 20(7), 14831510.

Javed K., Gouriveau R., Zerhouni N., Nectoux P., 2013. A feature extraction procedure based on trigonometric functions and cumulative descriptors to enhance prognostics modeling. IEEE Prognostics and Health Management (PHM) Conference, 1(7), 24–27.

Johnson G. R., Cook W. H., 1985. Fracture characteristics of three metals subjected to various strains, strain rates, temperatures and pressures. Engineering Fracture Mechanics, 21(1), 31–48.

Joshi N., Lambert J., 2011. Diversification of engineering infrastructure investments for emergent and unknown non systematic risks. Journal of Risk Research, 14 (4), 1466–4461.

Khawaja T., Vachtsevanos G., Wu B., 2005. Reasoning about uncertainty in prognosis: A confidence prediction neural network approach. Proceedings of the Annual Meeting of the North American Fuzzy Information Processing Society.

Kothamasu R., Huang S. H., VerDuin W. H., 2006. System health monitoring and prognostics a review of current paradigms and practices. The International Journal of Advanced Manufacturing Technology, 28(9–10), 1012–1024.

Ling Y., Mahadevan S., 2011. Quantitative model validation techniques: New insights. Department of Civil and Environmental Engineering, Vanderbilt University, TN 37235.

Mathew S., Mohammed A., Pecht M., 2012. Identification of failure mechanisms to enhance prognostic outcomes, Journal of Failure Analysis and Prevention, 12, 66–73.

Mati Z., Sruk V., 2014. The physics-of-failure approach in reliability engineering, faculty of electrical engineering and computing, University of Zagreb, Croatia.

Medjaher K., Tobon-Mejia D. A., Zerhouni N., 2012. Remaining useful life estimation of critical components with application to bearings. IEEE Transactions on Reliability, 61(2), 292–302.

Meselhy K. T., ElMaraghy H. A., ElMaraghy W. H., 2009. Comprehensive complexity-based failure modelling for maintainability and serviceability. Industrial Product Service Systems IPS2, 89–93.

Modarres M., 2019. Advances in Probabilistic Physics-of-Failure, Keynote Talk Well Engineering Reliability Workshop. Petrobras R&D Center, Rio de Janeiro, Brazil.

Montgomery N., Banjevic D., Jardine A. K. S., 2012. Minor maintenance actions and their impact on diagnostic and prognostic CBM models. Journal of Intelligent Manufacturing, 23(2), 303–311.

National Academy of Science, 2012. Assessing the reliability of complex models, Washington, D.C. www.nap.edu. Viewed: June 04, 2020.

Norwegian Oil & Gas, 2017. Black Swans, an enhanced perspective on understanding, assessing and managing risk. Stavanger, August 25, 2017. www.norskoljeoggass.no/contentassets/d3183372438841a180e14938177f6ec7/black-swans.pdf.

Okoh C., Roy R., Mehnen J., Redding L., 2014. Overview of remaining useful life prediction techniques in through-life engineering services. EPSRC Centre for Innovative Manufacturing in Through-Life Engineering Services Department of Manufacturing and Materials, Cranfield University, 158–163.

Pal S., Heyns P. S., Freyer B. H., Theron N. J., Pal S. K. 2011. Tool wear monitoring and selection of optimum cutting conditions with progressive tool wear effect and input uncertainties. Journal of Intelligent Manufacturing, 22, 491504.

Pecht M., 2009. Prognostics and health management of electronics, Center for Advanced Life Cycle Engineering (CALCE), University of Maryland, College Park, MD, USA.

Pecht M., Gu J., 2009. Physics-of-failure-based prognostics for electronic products, Transactions of the Institute of Measurement and Control, 31(3/4), 309–322.

Peng Y., Dong M., Zuo M. J., 2010. Current status of machine prognostics in condition-based maintenance: A review. The International Journal of Advanced Manufacturing Technology, 50(1–4), 297–313.

Pettersen K., A., 2013. Acknowledging the role of adductive thinking: A way out of proceduralization for safety management and oversight? In: Bieder C, Bourrier M, eds. Trapping safety into rules: How desirable or avoidable is proceduralization? Ashgate (pp. 107–117).

PSA, 2013. Hydrocarbon leak on Ula P platform 12 September 2012. Stavanger: Petroleum Safety Authority Norway.

Roman D. V., Dickie R. W., Flynn D., Robu V., 2018. A review of the role of prognostics in predicting the remaining useful life of assets, Heriot Watt University Smart Systems Group.

Roy C. J., Oberkampf W. L., 2011. A comprehensive framework for verification, validation, and uncertainty quantification in scientific computing. Aerospace and Ocean Engineering Department, Virginia Tech.

Sadiku M. N. O., Shadare A. E., Dada E., Musa S. M., 2016. Physics of failure: An introduction. International Journal of Scientific Engineering and Applied Science (IJSEAS), 2(9).

Saha B., Goebel K., 2008. Uncertainty management for diagnostics and prognostics of batteries using Bayesian techniques. IEEE Aerospace Conference, 1(8), 1–8.

Sande T., 2013. Misforstått risiko. Stavanger Aftenblad, p. 20. October 1, 2013.

Sankararaman S., 2012. Uncertainty quantification and integration in engineering systems. Ph.D. Dissertation, Vanderbilt University.

Sankararaman S., 2020. Remaining Useful Life Estimation in Prognosis: An Uncertainty Propagation Problem, SGT Inc., NASA Ames Research Center.

Saxena A., Celaya J., Saha B., Saha S., Goebel K., 2009. On applying the prognostic performance metrics. SGT Inc., NASA Ames Research Center, Intelligent Systems Division.

Schenkelberg F., 2020. Introduction to Physics of Failure Models. https://accendoreliability.com/introduction-physics-of-failure-models. Viewed: June 24, 2020.

Schwabacher M., Goebel K., 2007. A survey of artificial intelligence for prognostics, Working Notes of 2007 AAAI Fall Symposium: AI for Prognostics.

Simons J. W., Shockey D. A., 2006. Prognostics modeling of solder joints in electronic components. 2006 IEEE Aerospace.

Soh S. S., Radzi N. H. M., Habibollah H., 2012. Review on scheduling techniques of preventive maintenance activities of railway. Fourth International Conference on Computational Intelligence, Modelling and Simulation (CIMSiM) (pp. 310–315).

Taleb N., 2004. The roots of unfairness: The black swan in arts and literature, 2nd Draft, November 2004.

Taleb N., 2007. The Black Swan: The Impact of the Highly Improbable. Random House.

Taleb N., 2012. Anti fragile. Penguin.

Tobon-Mejia D. A., Medjaher K., Zerhouni N.,Tripot G., 2012. A datadriven failure prognostic method based on mixture of Gaussians hidden Markov models. IEEE Transactions on Reliability, 61(2), 491–503.

United Airlines, 2016. CRM/TEM training presentation. 2016. Internal company training document. United Airlines, Denver, Colorado.

Wikipedia, 2020a. Failure mode and effects analysis. https://en.wikipedia.org/wiki/Failure_mode_and_effec ts_analysis. Viewed: June 05, 2020.

Wikipedia, 2020b. Physics of failure. https://en.wikipedia.org/wiki/Physics_of_failure#:~:text=Physics%20 of%20failure%20is%20a,reliability%20and%20improve%20product%20performance. Viewed: June 24, 2020.

Yanga Z., Kanga R., Elsayed A., 2013. Reliability estimate of probabilistic-physics-of-failure degradation models, Guest Editors: E. Zio, P. Baraldi. AIDIC Servizi S.r.l.

6 Hybrid Prognostics Combining Physics-Based and Data-Driven Approaches

6.1 INFORMATION REQUIREMENTS FOR HYBRID MODELS IN PROGNOSIS

6.1.1 PROGNOSTICS TECHNOLOGY

Prognostics technology covers many aspects. For example, prognostics can give advance warning of impending failures and estimate remaining useful life (RUL), ultimately resulting in increased availability, reliability, and safety, as well as reduced maintenance costs. As defined in ISO 13381-1, prognostics is "an estimation of time to failure (TTF) and risk for one or more existing and future failure modes". In this understanding, prognostics is also called the "prediction of a system's lifetime" as its objective is to predict RUL before a failure occurs, given the present system condition and past operating profile (Jardine et al., 2006). RUL prediction is widely applied, including for military and aerospace systems, manufacturing equipment, structures, power systems, and electronics (Liao & Köttig, 2016).

There are several methods of monitoring the condition of equipment. The most common are physics-based and data-driven approaches for diagnosis and prognosis. Each has advantages and disadvantages, and they are therefore often used in combination (Fornlöf, 2016).

In general, RUL prediction models can be categorized into experience-based models, data-driven models, and physics-based models, as shown in Figure 6.1 (Liao & Köttig., 2014).

Prognostics focuses on predicting the future performance of a system, specifically the time at which the system will no longer perform its desired function, or in other words, its TTF. As an important aspect of prognostics, RUL prediction estimates the remaining usable life of a system. It is one of the key requirements in prognostics and health management (PHM) and is an essential element of maintenance decision-making and contingency mitigation. A system or component exhibits degradation during its lifecycle; a number of methods can predict its future performance and determine when it will no longer perform its desired function (Liao & Köttig, 2016).

6.1.1.1 Experience-Based Prognostic Models

In these methods, historical data collected over a significant period of time (failure times, maintenance data, operating data, etc.) are used to predict the TTF or the RUL. Their main advantage is that they are based on the use of simple reliability functions (exponential law, Weibull law, etc.) rather than complex mathematical models.

The prognostic results from these methods are less precise than those provided by physics-based and data-driven approaches, especially if the operating conditions are variable or if systems are new and lack useful failure data (Jammu & Kankar, 2011).

6.1.1.2 Data-Driven Models

In these methods, online data are captured with the help of sensors and converted into relevant information. The information is used to study degradation based on various models and

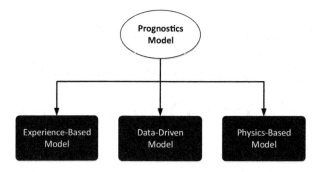

FIGURE 6.1 Categorical breakdown of prognostics models (Liao & Köttig, 2014).

tools such as neural networks, Bayesian networks (BNs), and Markovian processes or on statistical methods in order to predict the future health state and the corresponding RUL of the system.

Data-driven methods have an advantage over both physics-based prognostic methods and experience-based prognostic methods, as in real industrial applications, getting reliable data is easier than constructing physical or analytical behavior models. Moreover, the generated behavioral models from real monitoring data lead to more precise prognostic results than those obtained from historical data.

6.1.1.3 Physics-Based Prognostic Models

These methods employ an analytical model (set of differential or algebraic equations) to represent the system's dynamic behavior and degradation. Yu and Harris (2001) proposed a stress-based fatigue life model for ball bearings. The model delivers more accurate prognostic results, but real systems are often nonlinear and the degradation mechanisms are generally stochastic, so it is difficult to use analytical models. Consequently, the applicability of this approach may be limited in practice.

6.1.2 HYBRID PROGNOSTICS APPROACH

Hybrid models use a combination of several techniques to improve accuracy (Galar et al., 2012; Okoh et al., 2014). Very few researchers have considered hybrid modeling for fault diagnostics and maintenance decision-making. Ahmadzadeh and Lundberg (2014) reviewed three state-of-the-art models for RUL prediction, that is, experimental-based model, data-driven model, and physics-based model, as well as hybrid approaches. Jardine, Lin, and Banjevic (2006) reviewed machinery diagnostics and prognostics, implementing condition-based maintenance (CBM) to show that statistical, artificial intelligence, and physics-based prognostic approaches can be used to estimate life with improved accuracy. However, no work has specifically focused on hybrid prognostics approaches to leverage the advantages of different prognostics models (Liao & Köttig, 2014).

Hybrid approaches try to use the advantages of other approaches, such as physics-based, data-driven, and physics-based approaches, while avoiding their limitations. Hybrid approaches have the advantage of not requiring highly accurate models or an enormous amount of data; they operate in a complementary manner. They are flexible and useful for uncertainty management because they retain the intuitiveness of a model while explaining the observed data.

Hybrid approaches have the limitations of requiring both models and data. An incorrect model or noisy data or both are likely to lead to the inaccurate prediction of RUL. Therefore, there is a high probability of higher variance in error if not handled properly (Kabir et al., 2012).

When handled correctly, a hybrid approach can combine data-driven and physics-based approaches to get the best from each. A physics-based model can compensate for the lack of data,

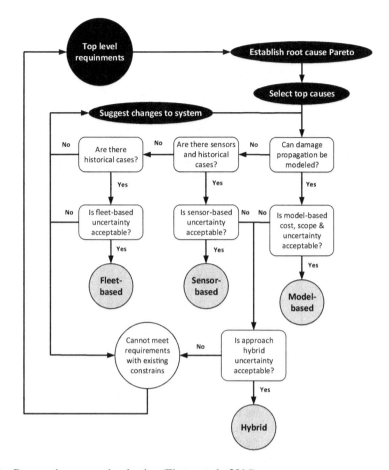

FIGURE 6.2   Prognostics approach selection (Elattar et al., 2016).

and a data-driven approach can compensate for the lack of knowledge about the system's physics. This fusion can be performed before RUL estimation; this is called pre-estimation. Fusion can also be performed after RUL estimation by fusing the results from each approach to obtain the final RUL. This is called post-estimation (Elattar et al., 2016). Goebel et al. (2006) used a fusion approach for aircraft engine bearings; the results showed this method has a more accurate and robust outcome than using either data-driven or physics-based approaches alone.

Although a hybrid approach is used to eliminate the drawbacks of physics-based and data-driven methods and gain their benefits, it still has the disadvantages of both to a certain extent.

Goebel (2007) created a flowchart (see Figure 6.2) as a guide to selecting a prognostics approach.

Prognostics models generally belong to one of three categories: experience-based models, data-driven models, and physics-based models (see Section 6.1.1). The approaches using different combinations of the three categories (i.e., hybrid models) are displayed in Figure 6.3. The figure is taken from work by Liao and Köttig (2014) and shows five combinations:

1. H1: Experience-based model + data-driven model.
2. H2: Experience-based model + physics-based model.
3. H3: Data-driven model + data-driven model (multiple data-driven models).
4. H4: Data-driven model + physics-based model.
5. H5: Experience-based model + data-driven model + physics-based model.

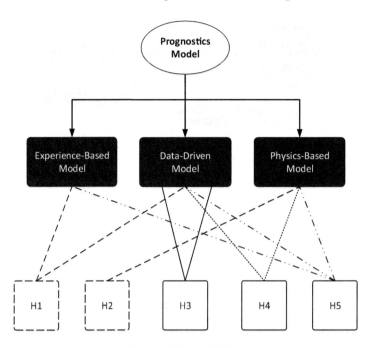

FIGURE 6.3 Hybrid prognostics models (Liao & Köttig, 2014).

6.1.2.1 Prognostics Application

Automated prognostics has been applied to several different types of systems, including actuators, aerospace structures, aircraft engines, batteries, bearings, clutch systems, cracks in rotating machinery, electronics, gas turbines, hydraulic pumps and motors, military aircraft turbofan oil systems, semiconductor manufacturing, heating, ventilation, and air conditioning, wheeled mobile robots, and unmanned aerial vehicle propulsion.

An example of a prognostics application in the aircraft industry is the Joint Strike Fighter (JSF) aircraft, currently under development (JSF, 2007). It will be used by the US Air Force, Navy, and Marines, and by certain US allies. The current plan is to have a PHM system that provides fault detection and isolation for every major system and subsystem on the aircraft, and prognostics for selected components. PHM is a key element in the justification for the choice of a single engine aircraft and is intended to improve safety and reduce maintenance costs. The proposed architecture includes an off-board PHM system, which will use data mining techniques (Schwabacher & Goebel, 2016).

In the last few years, a great deal of attention has been given to prognostics because of its potential to improve complex engineering systems' health management. For example, prognostics can make a contribution to the field of medicine; the future course and outcome of disease processes can be predicted after treatment. Another area of application is everyday weather forecasting. In fact, medicine and weather forecasting are mature prognostics applications that have already proved their value.

Prognostics applications can be online and work in real time or near real time whether onboard or off-board. Prognostics also can be applied off-line regardless of the operation time of the monitored system. Real-time prognostics takes online data from the data acquisition system to perform RUL estimation and gives a warning of impending failure to allow system reconfiguration and mission replanning. The off-line prognostics system uses fleet-wide system data and performs deep data mining processes that could not be performed onboard in real time because of a lack of resources and

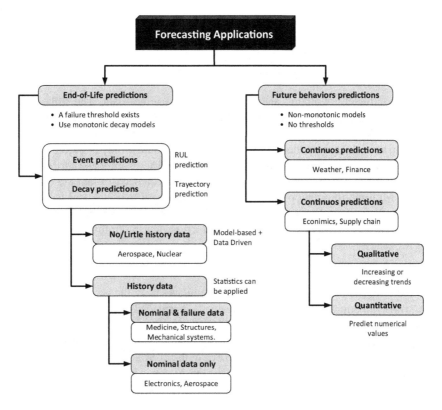

FIGURE 6.4 Forecast applications (Elattar et al., 2016).

time criticality. The results from an offline prognostics system can be used in maintenance planning and decision-making for logistics support management. Prognostics was originally a forecasting application (see Figure 6.4) (Elattar et al., 2016).

6.2 SYNTHETIC DATA GENERATION VS. MODEL TUNING

6.2.1 SYNTHETIC DATA GENERATION

Synthetic data refer to "any production data applicable to a given situation that are not obtained by direct measurement" (McGraw-Hill Dictionary, 2009) where production data, as defined by Mullins (2009), refer to "information that is persistently stored and used by professionals to conduct business processes".

Synthetic data generation (SDG) has become a focus in recent years because of the ability of synthetic data to support the validation of new algorithms and applications needing data which are not available or not accessible due to privacy concerns (Surendra & Mohan, 2017).

Synthetic data are randomly generated with constraints to hide sensitive private information and retain certain statistical information or relationships between attributes in the original data. The creation of synthetic data involves data anonymization; that is, synthetic data are a subset of anonymized data. Synthetic data are used in many different fields as a filter for information that would otherwise compromise the confidentiality of particular aspects of the data, including an individual's name, home address, Internet Protocol (IP) address, credit card number, telephone number, and social security number (Wikipedia, 2020).

6.2.1.1 Types of Synthetic Data

Synthetic data are broadly classified into three categories: fully synthetic, partially synthetic, and hybrid synthetic (Surendra & Mohan, 2017).

6.2.1.1.1 Fully Synthetic Data

These data are completely synthetic and do not contain original data. The fully synthetic data generators identify the density function of attributes in the original data and estimate the parameters of these density functions. Then for each attribute, privacy protected series are generated by randomly picking up the values from the estimated density functions. If only a few attributes of the original data are to be replaced with synthetic data, the protected series of these attributes are mapped with the other attributes of the original data to rank the protected series and the original series in the same order. Multiple imputation and bootstrap methods are two classical techniques used to generate fully synthetic data. Since the released data are completely artificially generated and do not contain original data, there is strong privacy protection, but the truthfulness of the data is lost (Surendra & Mohan, 2017).

6.2.1.1.2 Partially Synthetic Data

The method used to generate partially synthetic data replaces only the values of the selected sensitive attribute with synthetic values. The original values are replaced only if there is a high risk of disclosure. Masking the original values with synthetic values prevents re-identification, thus preserving privacy in the published data. Multiple imputation and model based techniques have been used to find the synthetic values for a selected attribute to avoid disclosure. These techniques are also useful for imputing missing values in the original data. Disclosure risk is higher for partially synthetic data than fully synthetic data, as they contain original data along with imputed synthetic data (Surendra & Mohan, 2017).

6.2.1.1.3 Hybrid Synthetic Data

Hybrid synthetic data are generated using both original and synthetic data. For each record of original data, a nearest record in the synthetic data is chosen, and both are combined to form hybrid data. The hybrid synthetic data have the advantages of both fully and partially synthetic data. Hence, they provide good privacy preservation with high utility compared to fully synthetic and partially synthetic data but require more memory and processing time (Surendra & Mohan, 2017).

6.2.1.2 Approaches to and Methods for Synthetic Data Generation

A variety of SDG methods is used across a wide range of domains. Generally speaking, five categories of synthetic generation can be identified. In the first category, data masking models replace personally identifiable data fields with synthetic data. In the second category, synthetic target data are embedded into recorded user data in a method known as signal and noise. In the third category, network generation approaches deliver relational or structured data. The fourth category contains truly random data generation approaches like the music box model. The fifth category includes probability weighted random generation models like the Monte Carlo, Markov chain, and Walker's alias methods.

Table 6.1 summarizes the types of synthetic data and gives an example of an application (McLachlan, 2017).

6.2.1.2.1 Data Masking

Data masking is also referred to as data scrambling, data binding, data anonymization, data sanitization, or data encoding. Data masking obfuscates or replaces sensitive data. Typical established methods of data masking include (McLachlan, 2017):

TABLE 6.1
Classification of Synthetic Data

True Synthetic Data	Data generated with no confidential or sensitive data directly used. Exemplar generation may rely on algorithms using models or frameworks to populate a dataset with generic seed data based on statistical probability or acute randomness. An example of true synthetic data can be seen in CoMSER.
Fully Synthetic Data	No real-world data are contained in the output data. Some synthetic approaches still use real or aggregate data in the input phase, but none of the real data is maintained across the generation method. Common methods for ensuring data are synthetic involve capturing and breaking up real-world data into much smaller components, rebuilding these components into new rows of data. Other methods use the real data to construct a database architecture, populating the new database architecture with synthetic data based on the observed real data.
Partially Synthetic Data	Partially synthetic data consist of some form of simulated or synthetic data intermixed or aggregated with unaltered real data. An example is the Outbreak-Detection system which uses simulated signals injected or superimposed on real background noise.
Anonymized-Only Data	These data are for projects which operate to identify and replace or scramble sensitive fields within a dataset, leaving the rest of the dataset largely unchanged.
Real Data	In real or observed data, no attempt has been made to anonymize, conceal or synthesize the values of any sensitive or confidential fields.

Source: McLachlan (2017).

- Randomization: Real data are replaced with randomly generated data that may be governed by rules to limit their scope to fall with a given range, variance, or percentage of the original value.
- Blocking: Also known as substitution. Original data are replaced entirely or partially with an artificial record, usually from a lookup table.
- Masking: Original data are fully or partially replaced with a masking character, such as the asterisk often used to replace digits in a credit card number or password.
- Scrambling: Data type and size are preserved; however, the value is entirely replaced.
- Shuffling: Substitution data are derived entirely from the value in the column itself. The data in the column are randomly moved or shifted between rows.

6.2.1.2.2   Signal and Noise
The signal and noise method involves the collection or creation of a large dataset of generally normal noise data, that is, data that would be seen in the target system when the issue, or signal, is not evident. This might consist of several days' worth of normal traffic seen on an interface that connects to a web server, application service or console, network or system firewall, or collected from a database engine such as that containing or processing messages such as emails (McHugh, 2000). This normal traffic may be used as it was captured or may constitute the seed data pool for some form of randomized or constrained generation method.

The signal is the element with which the method is mostly concerned. It is sought out in the eventual synthetic dataset, often being used to train other applications, systems, or analysts. The signal may be artificially or manually created by researchers or drawn from a set of data such as breach data captured in a manner similar to noise data (McLachlan, 2017).

6.2.1.2.3 Network Generation

The Merriam-Webster dictionary defines a network as a group or system of intersecting or connected people or things. Network generation is concerned with the creation of a dataset that describes a network of objects, whether a social network of similar people (e.g., a terrorist organization) or network types (e.g., generated in medicine for gene expression). The network generation dataset generally describes the nodes of the network, as well as the paths that interconnect or intersect each node (McLachlan, 2017).

6.2.1.2.4 Music Box Method

Using what they called the music box method, researchers de-identified real source health record data, broke the complete source data down into their respective components, and used these components to generate new random event records based on the pre-calculated collecting together of random segments matching an event type (Mwogi et al., 2014).

6.2.1.2.5 Markov Chain Method

The Markov chain method is a process by which each component of a record is generated by a random process constrained or dependent only on the value of the current state and the conditional state probability of the next step (Markov, 1971). Markov chain models use probabilities, in this case, the probability that the system will transition from the current state to any one of a number of random next states. In the research described above, Mowgi et al. (2014) compared the performance of a Markov chain method to the music box method. They analyzed existing health records, building a dataset of record segments and using the probabilities that a given segment would be followed by another given segment. Each had a random segment starting point, and probabilities were used to randomly select each segment selected until the new synthetic record was complete (Mwogi et al., 2014).

6.2.1.2.6 Monte Carlo Method

A Monte Carlo method is any method that solves the problem of generating suitable random numbers and observing a fraction of the numbers obeying some property or properties. The Monte Carlo method for SDG is probabilistic, in that it generates data with similar probability properties as those observed in real data (McLachlan, 2017).

6.2.1.2.7 Walker's Alias Method

Walker's alias method is an efficient two-step pseudo-random approach that uses defined frequency distributions with a finite number of outcomes to generate synthetic data (Walker, 1974, 1977; Davis, 1993). Over thousands or tens-of-thousands of records, the alias method has been found to produce synthetic data with a high degree of accuracy when compared to statistical distributions taken from observed data (McLachlan, 2017).

6.2.1.2.8 Distribution of Methods and Domains in Synthetic Data Generation (SDG)

Figure 6.5 shows the distribution of common SDG model types in the research (McLachlan, 2017).

Figure 6.6 shows the distribution of domains using SDG in the same collection of papers. Computer sciences were observed most frequently, followed by the energy and environmental sciences (McLachlan, 2017).

6.2.2 MODEL TUNING

6.2.2.1 Definition of Model Tuning

Tuning refers to trial-and-error process by which we change some hyper-parameters (e.g., the value of alpha in a linear algorithm for the number of trees in a tree-based algorithm), rerun the algorithm

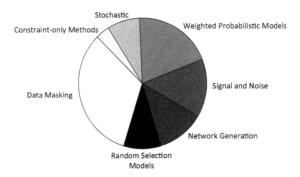

FIGURE 6.5   Distribution of synthetic data generation methods (McLachlan, 2017).

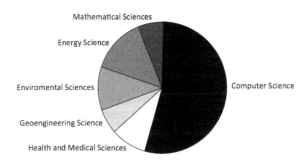

FIGURE 6.6   Distribution of synthetic data generation domains (McLachlan, 2017).

on the data, and then compare its performance on a validation set to determine which set of hyper-parameters yields the most accurate model.

All machine learning (ML) algorithms have a default set of hyper-parameters. They are external to the model, and their value cannot be estimated from the data. Different algorithms have different hyper-parameters: decision trees have a set number of branches, regularized regression models have coefficient penalties, and neural networks have a set number of layers. When they are building models, researchers choose the default configuration of the hyper-parameters after running the model on several datasets.

The generic set of hyper-parameters for each algorithm offers a starting point for analysis and usually results in a well-performing model. However, the generic set may not have the optimal configurations for a particular dataset or a specific problem. This is when tuning is required (Datarobot, 2020).

Many models have important parameters which cannot be directly estimated from the data. For example, in the k-nearest neighbor (k-NN) classification model, a new sample is predicted based on the k-closest data points in the training set. Figure 6.7 illustrates a five-nearest neighbor model. Here, two new samples (denoted by the solid dot and filled triangle) are being predicted. One sample (●) is near a mixture of the two classes; three of the five neighbors indicate the sample should be predicted as the first class. The other sample (▲) has all five points, indicating the second class should be predicted. The question remains as to how many neighbors should be used. A choice of too few neighbors may overfit the individual points of the training set, while too many neighbors

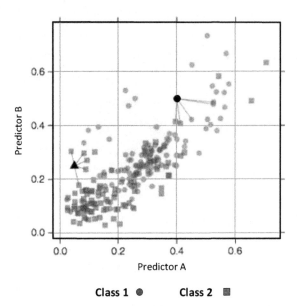

FIGURE 6.7 *K*-nearest neighbor classification model. Two new points, symbolized by filled triangles and solid dots, are predicted using the training set (Kuhn & Johnson, 2013).

may not be sensitive enough to yield reasonable performance. This type of model parameter is referred to as a tuning parameter because there is no analytical formula available to calculate an appropriate value (Kuhn & Johnson, 2013).

Several models discussed above have at least one tuning parameter. Since many of these parameters control the complexity of the model, poor choices for the values can result in overfitting. Figure 6.8 illustrates this point. A support vector machine (SVM) is used to generate the class boundaries in each panel. One of the tuning parameters for this model sets the price for misclassified samples in the training set and is generally referred to as the "cost" parameter. When the cost is large, the model will go to great lengths to correctly label every point (as in the left panel), while smaller values produce less aggressive models. The class boundary in the left panel is created by manually setting the cost parameter to a very high number. In the right panel, the cost value is determined using cross-validation (Kuhn & Johnson, 2013).

There are various approaches to searching for the best parameters. A general approach that can be applied to almost any model is to define a set of candidate values, generate reliable estimates of model utility across the candidate values, and then choose the optimal settings. A flowchart of this process is shown in Figure 6.9 (Kuhn & Johnson, 2013).

As Figure 6.9 shows, once a candidate set of parameter values has been selected, we must obtain trustworthy estimates of model performance. Next, the performance on the hold-out samples is aggregated into a performance profile used to determine the final tuning parameters. We then build a final model with all of the training data using the selected tuning parameters. Using the *k*-NN example to illustrate the procedure of Figure 6.5, the candidate set might include all odd values of *k* between 1 and 9 (odd values are used in the two-class situation to avoid ties). The training data will be resampled and evaluated many times for each tuning parameter value. These results will then be aggregated to find the optimal value of *k* (Kuhn & Johnson, 2013).

The procedure defined in Figure 6.9 uses a set of candidate models defined by the tuning parameters. Other approaches such as genetic algorithms or simplex search methods can also find optimal tuning parameters. These procedures algorithmically determine appropriate values for tuning parameters and iterate until they arrive at parameter settings with optimal performance. They

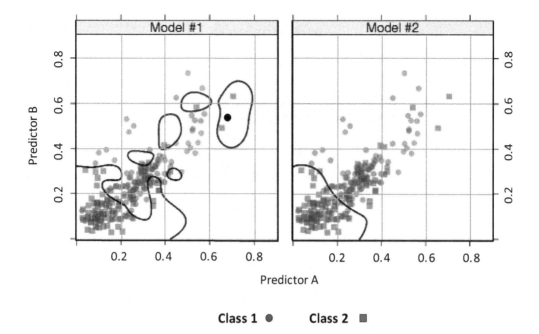

FIGURE 6.8 Example of a training set with two classes and two predictors. The panels show two different classification models and their associated class boundaries (Kuhn & Johnson, 2013).

tend to evaluate a large number of candidate models and can be superior to a defined set of tuning parameters when model performance can be efficiently calculated.

A more difficult problem is obtaining trustworthy estimates of model performance for these candidate models. The apparent error rate can produce extremely optimistic performance estimates. A better approach is to test the model on samples that were not used for training. Evaluating the model on a test set is the obvious choice, but to get reasonably precise performance values, the size of the test set may need to be large.

An alternate approach to evaluating a model on a single test set is to resample the training set. This process uses several modified versions of the training set to build multiple models and then employs statistical methods to provide honest estimates of model performance (i.e., not overly optimistic) (Kuhn & Johnson, 2013).

6.2.2.2   Why Model Tuning Is Important

Model tuning allows models to be customized so they generate the most accurate outcomes and yield valuable insights into data, thereby boosting the decision-making process (Datarobot, 2020).

6.3   HYBRID APPROACH INCORPORATING EXPERIENCE-BASED MODELS AND DATA-DRIVEN MODELS

Experience-based models incorporate domain knowledge into the reasoning, but they cannot directly deal with continuous variables, and their output is often a discrete event. This limits their ability to derive RUL.

By contrast, data-driven models easily handle continuous data and learn correlations and the underlying structure from the data.

Fuzzy logic models can deal with continuous variables by converting them into fuzzy representations using appropriately designed membership functions; then, the output can be converted

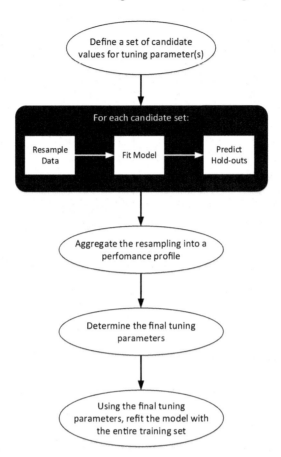

FIGURE 6.9   Schematic of parameter tuning process (Kuhn & Johnson, 2013).

into numerical results using another set of membership functions. However, they still need a human expert to specify the underlying system behavior to develop a complete fuzzy representation.

In a study of an industrial gearbox, Garga et al. (2001) proposed an automated reasoning method integrating explicit domain knowledge in the form of rules and a feedforward neural network trained with sensor data to assess gearbox condition. By integrating the rule-based domain knowledge with test and operational data from the gearbox to train the feedforward neural network, they combined redundant rules and identified inconsistencies. They trained two neural networks: one was trained to learn the explicit rule set from domain experts; another was trained to reason the health level of the gearbox, incorporating the operational data and measurements converted into fuzzy variables. Although Garga et al. did not give an exact RUL prediction method or quantify predictive performance, they argued that the health level estimated by the hybrid model can be used to extrapolate the RUL.

In their study, Satish and Sarma (2005) employed both neural networks and fuzzy logic to estimate state and predict RUL. They combined the neural network with fuzzy logic to create a fuzzy backpropagation (BP) network and used expert knowledge to select the membership function (LR-type, with L for left and R for right). They created two fuzzy BP networks: the first assessed the system health state; the second predicted the RUL based on the inputs plus the output (health indicator) of the first.

In work on drill bits, Chinnam and Baruah (2004) proposed a neuro-fuzzy inference model relying on experimental and domain expert knowledge and applied it to cutting tool monitoring and RUL estimation. More specifically, they used trained focused time-lagged feedforward neural networks (focused TLFNs) to forecast the thrust-force and the torque of drill bits. The focused TLFNs became the input for a fuzzy inference system (FIS), specifically a Sugeno FIS, and expert knowledge was incorporated into the degradation signal space to define failure models using fuzzy inference. Finally, the RUL was predicted based on the reliability estimation associated with the forecasted states from the focused TLFNs.

In their work on failure, Zhang et al. (2002) proposed an architecture for an integrated fault diagnostics and prognostics system and demonstrated it on a process simulator. They first developed a fuzzy logic module using a combination of direct user experience, simulated models, and experimental data to estimate the system's fault state and then proposed the wavelet neural network (WNN) method as an alternative. They subsequently built two modules to predict the RUL of a failing component: the first used a dynamic WNN to project the current state of the faulty component into the future to calculate the RUL; the second used a confidence prediction neural network to estimate the prediction uncertainties and confidence distributions.

Briefly stated, hybrid approaches using both experience-based and data-driven models integrate domain knowledge into data-driven models for prognostics. Domain knowledge (i.e., experience-based models) is mainly used to determine a system's faulty states, while data-driven models are used to refine the rules generated by expert knowledge and to perform the actual RUL prediction.

On the one hand, introducing domain knowledge into data-driven models sheds light on the dynamics of degradation under various operating conditions and failure modes, thus reducing the uncertainty of the predictions generated by purely data-driven models. In addition, it adapts machine-specific hypotheses and learns the system's nominal and faulty levels in the absence of failure data. On the other hand, the use of data-driven models results in a more parsimonious representation of the domain knowledge, with subsequent practical benefits for decision-making.

Despite the introduction of expert knowledge, we cannot guarantee that all main failure modes can be captured. A novelty detection mechanism may be needed in the data-driven prediction model to discover any unforeseen failure mode. Expert knowledge cannot capture intermediate states (i.e., neither new nor totally worn out) and thus may jeopardize the ability of the reasoning system to interpolate (Chinnam & Baruah, 2004). This issue may not be all that significant from a practical point of view, however, because the goal is to estimate the end of life (i.e., RUL) not the intermediate life residual time (Liao & Köttig, 2014).

6.3.1 Hybrid Approach

Simply stated, a hybrid approach combines physics-based and data-driven approaches to take advantage of both. The main idea is to achieve finely tuned prediction models with better ability to manage uncertainty and resulting in more accurate RUL estimates. Two categories of hybrid approaches are the series approach and the parallel approach (Javed, 2014).

A series approach, shown in Figure 6.10, combines a physics-based model which captures the failure mode or process being modeled through an understanding of the system and a data-driven approach that helps estimate the process parameters that are uncertain, using failure data from the field. The data model can be a simple parameter optimization method using classical optimization techniques when historical data are available. In cases when the degradation may not be observed directly, online parameter estimation techniques like Kalman filter, particle filter, and their variants can be used. These methods update the tunable parameters when new data are collected. The fundamental idea is that the predicted feature is not necessarily a direct outcome of the tuned parameters but could be a down-stream parameter. For example, in a system, the cooling effectiveness for metal temperature calculation could be tuned using the crack lengths observed from a borescope inspection (Pillai et al., 2016).

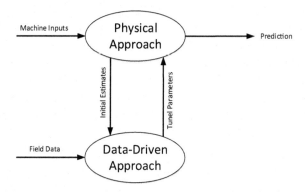

FIGURE 6.10 Series approach in hybrid modeling (Pillai et al., 2016).

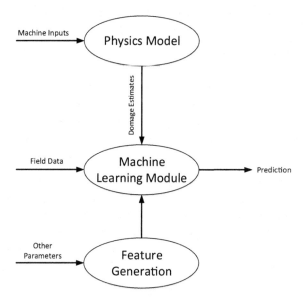

FIGURE 6.11 Parallel approach in hybrid modeling (Pillai et al., 2016).

In a parallel approach, the output from the physics model is combined with the data from other sources using data-based methods, as schematically shown in Figure 6.11. In this approach, the ML module can be trained to predict the errors in prediction that are not explained by the physics-based model, using other relevant features that cannot be modeled using the physics-based model, but impact the failure mode. These other parameters and associated features are estimates that account for un-modeled effects in the physics-based model (Javed, 2014). Such predictions could potentially capture the failure better than either of the models independently (Pillai et al., 2016).

6.3.2 Hybrid Approaches

Health condition monitoring algorithms are normally classified into physics-based models and data-driven models, and their respective benefits can be combined to obtain more robust algorithms called hybrid models. In one example, Orsagh et al. (2003) proposed a generic framework (Figure 6.12) using data-driven models for diagnostics and a physics-based model for prognostics of bearing spalling by combining sensor data for diagnostics with physics-based and historical data for

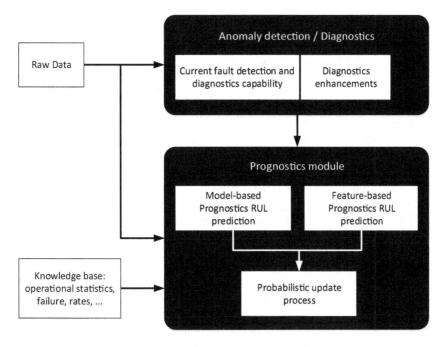

FIGURE 6.12 Hybrid model (Orsagh et al., 2003).

TABLE 6.2
Parameter Estimators

Linear estimators	Nonlinear estimators
Least square,	Extended Kalman filter,
Linear Kalman filter	Particle filter

Source: Cubillo et al. (2016).

prognostics. A similar approach was proposed by Zio and Di Maio (2012). They applied Paris' law for crack growth estimation (prognostics) and a relevance vector machine method for fault detection (diagnostics). In a third example, Pantelelis et al. (2000) detected faults in a naval turbocharger combining finite element models of the system and an artificial neural network (ANN) for fault identification using vibration analysis (Cubillo et al., 2016).

For prognostics, regardless of the approach (physics-based, data-driven, or hybrid), the health condition monitoring algorithm will provide an assessment of the current health of the system, normally by measuring a health indicator, and a prediction of the future health of the system. The prediction should be as accurate as possible, but there are always errors between the predicted value of a health indicator and the measured one. Thus, it is important to have the capability to adjust the model to minimize these differences based on the past errors between the predicted health indicator and the measured ones. Physics-based models can be combined with parameter estimators to minimize these errors in real time. A unified formulation proposed by Jaw and Wang (2006) can implement parameter estimators into physics-based models given a generic set of system equations.

The most common parameter estimators are shown in Table 6.2. Linear estimators are less computationally expensive, but if the system behavior is clearly more sophisticated, nonlinear algorithms are needed (Cubillo et al., 2016).

6.4 HYBRID APPROACH INCORPORATING EXPERIENCE-BASED MODELS AND PHYSICS-BASED MODELS

When experience-based and physics-based models are integrated, the output of the former is often used as an auxiliary to enhance the performance of the latter. The experience-based model can also estimate the system health state, thus allowing the RUL to be predicted (Liao & Köttig, 2014).

Swanson (2001) suggested a prognostics methodology using a Kalman filter and fuzzy logic, and applied it to the prognosis of crack growth in tension steel bands. With vibration mode frequencies used as responsive fault features, the Kalman filter performed feature tracking and forecasting; the RUL was estimated using Newton's method. While the Kalman filter facilitated feature trending and prediction, the failure threshold depended on the system's operational conditions. Swanson used fuzzy logic to adapt failure thresholds to the operational conditions.

Byington, Watson, and Edwards (2004) proposed a methodology to predict aircraft actuator components' RUL. They implemented a fuzzy logic process to quantify the level of damage (damage index) of the system using a predefined set of rules based on knowledge of the system and engineering expert judgment. They then applied a physics-based method (i.e., Kalman filtering) to predict the progression of the damage.

In their general framework and interface to integrate information from different types of models, the hybrid approaches integrating experience-based and physics-based models are similar to the approaches integrating experience-based and data-driven models. Expert knowledge is used to determine the system's faulty states and categorize damage levels; instead of using data-driven models, however, physics-based models are used for the actual RUL prediction. Hybrid approaches using experience-based and physics-based models share similar advantages and disadvantages with those combining experience-based and data-driven models (see Section 6.3) (Liao & Köttig, 2014).

6.4.1 MACHINE LEARNING VS. PHYSICS-BASED MODELING

A well-made physics-based model sheds light on complex processes and predicts future events. Such models have already been applied to vastly different processes, such as predicting the orbits of massive space rockets or the behavior of the nano-sized objects at the heart of modern electronics. The ability to make predictions is also one of the important applications of ML (Flovik, 2018).

A common key question is how to choose between a physics-based model and a data-driven machine learning model. The answer depends on what problem is being solved. In this setting, there are two main classes of problems (Flovik, 2018):

1. There is no theoretical knowledge about the system, but there are a lot of experimental data on how it behaves. If for instance, there is no direct knowledge about the behavior of a system, a mathematical model cannot be formulated to describe it and make accurate predictions.

   If there are a lot of example outcomes, an ML-based model could be used. Given enough example outcomes (the training data), an ML model should be able to learn any underlying pattern between the information there is about the system (the input variables) and the outcome that will be predicted (the output variables). An example is predicting the housing prices in a city. If there are enough examples of the selling prices of similar houses in the same area, it is possible to make a fair prediction of the price for a house put up for sale.

2. There is a good understanding of the system, and it can be described mathematically. If a problem can be well described using a physics-based model, this approach will often be a good solution. This does not mean ML is useless for any problem that can be described using

physics-based modeling. In fact, combining physics with ML in a hybrid modeling scheme is a valid prospect.

6.4.1.1   Hybrid Analytics: Combining Machine Learning and Physics-Based Modeling

Even if a system, at least in principle, can be described using a physics-based model, this does not mean that a ML approach will not work. The ability of ML models to learn from experience means they can also learn physics. Given enough examples of how a physical system behaves, the ML model can learn this behavior and make accurate predictions.

ML models – or algorithms – learn from experience, thus, in principle at least, resembling the way humans learn. A class of ML models called ANNs are computing systems inspired by how the brain processes information and learns from experience. This ability to learn from experience has inspired many researchers to try teaching physics to ML models rather than using mathematical equations (Flovik, 2018).

6.4.1.2   Why Use Machine Learning When Physics-Based Models Are Available?

An important question is why we want to implement ML-based approach physics-based models to describe the system in question. One of the key concerns is the computational cost of the model. We might be able to describe the system in detail using a physics-based model, but solving this model could be complicated and time-consuming. Thus, a physics-based approach might break down if we aim for a model that can make real-time predictions on live data.

In this case, a simpler ML-based model could be an option. The computational complexity of an ML model is mainly seen in the training phase. Once the model has finished training, making predictions on new data is straightforward. This is where the hybrid approach of combining ML and physics-based modeling becomes highly interesting (Flovik, 2018).

6.4.2   PHYSICS OF FAILURE-BASED PHM IMPLEMENTATION APPROACH

The general PHM methodology is shown in Figure 6.13. The first step involves a virtual (reliability) life assessment, where design data, expected lifecycle conditions, failure modes, mechanisms, and effects analysis (FMMEA), and physics of failure (PoF) models are the inputs. Based on virtual life assessments, the critical failure modes and failure mechanisms are prioritized. The existing sensor data, maintenance and inspection records, built-in-test results, and warranty data are used to identify abnormal conditions and parameters. Based on this information, the monitoring parameters and sensor locations for PHM can be determined (Pecht & Gu, 2009).

Based on the collected operational and environmental data, the health status of products can be assessed. Damage can also be calculated from the PoF models to determine the remaining life. Then, PHM information can be used for maintenance forecasting and decisions that minimize lifecycle costs and maximize availability or some other utility function (Pecht & Gu, 2009).

6.4.2.1   Failure Modes, Mechanisms, and Effects Analysis

The FMMEA used in the PoF-based PHM approach is shown in Figure 6.14. Design capture is the process of collecting structural (dimensional) and material information to generate a model. This step involves characterizing the product at all levels, that is, parts, systems, and physical interfaces. In most cases, this information is developed in the design process (Pecht & Gu, 2009).

The reliability assessment step involves identifying appropriate PoF models for the identified failure mechanisms. A load-stress analysis is conducted using material properties, product geometry, and lifecycle loads. With the computed stresses and the failure models, a damage analysis is conducted, and the accumulated damage is estimated using a damage model.

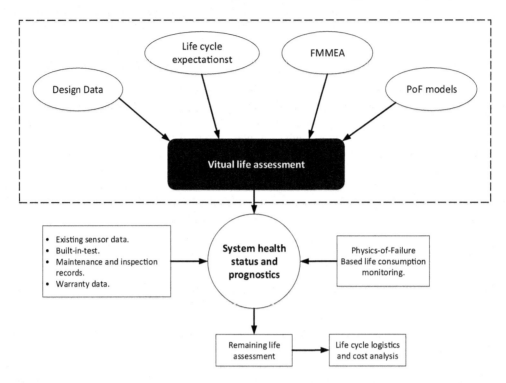

FIGURE 6.13   PoF-based PHM methodology (Pecht & Gu, 2009).

A failure mode is the effect by which a failure is observed to occur. All possible failure modes for each identified element should be listed. Potential failure modes may be identified using numerical stress analysis, accelerated tests to failure (e.g., highly accelerated life testing (HALT)), past experience, and engineering judgment. A failure cause is defined as the specific process, design, and/or environmental conditions that initiate a failure, whose removal will eliminate the failure. Knowledge of potential failure causes can help identify the failure mechanisms driving the failure modes for a given element.

Failure mechanisms are the physical, chemical, thermodynamic, or other processes that result in failure. Failure mechanisms are categorized as either overstress or wear out mechanisms. Overstress failure arises as a result of a single load (stress) condition, which exceeds a fundamental strength property. Wear-out failure arises as a result of cumulative damage related to loads (stresses) applied over an extended time. Within current technology, PHM can only be applied in the wear-out failure mechanisms.

Failure models help quantify the failure through the evaluation of TTF or likelihood of a failure for a given set of geometries, material construction, environmental, and operational conditions. For wear-out mechanisms, failure models use both stress and damage analysis to quantify the damage accumulated (Pecht & Gu, 2009).

6.4.2.2   Lifecycle Load Monitoring

In the lifecycle of a product, several failure mechanisms may be activated by different environmental and operational parameters acting at various stress levels, but in general, only a few operational and environmental parameters and failure mechanisms are responsible for the majority of the failures. High-priority mechanisms are those with high combinations of occurrence and severity. Prioritization of the failure mechanisms provides an opportunity for the effective use of resources.

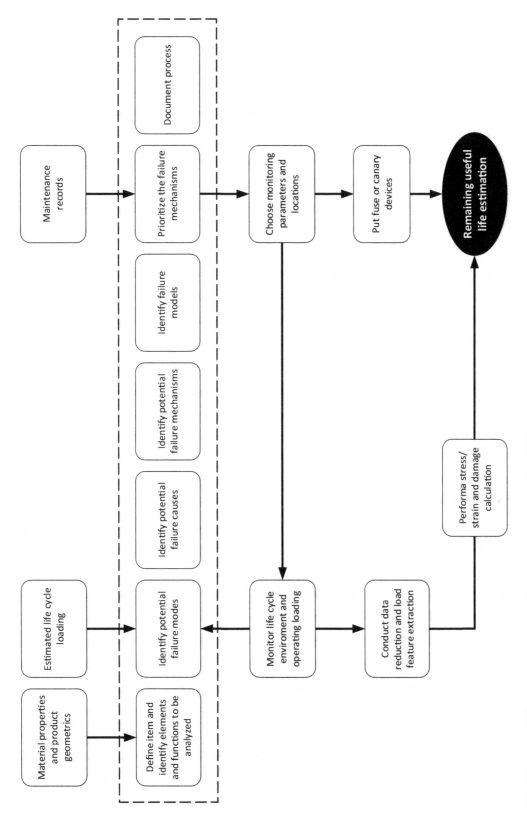

FIGURE 6.14 FMMEA analysis used in PHM approach (Pecht & Gu, 2009).

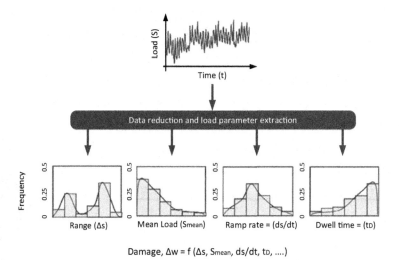

FIGURE 6.15   Load feature extraction (Pecht & Gu, 2009).

If loads can be measured *in situ*, the load profiles can be used in conjunction with damage models to assess degradation deriving from exposure to cumulative loads (Pecht & Gu, 2009).

6.4.2.3   Data Reduction and Load Feature Extraction

Experience shows even the simplest data collection systems can accumulate vast amounts of data quickly, requiring either a frequent download procedure or a large capacity storage device. Therefore, data reduction is necessary. Vichare and Pecht (2006) and Vichare et al. (2007) described the accuracy associated with several different data reduction methods: ordered overall range, peak counting, level crossing counting, rainflow cycle counting, range-pair counting, fatigue meter counting, and range counting. The efficiency measures of data reduction methods included: gains in computing speed and testing time; the ability to condense load histories without sacrificing important damage characteristics; and estimation of the error introduced by omitting data points.

As shown in Figure 6.15, a time-load signal can be monitored *in situ* using sensors and further processed to extract cyclic range (Δs), cyclic mean load (S_{mean}), ramp rate or rate of change of load (ds/dt), and dwell time (t_D) using embedded load extraction algorithms. The extracted load parameters can be stored in appropriately binned histograms to achieve further data reduction. After the binned data are downloaded, they can be used to estimate the distributions of the load parameters. This type of output can be inputted into fatigue damage accumulation models. Embedding the data reduction and load parameter extraction algorithms into the sensor modules as suggested by Vichare and Pecht (2006) can result in reduced on-board storage space, less power consumption, and longer periods of uninterrupted data collection (Pecht & Gu, 2009).

6.4.2.4   Damage Assessment and Remaining Life Calculation

PoF-based failure models use stress and damage analysis methods to evaluate the susceptibility to failure based on the TTF or likelihood of a failure for a given geometry, material construction, and environmental and operational conditions. The loading feature (e.g., cyclic range, cyclic mean, ramp rate, and dwell time) from raw data (e.g., temperature and vibration) after feature extraction can be the input of the failure model to calculate the damage. The damage is accumulated over a period until the item is no longer able to withstand the applied load. Remaining life prediction is the process of estimating the remaining life (e.g., the time in days and distance in miles) through which the product can function reliably, based on the damage accumulation information. Damage caused by temperature can be calculated in the time domain (e.g., using the Coffin–Manson model), while damage caused by vibration can be calculated in both the time and frequency domains (Pecht & Gu, 2009).

6.4.2.5   Uncertainty Implementation and Assessment

PoF models can be used to calculate the RUL, but it is necessary to identify the uncertainties in the prognostics approach and assess the impact of these uncertainties on the remaining life distribution to make risk-informed decisions. With uncertainty analysis, a prediction can be expressed as a failure probability.

Gu, Barker, and Pecht (2007) implemented uncertainty analysis of prognostics for electronics under vibration loading. These researchers identified the uncertainty sources and categorized them into four different types of uncertainty: parameter, measurement, failure criteria, and future usage uncertainty. They used sensitivity analysis to identify the dominant input variables influencing the model's output. With information on the input parameter variable distributions, they were able to perform a Monte Carlo simulation to determine the distribution of accumulated damage. Then, from the accumulated damage distributions, they were able to predict the remaining life with confidence intervals. They also presented a case study of an electronic board under vibration loading and a step-by-step demonstration of the uncertainty analysis implementation. The results showed the experimentally measured failure time was within the bounds of the uncertainty analysis prediction (Pecht & Gu, 2009).

6.4.3   Physics-Based Modeling Approaches to Engine Health Management

A physics-based model is a technically comprehensive modeling approach traditionally used to understand component failure mode progression. Physics-based models provide a means to calculate the damage to critical engine components as a function of operating conditions and assess the cumulative effects in terms of component life usage. By integrating physical and stochastic modeling techniques, the model can be used to evaluate the distribution of remaining useful component life as a function of uncertainties in component strength/stress properties, loading or lubrication conditions for a particular fault. Statistical representations of historical operational profiles serve as the basis for calculating future damage accumulation. The results from such a model can then be used for real-time failure prognostic predictions with specified confidence bounds.

A block diagram of this prognostics modeling approach is given in Figure 6.16. As illustrated at the core of this figure, the physics-based model uses the critical, life-dependent uncertainties, so current health assessment and future RUL projections can be examined with respect to a risk level (Roemer et al., 2006).

The results from such a model can be used to create a network or probabilistic-based autonomous system to obtain real-time failure prognostic predictions. Other information used as input to the prognostic model may include diagnostic results, current condition assessment data, and operational profile predictions. This knowledge-rich information can be generated from multi-sensory data fusion combined with in-field experience and maintenance information obtained from data mining processes.

Physics-based approaches to prognostics differ from feature-based approaches in that they can make RUL estimates in the absence of any measurable events, but when related diagnostic information is present, the model can often be calibrated based on this new information. A combination or fusion of the feature-based and physics-based approaches provides full prognostic ability over the entire life of the component, thus offering valuable information to plan which components to inspect during specific overhauls periods. While failure modes may be unique from component to component, this combined physics-based and feature-based methodology can remain consistent across different types of critical components.

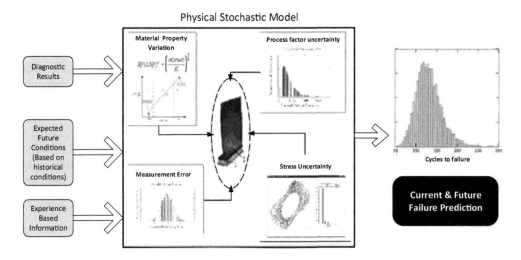

FIGURE 6.16   Physics-based modeling approach (Roemer et al., 2006).

To perform a prognosis with a physics-based model, an operational profile prediction must be developed using the steady state and transient loads, temperatures, or other on-line measurements. With this capability, probabilistic critical component models can be "run into the future" by creating statistical simulations of future operating profiles from the statistics of past operational profiles or expected future operating profiles (see Figure 6.17) (Roemer et al., 2006).

6.5   HYBRID APPROACH INCORPORATING MULTIPLE DATA-DRIVEN MODELS

In one hybrid approach, a data-driven model can be used to estimate the internal system state (e.g., crack growth rate) when it cannot be measured by sensors. The estimated system state can then be used to extrapolate the future system state to predict RUL using a second data-driven model. In another approach, different competing data-driven models can be developed for RUL prediction, and the results of different models aggregated to improve the prediction performance via a carefully designed fusion mechanism (Liao & Köttig, 2014).

Researchers have implemented a hybrid prognostics approach that combines two data-driven models, with one estimating and the other predicting the system state. For example, Liu et al. (2012b) combined least squares support vector regression (LSSVR) with hidden Markov models (HMMs) to predict the RUL of bearings. These researchers used features extracted from vibration sensor signals for offline and online training of the HMMs, with health states representing different health levels during system degradation. They used LSSVR to predict the trend of features to infer future health states by means of the HMMs and log-likelihood probabilities. They estimated RUL as the time until the HMMs' last health state (i.e., system failure) was reached.

Huang et al. (2007) predicted the RUL of bearings by combining two data-driven approaches. The researchers calculated the health index of the system state using the minimum quantization error (MQE) of established baseline self-organizing maps trained with vibration features. They then used the developed health index as an input into BP neural networks (BPNNs) for RUL prediction.

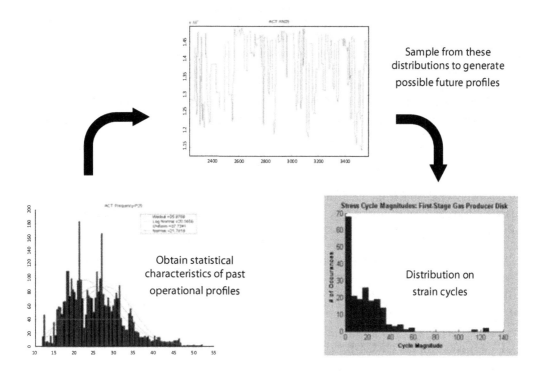

FIGURE 6.17 Operation profile and loading model for prognosis (Roemer et al., 2006).

Yan and Lee (2007) estimated the tool wear condition (i.e., state) in drilling operations via logistic regression analysis combined with a maximum likelihood technique for parameter estimation, based on features extracted from vibration signals using wavelet packet decomposition. They applied another data-driven model, an autoregressive moving average (ARMA) model, to predict the RUL based on the estimated tool wear.

Data-driven health state estimation methods and data-driven prediction methods are well-studied. They are an obvious choice to use sequentially, first for degradation detection and then for prediction. This type of hybrid approach is easily adopted, and sometimes does not have demanding data requirements. An unsupervised learning model can build a description of a system's baseline behavior using data collected during normal operation. Next, the level of degradation can be assessed by calculating a degradation index using deviations from the established baseline. Then, the extent of deviation can be used to predict future trends using a time series method. Finally, RUL can be defined when the predicted deviation exceeds a predefined threshold.

There is some difficulty involved in defining the threshold of the degradation index. For example, if the degradation is itself a deviation measure estimated using multiple sensor measurements, it is hard to determine the physical meaning of the degradation index, even with expert domain knowledge. Moreover, if the underlying physical degradation or fault propagation model is not understood, relying on the prediction model fitted using limited data may introduce unacceptable uncertainty to prediction (Liao & Köttig, 2014).

Some researchers have predicted RUL by fusing results from multiple data-driven models. In a study of bearings, Gebraeel et al. (2004) established the relations between the bearing's operating time and vibration signals using feedforward BPNNs. Each neural network was trained on data from one of 25 tested bearings. They predicted RUL by weighting the outputs of all 25 bearings using two methods: weight application to exponential parameters and weight application to failure time. Peel (2008) used a combination of multilayer perceptron and radial basis function neural networks to predict RUL. He created the neural network ensemble using a tournament style heuristic. The neural networks mapped sensor measurements in the state space in a nonlinear fashion, while a Kalman filter fused the multiple ensemble outputs over time, thus allowing RUL to be predicted. In a study of aircraft engines, Heimes (2008) applied a recurrent neural network trained by BP and incorporated an extended Kalman filter to filter the output. Heimes calculated the RUL for simulated aircraft engines by averaging the outputs of the three best models. Finally, in their work on bearing degradation, Du, Lv, and Xi (2012) proposed a selective neural network ensemble model to predict degradation time (from the appearance of the defect until failure). First, they used time and frequency domain features extracted from bearing vibration signals as input for several neural networks to map the features to operating time, employing a bagging method to create the neural networks. They then used particle swarm optimization with simulated annealing to select a neural network ensemble from the networks created by bagging. In a final step, they combined the outputs of the neural networks (i.e., the predictions of RUL) using a majority voting technique.

Building multiple data-driven models in this fashion can capture dynamics in failure modes and operating conditions and markedly reduce prediction error. Physics-based models can also be incorporated into the models to make more accurate predictions. Because the results will depend on the fusion method, however, it is essential to design the fusion mechanism very carefully. Moreover, building multiple models is computationally intensive and time consuming, thus limiting their use in certain applications (Liao & Köttig, 2014).

6.5.1 Hybrid LSSVR/HMM-Based Prognostics Approach

LSSVR can be combined with the HMM for prognostics. In this method, an LSSVR algorithm is used to predict the feature trends. The LSSVR training and prediction algorithms are modified by adding new data and deleting old data; the probabilities of the predicted features for each HMM are calculated using either forward or backward algorithms. It is possible to determine a system's future health state and estimate its RUL based on these probabilities (Liu et al., 2012a).

6.5.1.1 LSSVR/HMM-Based Prognostics

6.5.1.1.1 Hidden Markov Model

This section explains how to use the HMM-based method to describe a fault process. The fault process of a machine has a certain time-span. The fault process can be divided into several health states. Some fault states are unobservable but hidden in observable signals, such as vibration signals. The HMM-based method provides a way to detect the unobservable fault states, as HMM is a stochastic technique for classifying signals and modeling. An HMM describes a double stochastic process. One is the Markov chain, which describes state transitions. Another is the stochastic process, which describes the relations between states and observations.

Suppose the number of health states is N, which are $h_1, h_2,..., h_N$. The last state h_N represents that the system fails. This state sequence is a Markov chain. The structure of an N-state left-to-right HMM is commonly used in fault analysis as shown in Figure 6.18 (Liu et al., 2012a).

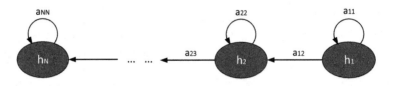

FIGURE 6.18 N-state left-to-right hidden Markov model (Liu et al., 2012a).

As shown in Figure 6.18, an HMM consists of a finite number of states, where a_{ij} $(i, j = 1,2,..., N)$ is the transition probability from state h_i to state h_j. Observation sequences from different conditions can be used to train different HMMs. For instance, observations from normal conditions can train an HMM for normal states. An HMM can be described as (Liu et al., 2012a):

$$\pi = (N, M, \pi, A, B)$$

where N is the number of states, M is the number of observations for each state, π is the initial probability vector, $A = (a_{ij})_{N \times N}$ is the probability transition matrix describing the transition relations between the states, $B = (b_{jk})_{N \times M}$ describes the observation probability distribution, and b_{jk} is the probability of the kth observation in state j. N and M are known based on history or prior knowledge. In other words, they are assumptions before HMM training. π, A, and B are learned from the data. As N and M are also included in A and B, the model can be abbreviated as (Liu et al., 2012a):

$$\lambda = (\pi, A, B). \tag{1}$$

As the observation sequence from a health state can train an HMM, N HMMs can be trained, written as $\lambda_1, \lambda_2,..., \lambda_N$, where λ_i represents the HMM trained based on the given observation sequence from the ith health state. Given an observation sequence $O = \{o_1, o_2,..., o_T\}$ where T is the length of the sequence, $P(O|\lambda_i)$ is the likelihood probability of observation sequence O occurring at the condition of model λ_i, which can be used to determine whether O belongs to the health state h_i.

HMMs can describe machine fault processes. According to the likelihood probabilities of present observations, HMMs can detect and diagnose faults. However, the HMM method cannot predict future health states and RUL, because the future observation sequence cannot be obtained (Liu et al., 2012a).

6.5.1.1.2 Framework of LSSVR/HMM Prognostics

The framework of the hybrid fault prognosis scheme is described in Figure 6.19. In this scheme, the LSSVR algorithm is used to predict the future fault features, and the HMMs are used to describe different health states (Liu et al., 2012a).

Figure 6.19 summarizes the process of the hybrid LSSVR/HMM prognostics approach in the following steps (Liu et al., 2012a):

1. Feature extraction: As most observation signals have random noise and uncertain interferences, features should be extracted from the signals before fault diagnosis and prognosis.
2. HMM training: Training can be carried out based on the fault features history. The essence of this step is to estimate the HMM parameters for each state. Based on observation sequences from N different conditions, optimal HMMs $\lambda_1, \lambda_2,..., \lambda_N$ are trained for the corresponding health state. This step is an off-line process.
3. Prediction of features based on LSSVR: Suppose the present time is t, and the feature vector after k time units is denoted $\hat{o}_{t+k}$ which can be predicted based on the data before time t using the LSSVR algorithm. Based on the feature vector $P(\hat{o}_{t+k}$, we can predict the system health state at the future time $t + k$.
4. Log-likelihood calculation: Calculating likelihood probabilities $P(\hat{o}_{t+k} | \lambda_i)$ according to forward or backward algorithms and comparing all of these probabilities allows us to find the maximum probability:

$$\text{i} = index\left(max\left(P_1, P_2,..., P_N\right)\right), \tag{2}$$

where $P_1, P_2,..., P_N$ are the likelihood probabilities for the corresponding HMMs.

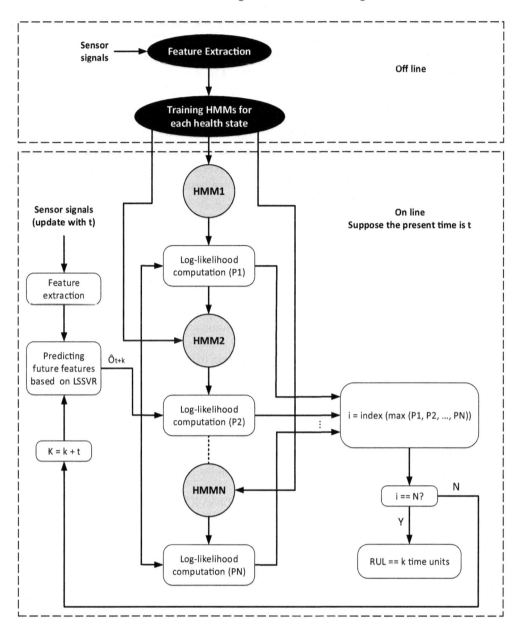

FIGURE 6.19 Framework of LSSVR/HMM fault prognostics scheme with two processes: HMM training (off line) and fault feature and RUL prediction process (on line) (Liu et al., 2012a).

The corresponding state h_i is considered the health state of the future time $t + k$. Therefore, this method can be used to predict future health states. If $k = 0$, the health state is the present state.

5. RUL prediction: The fault prognosis should run until the probability for the last state model is higher than for all other state models. Then, the probability $P(\hat{o}_{t+k} \mid \lambda_N)$ for the last state indicates the system will fail at the future time $t + k$. Therefore, the RUL is k time units.

6.6 HYBRID APPROACH INCORPORATING DATA-DRIVEN MODELS AND PHYSICS-BASED MODELS

There is no perfect prognostics model. Each has advantages and disadvantages, and it is better to think in terms of the right model for the case at hand. In work on power devices, for example, Celaya et al. (2011) proposed an approach for RUL prediction using both data-driven and physics-based prognostics algorithms. The researchers used the drain-source ON-state resistance as the health indicator. The data-driven side of the approach comprised Gaussian process regression. The physics-based approaches comprised an extended Kalman filter and a particle filter. They conducted accelerated aging tests on power devices and used prognostic performance metrics to compare the results of the methods. The particle filter method performed best in prognostics. The physics-based approach was better because the degradation model was exponential with two parameters estimated online using a Bayesian framework. The data-driven Gaussian process regression model could not make reasonable RUL predictions before a distinct (exponential) degradation behavior emerged. Alternatively, however, if the degradation did not follow an exponential model because of noisy data or different failure modes, the data-driven model would yield more accurate results because its results could be compared to the historical degradation patterns (Liao & Köttig, 2014).

Simply stated, many researchers prefer a hybrid approach to RUL analysis, electing to combine data-driven and physics-based models to leverage their respective strengths to improve prediction (Liao & Köttig, 2014).

6.6.1 PHYSICS-BASED PROGNOSTICS VS. DATA-DRIVEN PROGNOSTICS

Fault prognostics can be data-driven, physics-based, or hybrid. The first approach uses sensors' monitoring data capturing the evolution of the system's degradation. The data are pre-processed to extract features to learn models for health assessment and RUL prediction. Examples are neural networks, HMMs, regressions, and support vector regression.

The second approach requires a deep understanding of the physical system, including the evolution of the degradation. Physical laws are used to build a system model used for simulations and RUL prediction (Medjaher & Zerhouni, 2013). In work on the degradation of a car's suspension, Luo et al. (2003) used a mathematical method for physics-based prognostics. And in their study of damage evolution in a two-well magneto-mechanical oscillator, Chelidze and Cusumano (2004) took a similar approach proposing a method based on a dynamic systems approach. Note that in physics-based prognostics, the construction of the model supposes the availability of a model of the degradation, for example, of cracks related to fatigue, corrosion, or wear.

Figure 6.20 summarizes the strengths and weaknesses of each approach. As the figure suggests, the results of data-driven methods are less precise than those of physics-based methods. Another drawback of data-driven prognostics is the variability of the data used to learn the degradation models. As we mentioned in the previous paragraph, we need data representing the behavior of the degradation. In practice, however, the data on assets degrading in the same operating conditions will vary. The model learned from these data will represent the mean; consequently, RUL predictions will lack precision (Medjaher & Zerhouni, 2013).

Practitioners prefer data-driven methods for reasons of cost, applicability, and simplicity. Physics-based methods are not easily applied in industrial systems because of the difficulty involved in building a physical model of a system's degradation. Physics-based methods can be applied on systems for which the models are already known or on certain classes of systems (e.g., mechatronic systems). Even here, however, some experimentation is required to determine the behavior of the system degradation.

A possible solution is a hybrid approach. A hybrid approach to prognostics can take advantage of the strengths of both methods to improve prediction. At the same time, it can reduce their individual disadvantages (Medjaher & Zerhouni, 2013; Su & Chen, 2017).

Model-Based Prognostics		Data-Driven Prognostics	
Advantages	**Drawbacks**	**Advantages**	**Drawbacks**
• High precision. • Deterministic approach. • System-oriented approach: propagation of the failure in the whole system. • The dynamic of the states can be estimated and predicted at each time. • The failure thresholds can be defined according to the system perfomance (stability, precision). • Possibility to simulate several degradetions (drifts on the parameters).	• Need of degradation model. • High cost implementation. • Difficult to apply on complex systems.	• Simplicity implementation. • Low cost.	• Need of experimental data that represent the degradation phenomena. • Variability of test results even for a same type of component under same operating conditions. • Less precision. • Difficult to take into account the variable operating conditions. • Component-oriented approach rather than system-oriented approach. • Difficult to define the failure thresholds.

FIGURE 6.20   Data-driven prognostics vs. physics-based prognostics (Medjaher & Zerhouni, 2013).

Hybrid approaches combining physics-based and data-driven methods can be divided into four types (He et al., 2014):

- Using the data-driven method to infer the physical model: A complicated physics-based model can be replaced by a data-driven model when the degradation model is difficult to obtain or the system state cannot be measured directly.
- Using the data-driven method to estimate future measurements for the physics-based method: Predicted measurements from the data-driven method can be regarded as new measurements in the physics-based method when there is a lack of measurements for long-term prediction.
- Using the data-driven method to estimate/adjust the parameters of the physics-based method: The data-driven method describes the degradation-inherent relationships and trends based on data and uses them to estimate the parameters in the model.
- Using the filtering approach to estimate/adjust the parameters of the data-driven method: Filtering is often used to reduce noise in data and estimate the parameters of the model.

In the future, ensemble learning can be used for the fusion and integration of different data-driven prognostic methods. The fusion of online algorithms and uncertainty may be a key issue in a hybrid approach for RUL estimation (Su & Chen, 2017).

6.6.2   FUSION PROGNOSTICS FRAMEWORK OF DATA-DRIVEN AND PHYSICS-BASED METHODS

A fusion prognostics framework has the potential to estimate RUL by combining data-driven and physics-based methods.

The physics-based approach, also known as the physical model-based approach (Liao & Köttig, 2016), refers to an understanding of the physics of a system to make estimations of reliability (Okoh et al., 2016). The approach describes system degradation in the form of an analytical system equation (i.e., a degradation model). The degradation model should accurately describe the evolution of degradation, but in practice, there may be deviations from this model. The addition of data-driven prediction methods that incorporate historical data from comparable systems and from the system of interest can improve prediction accuracy and reduce the uncertainty boundaries.

Figure 6.21 illustrates the detailed interface between data-driven and physics-based methods (Liao & Köttig, 2016).

The fusion of prognostics introduces two data-driven methods into the classical physics-based particle filter framework to improve prediction accuracy. The novelty can be summarized as follows (Liao & Köttig, 2016):

- Introducing a data-driven method to estimate the measurement model.
- Introducing a data-driven method to predict future measurement in long-term prediction.

The proposed hybrid prognostics framework generalizes Bayesian state estimation by introducing two data-driven methods within a physics-based method, in this case particle filter. The internal system state X_k (e.g., degradation) of a complex system usually cannot be directly accessed by sensor measurements Y_k. This calls for an indirect estimate of the internal system state using a physics-based method. While the classical Bayesian state estimation relies on an analytical measurement model, $Y_k = h(X_k) + v_k$, in numerous cases, such analytical representation of the measurement model is unavailable. Consequently, a data-driven method (trained on historical data) is used instead (first data-driven method). With the estimated data-driven measurement model, state tracking can be performed as usual using the system degradation model $Y_k = h(X_{k-1}) + w_k$. During the state prediction phase, the classical physics-based particle filter applies the system degradation model to extrapolate the internal system state. In the proposed fusion prognostics framework, a second data-driven method is used at this stage to predict future measurements, $\hat{Y}_{k+1} = g(X_k, Y_{k-1}, \ldots) + u_{k+1}$. These are fed back into the particle filter algorithm. Using these predicted "measurements", the state prediction phase is accomplished in the same way as the state tracking phase, and the particles and their weights can be further updated (Liao & Köttig, 2016).

6.6.2.1 Data-Driven Methods

Two data-driven methods are included in the proposed method fusion framework. One uses a data-driven method to build a measurement model to map from the indirectly measurable internal state to the direct measurements. The other uses a data-driven method trained from historical data to predict future measurements (Liao & Köttig, 2016).

6.6.2.2 Physics-Based Method

The physics-based method (e.g., standard particle filter) is used to predict the future system state. It tracks system state in parallel with parameter identification of the system model during the state tracking phase when actual measurements are available. State is predicted based on the last estimated state by using the system equation (i.e., the degradation model).

The proposed fusion framework is similar to the standard particle filter during the state tracking, but the prediction of system state is different. In the fusion method, new measurements are provided by a data-driven prediction method. A similarity-based prediction method trained on run-to-failure datasets provides new measurements by predicting measurements during the prediction phase. Therefore, the particle weights can be updated during the prediction phase, thereby correcting an imperfect model-based degradation model or extremely variable measurements at the end of the

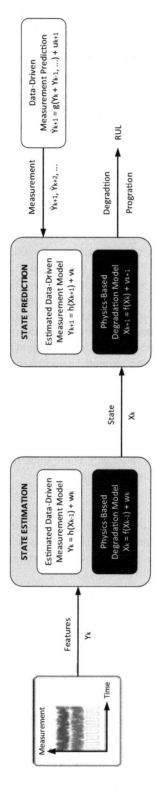

FIGURE 6.21 Data-driven and physics-based methods fusion prognostics framework (Liao & Köttig, 2016).

state tracking phase. If this is not done, the significant error in prediction of the internal state can lead to inaccurate or even incorrect RUL predictions.

The fusion framework builds on the one proposed by Liu et al. (2012), specifically in how a measurement model is introduced. It differs from the canonical representation $Y_k = X_k + v_k$, that is, the representation accounting for cases where the internal state is not directly accessible or measurable.

To sum up: physics-based methods rely on the potentially imperfect degradation model for state prediction, while purely data-driven prediction methods do not account for the physical degradation process. The use of a physics-based method with a data-driven prediction of new measurements and a data-driven measurement model can bridge the gap between them and provide more advanced and accurate prognostics (Liao & Köttig, 2016).

6.6.3 A New Hybrid Prognostic Methodology

Figure 6.22 provides a diagrammatic representation of the fields of diagnostics and prognostics. Prognostics involves two phases. The first is an assessment of the current health status – called health assessment, severity detection, or degradation identification. The second forecasts the system's future health level by propagating the current health level to a failure threshold to determine RUL. The first phase, the determination of health state, overlaps with diagnostics; the second, the prediction of failure, is the actual prognosis (Eker et al., 2019).

A prognostics model can be data-driven, physics-based, knowledge (or experience)-based, or hybrid.

The latter model type, hybrid prognostics, combined two or more approaches to capitalize on strengths of each, thus improving the prognostics outcome. Prognostics datasets that can be efficiently employed for data-based or physics-based models are hard to find. Most datasets either lack sufficient data, or details on physical modeling are missing. This means integrating them and using them together for the same system is a challenge. However, the hybrid prognostic approach has several advantages:

- Weaknesses of individual approaches can be compensated for.
- Prediction accuracy can be improved.

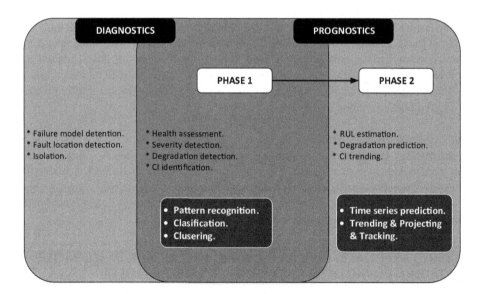

FIGURE 6.22 Prognostic and diagnostic phases (Eker et al., 2019).

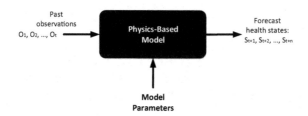

FIGURE 6.23   Input and outputs of physics-based model (Eker et al., 2019).

- Computation complexity because of the extensive data processing required in data-based models may be reduced by using physics-based models in combination with the data-based models.
- Combining approaches can compensate for lack of data.

Hybrid prognostic models are application-specific; one method can be used for health state estimation and another for RUL prediction. In Huang et al.'s (2007) hybrid methodology, for example, the health state of the system is estimated by a self-organizing map baseline, supported by MQE, while RUL is predicted by a trained BPNN. A number of researchers have used data-based models to infer the measurement model, and a physics-based model to predict RUL. The measurement model maps the sensory data to the underlying system state, which is not measured. To sum up, physics-based and data-based approaches can be used in different steps of the PHM process; one is useful for health state identification and the other is useful for forecasting (Eker et al., 2019).

Some studies combine data-based and physics-based models in the state forecasting process. They run the models independently and combine their results using fusion techniques such as Dempster–Shafer theory (Eker et al., 2019). Zhao et al. (2018) proposed an integrated method combining a Bayesian update with Archard's wear model (Eker et al., 2019).

6.6.3.1   Methodology
The hybrid method assumes the current health state has already been identified and focuses on forecasting the state progression (i.e., the evolution of failure). Physics-based and data-driven methods are integrated in the forecasting process.

The forecasting process is divided into two phases: short-term forecasting and long-term forecasting of the RUL. Physics-based methods are used for short-term forecasting, and data-driven for the longer term. In fact, data-driven methods rely on the forecasts of physics-based methods. The last forecasted time point using the physics-based model is assumed to be the current time by the data-driven model. The data-driven model's forecast starts at the time point after the end of the physics-based model's forecast.

6.6.3.1.1   Short-Term Forecasting
As we just mentioned, short-term forecasting uses a physics-based model. An equation defines the evolution of the health state from time $t-1$ to time t. Physics-based models are available for many different types of degradation; crack growth rate is a common one. Figure 6.23 shows the inputs and outputs of a physics-based model. Observations affected by the health states and model parameters are the input. The forecasted health states are the output (Eker et al., 2019).

6.7   HYBRID APPROACH INCORPORATING EXPERIENCE-BASED MODELS, DATA-DRIVEN MODELS, AND PHYSICS-BASED MODELS

The full strength of prognostics is achieved by incorporating the strengths of different prognostics models, that is, experience-based, data-driven, and physics-based models. Physics-based models characterize the degrading system by using analytical descriptions of the underlying physical principles; they make precise predictions when the physics is well known for all operating conditions and failure modes, but this is rarely the case in industrial applications. Data-driven prognostics models can mitigate this problem because of their ability to model from historical data without knowledge of the physics of degradation. Experience and expert knowledge-based models are extremely valuable and can enhance the prognostics of the other two.

Consequently, many suggest applying all three prognostics models in a hybrid approach (Liao & Köttig, 2014). For example, Bartram and Mahadevan (2012) proposed a general framework for diagnostics and prognostics incorporating heterogeneous information by using a dynamic BN (DBN). The DBN fused information from several sources to improve diagnostics and prognostics, including:

1. Expert opinion.
2. Operational data from the system.
3. Laboratory data, for example, parameter learning algorithms.
4. Published reliability data.
5. Mathematical behavior models, including physics-based (e.g., finite element, mathematical) and data-based (e.g., neural networks) models.

The authors tested their proposed framework on the example of a cantilever beam with a possible loose bolt at the connection or a crack in the middle of the span. The DBN tracked the system state of the cantilever beam and detected faults by monitoring the Bayes' factor of the system state estimate. The filtering and prediction used a particle filter method.

In another example, Xu and Xu (2011) proposed a prognostic fusion model built on an optimal linear combination (i.e., fusion) of the RUL estimates of single prognostic algorithms (i.e., knowledge-based, data-driven, and physics-based) by using absolute value and prediction error as the index of prognostic precision. The researchers then applied the model to the RUL prediction of avionics radar magnetrons. They incorporated failure mode and effects analysis, an ARMA model, and SVMs and fuzzy logic combined with neural networks.

Orsagh et al. (2003) used physics-based information to predict the RUL of bearings using spall initiation and propagation models. They started by examining vibration using the Dempster–Shafer approach, then switched from the spall initiation model to the spall propagation model. Their proposed framework fused data-driven RUL prediction results, physics-based RUL prediction results, operational statistics, failure rates, and experience-based information using a probabilistic update process to predict future failure probability.

Gola and Nystad (2011) used their proposed hybrid method to assess the health state of an eroding choke valve and estimate its RUL. The system's health state, in this case, the valve erosion, could be accurately determined using a physics-based model, in this case, a generic choke valve fluid dynamic model which accounted for all relevant physical process parameters and incorporated them into an analytical expression. This was a more complicated process than it sounds because the measurements were not accessible under normal operating conditions, so estimates of health state were inaccurate. Accordingly, Gola and Nystad developed a concept called Virtual Sensor: an ensemble of feedforward ANNs was trained with appropriate input (i.e., choke opening, pressure drop, and fluid and gas flow rates) and output (i.e., choke valve erosion state) data. They used moving average and moving maxima filters to filter the median of the individual ensemble outputs and then fed the filtered result to a statistics-based approach based on the gamma process for degradation prediction and RUL estimation. Meanwhile, for the gamma process, they used a state-based approach so they could incorporate a predefined list of health states based on expert analysis and domain knowledge to reduce prediction uncertainty.

While each type of model (experience-based, physics-based, and data-driven) may encounter difficulties when combined, making the process of creating and using a hybrid model time-consuming, complex, and computationally expensive, fusing all types of information (e.g., domain knowledge, maintenance history, condition data, and physical knowledge) and thus leveraging the models' individual strengths can be beneficial. Remaining challenges include how to aggregate results from the different (sometimes competing) models, how to design a fusing mechanism to integrate heterogeneous information, and how to use data-driven models to reduce prediction uncertainty (Liao & Köttig, 2014).

6.7.2 PROPOSED PROGNOSIS FRAMEWORK

A major challenge in prognostics is how to minimize uncertainty in the estimated RUL given constraints on available information about the system or the operating environment, computational resources, and time horizon. To this end, some researchers have proposed the use of a DBN-based prognostics framework (Bartram & Mahadevan, 2013). The first step is the construction of a DBN-based system model using heterogeneous information sources: reliability data, expert opinion, mathematical models, operational data, and laboratory data. The inclusion of physics-based models is particularly important at this stage, as is knowledge of the evolution of the degradation of interest in the system, such as wear, corrosion, or cracking. The system model is then used for diagnosis: that is, it defines the current state of the system. In the next step, a sequential Monte Carlo process predicts future system states and estimates RUL. In the final step, the prognostics, diagnostics, and predictive algorithms are validated using a four-step hierarchical procedure. The prognosis procedure is shown in Figure 6.24 (Bartram & Mahadevan, 2013).

6.7.2.1 Dynamic Bayesian Networks

A DBN is the temporal extension of a static BN. A static BN, sometimes called a directed acyclic graph (DAG) or a belief network, is a probabilistic graphical representation of a set of random variables and their conditional dependencies. In a BN, variables are represented by nodes (vertices of the graph), and conditional dependence is represented by directed edges. Unconnected nodes are conditionally independent of each other. The acyclic requirement means when starting at node x_i, it is impossible to return to node x_i; that is, there are no paths connecting them.

By contrast, the variables in a DBN may be all discrete, all continuous, or a combination of both. The continuous variables are typically modeled as continuous and the faults as discrete To create the network, a conditional probability distribution is chosen for each variable; these can be tabular (multinomial), softmax, Gaussian, deterministic, and so on.

A hybrid system such as the DBN is particularly useful when modeling systems with faults, as it integrates various types of information: expert opinion, mathematical models (including system state space and physics-based models), reliability data, operational data, laboratory data, and online measurement information from sensors (Bartram & Mahadevan, 2013).

6.7.2.2 Physics-Based Models

A prognostics model can estimate the future evolution of damage while a diagnostics model infers the current state of damage, often based on fault signatures or pattern recognition. While they may be able to detect and isolate damage and then quantify it using least squares-based estimation, they

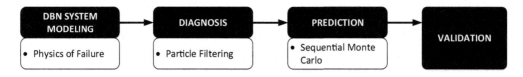

FIGURE 6.24 Proposed prognosis methodology (Bartram & Mahadevan, 2013).

can't always model the evolution of the damage, for example, wear, crack growth, or corrosion. In fact, it is a challenge to develop accurate and comprehensive physics-based models of these complex damage mechanisms, as damage will evolve differently with different system designs and dynamics, and they may also interact (Bartram & Mahadevan, 2013).

6.7.2.3   Diagnosis

When we perform diagnostics, our goal is to detect and isolate damage in a system and quantify its magnitude. In the context of prognostics, diagnosis specifies the initial conditions of damage for later prognosis of a unit under test. Its accuracy strongly impacts the accuracy of the RUL estimation, making it essential to account for any uncertainty.

Uncertainty in diagnosis has a number of possible causes, including measurement error, natural variability, model error, hypothesis testing error, and error in inference. Accounting for uncertainty is an integral part of the diagnosis procedure (Bartram & Mahadevan, 2013).

6.7.2.4   Fault Diagnosis and Diagnosis Uncertainty Quantification

A particle filter, a physics-based approach, is often used in the context of diagnosis uncertainty. In this case, the belief state provides the information necessary for fault detection, isolation, and damage quantification. The marginal distribution over combinations of the discrete fault indicator variables is multinomial; its parameters are calculated from the particles representing the current belief state (Bartram & Mahadevan, 2013).

6.7.2.5   Prognosis Validation

It is necessary to validate the outcome of prognosis to generate confidence in the RUL estimate. Many of the sources of uncertainty we have mentioned to this point, including modeling errors, sensor faults, data noise, unpredictable loading conditions, and diverse operating environments, will affect prognostics. Carefully selected performance metrics can validate prognostics. Their selection requires some thought, however, given the many possible problems that may arise when evaluating prognostics algorithms, including, for example, the need to improve accuracy as more data are acquired (Saxena et al., 2010). A possible solution is the use of a standard off-line four-metric hierarchical test proposed by Saxena et al. (2010) to evaluate a prognosis algorithm. This hierarchical test assumes prognostics improve as more measurements become available. The combination of four metrics permit the testing and comparison of prognostic algorithms.

As mentioned above, DBNs are able to handle uncertainty and can integrate many types of information, both in the off-line model construction phase and the online belief state updating phase. For prognostics, it is especially important to model complex physics-based phenomena and integrate the model into the DBN. Once the DBN model is built, it can be used to track the state of the system of interest, while particle filtering is used to update the belief state as new measurements are obtained. Uncertainty in the state estimate (i.e., diagnostics) is quantified, and when a fault is detected, estimation of RUL begins (i.e., prognostics). The result (i.e., output) is an estimate of the RUL distribution (Bartram & Mahadevan, 2013).

6.7.3   PROGNOSTICS AND HEALTH MONITORING IN THE PRESENCE OF HETEROGENEOUS INFORMATION

Diagnosis and prognosis methodologies use DBNs to fuse many types of information. These methodologies, however, fuse problem-specific information and focus only on a subset of information types. By using a subset of information, the interactions between or individual behaviors of subsystems, components, and faults may not be fully realized.

Diagnosis and prognosis have become increasingly important in the quest for safer, more intelligent, more efficient, and more cost-effective systems. Accurate diagnosis and prognosis are integral parts of system maintenance, operation, and design. CBM using automated diagnosis has gained traction because such a program may improve the safety and minimize the maintenance costs of a system. Mission-level decision-making can benefit from improved diagnosis and prognosis capabilities. Systems designed for PHM also stand to benefit from improved diagnosis and prognosis.

One important requirement for a PHM methodology is the ability to utilize existing information about a system. Such information may be found in the form of expert opinion, operational and laboratory data, reliability data, and mathematical models. This information needs to be integrated into a system-level approach to better understand interactions between subsystems and components and make all of the information available for system-level diagnosis and prognosis procedures.

This section proposes a methodology for performing diagnosis and prognosis of a mechanical system in the presence of heterogeneous information. First, it constructs a DBN model of the system and uses it in conjunction with a particle filter to track the system. Next, binary faults are detected based on Bayes' factors and isolated using the state estimate of the system. Finally, the future trajectory of the system is estimated via particle filtering (Bartram & Mahadevan, 2012).

6.7.3.1 Bayesian Networks and Dynamic Bayesian Networks

As we mentioned previously, a static BN, sometimes called a DAG or a belief network, is a probabilistic graphical representation of a set of random variables and their conditional dependencies. Variables are represented by nodes, and conditional dependence is represented by directed edges. Unconnected nodes are conditionally independent. The acyclic requirement means when starting at node X_i, it is impossible to return to node X_i (see Section 6.7.2.1).

A BN can be formulated as the joint probability space U containing variables $X_1,\ldots,X_n$ and represented in factored form as the following:

$$p\left(X_1,\ldots,X_2\right) = \prod_{i=1}^{n} p\left(X_i \# _i\right)$$

where Π_i is the set of nodes on which X^ϱ is conditionally dependent, so $p\left(X_i \#\ X_1,\ldots,X_{i-1}\right) = p\left(X_i \#_i\right)$. The nodes in Π_i are commonly referred to as the parents of node i. Note that this definition of a BN depends on the ordering of variables.

The factored formulation is readily extended to handle different types of multivariate continuous distributions, such as the Gaussian or distributions consisting of Gaussian and discrete variables.

In the factored formulation of a DBN, the variable $X[t]$ is the value of X at time t. The probability distribution describing X on the interval $[0, \infty)$ is very complex, as it is over $\bigcup_{t=0}^{t=\infty} X[t]$. Using the Markov assumption simplifies this distribution by assuming that only the present state of the variable $X[t]$ is necessary to estimate $X[t+1]$, and thus $p\left(X[t+1]X[0]\ldots X[t]\right) = p\left(X[t+1]X[t]\right)$. In addition, the process is assumed to be stationary, so $p\left(X[t+1]X[t]\right)$ is independent of t.

As mentioned in the previous section, a DBN may be composed of all discrete or all continuous variables or a combination of the two. In hybrid systems, the continuous variables are modeled as Gaussian and the faults as discrete (Bartram & Mahadevan, 2012).

6.7.3.2 Heterogeneous Information

An important feature of Bayesian analysis is the ability to integrate mixed information types. The graphical and probabilistic nature of BNs provides many opportunities to integrate information into the model, allowing them to benefit from heterogeneous information, that is, information from a variety of sources and formats. These sources include observational and experimental data, published reliability data, mathematical behavior models of components or subsystems, and expert opinion (see Section 6.7, opening paragraphs of the section). Research has not fully exploited heterogeneous information, especially with respect to building DBN models for diagnostics and prognostics.

6.7.3.3   Learning Bayesian Networks and Dynamic Bayesian Networks

Learning BNs and DBNs consists of learning the structure of a network which defines the conditional independence relationships between variables and learning the distribution parameters of the network. Algorithms can learn the structure and parameters for networks composed of discrete and continuous variables, as well as hidden variables and missing data (Bartram & Mahadevan, 2012).

6.7.3.3.1   Learning a Static BN

It is initially assumed that all variables are independent with unknown distributions. After gathering all available expert opinion about the variables (distribution types, constraints on the network structure), the structure learning algorithm evaluates any laboratory and experimental data available to determine the system structure of the system BN. The laboratory and experimental data, along with any reliability data or mathematical models, are used to determine the distribution parameters of the network.

6.7.3.3.2   Learning a Dynamic BN

DBN learning generally consists of learning the structure of a transition network which defines the relationships between variables in different time slices and learning the parameters of this network. Variations of several algorithms can handle various combinations of discrete and continuous variables.

Structure learning in a DBN is similar to that in a static BN. The observations for a dynamic system may include multiple time histories of instantiations of the system. By lagging each time history individually, data are obtained for a variable at t and $t + 1$. After each time history is lagged, all the time histories can be combined into a database. In this database, a data case represents simultaneous observations of all the variables at a discrete point in time. Note that for the purpose of learning, it is assumed that all variables are observable. In general, this may not be true.

6.7.3.4   Diagnosis

Diagnosis is the process of detecting and isolating faults in a system and quantifying their magnitudes. Detection and isolation can be performed using a BN or DBN model of the system which treats faults as binary (true or false).

In detection, it is desirable to determine the state of the system, including the states of potentially unobservable variables (faults) (Bartram & Mahadevan, 2012).

6.7.3.4.1   Diagnosis of a Static System

Diagnosis of a static system using a BN is a relatively straightforward problem. The purpose of diagnosis is to determine values of the unknown parameters (hidden state) from the measurements. Given observations of observable variables, the state of unobservable variables, which are hidden, can be inferred using Bayes' theorem. The state of the fault values that best explain the observations is taken as the fault state. Inference is automated using algorithms, such as the junction tree algorithm for exact inference or Monte Carlo-based approaches for approximate inference. Static BN diagnosis has been implemented in medical diagnosis.

6.7.3.4.2   Diagnosis of a Dynamic System

Diagnosis of a dynamic system is built around the concept of tracking. Prognosis makes an inference about the future distribution of the state of a system given its current state and possible future scenarios. As no new measurements are available, a Bayesian recursive filter which integrates measurement data is no longer necessary. The last set of state estimates and measurements are propagated through the DBN. The result is an estimate of the future distribution of the state variables (Bartram & Mahadevan, 2012).

REFERENCES

Ahmadzadeh F., Lundberg J., 2014. Remaining useful life estimation: Review. International Journal of Systems Assurance Engineering and Management, 5(4), 461–474.

Bartram G., Mahadevan., 2012. Prognostics and health monitoring in the presence of heterogeneous information. In Proceedings of the Annual Conference of the Prognostics and Health Management Society, 3.

Bartram G., Mahadevan S., 2013. Dynamic Bayesian networks for prognosis. Vanderbilt University. Annual Conference of the Prognostics and Health Management Society.

Byington C. S., Watson M., Edwards D., 2004. Data-driven neural network methodology to remaining life predictions for aircraft actuator components, Proceedings IEEE Aerospace Conference, 6, 3581–3589.

Celaya J., Saxena A., Saha S., Goebel K., 2011. Prognostics of power MOSFETS under thermal stress accelerated aging using data-driven and model-based methodologies. Proceedings International Conference on Prognostics and Health Management.

Chelidze D., Cusumano J., 2004. A dynamical systems approach to failure prognosis. Journal of Vibration and Acoustics, 126, 2–8.

Chinnam R., Baruah P., 2004. A neuro-fuzzy approach for estimating mean residual life in condition-based maintenance systems. International Journal of Materials and Product Technology, 20(1), 166–179.

Cubillo A., Perinpanayagam S., Esperon-Miguez M., 2016. A review of physics-based models in prognostics: Application to gears and bearings of rotating machinery. Advances in Mechanical Engineering. 8(8), 1–21.

Datarobot, 2020. Model Tuning. www.datarobot.com/wiki/tuning. Viewed: June 11, 2020.

Davis C., 1993. The computer generation of multinominal variants. Computational Statistics and Data Analysis, 16.

Du S., Lv J., Xi L., 2012. Degradation process prediction for rotational machinery based on hybrid intelligent model. Robotics and Computer-Integrated Manufacturing, 28(2), 190–207.

Eker O. F., Camci F., Jennions I. K., 2019. A new hybrid prognostic methodology. International Journal of Prognostics and Health Management.

Elattar H. M., Elminir H. K., Riad A. M., 2016. Prognostics: A literature review. Complex & Intelligent Systems, 2, 125–154.

Flovik V., 2018. How do you teach physics to machine learning models? Hybrid analytics: Combining the best of two worlds. https://towardsdatascience.com/how-do-you-combine-machine-learning-and-physics-based-modeling-3a3545d58ab9. Viewed: June 12, 2020.

Fornlöf V., Galar D., Syberfeldt A., Almgren T., 2016. RUL estimation and maintenance optimization for aircraft engines: A system of system approach. International Journal of System Assurance Engineering and Management, 7(4), 450–461.

Galar D., Palo M., Van Horenbeek A., Pintelon L., 2012. Integration of disparate data sources to perform maintenance prognosis and optimal decision making. Insight: Non-Destructive Testing and Condition Monitoring, 54 (8), 440–445.

Garga A., McClintic K., Campbell R., Yang C., Lebold M., Hay T., Byington C., 2001. Hybrid reasoning for prognostic learning in CBM systems. Proceedings IEEE Aerospace Conference, 6, 2957–2969.

Gebraeel N., Lawley M., Liu R., Parmeshwaran V., 2004. Residual life predictions from vibration-based degradation signals: A neural network approach. IEEE Transactions on Industrial Electronics, 51(3), 694–700.

Goebel K., 2007. Prognostics and Health Management. Guest lecture, ENME 808A University of Maryland.

Goebel K., Eklund N., Bonanni P., 2006. Fusing competing prediction algorithms for prognostics. Proceedings of 2006 IEEE Aerospace Conference, Big Sky, MT, USA, 4–11 March 2006.

Gola G., Nystad B., 2011. From measurement collection to remaining useful life estimation: Defining a diagnostic-prognostic frame for optimal maintenance scheduling of choke valves undergoing erosion. Proceedings Annual Conference Prognostics and Health Management Society, 26–29.

Gu J., Barker D., Pecht M., 2007. Uncertainty assessment of prognostics implementation of electronics under vibration loading. AAAI Fall Symposium on Artificial Intelligence for Prognostics, Arlington, VA, 50–57.

He W., Williard N., Chen C., Pecht M., 2014. State of charge estimation for li-ion batteries using neural network modeling and unscented Kalman filter-based error cancellation, International Journal of Electrical Power & Energy Systems, 62, 783–91.

Heimes F., 2008. Recurrent neural networks for remaining useful life estimation. Proceedings International Conference on Prognostics and Health Management, 1–6.

Huang R., Xi L., Li X., Richard Liu C., Qiu H., Lee J., 2007. Residual life predictions for ball bearings based on self-organizing map and back propagation neural network methods, Mechanical Systems and Signal Processing, 21(1), 193–207.

Jammu N. S., Kankar P. K., 2011. A review on prognosis of rolling element bearings. International Journal of Engineering Science and Technology (IJEST), 3(10), 7497.

Jardine A. K. S., Lin D., Banjevic D., 2006. A review on machinery diagnostics and prognostics implementing condition-based maintenance. Mechanical Systems and Signal Processing, 20(7), 1483–1510.

Javed K., 2014. A robust & reliable data-driven prognostics approach based on extreme learning machine and fuzzy clustering. Doctoral dissertation, Universite de Franche-Comte.

Jaw L. C., Wang W., 2006. Mathematical formulation of model-based methods for diagnostics and prognostics. Proceedings of the 2006 ASME 51st Turbo Expo: Power for Land, Sea, and Air, Barcelona, 8–11 May 2006 (pp. 691–697).

JSF, 2007. Joint Strike Fighter. www.jsf.mil. Viewed: June 7, 2020.

Kabir A., Bailey C., Lu H., Stoyanov S., 2012. A review of data–driven prognostics in power electronics. IEEE 189 35 International Spring Seminar on Electronics Technology.

Kuhn M., Johnson K., 2013. Over-fitting and model tuning. In: Applied Predictive Modelling, 61–92. Springer. https://link.springer.com/chapter/10.1007/978-1-4614-6849-3_4#Bib1. Viewed: June 11, 2020.

Liao L., Köttig F., 2014. Review of hybrid prognostics approaches for remaining useful life prediction of engineered systems, and an application to battery life prediction. IEEE Transactions on Reliability, 63(1).

Liao L., Köttig F., 2016. A hybrid framework combining data-driven and model-based methods for system remaining useful life prediction. Applied Soft Computing, 44(C).

Liu J., Wang W., Ma F., Yang Y., Yang C., 2012. A data-model-fusion prognostic framework for dynamic system state forecasting, Engineering Applications of Artificial Intelligence, 25(2012), 814–823.

Liu Z., Li Q., Mu C., 2012a. A hybrid LSSVR-HMM-based prognostics approach. Proceedings of the 4th International Conference on Intelligent Human-Machine Systems and Cybernetics, 2, 275–278.

Liu Z., Li Q., Mu C., 2012b. A hybrid LSSVR-HMM based prognostics approach. Proceedings of the 4th International Conference on Intelligent Human-Machine Systems and Cybernetics, 2, 275–278.

Luo J., Pattipati K. R., Qiao L., Chigusa S., 2003. Model-based prognostic techniques applied to a suspension system. Transactions on Systems, Man, and Cybernetics, 38, 1156–1168.

Markov A., 1971. Extension of the limit theorems of probability theory to a sum of variables connected in a chain. Reprinted in Appendix B of R. Howard, Dynamic Probabilistic Systems, Vol 1: Markov Chains. John Wiley and Sons.

Medjaher K., Zerhouni N., 2013. Framework for a Hybrid Prognostics. FEMTO-ST Institute, UMR CNRS 6174 - UFC/ENSMM/UTBM, Automatic Control and Micro-Mechatronic Systems Department. 24, rue Alain Savary, 25000 Besançon, France. Vol. 33, 2013. Guest Editor: E. Zio.

McGraw-Hill Dictionary, 2009. Synthetic data. McGraw-Hill Dictionary of Scientific and Technical Terms. November 29, 2009.

McHugh J., 2000. Testing intrusion detection systems: A critique of the 1998 and 1999 DARPA Intrusion detection system evaluations as performed by Lincoln Laboratory. ACM Transactions on Information and Systems Security, 3(4).

McLachlan S., 2017. Realism in synthetic data generation. A thesis presented in fulfilment of the requirements for the degree of: Master of Philosophy in Science. School of Engineering and Advanced Technology. Massey University. Palmerston North, New Zealand.

Mullins C. S., 2009. What is production data? NEON Enterprise Software, Inc. Archived from the original on July 21, 2009.

Mwogi T., Biondich P., Grannis S., 2014. An evaluation of two methods for generating synthetic HL7 segments reflecting real-world health information exchange transactions. AMIA Annual Symposium Proceedings.

Okoh C., Roy R., Mehnen J., 2016. Predictive maintenance modelling for through-life engineering services. The 5th International Conference on Through-life Engineering Services (TESConf 2016). ScienceDirect. Procedia CIRP 59 (2017) 196–201.

Okoh C., Roy R., Mehnen J., Redding L., 2014. Overview of remaining useful life prediction techniques in through-life engineering services. Procedia CIRP, 16, 158–163.

Orsagh R. F., Sheldon J., Klenke C. J., 2003. Prognostics/diagnostics for gas turbine engine bearings. Proceedings of the 2003 IEEE Aerospace Conference, Big Sky, MT, 8–15 March 2003, pp. 3095–3103.

Pantelelis N. G., Kanarachos A. E., Gotzias N. 2000. Neural networks and simple models for the fault diagnosis of naval turbochargers. Mathematics and Computers in Simulation, 51: 387–397.

Pecht M., Gu J., 2009. Physics-of-failure-based prognostics for electronic products. Transactions of the Institute of Measurement and Control, 31(3/4), 309–322.

Peel L., 2008. Data driven prognostics using a Kalman filter ensemble of neural network models. Proceedings International Conference on Prognostics and Health Management, 1–6.

Pillai P., Kaushik A., Bhavikatti S., Roy A., Kumar V., 2016. A hybrid approach for fusing physics and data for failure prediction. International Journal of Prognostics and Health Management.

Roemer M. J., Byington C. S., Kacprzynski G. J., Vachtsevanos G., 2006. An overview of selected prognostic technologies with application to engine health management. Proceedings of GT2006. ASME Turbo Expo 2006: Power for Land, Sea, and Air May 8-11, 2006, Barcelona, Spain.

Satish B., Sarma N., 2005. A fuzzy BP approach for diagnosis and prognosis of bearing faults in induction motors. Proceeding IEEE Power Engineering Society General Meeting, 2291–2294.

Saxena A., Celaya J., Saha B., Saha S., Goebel K., 2010. Metrics for offline evaluation of prognostic performance. International Journal of Prognostics and Health Management.

Schwabacher M., Goebel K., 2016. A survey of artificial intelligence for prognostics. NASA Ames, Research Center, MS 269-3. Moffett Field, CA 94035.

Su C., Chen H. J., 2017. A review on prognostics approaches for remaining useful life of lithium-ion battery. IOP Conference Series: Earth and Environmental Science, 93, 012040.

Surendra H., Mohan H. S., 2017. A review of synthetic data generation methods for privacy preserving data publishing. International Journal of Scientific and Technology Research, 6(3).

Swanson D. C., 2001. A general prognostic tracking algorithm for predictive maintenance. Proceedings of the IEEE Aerospace Conference, 6, pp. 2971–2977.

Vichare N., Pecht M., 2006. Prognostics and health management of electronics. IEEE Transactions on Components and Packaging Technologies 29, 222–29.

Vichare N., Rodger P., Eveloy V., Pecht M., 2007. Environment and usage monitoring of electronic products for health assessment and product design. International Journal of Quality Technology and Quantitative Management 4, 235–50.

Walker A., 1974. New fast method for generating discrete random numbers with arbitrary frequency distributions. Electronic Letters, 10(8).

Walker A., 1977. An efficient method for generating discrete random variables with general distributions. ACM Transactions on Mathematical Software, 3(3).

Wikipedia, 2020. Synthetic data. May 13, 2020. https://en.wikipedia.org/wiki/Synthetic_data#cite_note-Mullins-2. Viewed: June 12, 2020.

Xu J., Xu L., 2011. Health management based on fusion prognostics for avionics systems. Journal of Systems Engineering and Electronics, 22(3), 428–436.

Yan J., Lee J., 2007. A hybrid method for on-line performance assessment and life prediction in drilling operations. Proceedings IEEE International Conference on Automation and Logistics, 2500–2505.

Yu W. K., Harris T. A., 2001. A new stress-based fatigue life model for ball bearings. Tribology Transactions, 44(1), 11–18.

Zhang G., Lee S., Propes N., Zhao Y., Vachtsevanos G., Thakker A., Galie T., 2002. A novel architecture for an integrated fault diagnostic/ prognostic system. Proceedings of the AAAI Symposium, Stanford, California.

Zhao F., Tian Z., Liang X., Xie M., 2018. An integrated prognostics method for failure time prediction of gears subject to the surface wear failure mode. IEEE Transactions on Reliability, 67(1), 316–327.

Zio E., Di Maio F., 2012. Fatigue crack growth estimation by relevance vector machine. Expert Systems with Applications, 39, 10681–10692.

7 Prognosis in Prescriptive Analytics

7.1 EVOLUTION FROM DESCRIPTION TO PREDICTION AND PRESCRIPTION

7.1.1 EVOLVING ANALYTICS: DESCRIPTIVE TO PRESCRIPTIVE TO PREDICTIVE

With the flood of data available today, companies are turning to analytics to improve decision-making. By analyzing their historical data, they will be able to forecast what could happen in the future. Becoming a data-driven organization stands to yield a big return on investment, lowering operating costs, increasing revenues, and improving customer service and product mix (Bachar, 2020).

There are several types of analytics; no one type is better than another. They coexist with and complement each other:

1. Descriptive analytics: Using data aggregation and data mining to gain insight into the past and answer: "What has happened?"
2. Prescriptive analytics: Using optimization and simulation algorithms to suggest possible outcomes and answer: "What should we do?"
3. Predictive analytics: Using statistical models and forecasting techniques to get a glimpse of the future and answer: "What could happen?"

7.1.1.1 Descriptive Analytics: Understanding the Past

Descriptive analysis does exactly what the name implies: it describes or summarizes raw data and makes them interpretable by humans. These analytics describe the past. The past refers to any point of time that an event has occurred, whether one minute ago, or one year ago. Descriptive analytics are useful because they allow us to learn from past behaviors and understand how they might influence future outcomes.

Most statistics fall into this category (e.g., basic arithmetic like sums, averages, percent changes). The underlying data are usually a count or aggregate of a filtered column of data to which basic math is applied. For all practical purposes, there is an infinite number of these statistics. Descriptive statistics are useful to show things like total stock in inventory, average dollars spent per customer, or year-over-year change in sales. Common examples are reports providing historical insights into production, financials, finance, inventory, operations, sales, and customers. We use descriptive analytics when we want to understand at an aggregate level what is going on and to summarize and describe different aspects of the business enterprise (Bachar, 2020).

Most social analytics are descriptive. They summarize certain groupings based on simple counts of certain events. The number of followers in Twitter, likes, posts, and fans are event counters. These metrics are used for social analytics like average response time, average number of replies per post, %index, or number of page views, using basic arithmetic operations. An organization can use the results it gets from a web server through Google Analytics tools to help understand what actually happened in the past. For example, it can validate if a promotional campaign was successful based on basic parameters like page views (DeZyre, 2019).

DOI: 10.1201/9781003097242-7

The components and the output of the system itself fall into descriptive analysis. Through the data warehouse, the engines and reporting tools analyze past data, categorize, filter, and aggregate those data, and apply mathematical or statistical functions to them. These engines perform many operations but their analysis describes the past (Data Skills, 2018).

Many tools have the ability to perform prescriptive and predictive analytics, but their use requires a contextual understanding of the underlying dataset and custom coding. Furthermore, descriptive systems require the user to sort through volumes of information to determine what actions to take. Codifying the rule sets into a prescriptive analytics system reduces the effort required by the user to determine what actions need to be taken. In today's fast-moving operations environments, simplifying the process can have substantial rewards.

Despite their disadvantages, descriptive analytics will likely remain popular because they provide useful insights into performance.

7.1.1.2 Predictive Analytics: Understanding What to Do

Predictive analytics are all about understanding the future. This type of analytics gives organizations actionable insights based on data and estimates the likelihood of a future outcome. No statistical algorithm can predict the future with total certainty, however, and predictive analytics is based on probabilities.

These statistics take existing data and fill in missing data with best guesses. They combine historical data found in various systems to identify patterns and apply statistical models and algorithms to capture relationships between various datasets. Predictive analytics can be used throughout an organization, from forecasting customer behavior and purchasing patterns to identifying trends in sales activities or forecasting demand for inputs from the supply chain, operations, and inventory.

Retail organizations like Amazon and Walmart leverage predictive analytics to identify trends in sales based on customers' purchase patterns. They then use these identified trends to forecast customer behavior, forecast inventory levels, predict the products customers are likely to purchase together so they can make personalized recommendations, and predict quarterly or yearly sales. A common use of predictive analytics is to produce credit scores. Credit scores help financial institutions determine the probability of a customer paying his or her credit bills on time (DeZyre, 2019).

Simply stated, predictive analytics can be used whenever we need to know something about the future or fill in information we do not actually have (Bachar, 2020). Predictive analytics include the following:

1. Predictive modeling: Deciding what is likely to happen next.
2. Root cause analysis: Determining why something happened.
3. Data mining: Identifying correlated data.
4. Forecasting: Determining what will happen if current trends continue.
5. Monte Carlo simulation: Determining what could happen.
6. Pattern identification and alerts: Identifying when an action should be taken to correct a process.

Examples of predictive analytics applied to business are (Data Skills, 2018):

- Churn analysis: This method involves analyzing the customer pool to determine which customers have a high probability of migrating to the competition. This allows the organization to intervene early and prevent their departure.
- Searching for anomalies: In this method, machine learning (ML) algorithms identify abnormal situations, such as the fraudulent use of credit cards or insurance fraud.
- Cross selling/up selling: This method involves the use of predictive techniques to determine which customers are most likely to buy a certain product. This allows the company to target a specific group of potential buyers.

Predictive analytics algorithms can be categorized as the following (Data Skills, 2018):

- Classification algorithms: These algorithms allow users to assign an object to a category. An example is the identification of churners (see churn analysis above); in this case, the assigned categories are churners vs. loyal customers.
- Clustering algorithms: Clustering techniques are based on similarity between elements. Accordingly, these algorithms group homogeneous elements in a dataset.
- Frequent pattern mining: These algorithms identify sets of recurring elements even in extremely large datasets (i.e., Big Data).

7.1.1.3 Prescriptive Analytics: Seeing the Possible Future

Prescriptive analytics models essentially belong to predictive analytics but they add the ability to explain the reasons for a certain event. Predictive models tell us what is likely to occur, while prescriptive models use a set of rules to explain the reasons (Data Skills, 2018). Simply stated, prescriptive analytics use historical data to predict possible future outcomes. Big Data and cutting-edge software tools make this achievable.

Prescriptive analytics allow users to prescribe a plethora of possible actions and guide them toward selecting one. In a way, these analytics are all about providing advice. Prescriptive analytics attempt to quantify the effect of future decisions in order to give advice on possible outcomes before a decision is actually made (Bachar, 2020). In other words, prescriptive analytics are all about optimization: how to achieve the best outcome and identify data uncertainties to make better decisions (DeZyre, 2019).

Prescriptive analysis explores several possible actions and suggests actions depending on the results of descriptive and predictive analytics of a given dataset (DeZyre, 2019). These analytics go beyond descriptive and predictive analytics, however, by recommending one or more possible courses of action. Essentially, they predict multiple futures and allow companies to assess a variety of different outcomes.

Prescriptive analytics use a combination of techniques and tools, including business rules, algorithms, ML, and computational modeling. These techniques are applied against input from many different datasets including historical and transactional data, real-time data feeds, and Big Data.

Large companies are successfully using prescriptive analytics to optimize production, scheduling, and inventory in the supply chain to make sure they are delivering the right products at the right time and optimizing the customer experience. When implemented correctly, they can have a large impact on how the company makes decisions, and on the company's bottom line. But prescriptive analytics are relatively complex to administer, and most companies are not using them in their everyday business transactions (Bachar, 2020).

Some possible predictive outcomes based on currently available data include the following:

- Labor forecasting: Correlating historical volume cycles and current employees to forecast staffing needs.
- Equipment breakage: Analyzing vehicle telematics and comparing this to part failures to predict vehicle breakdown before it occurs.
- Employee turnover: Analyzing employee performance trends correlated with time and attendance to identify employees who may quit.

Figure 7.1 shows how prescriptive analytics work. The prescriptive model enables decision-makers to take immediate action, based on probabilistic forecasts and on comprehensible rules coming from the model itself. Note that an important aspect of prescriptive analytics is the ability to analyze the feedback coming from using the rules and to take action based on that analysis (Data Skills, 2018).

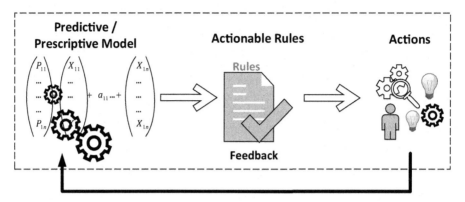

FIGURE 7.1 Prescriptive model (Data Skills, 2018).

Predictions are not valid forever, but if the rules produced by the model are simple and easy to interpret, the time necessary for the action to take place will be short, allowing actions to occur while the prediction is still valid. Algorithms that create immediately usable rules belong to three classes (Data Skills, 2018):

1. Decision trees: Decision trees are simple to use but do not have a particularly good predictive performance. They are widely used, however, and are implemented in many data mining tools, as they can produce rules.
2. Fuzzy rule-based systems: Fuzzy systems are the best from a predictive view but are not very common.
3. Switching neural networks (logic learning machine): Switching neural networks is implemented only in one ML tool (Rulex) and has high predictive ability.

Other systems may be effective from the point of view of the sensitivity, accuracy, and specificity of the model but do not provide any insight into how they reach a certain prediction. Examples include neural networks and support vector machines (Data Skills, 2018).

7.1.2   Categories of Methods for Predictive Analytics and Prescriptive Analytics

Figure 7.2 depicts the classification of the methods used in predictive analytics: probabilistic models, ML/data mining, and statistical analysis (Lepenioti et al., 2020b).

Figure 7.3 shows the classification of the methods used for prescriptive analytics: probabilistic models, ML/data mining, mathematical programming, evolutionary computation, simulation, and logic-based models (Lepenioti et al., 2020b).

7.1.3   Toward Prescriptive Analytics

Prescriptive analytics is the most sophisticated type of business analytics and can bring the greatest intelligence and value. It suggests (prescribes) the best decision options to take advantage of the predicted future utilizing large amounts of data. To do this, it incorporates the predictive analytics output and utilizes artificial intelligence (AI), optimization algorithms, and expert systems in a probabilistic context to provide adaptive, automated, constrained, time-dependent, and optimal decisions (Lepenioti et al., 2020b).

Prescriptive analytics has two levels of human intervention (Lepenioti et al., 2020b):

1. Decision support, for example, providing recommendations.
2. Decision automation, for example, implementing the prescribed action.

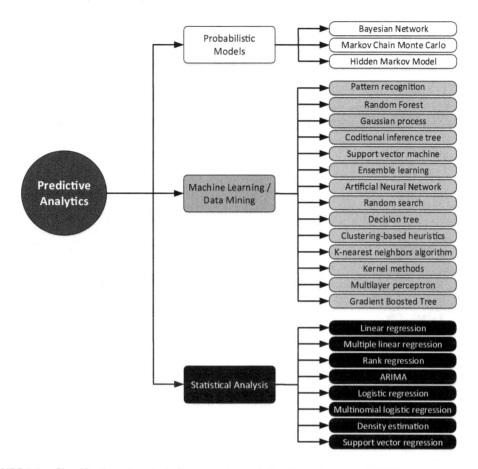

FIGURE 7.2 Classification of methods for predictive analytics (Lepenioti et al., 2020b).

The effectiveness of the prescriptions depends on how well these models incorporate a combination of structured and unstructured data, represent the domain under study, and capture impacts of decisions being analyzed.

The business value of the three stages of business analytics over time is depicted in Figure 7.4. Beginning from the left side, descriptive analytics aims to determine what is happening now by gathering and analyzing parameters related to the root causes of the event to be eliminated or mitigated. Descriptive analytics is able to detect patterns that indicate a potential problem or a future opportunity for the business. On this basis, predictive analytics is able to predict whether an event will happen, when it is about to happen, and the reason why it will happen. As shown in Figure 7.4, predictive analytics contributes significantly to the business value. However, this is closely related to the decisions that are taken and the actions that are implemented. In case of human decisions, it very much depends on their knowledge and experience (Lepenioti et al., 2020b).

Full exploitation of predictive analytics can be achieved in conjunction with prescriptive analytics for optimized decision-making ahead of time. However, even in this case, it is not possible to achieve the maximum potential of predictive analytics since there is a time interval until the decision. Therefore, there is an inevitably lost business value between the event prediction and the proactive decision. The time scale of this interval may vary depending on the computational environment, for example, real-time vs. offline, and the application domain. In any case, the minimization of this interval is of utmost importance.

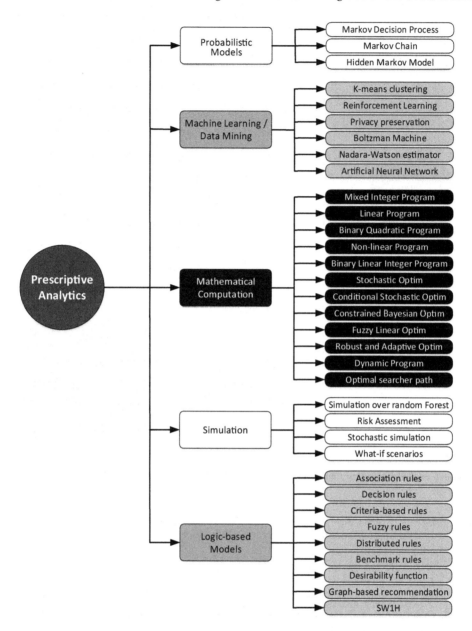

FIGURE 7.3 Classification of methods for prescriptive analytics (Lepenioti et al., 2020b).

As shown in Figure 7.4, prescriptive analytics generates proactive decisions on the basis of the predictive analytics outcomes. Between the decision and the implementation of the action, there is a time interval for the preparation of the action. Moreover, an action may be better implemented at a specific time before the predicted event occurrence when the expected utility/loss is optimized. When the event actually occurs, descriptive analytics is applied to derive insights about what happened and why it happened. In this case, descriptive analytics may be applied in a different timescale. In this sense, it may deal with near real-time reactive actions or with more long-term actions. Therefore, the timely detection of the current state of a business and the timely prediction of emerging events are crucial in terms of potential loss of business value (Lepenioti et al., 2020b).

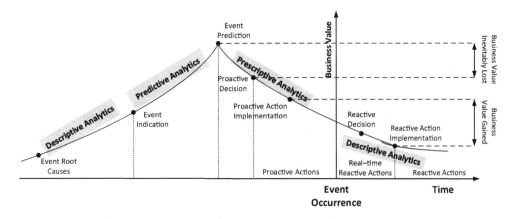

FIGURE 7.4 Business value of analytics with respect to time (Lepenioti et al., 2020b).

7.1.3.1 How Prescriptive Analytics Works

Prescriptive analytics relies on AI, specifically the field of ML. In ML, algorithms and models allow computers to make decisions based on statistical data relations and patterns.

The Bayes' classifier is a common ML algorithm that uses a statistical model (Bayes' theorem) to determine the conditional probability of an event. Another common but non-statistical algorithm is ID3; the algorithm creates a decision tree that structures a graph of possible outcomes from a dataset. In both cases, the goal is to create a model from historical data that can accept new inputs and predict outcomes.

Different algorithms make different assumptions about the structure and completeness of data so it is necessary to experiment with ML algorithms and features to create a prescriptive analytics system. For example, a linear regression analysis assumes the prediction variable can be modeled as a weighted sum of the descriptive features. Not all data are linearly related, however, so linear regression is not suitable for every problem.

Despite their many benefits, prescriptive analytics systems are not perfect. Data quality issues such as missing or incorrect information can lead to false predictions, while overfitting in models can lead to inflexible predictions that cannot handle changes over time. In other words, the models need close monitoring and maintenance (Stitch, 2020).

7.1.4 FIVE PILLARS OF PRESCRIPTIVE ANALYTICS SUCCESS

As the Big Data analytics space continues to evolve, prescriptive analytics will become increasingly important. The promise of prescriptive analytics is certainly alluring: it enables decision-makers to not only look into the future of their mission critical processes and see the opportunities (and issues) that are potentially out there, but it also presents the best course of action to take advantage of that foresight in a timely manner (Basu, 2013).

There are five pillars to prescriptive analytics success (Basu, 2013).

7.1.4.1 Hybrid Data

Most businesses run on structured data – numbers and categories. According to IBM, 80 percent of the data currently produced are unstructured – text, image, video, and audio. While some businesses may choose to run the same way in the future as they do today, doing so could render them unproductive and noncompetitive. These businesses may not survive as their customers, suppliers, and competitors move beyond them by taking full advantage of hybrid data, a combination of unstructured and structured data. Hybrid data empower businesses to use all the available data to make the best decisions possible.

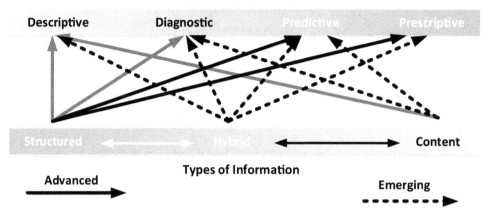

FIGURE 7.5   Evolution of analytics (Basu, 2013).

For a prescriptive analytics technology to be transformative, it must be able to process hybrid data. Without incorporating hybrid data, decision-makers are making their decisions based on just 20 percent of the available data. Figure 7.5 showcases the evolution of analytics, culminating in prescriptive analytics with hybrid data.

Processing hybrid data brings into the mix new technologies that are essential ingredients in prescriptive analytics software, including computer vision, speech recognition, image processing, natural language processing, signal processing, and more. While traditional disciplines such as applied statistics remain invaluable, they aren't designed to process image, video, audio, and text.

7.1.4.2   Integrated Predictions and Prescriptions

Prescriptive analytics is about seeing and then shaping the future. Common sense tells us that we need to see the future before we can shape it. The functions – predictions and prescriptions – must work synergistically for prescriptive analytics to deliver on its promise. The symbiotic integration of predictions and prescriptions is the key to the widespread adoption and inherent value of prescriptive analytics.

Assume a scenario where predictions and prescriptions are coming from two different systems that have been cobbled together. Say this software combination produced a prescription that turned out to be faulty. If this is a software issue, is this error due to a bug in the prediction software or the prescription software or both? Imagine the disruption to business as you investigate the root cause and attempt to preempt similar erroneous prescriptions in future.

7.1.4.3   Prescriptions and Side Effects

Prescriptions, that is, recommended, time-dependent actions to improve the future, in prescriptive analytics technology are generated using several methods. A prevalent method of coming up with prescriptions is through a guided framework of business rules. This rule framework can be simple or complex, depending on the business process or the initiative that is being governed by prescriptive analytics.

A more scientific and rigorous way to produce prescriptions to improve the future is through operations research, the science of data-driven decision-making. Operations research takes into account the objectives, the constraints, and the decision variables to produce the best course of action – a prescription – that doesn't lead to undesirable side effects. Both optimization and simulation technologies, two prominent branches of operations research, can be used to generate effective prescriptions.

For a prescriptive analytics technology to scale, the solution should use both business rules and operations research and use them synergistically. At this point, this technology will be able to generate the most effective and timely prescriptions that the available data allow. For the Internet of Everything (or the Industrial Internet) to reach its true potential, prescriptive analytics – and the resulting decision automation – must play a pivotal role.

7.1.4.4   Adaptive Algorithms

In a world of growing data volume, velocity and variety, the prescriptive analytics technology must be able to automatically recalibrate all its built-in algorithms, plus automatically create new algorithms to remain relevant. This total recalibration also needs to be adaptive – dynamic and/or continual – to successfully manage the business process in an ongoing fashion. Recalibration of the algorithms in the prescriptive analytics software can be triggered in several ways: with new data arrival, with data change(s), after a specified time period, and more.

Another important factor is the "action-ability" of the prescriptions; it is, of course, different for different business processes. At times, it may not be helpful to automatically generate newer prescriptions if the old ones haven't been acted upon properly.

7.1.4.5   Feedback Mechanism

How will the prescriptive analytics software know if its prescriptions are being acted upon? Prescriptions are, generally speaking, time-sensitive action plans involving changes to some actionable influencers to preempt one or more predicted issues or to benefit from one or more predicted opportunities. If a business manager decides to ignore a prescription from the software, this inaction would at some point be reflected in the incoming data being collected on the actionable influencers. The consequence of inaction, if any, will be reflected in the upcoming predictions and prescriptions. For example, due to lost time, inaction on a valid prescription could lead to additional expenses in preempting an upcoming issue that has been flagged (via prediction) and addressed (via prescription) in the last round.

While this may change in the near future, there is still a difference between prescriptive analytics software and prescriptive automation. Prescriptive automation (e.g., Google Car) has elaborate, built-in process control (software, hardware, firmware, etc.) to automatically "action" the prescriptions coming from the software side. Prescriptive analytics software still requires human assistance to carry out these prescriptions.

While Google can probably outfit a car any way it wants to, companies with highly sophisticated prescriptive analytics software are dependent on humans to act on the prescriptions coming out of the software. In the future, this distinction will disappear, and prescriptive analytics software will become a fully integrated and embedded component of the business process it is improving.

7.1.5   Maintenance Analytics Concept

For maintenance decision-making to be supported smoothly and efficiently in an era when the technological revolution, including Internet of Things (IoT), Industrial IoT (IIoT), Industry 4.0, and smart factories, enables an enhanced availability of data and information, we need a structured approach and concept to improve information extraction and knowledge discovery. Maintenance analytics (MA) has been proposed to deal with analytics in maintenance from methodological and technological perspectives. MA is based on four interconnected time-lined phases. The goal is to facilitate maintenance through better understanding of data and information.

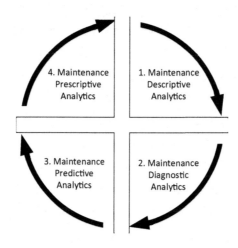

FIGURE 7.6   Constitution phases of maintenance analytics (Karim et al., 2016).

The MA phases are shown in Figure 7.6 and listed below (Karim et al., 2016):

1. Maintenance descriptive analytics.
2. Maintenance diagnostic analytics.
3. Maintenance predictive analytics.
4. Maintenance prescriptive analytics.

In maintenance decision-making, it is essential to deal with faults (inability to perform as required, due to an internal state) and failures (the termination of the ability of an item to perform a required function). In effect, a fault is the result (state) of an associated failure (event). To deal with both event and state in a system, the MA phases are in a time-related order (Karim et al., 2016).

7.1.5.1   Maintenance Descriptive Analytics

What happened? The descriptive phase of MA tries to figure out what has happened, making access to data on system operation, system condition, and expected condition key. To understand the relations of events and states during descriptive analytics, we need to understand the time and time frame associated with each specific log. In addition, events and states need to be associated with the system configuration for the time. Thus, time synchronization is an important aspect of MA (Karim et al., 2016).

7.1.5.2   Maintenance Diagnostic Analytics

Why did it happen? The diagnostic phase of MA asks what caused the event of interest. The outcome from maintenance descriptive analytics is used to frame these analytics, but reliability data are also needed (Karim et al., 2016).

7.1.5.3   Maintenance Predictive Analytics

What will happen in the future? The predictive phase of MA tries to predict what may happen. Like maintenance diagnostic analytics, this phase uses the outcome from maintenance descriptive analytics. Reliability data and maintainability data are both required as well. To predict upcoming failures and faults, business data such as planned operation and planned maintenance are useful (Karim et al., 2016).

7.1.5.4 Maintenance Prescriptive Analytics

What needs to be done? The prescriptive phase of MA uses the outcome from maintenance diagnostic analytics and maintenance predictive analytics to decide what should be done. Resource planning data and business data are also required to predict failures and faults (Karim et al., 2016).

7.1.6 MAINTENANCE ANALYTICS AND eMAINTENANCE

Appropriate information logistics are essential for MA. The main aim is to provide just-in-time information to users and optimize the information supply process, simply stated, to make the right information available at the right time and in the right location. Solutions for information logistics need to deal with the following (Karim et al., 2016):

- Time management: When to deliver.
- Content management: What to deliver.
- Communication management: How to deliver.
- Context management: Where and why to deliver.

eMaintenance is defined as the use of information logistics to support maintenance decision-making (Karim, 2008; Kajko-Mattson et al., 2011). eMaintenance solutions supporting the MA concept require an overarching approach to combine the modeling of data, knowledge, and context. When dealing with information integration, two essential issues need to be considered (Karim et al., 2016):

- Syntactic problem: Format and structure of content.
- Semantic problem: Meaning of content.

In an application area, such as maintenance, it is necessary to apply an ontology, that is, a representational vocabulary for a shared domain that includes definitions of classes, relations, functions, and other objects. Ontology is an explicit specification of a conceptualization (i.e., abstract and simplified view of the world), including the objects, concepts, and other entities presumed to exist in some area of interest and the relationships that connect them (Karim et al., 2016).

eMaintenance includes tools, technologies, and methodologies aimed at maintenance decision-making, including analytics. As Figure 7.7 shows, eMaintenance is a concept through which MA can be materialized (Karim et al., 2016).

7.2 ROLE OF PROGNOSIS IN A DYNAMIC ENVIRONMENT

7.2.1 PROCEDURE FOR PROGNOSTICS OF DYNAMIC SYSTEMS

7.2.1.1 Dynamic Bayesian Networks

Prognostics are used to estimate the remaining useful life (RUL) of a system. Various methods can be used, but they are restricted by the type and amount of available data, the need to build a model, and so on. Prognostics tools must consider the inherent uncertainty of prognostics to enable the confident choice of timely and accurate maintenance actions.

Some techniques are well suited for this. For example, neuro-fuzzy systems are able to forecast applications under imprecision and uncertainty (Wang et al., 2004). But these tools make no assumption of the system's dynamics (they are based on a learning phase) and do not explicitly take into account cause-effect relations.

It may not be necessary, but it can be useful to represent events' causal dependencies to improve prognostics by considering diagnosis outputs. In this case, dynamic Bayesian networks (DBNs) (Murphy, 2002) are applicable: their graphical representation and inference capability make them

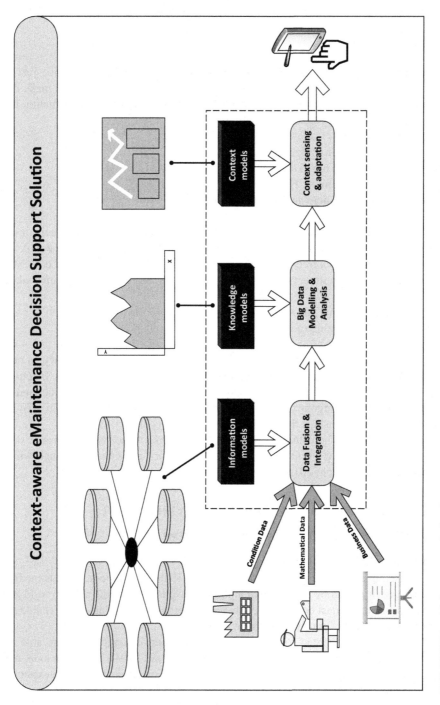

FIGURE 7.7   eMaintenance solution for maintenance analytics (Karim et al., 2016).

appropriate for fault diagnosis (Lerner et al., 2000) and prognostics (Muller et al., 2008). A DBN prognostics model for dynamic systems would take into account the dynamics of the system, sensors, and monitoring data, diagnosis outputs, and online introduced maintenance actions (Medjaher et al., 2009).

7.2.1.2 Prognostic Procedure for Dynamic Systems

When studying a dynamic system, it is useful to represent its physical knowledge and dynamic behavior using a bond graph (BG) tool. DBN is another suitable prognostics tool. The approaches can be linked for prognostics purposes (see Figure 7.8). This can be done in five steps (Medjaher et al., 2009):

1. Generate the BG model in integral causality.
2. Generate residuals from the BG model in derivative causality.
3. Construct the temporal causal graph (TCG) from the integral BG model.
4. Generate the DBN by using the structure of the TCG and the residuals' information.
5. Simulate the DBN to estimate the future state of the dynamic system and provide prognostic metrics.

The BG model shown in Figure 7.8 is constructed from the dynamic system by representing the power transfer between the physical elements composing it (Karnopp et al., 1990; Paynter, 1961). The mathematical model of the system's behavior can be obtained from the BG model in the form of state space equations or the transfer function. Residuals are generated from the derivative BG form. The BG form is preferred to the integral form for two reasons (Medjaher et al., 2009). First, in the detection stage, initial conditions are not always known, making integral calculation impossible. Second, the diagnosis phase involves observing the effect and trying to identify the cause (an inversion of causal relations). In this prognostics procedure, the generated residuals are used as monitoring features.

A TCG is obtained from the BG model in integral causality of the dynamic system (Mosterman & Biswas, 1997). This topological representation captures local dynamic relations between variables and provides a more explicit representation of relations between the system's parameters and the behavior variables (Lerner et al. 2000). It also indicates the algebraic and temporal constraints between the BG model's effort and flow variables. The TCG can be used to build the DBN.

A DBN is a way to extend Bayes' nets to model probability distributions over semi-infinite collections of random variables (U_t, X_t, Y_t) representing the input variables and the hidden and output variables of a state-space model (Murphy, 2002). DBNs generalize Kalman filters, hidden Markov

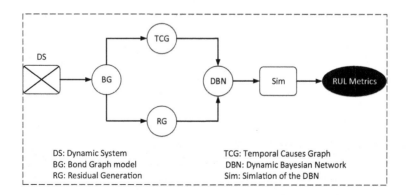

FIGURE 7.8 Steps of prognostics procedure (Medjaher et al., 2009).

models (HMMs), and hierarchical HMMs; users can thus monitor the system and predict its future state. The DBN related to a dynamic system is generated from the TCG as explained above (Lerner et al., 2000). In the simulation step, the DBN is parametrized by specifying the initial probability distributions of the nodes; inference algorithms can be implemented to predict the system's future states. Finally, the simulation results can be processed to provide prognostics metrics (Medjaher et al., 2009).

7.3 PROBABILISTIC MODELS FOR PRESCRIPTION

A probabilistic model quantifies uncertainty by integrating first principle knowledge with data to capture the dynamics in a distribution over model predictions for state transitions between samples in a batch run. In this sense, this category includes models that represent uncertain causal relationships (i.e., between causes and effects). In predictive and prescriptive analytics, probabilistic models are used to calculate the likelihood of certain events occurring instead of monitoring actual data in search of events and data points that conform to a set of rules defined by historical analysis (Lepenioti et al., 2020b).

As shown in Figure 7.3 and explained at greater length below, the probabilistic models for prescriptive analytics are classified as:

1. Markov decision process (MDP).
2. HMM.
3. Markov chain (MC).

7.3.1 MARKOV DECISION PROCESS

A mathematical representation of a complex decision-making process is the MDP. MDP is defined as the following (Cavaioni, 2017):

- State S: This represents every state we could be in, within a defined world.
- Model or transition function T: This is a function of the current state, the action taken and the state where it ends up. This transition produces a certain probability of ending up in state S, starting from state S and taking action A.
- Actions: These are things that can be done in a particular state.
- Reward: A reward is a scalar value for being in a state. It tells us the usefulness of entering the state.

The final goal of the MDP is to find a policy that can tell us, for any state, which action to take. The optimal policy maximizes the long-term expected reward (Cavaioni, 2017).

The Markovian process is based on two properties (Cavaioni, 2017):

1. Only the present matters: The transition function only depends on the current state S and not any of the previous states.
2. Things are stationary: Therefore, rules do not change over time.

7.3.2 HIDDEN MARKOV MODEL

HMMs are computationally straightforward and underpinned by powerful mathematical formalism. Thus, they provide a good statistical framework for solving a wide range of time-series problems. They have been successfully applied to pattern recognition and classification for many years.

The study of MCs was initiated in the early 1900s by Markov, who laid the foundation for the theory of stochastic processes. From the 1940s to 1960s, HMMs were investigated as a representation

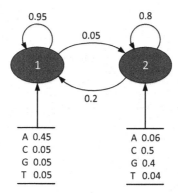

FIGURE 7.9 Simple HMM model for generating DNA sequences (Choo et al., 2004).

of stochastic functions of MCs. Their initial development was predominated by theoretical reasoning that attempted to solve problems pertaining to the issues of uniqueness and identifiability. HMMs did not gain popularity until the early 1970s when the technique was applied to speech recognition using an efficient training algorithm for HMMs.

In the late 1980s and early 1990s, HMMs were introduced to computational sequence analysis and protein structural modeling in molecular biology. However, HMMs became popular in computational biology only after researchers explored HMM-based profile methods for sequence alignment (Choo et al., 2004).

In a nutshell, HMMs are composed of two components. Associated with each HMM is a discrete-state, time-homologous, first-order MC with suitable transition probabilities between states and an initial distribution. In addition, each state emits symbols according to a pre-specified probability distribution over emission symbols or values. Emission probabilities are dependent only on the present state of the MC, regardless of previous states. Starting from some initial states with the initial probability, a sequence of states is generated by moving from one state to another following the state-transition probabilities until a final state is reached. Each state emits a symbol when it is visited, thus creating an observable sequence of symbols (Rabiner, 1989).

The key idea is that an HMM is a sequence generator describing a probability distribution over a set of possible sequences. The set of sequences ultimately forms a finite model. A simple HMM for generating a DNA sequence is specified in Figure 7.9 (Choo et al., 2004).

In Figure 7.9, state transitions and their associated probabilities are indicated by arrows; symbol emission probabilities for A, C, G, T at each state are indicated below the state. For clarity, the initial and final states are omitted, as well as the initial probability distribution. For instance, this model can generate the state sequence given in Figure 7.10, and each state emits a nucleotide according to the emission probability distribution (Choo et al., 2004).

When sequences of emissions are produced, only the output symbols can be observed. The sequences of states underlying MC are hidden and cannot be observed, hence the name HMM. Any sequence can be represented by a state sequence in the model. The probability of any sequence, given the model, is computed by multiplying the emission and transition probabilities along the path (Choo et al., 2004).

7.3.3 MARKOV CHAIN

An MC is a stochastic model describing a sequence of possible events in which the probability of each event depends only on the state attained in the previous event. In continuous-time, it is known as a Markov process. It is named after the Russian mathematician Andrey Markov. The adjective Markovian is used to describe something that is related to a Markov process.

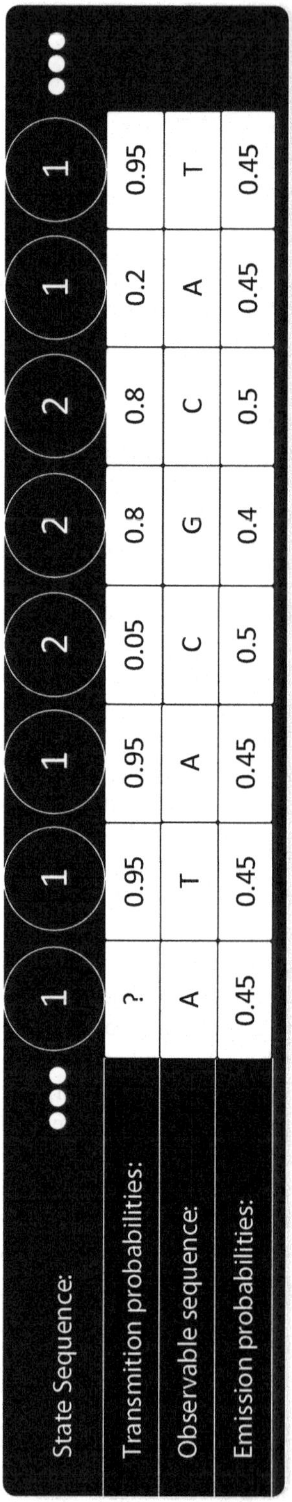

State Sequence:	1	1	1	2	2	2	2	1	1	
Transmition probabilities:	?	0.95	0.95	0.05	0.8	0.8	0.05	0.2	0.95	
Observable sequence:	A	T	A	C	G	C	C	A	T	
Emission probabilities:	0.45	0.45	0.45	0.5	0.4	0.5	0.5	0.45	0.45	

FIGURE 7.10 Generated state sequence and associated DNA sequence (Choo et al., 2004).

Markov processes are the basis for general stochastic simulation methods known as MC Monte Carlo, used for simulating sampling from complex probability distributions, and have found application in Bayesian statistics and AI.

A Markov process is a stochastic process that satisfies the Markov property (sometimes characterized as "memorylessness"). In simple terms, it is a process for which predictions can be made regarding future outcomes based solely on the present state of the system. Importantly, such predictions are just as good as the ones that could be made knowing the process's full history. In other words, conditional on the present state of the system, its future and past states are independent.

An MC is a type of Markov process that has either a discrete state space or a discrete index set (often representing time), but the precise definition of an MC varies. For example, it is common to define an MC as a Markov process in either discrete or continuous time with a countable state space (thus regardless of the nature of time), but it is also common to define an MC as having discrete time in either countable or continuous state space (thus regardless of the state space) (Wikipedia, 2020).

MCs have many applications as statistical models of real-world processes, such as studying cruise control systems in motor vehicles, queues or lines of customers arriving at an airport, currency exchange rates, and animal population dynamics.

7.3.3.1 Types of Markov Chains

The system's state space and time parameter index need to be specified. Table 7.1 gives an overview of the different instances of Markov processes for different levels of state space generality and for discrete time vs. continuous time (Wikipedia, 2020).

There is no definitive agreement on the use of some of the terms that signify special cases of Markov processes. Usually the term MC is reserved for a process with a discrete set of times, that is, a discrete-time MC (DTMC), but a few authors use the term Markov process to refer to a continuous-time MC (CTMC) without explicitly saying so. Other extensions of Markov processes are referred to as such but do not necessarily fall within any of the four categories. Moreover, the time index need not necessarily be real-valued; like with the state space, there are conceivable processes that move through index sets with other mathematical constructs. The general state space CTMC is general to such a degree that it has no designated term.

While the time parameter is usually discrete, the state space of an MC does not have any generally agreed-on restrictions: the term may refer to a process or an arbitrary state space. However, many applications of MCs employ finite or countably infinite state spaces, which have a more straightforward statistical analysis. Besides time-index and state-space parameters, there are many other variations, extensions and generalizations (Wikipedia, 2020).

7.3.3.2 Transitions

The changes of state of the system are called transitions. The probabilities associated with various state changes are called transition probabilities. The process is characterized by a state space, a

---

TABLE 7.1

Different instances of Markov Processes

	Countable State Space	Continuous or General State Space
Discrete-time	(Discrete-time) Markov chain on a countable or finite state space.	Markov chain on a measurable state space (e.g., Harris chain).
Continuous-time	Continuous-time Markov process or Markov jump process.	Any continuous stochastic process with the Markov property (e.g., the Wiener process).

Source: Wikipedia (2020).

transition matrix describing the probabilities of particular transitions, and an initial state (or initial distribution) across the state space. By convention, we assume all possible states and transitions have been included in the definition of the process, so there is always a next state, and the process does not terminate.

A discrete-time random process involves a system which is in a certain state at each step, with the state changing randomly between steps. The steps are often thought of as moments in time, but they can equally well refer to physical distance or any other discrete measurement. Formally, the steps are the integers or natural numbers, and the random process is a mapping of these two states. The Markov property states that the conditional probability distribution for the system at the next step (and in fact at all future steps) depends only on the current state of the system and not additionally on the state of the system at previous steps.

Since the system changes randomly, it is generally impossible to predict with certainty the state of an MC at a given point in the future. However, the statistical properties of the system's future can be predicted. In many applications, it is these statistical properties that are important (Wikipedia, 2020).

7.4 MACHINE LEARNING AND DATA MINING IN PRESCRIPTIVE ANALYTICS

ML refers to algorithms that rely on models and inference based on data processing without explicit instructions (Nasrabadi, 2007). ML algorithms build a mathematical model of sample data, known as "training data", in order to make predictions or decisions without being explicitly programmed to perform the task. It has been considered a subset of AI.

Data mining looks for patterns in large datasets. The goal is to extract information and transform it into a comprehensible structure for further use. It involves analysis, database and data management aspects, data pre-processing, interestingness metrics, complexity considerations, model and inference considerations, post-processing of discovered structures, visualization, and online updating.

ML and data mining are closely interrelated (Chakrabarti et al., 2006) and are treated as one category of methods. Essentially, using ML and data mining techniques, we can build algorithms to extract data and see important hidden information. In predictive analytics, information is sought that can predict the future outcome from data based on previous patterns while prescriptive analytics looks for information that can help find the best course of action for a given situation (Lepenioti, 2020b).

ML/data mining algorithms have been widely used in predictive analytics. However, their exploitation in the context of prescriptive analytics is scarce. One use of ML and data mining algorithms is based on k-means clustering to address the problem of adaptive data placement across distributed nodes in a secure way by considering the sensitivity and security of the underlying data (Revathy & Mukesh, 2018). In another application, Shroff et al. (2014) proposed a framework based upon reinforcement learning (RL), that is, learning the behavior of an agent through trial-and-error interactions within a dynamic environment, integrated with simulation and optimization in an online learning scenario.

Prescriptive analytics takes the output from ML and deep learning to predict future events (predictive analytics) and then uses the prediction to initiate proactive decisions. These can be outside the bounds of human interaction. An autonomous car transports people safely to a predetermined destination after selecting a route based on current data (e.g., traffic congestion). But an autonomous car equipped with prescriptive analytics does more. It can transport people to a location based on deep learning analytics, for example, going to the grocery store based on dinner plans pulled from social media channels or going to the doctor's office based on external inputs recognizing an individual has been exposed to a virus and may need medical attention (such inputs could come from personal devices based heart rate monitor, social media, proximity tracking etc.).

In another example of prescriptive analytics, an enterprise application might be autonomously moved from one server to another based on the predicted failure of the server or the data center.

In the very near future, prescriptive analytics will be incorporated into a host of applications, including stock market transactions, medical diagnosis and treatment, design manufacturing, power generation, and more (Elliott, 2018).

7.4.1 MACHINE LEARNING FOR PRESCRIPTIVE ANALYTICS

Prescriptive analytics aims at answering the questions "What should I do?" and "Why should I do it?" It is able to bring business value through adaptive, time-dependent, and optimal decisions on the basis of predictions about future events. There is an increasing interest in prescriptive analytics for smart manufacturing; it is considered to be the next step in the evolution of data analytics maturity for optimized decision-making, ahead of time (Lepenioti al., 2020a).

The most important challenges of prescriptive analytics include the following (Lepenioti et al., 2020a):

1. Addressing the uncertainty introduced by the predictions, the incomplete and noisy data, and the subjectivity in human judgment.
2. Combining the "learned knowledge" of ML and data mining methods with the "engineered knowledge" elicited from domain experts.
3. Developing generic prescriptive analytics methods and algorithms utilizing AI and ML instead of problem-specific optimization models.
4. Incorporating adaptation mechanisms capable of processing data and human feedback to continuously improve the decision-making process over time and to generate non-intrusive prescriptions.
5. Recommending optimal plans out of a list of alternative (sets of) actions.

A novel method of prescriptive analytics is RL, which is considered to be a third ML paradigm, alongside supervised learning and unsupervised learning. RL shows increasing use as a tool for optimal policies in manufacturing problems. In RL, the problem is represented by an environment consisting of states and actions and learning agents with a defined goal state. The agents aim to reach the goal state while maximizing the rewards by selecting actions and moving to different states. Interactive RL has the additional capability of incorporating evaluative feedback from a human observer, so the RL agent learns from both human feedback and environmental reward. Another extension is multi-objective RL (MORL), a sequential decision-making problem with multiple objectives. MORL requires a learning agent to obtain action policies that can optimize multiple objectives at the same time (Lepenioti et al., 2020a).

7.4.1.1 Multi-Objective Reinforcement Learning for Prescriptive Analytics

When MORL is incorporated into prescriptive analytics, the process is able to (Lepenioti et al., 2020a):

- Recommend (prescribe) both perfect and imperfect actions (e.g., maintenance actions with various degrees of restoration.
- Model the decision-making process under uncertainty instead of the physical manufacturing process, thus making it applicable to various industries and production processes.
- Incorporate the preference of domain experts into the decision-making process (e.g., according to their skills, and experience), in the form of feedback on the generated prescriptions.

Most multi-objective optimization approaches result in a Pareto set of optimal solutions, but this approach provides a single optimal solution (prescription), thus generating more concrete insights to the user. The algorithm consists of three steps: prescriptive model building, prescriptive model solving, and prescriptive model adapting, described in detail below (Lepenioti et al., 2020a).

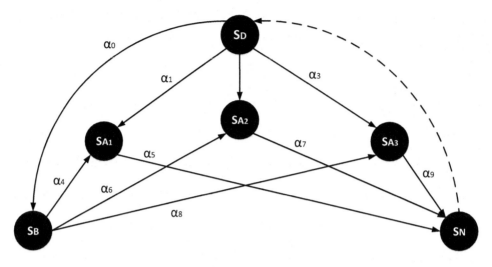

FIGURE 7.11 Example of prescriptive analytics models with three alternative (sets of) actions (Lepenioti et al., 2020a).

7.4.1.1.1 Prescriptive Model Building

The prescriptive analytics model representing the decision-making process is defined by a tuple (S, A, T, R), where S is the state space, A is the action space, T is the transition function $T : S \times A \times S \rightarrow \mathbb{R}$, and R is the vector reward function $R : S \times A \times S \rightarrow \mathbb{R}^n$ where the n-dimensions are associated with the objectives to be optimized, O_n.

The model has a single starting state S_N, from which the agent starts the episode, and a state S_B that the agent tries to avoid. Each episode of the training process of the RL agent will end when the agent returns to the normal state S_N or when it reaches S_B.

Figure 7.11 depicts an example including three alternative (perfect and/or imperfect maintenance) actions (or sets of actions), S_{A_i}, $i = 1, 2, 3$, each one of which is assigned to a reward vector. The prescriptive analytics model is built dynamically. In this sense, the latest updates on the number of the action states S_{A_i} and the estimations of the objectives' values for each state S_k are retrieved through application programming interfaces (APIs) from the predictive analytics. Each action may be implemented either before the breakdown (to eliminate or mitigate its impact) or after the breakdown (if this occurs before the implementation of mitigating actions). After the implementation of each action, the equipment returns to its normal state SN. The solid lines in the figure represent the transitions that have non-zero reward with respect to the optimization objectives and move the agent from one state to another (Lepenioti et al., 2020a).

7.4.1.1.2 Prescriptive Model Deployment

On the basis of event triggers for predicted abnormal situations (e.g., about the time of the next breakdown) received through a message broker, the model moves from the normal state S_N to the dangerous state S_D. For each objective, the reward functions are defined according to whether the objective is to be maximized or minimized. On this basis, the optimal policy $\pi_{O_i}(s,a)$ for each objective O_i is calculated with the use of the actor-critic algorithm, a policy gradient algorithm aiming at searching directly in (some subset of) the policy space starting with a mapping from a finite-dimensional (parameter) space to the space of policies. Assuming independent objectives, the multi-objective optimal policy is derived from: $\pi_{opt}(s,a) = \prod_{i \in I} \pi_{O_i}(s,a)$. The time constraints of the optimal policy (prescription) are defined by the prediction event trigger. The prescription is exposed to operators on the shop-floor (e.g., through a mobile device) providing them the capability to accept or reject it. If accepted, the prescribed action is added to the action plan (Lepenioti et al., 2020a).

7.4.1.1.3 Prescriptive Model Adaptation

The prescriptive analytics model is able to adapt according to feedback from experts on the generated prescriptions. This approach learns from the operators whether the prescribed actions converge with their experience or skills and incorporates their preference into the prescriptive analytics model. In this way, it provides non-disruptive decision augmentation and, thus, achieves an optimized human-machine interaction, while, at the same time, optimizing manufacturing key performance indicators (KPIs). To do this, it implements the policy shaping algorithm, a Bayesian approach that attempts to maximize the information gained from human feedback by utilizing it as a direct label on the policy. For each prescription, optional human feedback is received as a signal of approval or rejection, numerically mapped to the reward signals, and interpreted as a step function. The feedback is converted into a policy $\pi_{feedback}(s,a)$, the distribution of which relies on the consistency, expressing the user's knowledge of the optimality of the actions, and the likelihood of receiving feedback. Assuming the feedback policy is independent of the optimal multi-objective policy, the synthetic optimal policy for the optimization objectives and the human feedback is calculated as: $\pi_{opt}(s,a) = \pi_{opt}(s,a) * \pi_{feedback}(s,a)$ (Lepenioti et al., 2020a).

7.4.2 Data Mining for Prescriptive Analytics

Predictive analytics and data mining use algorithms to discover knowledge and find the best solutions. These algorithms can extract and analyze useful information and automatically discover hidden patterns and relationships in data. Predictive analytics uses data patterns to make predictions. In this aspect, it is much like ML, in which machines take historical and current information and apply it to a model to predict future trends (Expert System, 2016).

When data mining and predictive analytics are done right, the analyses aren't a means to a predictive end; rather, the desired predictions become a means to analytical insight and discovery. A better job can be done analyzing what really is necessary to analyze and predicting what we really want to predict (SAS, 2020).

7.4.2.1 Predictive and Prescriptive Analytics

7.4.2.1.1 Predictive Analytics

Predictive analytics makes future estimations based on past measurements or observation of the process (see Section 7.1.1.2). These analytics use statistics integrated with mathematical programming and operations research techniques. Predictions on manufacturing yields and properties, raw material and product demands, revenues, etc. are the main determinations. For Smart Manufacturing (SM) in the Information Age (IA), the gains and biases determined by predictive analytics must be updated online in iterative prescriptions to reduce the uncertainties in the prescribed near to short future decisions (Menezes et al., 2019).

7.4.2.1.2 Prescriptive Analytics

Prescriptive analytics is the application of mathematical programming and operations research techniques to automate decisions on given data and proposed modeling (see Section 7.1.1.3). Exact methods found in the optimization, such as branch-and bound, solve thousands of logic variables. However, more significant and complex problems are forcing the industry to use heuristics approaches (non-exact methods) to find good but perhaps not optimal solutions of what to do next (Menezes et al., 2019).

7.4.2.1.3 Prescriptive with Predictive Analytics for Smart Manufacturing

Combining predictive and prescriptive analytics is essential for smarter decisions in manufacturing. Strategies toward online automated decisions for prescriptive analytics such as scheduling bring new modeling and solving aspects to factually correct the scenarios and input data (or even output data) in a re-decision cycle. Hence, both activation and frequency must be determined for the manufacturing process characteristics.

Brand-new information on the process state (different from the last one) makes the prescription about what to do suboptimal or infeasible, motivating the need for re-calculation of the decisions, even considering an open-loop strategy by only updating the inputs of the problem. The inherent uncertainty of the process can be reduced, for example, by better prediction of the model by including the valid state or measured outputs using parameter feedback and operational updates from the field in a closed-loop strategy. In this case, using data reconciliation and parameter estimation techniques, gains and biases can be updated and integrated into the next online re-decision cycle. Parameter-feedback considers bias or corrections in the outputs (results) of the next re-calculation for a new solution considering measurements from the near past time. Other feedbacks are the gradient-feedback for updates related to the process and objective function gains and the variable-feedback to refresh the input data at each cycle (Menezes et al., 2019).

7.4.2.2 Prescriptive Analytics in the Information Age (IA)

Prescriptive analytics uses optimization techniques with exact methods (or even non-exact) to automatically find solutions of a constrained maximization or minimization problem with model and data to be matched to the physical system. This is supported by mathematical programming solvers, such as CPLEX, GUROBI, and XPRESS for mixed-integer linear programming, and IPOPT, CONOPT, or any sequential linear programming (SLP) technology, etc., for nonlinear programming. In this type of analytics, the model and data are transformed into smarter decisions to prescribe future activities within a near-, short-, medium-, and long-term horizon. This enables companies to efficiently and effectively expand the scope and scale of the decision domain rather than counting on manual calculations.

In prescriptive applications, the veracity of the data is a critical body to be shaped and explored to integrate proper data (timely and good quality) to high-performance decision-making engines. When comparing cyber-physical systems (CPSs) from the so-called process mining farmland, with decisions traced exclusively from data (the data mining territory), the consistency of the data is more critical in the latter, as what is discovered and drawn from the data guides the decision to be made. The exact and non-exact solutions of rigorous optimization and heuristic algorithms respectively are an intricate part of the whole. Ultimately, if the data are not updated or treated enough, the decisions are unreliable and inconsistent (Menezes et al., 2019).

7.5 SIMULATION AND LOGIC-BASED METHODS FOR RUL ESTIMATION IN PRESCRIPTIVE ANALYTICS

7.5.1 Simulation

Another important category of methods for prescriptive analytics is simulation. Giurgiu et al. (2017) applied simulation over the random forest model to prescribe the required improvement actions to servers by using the information available from incident tickets. Jank, Dölle, and Schuh (2018) exploited stochastic simulation to maximize the benefits of a company's product portfolio in accordance with the corporate objectives. Meanwhile, Wang, Cheng, and Deng (2018) proposed a method based on stochastic simulation over a Bayesian belief network to conduct fact-based decision-making based on KPIs in a data-driven way. In this way, they moved beyond expert-oriented or multi-criteria decision-making (Lepenioti et al., 2020b).

7.5.2 Logic-Based Methods

Logic-based models, especially rule-based systems, have been used to incorporate expert knowledge into prescriptive analytics models. Ceravolo and Zavatarelli (2015) developed a knowledge acquisition process as an investigation of process executions. The prescriptive knowledge base evaluates the achievement of the business rules or the objectives associated with a process and identifies unexpected patterns. Matyas et al. (2017) proposed a procedural approach to prescriptive maintenance planning. The approach is based on rules represented by mathematical functions for each machine component, taking into account the prognoses of the wear reserve for machine components, condition-based monitoring, and variations in product quality.

Ramannavar and Sidnal (2018) proposed a distributed architecture integrating advanced analytics to map a job to resumes using semantic technologies. For a given job description, every resume is ranked according to two measures, coverage and comprehensibility, denoting the number of concepts from a predefined class and the number of sections and subsections covered in a resume, respectively. Du et al. (2016) focused on the presentation and explanation of recommendations of temporal event sequences using a prescriptive analytics interface. They developed the Event Action system, which identifies similar records, estimates the current record's potential outcomes based on the outcome distribution of the similar archived records, and recommends actions by summarizing the activities of those who achieved the desired outcome. This approach was applied in an education scenario and in a digital marketing scenario (Du et al., 2018). In the latter, the authors evaluated the effectiveness of their tool in helping marketers by prescribing personalized marketing interventions.

Prescriptive analytics have also been utilized in the 5W1H (what to do, with whom, where, how/when) methodology embedded in an information system for the purpose of providing recommendations and advice on research directions to researchers (Lepenioti et al., 2020b).

As shown in Figure 7.3 (see Section 7.1.2), the logic-based methods for prescriptive analytics are classified as:

1. Association rules.
2. Decision rules.
3. Criteria-based rules.
4. Fuzzy rules.
5. Distributed rules.
6. Benchmark rules.
7. Desirability functions.
8. Graph-based recommendations.
9. 5W1H.

7.5.2.1 Association Rules

Association rules are the result of process analytics that specify patterns of relationships among items. An example would be:

{charcoal, lighter, chicken wings} → {barbecue sauce}

To explain what we see here, charcoal, lighter, and chicken wings imply barbecue sauce. Those curly brackets indicate a set. Items in a set are called elements. When an item-set like {charcoal, lighter, chicken wings, barbecue sauce} appears in our data with some regularity, this means we have discovered a pattern (SOCR, 2017).

Association rules are commonly used for unsupervised discovery of knowledge, not the prediction of outcomes. In biomedical research, association rules are widely used to (SOCR, 2017):

- Search for interesting or frequently occurring DNA patterns.
- Search for protein sequences in analyses of cancer data.

Elsewhere in the medical field, association rules can be used to find patterns of medical claims that occur in combination with fraudulent credit card or insurance use.

7.5.2.1.1   Mining Association Rules

Association rule mining is a technique used to discover relationships among very large sets of variables in databases (Lin et al., 2019). Basic association analysis deals with the occurrence of one item together with another. More complicated analysis can also consider the quantity of occurrence, price, sequence of occurrence, and so on. The method for finding association rules through data mining involves the following sequential steps (Kotu & Deshpande, 2016):

- Step 1: Prepare the data in transaction format. An association algorithm needs input data to be formatted in a particular format.
- Step 2: Short-list frequently occurring item sets. Item sets are combinations of items. An association algorithm limits the analysis to the most frequently occurring items, so the final rule set extracted in the next step is more meaningful.
- Step 3: Generate relevant association rules from item sets. The algorithm generates and filters the rules based on the interest measure.

At times, the associations between the data in the database are not known. Association rule mining (also called association rule learning) is a common technique to find the hitherto unknown interdependence of the data and discover the rules between those items. In other words, it involves finding the rules that govern associations and causality (antecedent and consequent) between sets of items. Simply stated, the rules for association rule analysis refer to $X \Rightarrow Y$, where X is called the antecedent, and Y is the consequent. That is, the rule of "if X then Y".

Agrawal, Imielinski, and Swami (1993) first proposed the framework for association rule mining. The rule's basic measurements have the following:

- Minimum transactional support s: The union of items in the consequent and antecedent of the rule is present in a minimum of $s\%$ of transactions in the database.
- Minimum confidence c: At least $c\%$ of transactions in the database that satisfy the antecedent of the rule also satisfy the consequent of the rule.

Association rules mining is usually required to have support and confidence greater than or equal to the user-specified minimum support and minimum confidence respectively (Lin et al., 2019).

7.5.2.2   Decision Rules

Rule-based decision techniques are used in prescriptive analytics to make decisions (MathWorks, 2020). One example is opting to shut down equipment for maintenance when sensor readings exceed certain established thresholds. Another is deciding to accept a financial transaction when a credit score is high enough. Prescriptive analytics using rule-based decision techniques includes consideration of uncertainty so that decisions are robust against a range of outcomes. Monte Carlo simulation is commonly used for this analysis.

Examples of rule-based decision techniques are: criteria-based rules, fuzzy rules, distributed rules, and benchmark rules.

REFERENCES

Agrawal R., Imielinski T., Swami A., 1993. Mining association rules between sets of items in large databases. Proceedings of the 1993 ACM SIGMOD Conference. https://rakesh.agrawal-family.com/papers/sigmod93assoc.pdf

Bachar D., 2020. Descriptive, predictive and prescriptive analytics explained. Logility: Planning Optimized. www.logility.com/blog/descriptive-predictive-and-prescriptive-analytics-explained. Viewed: June 22, 2020.

Basu A., 2013. Five pillars of prescriptive analytics success. June 3, 2013 in Executive Edge. https://pubsonl ine.informs.org/do/10.1287/LYTX.2013.02.07/full/. Viewed: July 1, 2020.

Cavaioni M., 2017. Machine learning: Reinforcement learning – Markov decision processes. February 7, 2017. https://medium.com/machine-learning-bites/machine-learning-reinforcement-learning-markov-decis ion-processes-431762c7515b. Viewed: July 1, 2020.

Ceravolo P., Zavatarelli F., 2015. Knowledge acquisition in process intelligence. 2015 International Conference on Information and Communication Technology Research.

Chakrabarti S., Ester M., Fayyad U., Gehrke J., Han J., Morishita S., Piatetsky-Shapiro G., Wang W., 2006. Data mining curriculum: A proposal (version 1.0). Intensive Working Group of ACM SIGKDD Curriculum Committee. April 30, 2006.

Choo K. H., Tong J. C., Zhang L., 2004. Recent applications of hidden Markov models in computational biology. Genomics Proteomics Bioinformatics. 2(2), 84–96.

DataSkills, 2018. From predictive analytics to prescriptive analytics. Understanding the World. www.dataskills. ai/from-predictive-analytics-to-prescriptive-analytics/#gref. Viewed: June 23, 2020.

DeZyre, 2019. Types of analytics: Descriptive, predictive, prescriptive analytics. August 01, 2019. www.dezyre. com/article/types-of-analytics-descriptive-predictive-prescriptive-analytics/209. Viewed: June 23, 2020.

Du F., Malik S., Koh E., Theocharous G., 2018. Interactive campaign planning for marketing analysts Extended abstracts of the 2018 CHI Conference on Human Factors in Computing Systems, LBW006.

Du F., Plaisant C., Spring N., Shneiderman B., 2016. Event action: Visual analytics for temporal event sequence recommendation. IEEE Conference on Visual Analytics Science and Technology, 61–70.

Elliott M., 2018. Machine learning, deep learning and prescriptive analytics: What's the difference? January 10, 2018. https://blog.netapp.com/machine-learning-deep-learning-and-prescriptive-analytics-whats-the-dif ference/#:~:text=Prescriptive%20analytics%20takes%20the%20output,a%20destination%20that%20 you%20determine. Viewed: June 20, 2020.

Expert System, 2016. Data mining and predictive analytics: What is the difference? December 2, 2016. https:// expertsystem.com/data-mining-predictive-analytics-difference. Viewed: July 19, 2020.

Giurgiu I., Wiesmann D., Bogojeska J., Lanyi D., Stark G., Wallace R. B., Hidalgo A. A., 2017. On the adoption and impact of predictive analytics for server incident reduction IBM Journal of Research and Development, 61(1), 9–98.

Jank M. H., Dölle C., Schuh G., 2018. Product portfolio design using prescriptive analytics. Congress of the German Academic Association for Production Technology, 584–593.

Kajko-Mattsson M., Karim R., Mirjamdotter A., 2011. Essential components of eMaintenance. International Journal of Performability Engineering, 7(6), 505–521.

Karim R., 2008. A service-oriented approach to eMaintenance of complex technical systems. Doctoral thesis, Luleå University of Technology, Luleå, Sweden.

Karim R., Westerberg J., Galar D., Kumar U., 2016. Maintenance analytics: The new know in maintenance. Division of Operation, Maintenance Engineering, Luleå University of Technology. SE-971 87, Luleå, Sweden.

Karnopp D., Margolis D., Rosenberg R., 1990. Systems dynamics: A unified approach. John Wiley, 2nd edition.

Kotu V., Deshpande B., 2016. Predictive analytics and data mining. Chapter 6 – Association analysis. Concepts and Practice with Rapidminer. 195–216.

Lepenioti K., Pertselakis M., Bousdekis A., Louca A., Lampathaki F., Apostolou D., Mentzas G., Anastasiou S., 2020a. Machine learning for predictive and prescriptive analytics of operational data in smart manufac turing. Dupuy-Chessa and H. A. Proper (Eds.): CAiSE 2020 Workshops, LNBIP 382, 5–16.

Lepenioti K., Alexandros Bousdekisa A., Apostoloua D., Mentzasa G., 2020b. Prescriptive analytics: Literature review and research challenges. International Journal of Information Management 50, 57–70.

Lerner U., Parr R., Koller D., Biswas G., 2000. Bayesian fault detection and diagnosis in dynamic systems. Proc. of American Association for Artificial Intelligence.

Lin H-K., Hsieh C-H., Wei N-C., Peng Y-C., 2019. Association rules mining in R for product performance management in industry 4.0. 11th CIRP Conference on Industrial Product-Service Systems. Procedia CIRP, 83(2019), 699–704.

MathWorks, 2020. Prescriptive Analytics. www.mathworks.com/discovery/prescriptive-analytics.html. Viewed: July 1, 2020.

Matyas K., Nemeth T., Kovacs K., Glawar R., 2017. A procedural approach for realizing prescriptive mainten ance planning in manufacturing industries. CIRP Annals, 66(1), 461–464.

Medjaher K., Gouriveau R., Zerhouni N., 2009. A procedure for failure prognostic in dynamic systems. FEMTO-ST Institute, CNRS - UFC/ENSMM/UTBM, France. Proceedings of the 13th IFAC Symposium on Information Control Problems in Manufacturing Moscow, Russia, June 3–5, 2009.

Menezes B. C., Kelly J. D., Leal A. G., Le Roux G. C., 2019. Predictive, prescriptive and detective analytics for smart manufacturing in the information age. IFAC Papers OnLine 52-1 (2019) 568–573.

Mosterman P., Biswas G., 1997. Monitoring, prediction and fault isolation in dynamic physical systems. Proceedings of AAAI, 100–105.

Muller A., Suhner M., Iung B., 2008. Formalization of a new prognosis model for supporting proactive maintenance implementation on industrial system. Reliability Engineering and System Safety, 93, 234–253.

Murphy P., 2002. Dynamic Bayesian networks: Representation, inference and learning. Ph.D. thesis, University of California, Berkeley.

Nasrabadi N. M., 2007. Pattern recognition and machine learning. Journal of Electronic Imaging, 16(4), 049901.

Paynter H., 1961. Analysis and design of engineering systems. M.I.T. Press.

Rabiner L. R., 1989. A tutorial on hidden Markov models and selected applications in speech recognition. Proceedings IEEE 1989, 77:257–286.

Ramannavar M., Sidnal N. S., 2018. A proposed contextual model for Big data analysis using advanced analytics Big data analytics. Springer, 329–339.

Revathy P., Muskesh R., 2018. HadoopSec: sensitivity-aware secure data placement strategy for Big Data/Hadoop Platform using Prescriptive Analytics. Journal on Computing, 6(1).

SAS, 2020. Data mining. What it is and why it matters. www.sas.com/en_us/insights/analytics/data-mining.html. Viewed: July 18, 2020.

Shroff G., Agarwal P., Singh K., Kazmi A. H., Shah S., Sardeshmukh A., 2014. Prescriptive information fusion. TCS Research. October 2014.

SOCR, 2017. Data Science and Predictive Analytics (UMich HS650). Apriori Association Rules Learning. SOCR/MIDAS (Ivo Dinov). SOCR Resource. June 2017. www.socr.umich.edu/people/dinov/2017/Spring/DSPA_HS650/notes/11_Apriory_AssocRuleLearning.html#2_association_rules. Viewed: July 1, 2020.

Stitch, 2020. Using prescriptive analytics. Stitch: A talent company. www.stitchdata.com/resources/prescriptive-analytics. Viewed: June 24, 2020.

Wang C. H., Cheng H. Y., Deng Y. T., 2018. Using Bayesian belief network and time-series model to conduct prescriptive and predictive analytics for computer industries. Computers & Industrial Engineering, 115, 486–494.

Wang W., Goldnaraghi M., Ismail F., 2004. Prognosis of machine health condition using neuro-fuzzy systems. Mechanical Systems and Signal Processing, 18, 813–831.

Wikipedia, 2020. Markov Chain. https://en.wikipedia.org/wiki/Markov_chain. Viewed: July 3, 2020.

8 Uncertainty Management and the Confidence of RUL Predictions

8.1 UNCERTAINTY REPRESENTATION AND INTERPRETATION

Uncertainty representation and the interpretation of uncertainty are affected by which modeling and simulation frameworks are being used. Methods for uncertainty representation vary in both granularity and detail. Common theories include the following: classical set theory, probability theory, fuzzy set theory, fuzzy measure theory (plausibility and belief), and rough set theory (upper and lower approximations).

Probability theory is widely used for prognostics and health management (PHM) (Celaya, 2012), where the representation of uncertainties is dominated by probabilistic measures. This approach is mathematically rigorous, but it assumes the availability of a statistically sufficient database. Other approaches, such as possibility theory (fuzzy logic) and Dempster–Shafer theory (DST), can be employed when data are scarce or incomplete (Wang, 2011). Note that the choice of a probability density function (PDF) affects the quality of the prognostics. Several approaches assume normal PDFs (Celaya, 2012).

Probabilistic methods can interpret uncertainty as physical or frequentist (classical) or subjectivist (Bayesian) (Sankararaman & Goebel, 2013b). The former interpretation implies uncertainty exists only in the context of natural randomness across multiple nominally identical experiments; the latter associates uncertainty with non-random events. In other words, uncertainty reflects the analyst's beliefs about the occurrence or non-occurrence of events (Sankararaman & Goebel, 2015).

8.1.1 UNCERTAINTY INTERPRETATION

Even though the mathematical axioms and theorems of probability are well-established, researchers disagree on the interpretation of probability. As mentioned above, there are two major interpretations, one based on physical probabilities and the other on subjective probabilities (Sankararaman & Goebel, 2014). It is essential to understand the difference before attempting to interpret uncertainty to make remaining useful life (RUL) predictions and perform prognostics (Sankararaman, 2015).

Physical probabilities (alternatively termed objective or frequentist probabilities) are related to random physical systems. To give an example, each trial in an experiment leads to a certain event (a subset of the sample space), and over repeated trials, each event tends to occur at more or less the same rate, called the relative frequency. The relative frequency is expressed in terms of physical probability. Thus, physical probabilities are defined in the context of random experiments.

Any statement can have a subjective probability. The statement does not need to concern an event that is a possible outcome of an experiment. Bayesian methodology is based on subjective probabilities that are simply degrees of belief. They quantify the extent to which the statement is supported by available knowledge and evidence. In fact, even deterministic quantities can be represented using probability distributions that reflect the analyst's subjective beliefs.

DOI: 10.1201/9781003097242-8

Subjective probability analysis is not really suitable for condition-based monitoring or online health monitoring. Only one asset or system is being monitored, and at any given time, there is no physical randomness. Therefore, despite uncertainty, the asset or system cannot be represented using a probability distribution. At the same time, however, system state estimation during health monitoring frequently uses Kalman or particle filters (PFs), and these approaches compute the state variables' probability distributions. Particle and Kalman filtering approaches use Bayes' theorem, and they deal with subjective probability. This suggests the uncertainty estimated by filtering algorithms reflects the analyst's beliefs and, as such, is not related to actual physical probabilities (Sankararaman & Goebel, 2014).

8.1.2 Representing Uncertainty in Prognostic Tasks

When modeling the degradation and failure behavior of a structure, system, or component (SSC), uncertainty can be considered essentially of two different types: randomness because of inherent variability in physical behavior (aleatory uncertainty) and imprecision due to lack of knowledge and information (epistemic uncertainty):

- Aleatory uncertainty arises from the intrinsic variability of the process under study and cannot be reduced by further measurements. It is usually due to the random nature of the input data, which can be mathematically represented by a probability distribution once enough experimental data are available.
- Epistemic uncertainty arises from the lack of knowledge about the parameters characterizing the physical system. It is due to ignorance, lack of knowledge, or incomplete information.

The distinction between aleatory and epistemic uncertainty is relevant to prognostics. The aleatory uncertainty affects the time evolution of the degradation process of the SSC, whereas epistemic uncertainty arises from incomplete knowledge of fixed but poorly known parameter values which enter the models used to evaluate the SSC reliability and RUL.

While the aleatory uncertainty is appropriately represented by probability distributions, current scientific discussions dispute the advantages and disadvantages of two possible representations of the epistemic uncertainty: probabilistic and possibilistic (Baraldi, 2012).

8.1.2.1 Probabilistic Representation of Epistemic Uncertainty

In current reliability, availability, maintainability, safety (RAMS) practice, epistemic uncertainty is often represented using probability distributions as is done for aleatory uncertainty. When sufficient data are not available for statistical analysis, it is possible to adopt a subjective view of probability based on expert judgment. For example, in risk assessments and RAMS analyses, it is common practice to use lognormal distributions to represent epistemic uncertainty on the parameters of the probability distributions of the time to failure (TTF) and time to repair of a component.

However, there might be some limitations to the probabilistic representation of epistemic uncertainty under limited knowledge (Baraldi, 2012).

8.1.2.2 Possibilistic Representation of the Epistemic Uncertainty

Given the potential limitations associated with a probabilistic representation of epistemic uncertainty under limited information, several alternative frameworks have been proposed, including evidence theory, fuzzy set theory, possibility theory, imprecise probability, and interval analysis. Possibility theory is particularly attractive for prognosis because of its representation power and relative mathematical simplicity. It is similar to probability theory in that it is based on set functions but differs in that it relies on two cumulative functions, called possibility (Π) and necessity N measures, instead of only one, to represent uncertainty.

The basic notion upon which the theory is founded is the possibility distribution of an uncertain variable (not necessarily random) which assigns to each real number x in a range X (universe of discourse a degree of possibility π x) $\in[0,1]$ of being the correct value.

A possibility distribution is thus an upper, semi-continuous mapping from the real line to the unit interval describing what an analyst knows about the more or less plausible values x of the uncertain variable ranging on X. These values are mutually exclusive, as the uncertain variable can take on only one true value (Baraldi, 2012).

8.1.2.3 Transformation from a Possibilistic Distribution to a Probabilistic Distribution

For the purposes of comparing probability and possibility representations of uncertainty, it is useful to consider a method for transforming a possibility distribution into a probability distribution. Obviously, the probabilistic and the possibilistic representations of uncertainty are not equivalent: the possibilistic representation is weaker than the probabilistic one because it explicitly handles imprecision (e.g., incomplete knowledge) and because its measures are based on an ordinal structure rather than an additive one. Remembering this, given that a possibility measure encodes a family of probability measures, the transformation aims at finding the probability measure which preserves as much as possible of the uncertainty in the original possibility distribution (Baraldi, 2012).

8.1.3 UNCERTAINTY WITHIN THE CONTEXT OF RISK ANALYSIS

As mentioned, uncertainty is commonly classified as aleatory or epistemic. Aleatory uncertainty can be mathematically represented by a probability distribution once enough experimental data are available, but epistemic uncertainty stems from incomplete information. Representing epistemic uncertainty by probabilistic means is questionable because there is no reason *a priori* to prefer one probability distribution function over another. Researchers have proposed a number of theoretical alternatives to the probabilistic approach. Of these, possibility theory and evidence theory (Shafer, 1976), also known as DST, are considered the most promising methodologies to deal with epistemic uncertainty.

Moreover, uncertainty modeling can be important to understand how the input uncertainty has propagated to output uncertainty. Formally, it is sufficient for a model to link the important output variables of interest to a number of continuous or discrete inputs that can be uncertain, subject to randomness, lack of knowledge, errors or any other sources of uncertainty, or fixed, namely considered to be known. In addition to uncertainty handling and modeling, reducing data uncertainty as much as possible is necessary for precise and reliable results.

Sensitivity analysis plays a key role in communicating the shortcomings and limitations of probabilities and expected values. It is usually done to examine the standard error caused by the estimation process. The approach is based on a direct decomposition of the model output variance into factorial terms called "importance measures" (Ratto et al., 2001). Sensitivity indices are computed by dividing the importance measures by the total output variance. These indices represent the expected amount of variance that would be removed from the total output variance if the true value of an individual input variable is known (within its uncertainty range).

Uncertainty in reliability characterization is also associated with the censoring point and initial sample size. This situation can be solved by using Big Data or by generating simulated values. Monte Carlo simulation (MCS) is usually applied to determine the minimum sample size, as well as the earliest censoring point for accurate reliability assessment. This simulation is designed to determine the earliest censoring point associated with test duration according to initial sample sizes which provides the right balance between censored and failed sample sizes without compromising uncertainty in reliability assessment (González-Prida et al., 2019).

8.2 UNCERTAINTY QUANTIFICATION, PROPAGATION, AND MANAGEMENT

8.2.1 Uncertainty Quantification

Uncertainty quantification tries to determine the likelihood of certain outcomes if some aspects of the system are not known with any degree of exactitude. Uncertainty is the source of many problems in both natural science and engineering fields. Computer simulations are frequently used to study these types of problems (Wikipedia, 2020a).

The discipline of uncertainty quantification is devoted to the problem of characterizing and properly addressing uncertainties in the use of mathematical and computational models of complex processes and data. This literature sprang from the recognition of the need to address uncertainty in computer models and simulations of highly complex physical processes, such as petroleum reservoir prediction, groundwater flows, or nuclear reactor safety, and to study how uncertainty propagates through complex models to affect conclusions drawn from the models (Villiers et al., 2015).

Real life applications have both aleatory and epistemic uncertainties. The goal of uncertainty quantification is to reduce epistemic uncertainties to aleatoric ones, as the quantification of the latter can be relatively straightforward, depending on the application. The Monte Carlo method is a common choice for such analysis. A probability distribution can be represented by its moments; more recent techniques include Karhunen–Loève and polynomial chaos expansions. The evaluation of epistemic uncertainties requires knowledge of the system or process. Fuzzy logic, probability bounds analysis, and evidence theory, for example, DST, a generalization of the Bayesian theory of subjective probability (Shafer, 1976), are some of the methods commonly used (Wikipedia, 2020a).

Uncertainty quantification is important to identify and characterize the sources of uncertainty affecting RUL estimation and prognostics. These sources of uncertainty must be incorporated into models and simulations as accurately as possible. Common sources of uncertainty in a PHM application include modeling errors, model parameters, sensor noise and measurement errors, state estimate errors, and future loading, operating, and environmental conditions. Each source of uncertainty is addressed separately and quantified using probabilistic/statistical methods. As noted previously, the Kalman filter is a Bayesian tool to quantify uncertainty wherein the uncertainty in the states is estimated continuously as a function of time, based on data which are typically available continuously as a function of time (Sankararaman & Goebel, 2013b).

The goal of uncertainty quantification is to characterize a source of uncertainty and thus allow its incorporation into models and simulations as correctly as possible. Accordingly, a characterization/ quantification step is likely to involve actual systems observed in realistic and relevant environments. Even with carefully designed experimentation, an accurate quantification of uncertainties is challenging (Engel, 2009), but some researchers have attempted it because quantification allows the most critical uncertainties to be prioritized (Celaya, 2012). In one case, Sankararaman et al. (2011) performed sensitivity analysis to identify which input uncertainty contributed most to the output uncertainty in prognostics for fatigue crack damage.

In addition to difficulties with dimensionality scaling, there can be a problem of identifiability. Multiple combinations of unknown parameters and discrepancy functions may yield the same experimental predictions, or in other words, they may all fit well to the data. In this case, different values of parameters or different model functions cannot be distinguished or identified. Explicitly characterizing the uncertainty of the parameters and/or models by probability distributions, belief functions, fuzzy sets, and so on may assist in quantifying such ambiguities (Villiers et al., 2015).

8.2.1.1 Uncertainty Quantification Classification

To sum up the discussion thus far: aleatory uncertainty or statistical uncertainty is representative of unknowns that differ each time we run the same experiment. Epistemic uncertainty or systematic uncertainty is due to things we could in principle know but do not in practice (Wikipedia, 2020a).

8.2.1.2 Two Types of Uncertainty Quantification Problems

Successful uncertainty quantification encounters two major problems.

1. The forward propagation of uncertainty, where various sources of uncertainty are propagated through the model to predict the overall uncertainty in the system response (Wikipedia, 2020a) and quantify the effect of uncertainty in model parameters and input variables on the model's output variables (Villiers et al., 2015).
2. The inverse assessment of model uncertainty and parameter uncertainty, where the model's parameters are calibrated simultaneously using test data.

The first problem is widely researched, and many uncertainty analysis techniques have been developed to address it. The second is attracting attention in the engineering design field, as uncertainty quantification of a model and the subsequently more accurate predictions of system responses are of interest to those designing robust systems (Wikipedia, 2020a).

8.2.1.2.1 Forward Uncertainty Propagation

Uncertainty propagation is the quantification of uncertainties in system output(s) propagated from uncertain inputs. It focuses on the influence on the outputs from the parametric variability listed in the sources of uncertainty (Wikipedia, 2020a). In other words, this typically involves how measurement errors propagate through the mathematical model and how they influence the output variables. Typical methods to perform forward uncertainty propagation analyses are random and deterministic sampling methods. Sensitivity analysis and response surface methods are other ways in which the effect of perturbations of input variables on the output can be quantified. In the information fusion domain, forward uncertainty propagation may involve how uncertainty in certain inputs affects the fusion results and ultimately the decisions and outcomes (Villiers et al., 2015).

Uncertainty propagation analysis is useful for the following (Wikipedia, 2020a):

- Evaluating low-order aspects of the outputs, for example, the mean and variance.
- Evaluating the reliability of the outputs. This is especially useful in reliability engineering, as a system's outputs are usually closely related to its performance.
- Assessing the probability distribution of the outputs as a whole. This is useful in utility optimization where the complete distribution is used to calculate the utility.

8.2.1.2.2 Inverse Uncertainty Quantification

Inverse uncertainty quantification is a generalization of parameter estimation error analysis and falls within the category of inverse problems. The objective of inverse uncertainty quantification is to study or estimate discrepancies between the model (i.e., bias correction) and its parameters by estimating the values of any unknown parameters in the model, such as parameter calibration, on the one hand, and estimating observed outcomes on the other. Methods for performing inverse uncertainty quantification include frequentist, modular Bayesian, or fully Bayesian approaches (Villiers et al., 2015). While this is generally a much more difficult problem than forward uncertainty propagation, it is extremely important, as it is often used in a model updating process. Inverse uncertainty quantification appears in the following scenarios (Wikipedia, 2020a):

- Bias correction only.
- Parameter calibration only.
- Bias correction and parameter calibration.

8.2.1.3 Uncertainty Quantification in RUL Prediction

In this section, we present a general computational framework for uncertainty quantification in prognostics and RUL prediction in the context of online, condition-based health monitoring. We explain how computing uncertainty in RUL predictions can be viewed as an uncertainty propagation problem, discuss the need for rigorous mathematical algorithms for uncertainty quantification in RUL, and suggest a variety of statistical methods for uncertainty propagation (Sankararaman, 2015).

8.2.1.3.1 Computational Framework for Prognostics

If we want to perform prognostics and predict the RUL at a generic time-instant, t_p, it is important to employ an appropriate framework or architecture (Daigle & Goebel, 2011). In the architecture shown in Figure 8.1, prognostics can be subdivided into the following three sub-problems (Sankararaman, 2015):

1. Present state estimation.
2. Future state prediction.
3. RUL computation.

1. State estimation: The first step of estimating the state at t_p serves as the precursor to prognosis and RUL computation. Consider the state space model used to continuously predict the state of the system as

$$\dot{x}(t) = f\left(t, x(t), \theta(t), u(t), v(t)\right), \tag{1}$$

where $x(t) \in \mathbb{R}^{n_x}$ is the state vector, $\theta(t) \in \mathbb{R}^{n_\theta}$ is the parameter vector, $u(t) \in \mathbb{R}^{n_u}$ is the input vector, $v(t) \in \mathbb{R}^{n_v}$ is the process noise vector, and f is the state equation. The state of the system uniquely defines the amount of damage in the system.

The state vector at time t_p, that is, $x(t_p)$, as well as the parameters $\theta(t_p)$, if they are unknown, is estimated using output data collected until t_p. Let $y(t) \in \mathbb{R}^y$, $n(t) \in \mathbb{R}^{n_n}$, and h denote the output vector, measurement noise vector, and output equation respectively. Then,

$$y(t) = h\left(t, x(t), \theta(t), (t), n(t)\right). \tag{2}$$

Filtering approaches such as Kalman filtering and particle filtering are typically used for such state estimation. As we noted in a previous section, these methods are considered Bayesian tracking methods for two reasons: they use Bayes' theorem for state estimation, and they rely on the subjective interpretation of uncertainty. In other words, at any time instant, there is nothing uncertain about the true states. Yet the true states are not known precisely; therefore, filtering is used to estimate the probability distributions of these state variables. The estimated probability distributions reflect the analyst's subjective knowledge of the state variables. Note that Bayes is used for only state estimation and not thereafter.

A number of alternatives are available to estimate state. These alternatives are based on least squares-based regression techniques and include such methods as moving least squares, total least squares, and weighted least squares. They are based on classical statistics and express uncertainty in the state using confidence intervals.

2. State prediction: Having estimated the state at time t_p, the next step is to predict the future states of the component/system. As the focus is on predicting the future, no data are available, and it is necessary to rely on Equation (1). Note that when it is discretized, this differential equation can be used to predict state at any future time instant, $t > t_p$, as a function of the state at time t_p.

3. RUL computation: RUL computation is concerned with the performance of the component lying outside a given region of acceptable behavior. The desired performance is expressed

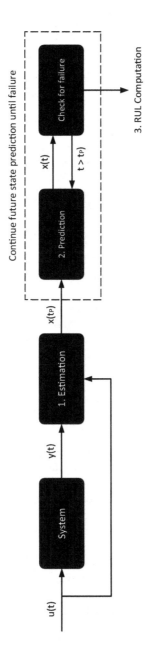

FIGURE 8.1 Model-based prognostics architecture (Sankararaman, 2015).

through a set of n_c constraints, $C_{EOL} = \{C_i\}_{i=1}^{n_c}$, where $C_i = \mathbb{R}^{n_x} \times \mathbb{R}^{n_\theta} \times \mathbb{R}^{n_u} \to B$ maps a given point in the joint state-parameter space given the current inputs, $(x(t), \theta(t), u(t))$, to the Boolean domain $B \triangleq [0,1]$, where $C_i(x(t), \theta(t), u(t)) = 1$ if the constraint is satisfied, and 0 otherwise. These individual constraints may be combined into a single threshold function:

$$T_{EOL} = \mathbb{R}^{n_x} \times \mathbb{R}^{n_\theta} \times \mathbb{R}^{n_u} \to B. \tag{3}$$

T_{EOL} is equal to 1 when any constraint is violated. Then, end of life (EOL), denoted by E, at any time instant t_p, is defined as the earliest time point at which the value of T_{EOL} becomes equal to 1. Thus,

$$E(t_P) \triangleq \inf\left\{t \in \mathbb{R} : t \geq t_P \quad T_{EOL}(x(t), \theta(t), u(t)) = 1\right\}. \tag{4}$$

RUL, denoted by R, at time instant t_p is expressed as

$$R(t_P) \triangleq E(t_P) - t_P. \tag{5}$$

Note that the output equation (Equation (2)) or output data ($y(t)$) is not used in the prediction stage, and EOL and RUL are dependent only on the state estimates at time t_p; although these state estimates are obtained using the output data, the output data are not used for EOL/RUL calculation after state estimation.

For the purpose of implementation, f in Equation (1) is transformed into the corresponding discrete-time version. Discrete time is indexed by k, and there is a one-to-one relation between t and k depending on the discretization level. While the time at which prediction needs to be performed is denoted by t_p, the corresponding index is denoted by k_p. Similarly, k_E denotes the time index corresponding to the EOL.

8.2.1.3.2 RUL Prediction through Uncertainty Propagation

RUL is predicted at time t_p. In other words, $R(t_p)$ depends on the following (Sankararaman, 2015):

1. Present state estimate ($x(k_p)$): The present state estimate and the state space equation in Equation (1) can be used to calculate the future states ($x(k_p)$, $x(k_p) + 1$, $x(k_p) + 2, ..., x(k_E)$).
2. Future loading ($u(k_p)$, $u(k_p) + 1$, $u(k_p) + 2, ..., u(k_E)$): These values are needed to calculate the future state values using the state space equation.
3. Parameter values from time-index k_p until time-index k_E (denoted by $\theta(k_p)$, $\theta(k_p) + 1, ..., \theta(k_E)$).
4. Process noise ($v(k_p)$, $v(k_p) + 1$, $v(k_p) + 2, ..., v(k_E)$).

In RUL prediction, these are all independent quantities, and RUL becomes a dependent quantity. Let $X = \{X_1, X_2, ..., X_i, ..., X_n\}$ denote the vector of all of the above independent quantities, where n is the length of the vector X and therefore the number of uncertain quantities that influence the RUL prediction. Then, the calculation of RUL (denoted by R) can be expressed as (Sankararaman, 2015):

$$R = G(X) \tag{6}$$

The functional relation in Equation (6) is graphically explained in Figure 8.2. When we know the values of X, we can compute the corresponding value of R. The quantities in X are uncertain; simply stated, prognostics attempts to compute their combined effect on the RUL prediction, and thereby compute the probability distribution of R. The problem of estimating the uncertainty in R

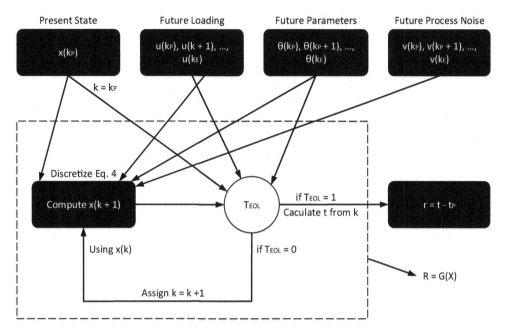

FIGURE 8.2 Definition of G (Sankararaman, 2015).

is equivalent to propagating the uncertainty in X through G; this requires the use of computational methods (Sankararaman, 2015).

8.2.1.3.3   Need for Computational Approaches

Estimating the uncertainty in R using uncertainty propagation techniques calls for rigorous computational approaches. This involves estimating the PDF of R, denoted by $f_Y(y)$, or the cumulative distribution function (CDF) of R, denoted by $F_Y(y)$. In a few rare cases, it is possible to analytically obtain the distribution of R (Sankararaman, 2015):

1. Each quantity contained in X follows a normal (Gaussian) distribution, and the function G can be expressed as a weighted linear combination of the quantities in X. In this case, R follows a normal distribution and can be calculated analytically.
2. Each quantity contained in X follows a lognormal distribution. If the logarithm of the function G can be expressed as a weighted combination of the quantities in X, then $\log(R)$ follows a normal distribution whose statistics can be analytically calculated. In this case, R follows a lognormal distribution.

Note that while Gaussian distributions and linear state-space models (linear f in Equation (1)) are used in PHM, using linear state-space models is not equivalent to G being linear. The use of the threshold function along with the linear state-space models automatically renders G nonlinear (Sankararaman, 2015).

8.2.1.3.4   Uncertainty Propagation Methods for RUL Prediction

Practical problems in PHM include (Sankararaman, 2015):

1. Non-Gaussian random variables affecting RUL prediction.
2. A nonlinear multi-dimensional state-space model.

3. Uncertain future loading conditions.
4. A complicated threshold function defined in multi-dimensional space.

The RUL depends on the quantities indicated in Figure 8.2. Thus, it is technically inaccurate to artificially assign the probability distribution type to RUL. As we explained above, RUL is simply a dependent quantity, and the probability distribution of R needs to be accurately estimated using computational approaches.

Calculating the uncertainty in R and estimating the PDF of R call for the use of rigorous computational methodologies developed for the purpose of uncertainty propagation. These include sampling-based, analytical, and hybrid methods.

Some methods calculate the CDF of R, while others generate samples from the probability distribution of R (Sankararaman, 2015).

8.2.1.4   Prognostics Challenges in Using Uncertainty Quantification

There are several challenges in using the various types of uncertainty quantification methods for prognostics, health management, and decision-making. It is important to understand these challenges and the need for PHM systems to integrate efficient uncertainty quantification with prognostics to facilitate risk-informed decision-making. An uncertainty quantification methodology for prognostics needs to do the following (Sankararaman & Goebel, 2014):

1. Make timely calculations: The methodology needs to be computationally feasible for implementation in online health monitoring. This requires quick calculations. Uncertainty quantification methods have traditionally been considered time-consuming and computationally intensive.
2. Characterize uncertainty: In many practical applications, it is difficult to assess each individual source of uncertainty. For example, the future loading uncertainty, an important contributor of uncertainty to prognostics, is highly uncertain, and it may not even be possible to characterize it. The model used for prediction may also be uncertain; it may be difficult to estimate the future values of the model parameters and the model errors in advance.
3. Perform uncertainty propagation: After each source of uncertainty has been characterized, it is not straightforward to compute their combined effect on prognostics and the RUL prediction. This computation must be the result of rigorous uncertainty propagation, resulting in the entire probability distribution of RUL prediction.
4. Capture distribution properties: The probability distribution of RUL in prognostics may be multi-modal; the uncertainty quantification methodology needs to be able to accurately capture such distributions.
5. Ensure accuracy: The method needs to be accurate. The entire probability distribution of X needs to be correctly accounted for, as does the functional relation defined by G in Figure 8.2. Some methods use only a few statistics (e.g., mean and variance) of X, while others make approximations (e.g., linear) of G. It is important to propagate uncertainty to compute the entire probability distribution of RUL, without making significant assumptions on the distribution types and functional shapes.
6. Obtain uncertainty bounds: It is important to be able to calculate the entire probability distribution of RUL, but it is equally important to quickly obtain bounds on RUL to assist in online decision-making.
7. Make deterministic calculations: Existing verification, validation, and certification protocols require algorithms to produce deterministic (i.e., repeatable) calculations. Several sampling-based methods produce different results (albeit, only slight differences if methods are implemented well) on repetition.

Each uncertainty quantification method addresses one or more of the above, suggesting the need to use different methods to achieve different goals. Future research needs to investigate this more fully, analyzing different types of uncertainty quantification methods and studying their applicability to prognostics (Sankararaman & Goebel, 2014).

8.2.2 UNCERTAINTY PROPAGATION

Uncertainty propagation accounts for all the previously quantified uncertainties and uses this information to predict:

1. Future states and the associated uncertainty.
2. RUL and the associated uncertainty.

The former is computed by propagating the various sources of uncertainty through the prediction model. The latter is computed using the estimated uncertainty in the future states along with a Boolean threshold function to predict EOL. As the predictions of future states and RUL depend on the uncertainties characterized in the previous step, the choice of distribution type and distribution parameters of future states and RUL should not be arbitrary. At times, a normal (Gaussian) distribution is assigned to RUL predictions; this is a mistake – the probability distribution of RUL needs to be estimated through the rigorous uncertainty propagation of the various sources of uncertainty through the state-space model and the EOL threshold function, and both could be nonlinear (Sankararaman & Goebel, 2013b).

In statistics, uncertainty propagation (or propagation of error) is the effect of variables' uncertainties (or errors, more specifically, random errors) on the uncertainty of a function based on them. When the variables are the values of experimental measurements, they have uncertainties because of measurement limitations (e.g., instrument precision) which propagate because of the combination of variables in the function.

The uncertainty u can be expressed in a number of ways. It may be defined by the absolute error Δx. Uncertainties can also be defined by the relative error $(\Delta x)/x$, usually written as a percentage. Most commonly, the uncertainty on a quantity is quantified in terms of the standard deviation, σ, the positive square root of the variance. The value of a quantity and its error are then expressed as an interval $x \pm u$. If the statistical probability distribution of the variable is known or can be assumed, it is possible to derive confidence limits to describe the region within which the true value of the variable may be found. For example, the 68% confidence limits for a one-dimensional variable belonging to a normal distribution are approximately ± one standard deviation σ from the central value x, which means the region $x \pm \sigma$ will cover the true value in roughly 68% of cases.

If the uncertainties are correlated, covariance must be taken into account. Correlation can arise from two different sources. First, the measurement errors may be correlated. Second, when the underlying values are correlated across a population, the uncertainties in the group averages will be correlated (Wikipedia, 2020b).

8.2.2.1 Uncertainty Propagation Methods

Monte Carlo sampling is the most commonly used uncertainty propagation technique. It is based on drawing random samples of independent quantities and computing corresponding realizations of the dependent quantity in this case, RUL.

To give an example, consider a generic engineering component whose health state at any time instant is $x(t)$. If $x(0)$ follows a Gaussian distribution with mean and standard deviation equal to 975 and 50 respectively, and:

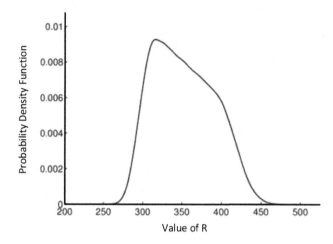

FIGURE 8.3   RUL: conceptual example (Sankararaman & Goebel, 2015).

1. Follows a uniform distribution (with lower and upper bounds of 0.990 and 0.993).
2. Follows a uniform distribution (with lower and upper bounds of -0.005 and 0).

Then, the RUL (defined by Equation (7), where the critical lower bound (l) follows a Gaussian distribution with mean deviation equal to 50 and standard deviation equal to 5, thus reflecting the presence of uncertainty in the EOL threshold definition) can be calculated as a probability distribution, using Monte Carlo sampling (note that 5,000 random samples were used for this illustration). If we use unit discretization (i.e., the time interval between the kth and $(k + 1)^{th}$ instants is equal to 1 second) to obtain the solution, we arrive at the PDF shown in Figure 8.3. This is clearly not a typical parametric distribution (e.g., normal and lognormal). This is why rigorous uncertainty propagation methods are necessary to accurately estimate this PDF (Sankararaman & Goebel, 2015).

In Equation (7), the RUL (where r is an instance of random variable R) is equal to the smallest n, such that $x(n) < 1$:

$$r = inf\left\{n : x(n) < l\right\}. \tag{7}$$

Monte Carlo sampling can be accurate, but it is time consuming and computationally expensive. Therefore, researchers have focused on developing computationally cheaper methods, including Latin hypercube sampling, adaptive sampling, importance sampling, and unscented transform sampling. Alternative analytical methods include the first-order second moment method, the first-order reliability method (FORM), and the second-order reliability method (SORM). There are also methods, such as the efficient global reliability analysis method, which involve both sampling and the use of analytical techniques. All of these methods empirically calculate the probability distribution of RUL; while some calculate the PDF ($f_R(r)$) of RUL, others calculate the CDF ($F_R(r)$), and still others directly generate samples from the desired PDF ($f_R(r)$).

Admittedly, each of these methods has limitations and thus may not accurately calculate the actual probability distribution of R. Accurate calculation requires the use of infinite samples for Monte Carlo sampling. Any other method will lead to uncertainty in the estimated probability distribution, commonly called prediction-method uncertainty. It is possible to decrease or even eliminate prediction-method uncertainty by using advanced probability techniques or employing powerful computing power.

It is necessary to look more closely at uncertainty propagation methods and determine whether they can be applied to prognostics and health monitoring applications. A few recent studies have investigated the use of Monte Carlo sampling, unscented transform sampling, and FORMs, with promising results. This is only a start, however; we need to continue to develop methods for quantifying uncertainty in the context of online PHM (Sankararaman & Goebel, 2015).

8.2.3 Uncertainty Management

The early development of diagnostic/prognostic algorithms concentrated on logical systems, interacting with the world via "if then" statements. The focus on probabilities began when researchers realized logical systems could not anticipate all possible outcomes. Consequently, Bayesian techniques (among others) were assimilated into newer approaches.

Simply put, Bayes' theory defines the concept of probability as the degree of belief that a proposition is true. Thus, Bayes' theorem can be used to infer or update the degree of belief in light of new data – in fact, the more data, the better the predictions. Importantly, Bayesian models are self-correcting, so the predictions evolve with change in data trends (Saha & Goebel, 2008).

The following are some techniques based on Bayes:

1. Support vector machines (SVMs): These are a set of related supervised learning methods used for classification and regression that belong to a family of generalized linear classifiers.
2. Kalman filter: Bayesian techniques are used in the Kalman filter. These techniques provide a general but rigorous framework for dynamic state estimation problems. The basic idea is to construct a PDF of the state based on available information. For a linear system with Gaussian noise, the method reduces to the Kalman filter. The state space PDF remains Gaussian at every iteration, and the filter equations propagate and update the mean and covariance of the distribution (Saha & Goebel, 2008).
3. Relevance vector machine (RVM): The RVM is a Bayesian form representing a generalized linear model whose functional form is identical to the SVM. In a given classification problem, the data points may be multidimensional, say n_{dim}. The task is to separate them by an $n_{dim} - 1$ dimensional hyper-plane. This is a typical form of linear classifier. While many linear classifiers might satisfy this property, an optimal classifier will create the maximum separation (margin) between the two classes. Such a hyper-plane is known as the maximum-margin hyper-plane, and such a linear classifier is known as a maximum-margin classifier.

   Note that nonlinear kernel functions can be used to create nonlinear classifiers. This allows the algorithm to fit the maximum-margin hyper-plane in the transformed feature space, even though the classifier is nonlinear in the original input space.

   Although SVM is frequently used for classification and regression, it has some disadvantages, including the lack of probabilistic outputs useful in health monitoring applications. RVM addresses the issues in a Bayesian framework. Besides offering a probabilistic interpretation of its output, it uses fewer kernel functions in a comparable generalization performance (Saha & Goebel, 2008).
4. PF: In the PF approach, the state PDF is approximated by a set of particles (points) representing sampled values from the unknown state space and a set of associated weights denoting discrete probability masses. The particles are generated and recursively updated from a nonlinear process model describing the evolution in time of the system under analysis, a measurement model, a set of available measurements, and an *a priori* estimate of the state PDF. PF implements a recursive Bayesian filter using MCSs and, as such, is known as a sequential Monte Carlo method. Particle methods assume state equations can be modeled as first-order Markov processes with conditionally independent outputs (Saha & Goebel, 2008).

8.2.3.1 Uncertainty Management in Prognostics

Uncertainty management of prognostics is necessary for prognostics to become a key enabler of health management in industrial applications. Techniques to manage the uncertainty in the many factors contributing to current health state estimation – such as signal-to-noise ratio on diagnostic features, optimal features with respect to detection statistics, and ambiguity set minimization – have received attention because of the maturity of the diagnostics domain. By contrast, uncertainty management in prognostics is underdeveloped (Orchard et al., 2008).

Even so, a number of approaches have been suggested for uncertainty representation and management in prediction. These include probabilistic methods, soft computing methods, and tools derived from evidential theory or DST. Probabilistic methods are mathematically rigorous assuming that a statistically sufficient database is available to estimate the required distributions. Possibility theory (fuzzy logic) offers an alternative when data are scarce or incomplete or contradictory data are available. Dempster's rule of combination and such concepts from evidential theory as belief on plausibility (as upper and lower bounds of probability) based on mass function calculations can support uncertainty representation and management tasks (Orchard et al., 2008).

In one area of application, uncertainty management is used to refer to activities which help to manage uncertainty in condition-based maintenance (CBM) during real-time operation. For optimal maintenance, it is a good idea to look for ways to constantly improve the uncertainty estimates. For example, it is useful to identify which sources of uncertainty are significant contributors to the uncertainty in the RUL prediction. If the quality of the sensors can be improved, it may be possible to obtain a better state estimate (with less uncertainty) during Kalman filtering, and this, in turn, may lead to a less uncertain RUL prediction. Another aspect of uncertainty management deals with the use of uncertainty-related information in the decision-making process (Sankararaman & Goebel, 2013b).

Most of the PHM research pertains to uncertainty quantification and propagation; little work directly addresses uncertainty management. Even within the area of uncertainty quantification and propagation, estimates of uncertainty are sometimes misinterpreted. For example, when statistical principles are used to estimate a parameter, many seek to calculate the estimate with the minimum variance. When this principle is applied to RUL estimation, it is important not to arbitrarily reduce the variance of the RUL itself (Sankararaman & Goebel, 2013). Celaya (2012) explored this idea and argued the RUL variance needs to be carefully calculated by accounting for the different sources of uncertainty.

The goal of prognosis uncertainty management is to analyze uncertainty sources, assign a PDF to the RUL estimate, and calculate confidence bounds for important system reliability measures, such as probability of failure. Prognosis uncertainty management is important because it can tell decision-makers when to take action, either offline or online, based on the current risk tolerance. Advanced prognosis uncertainty management solutions include algorithms that reduce uncertainty as more data become available (Tang et al., 2008).

Researchers have developed a variety of techniques for uncertainty management in many different areas. For example, one study addressed uncertainty management for diagnostics and prognostics of batteries by applying Bayesian techniques, including the RVM and particle filtering (Saha & Goebel, 2008). Usynin and Hines (2007) used an optimal principal component analysis (PCA) transformation to reduce the uncertainty effects imposed by degradation data obtained in the presence of unobservable failure mechanisms. Other researchers developed prognostic fusion techniques to fuse competing prediction algorithms and reduce prognostic uncertainty (Goebel et al., 2006; Goebel & Eklund, 2007). A particle filtering-based uncertainty representation and uncertainty management approach was tested with real vibration data from a critical aircraft component (Orchard et al., 2008), where uncertainty management was accomplished by making parametric adjustments in a feedback correction loop of the state model and its noise distributions. Using a Bayesian approach to obtain a posterior equivalent pre-crack size (EPS) distribution, other authors achieved more accurate estimates of probability of failure of an FA/18 bulkhead and demonstrated that the tail of the EPS distribution had a significant effect on probability of failure (Macheret & Teichman, 2008).

Engel (2008) suggested methods to create and validate verifiable requirements for accurate and safe prognostic predictions. Overall, researchers have made significant efforts to address prognostics and uncertainty management issues by standardizing research methods for prognostics, defining prognostics performance metrics, and developing new technologies (Tang et al., 2008).

Uncertainty management requires the incorporation of all relevant and/or significant sources of uncertainty into prognostic models and simulations. Therefore, the problem formulation stage is key. Uncertainty management is only effective when all relevant sources of uncertainty are identified and included. Moreover, various measures of uncertainty must be appropriately combined in the prognostic model as the input variability filters through a complex (possibly nonlinear) system model.

Once all sources of uncertainties are identified, modeled, and managed correctly, the output PDF for random variables like RUL or EOL should match the true spread. In practice, however, this is impossible because no model is perfect, and not all sources of uncertainties can be characterized. Furthermore, an exhaustive sampling-based method (e.g., MCS) is computationally prohibitively expensive. Consequently, researchers have worked to develop intelligent sampling-based algorithms and mathematical transformations (e.g., support vectors and PCA) to capture most details of the true variability. It may not be possible to identify and characterize all sources of uncertainty; sensitivity analysis is recommended to isolate the most important factors.

Through effective uncertainty management practices, we can at least attempt to bring the predicted estimate close to the true spread and not arbitrarily reduce the spread of RUL itself. We can minimize the variability in the estimate of a given parameter of interest, not the variability in the parameter of interest itself (Celaya, 2012).

8.2.3.2 Uncertainty Management in Long-Term Predictions

The issue of uncertainty management in a particle filtering-based prognosis framework is basically related to a set of techniques aimed to improve the estimate at the current time instant, since the expectation of the predicted trajectories for particles and the bandwidth of Epanechnikov kernels (widely used to model uncertainty in stochastic processes) both depend on that probability density function estimate (Orchard et al., 2008).

In this sense, it is important to distinguish between two main types of adjustments that may be implemented to improve the current representation of uncertainty for future time instants (Orchard et al., 2008):

- Adjustments in unknown parameters in the state equation.
- Adjustments in the parameters that define the noise PDFs embedded in the state equation. These parameters are "hyper-parameters".

The accuracy of long-term predictions is directly related to the estimates of x_t and the model hyper-parameters that affect $E[x_t \mid x_{t-1}]$. Precision in long-term predictions is directly related to the hyper-parameters that describe the variance of the noise structures considered in the state equation. In this sense, any uncertainty management system for future time instants within a particle filtering-based framework should follow the general structure presented in Figure 8.4, where the performance of the algorithm is evaluated in terms of the short-term prediction error (which depends on the PF-based PDF estimate). Whenever the performance criteria for the short-term prediction error are not met, an outer correction loop directly modifies both model parameters and hyper-parameters (Orchard et al., 2008).

When the hyper-parameters are modified via an outer correction loop, short-term predictions may be used to improve the efficiency of the particle filtering-based estimate and, thus, the subsequent generation of long-term predictions. A number of outer correction loops may be applied for

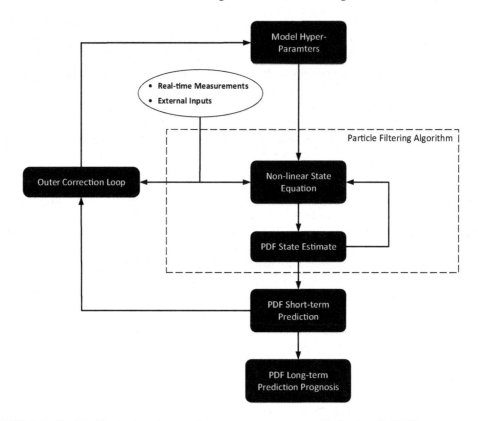

FIGURE 8.4   Particle filtering-based uncertainty management system (Orchard et al., 2008).

this purpose. One of these aims to modify the variance of the noise term in the state equation. In this case, let $\omega_2(t)$ represent the model uncertainty. Then, for the example detailed in what follows, the recommended parameters are (Orchard et al., 2008):

$$\begin{cases} var\left\{\omega_2\left(t+1\right)\right\} = p.var\left\{\omega_2\left(t\right)\right\}, if\ \dfrac{Pred\_error(t)}{Feature(t)} < Th \\[4mm] var\left\{\omega_2\left(t+1\right)\right\} = p.var\left\{\omega_2\left(t\right)\right\}, if\ \dfrac{Pred\_error(t)}{Feature(t)} > Th \end{cases} \tag{8}$$

where $Pred\_error(t)$ is the short-term prediction error computed at time t, is any well-defined norm (usually L_2- norm), $0 < p < 1$, $q > 1$, and $0 < Th < 1$ are scalars. In particular, $p \in [0.925, 0.975]$, $q \in [1.10, 1.20]$, and $Th = 0.1$. These values have been determined through exhaustive analysis of simulations considering scenarios with different combinations of values for the parameters p, q, and Th. The range for short-term predictions depends on the system under analysis, although a 5-step process is recommended to ensure rapid adaptation of the scheme.

Outer correction loops may be also implemented using neural networks, fuzzy expert systems, and proportional–integral–derivative (PID) controllers, among others. Additional correction loops include the modification of the number of particles used for 1-step or long-term prediction purposes and the reduction of the threshold for the use of the importance resampling algorithm (Orchard et al., 2008).

8.2.3.3 Implications for Uncertainty Management

The definition of RUL as a random variable or random process has implications for uncertainty management and the representation of uncertainty in a model-based prognostics methodology. If RUL is not modeled within a probability framework, like a fuzzy variable or just a deterministic variable, uncertainty management activities will differ. As an illustration, consider a simple point estimate example from basic mathematical statistics.

Let us assume we can perform a set of run to failure experiments with a high level of control, ensuring same usage and operating conditions. RUL at time t_p is computed by measuring the elapsed time from t_p until failure for all the n samples $(R_1,\ldots, R_n)$ on the set of run to failure experiments. Let us also assume these random samples come from a PDF $f_R(r)$, with expected value $E(R) = \mu$ and variance $Var(R) = \sigma^2$.

Let θ_1 be a parameter estimator of the mean μ of f_R, with expected value $E(\theta_1) = \mu_{\theta 1}$ and variance $V(\theta_1) = \sigma^2_{\theta_1}$. This estimator will be a function of all the sample values and have a PDF $f_{\theta 1}$. θ_1 is a point estimate of the random variable R, such as the sample mean, median, or some other location statistic. Now, from the uncertainty management perspective in prognostics, it is necessary to judge the ability of the algorithm to compute the point estimate of the process, in this case, to properly estimate μ. We expect that this estimate θ_1 has the least variability, the least variance possible, making θ_1 less uncertain. As a result, $\sigma^2_{\theta_1}$ should be as small as possible. It is, however, incorrect to expect the estimation process to reduce σ^2 itself.

This is often misinterpreted in prognostics methodologies based on computational statistics that do not directly focus on a point estimate, concentrating instead on generating an approximation of the distribution of R. Variability can be assessed by a measure of spread like the sample standard deviation computed directly from the sample distribution of R. Therefore, this variation should not be arbitrarily decreased by tuning the algorithm, as it is intended to represent the real statistical uncertainty of the process. This applies to RUL predictions without loss of generality if they are modeled as random variables, and this is typically the case.

The concept can be further described considering the sample average $\bar{R}$ as the estimator $(\theta_1 = \bar{R})$. From basic probability theory, we observe that $\mu_{\theta_1} = \mu$ and $\sigma^2_{\theta_1} = \sigma^2/n$. This estimator is unbiased, and its variance $\sigma^2_{\theta_1}$ can be reduced by increasing the sample size. But σ^2 cannot be reduced because it represents the inherent variability in the random variable R (Celaya, 2012).

8.3 SOURCES OF UNCERTAINTY AND MODELING UNCERTAINTY

Several different types of sources of uncertainty must be accounted for in the formulation of a prognostics system. These sources may be slotted into three categories and require separate representation and management methods (Celayaa, 2012):

1. Aleatoric or statistical uncertainties: These are caused by inherent variability in any process and cannot be eliminated. They can be characterized by multiple experimental runs but cannot be reduced by improved methods or measurements. Sampling fluctuations from the characterized PDF of a source of aleatoric uncertainty can result in different predictions every time. Examples include manufacturing variations and material properties.
2. Epistemic or systematic uncertainties: These occur because of unknown details that cannot be identified and thus are not incorporated into a process. With improved methods and deeper investigations, these uncertainties may be reduced but are rarely eliminated. Modeling uncertainties falls into this category and includes modeling errors due to unmolded phenomena in both the system and fault propagation models.
3. Prejudicial uncertainties: These occur because of the way a process is set up and are expected to change if the process is redesigned. They can be considered a type of epistemic uncertainty, but it is possible to control them to a better extent. Examples include sensor noise,

sensor coverage, information loss due to data processing, numerical errors, and various approximations and simplifications.

While it is possible to reduce some of these uncertainties, they cannot be entirely eliminated. Representing them and accounting for them in prognostic outputs is extremely important, as they directly affect the associated decision-making process and are typically expressed by the concept of risk generated by unwanted outcomes. Several PHM approaches quantify risk based on uncertainty quantification in an algorithm's output and incorporate it into a corresponding cost-benefit equation using monetary concepts (Celaya, 2012).

From a technical point of view, approaches to prognostics for integrated vehicle health management (IVHM) are commonly categorized as model-based, data-driven, or hybrid approaches:

- Model-based approaches: These approaches attempt to incorporate a physical understanding of the system into the estimation of RUL. Multiple models might be needed to address different failure modes for the same component; for example, the integrity of an aircraft's structure can be affected by many different factors, including corrosion and fatigue, each governed by a different physics of failure (PoF) mechanism. Model-based approaches also need to account for the assumptions and simplifications associated with the underlying prognostics model which involve both the parameters and the form (or structure) of the model.
- Data-driven approaches: These approaches use monitored operational data to infer and predict system health. A main uncertainty with data-driven approaches is data coverage. Do the data used for training cover all the failure scenarios?

Many other factors affect prognostics output, including sensor measurement uncertainty, future load profile uncertainty, and fault detection (diagnostics) uncertainty. Figure 8.5 summarizes the major contributors to prognostics uncertainty for IVHM applications. Their individual contributions and combined impact form the basis for uncertainty representation and management (Tang et al., 2008).

A more analytical approach to uncertainty sources and their contributions requires a mathematical formulation of the prognosis process. Without a generic mathematical formulation able to accommodate all possible approaches (likely impossible), prognosis can generally be considered a function of prediction models, model parameters, future load profile, current and historic health condition, and so on (Tang et al., 2008). Thus,

$$P_r = f(M, P, L, H, t) + \omega \tag{9}$$

where P_r is the prognosis, that is, either the RUL or PoF at a future time; $M \in \{M_1, M_2, ..., M_N\}$ is the prognostic model; N is the number of models available; $P \in \mathbb{R}^n$ is the model parameter vector; n is the number of model parameters; $L \in \mathbb{R}^m$ is a vector representing the future load profile factors; $H \in \mathbb{R}^s$ is the health/damage indicator; s is the number of indicators (or features). When a Markov process assumption is made, H represents current health/damage condition; otherwise, H can include the trajectories of health indicators leading to the current time; t is the time or load cycle; ω is the process noise term that captures the rest of the modeling errors.

For uncertainty management, it is important to quantify the influence of each uncertainty source to identify the most significant uncertainty contributors to focus on to improve not only the accuracy but also the precision of the prognosis. This sensitivity analysis can often be calculated either analytically or via an MCS. Sensitivity analysis can be conducted at the parameter level, for example, $\partial Pr/\partial P$, or at the hyper-parameter level, for example, $\partial \mu_{Pr}/\partial \mu_P$, $\partial \mu_{Pr}/\partial \sigma_P$, where μ and σ represent the mean and standard deviation, respectively. Both deterministic and probabilistic sensitivity analysis should be conducted to identify the most significant uncertainty sources. The output from a

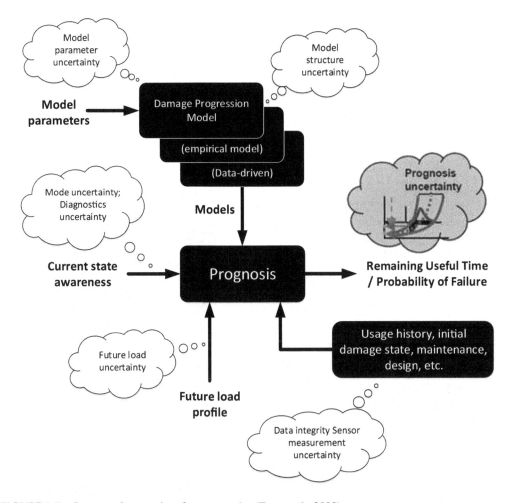

FIGURE 8.5 Sources of uncertainty for prognostics (Tang et al., 2008).

sensitivity analysis is typically presented as a Pareto chart as shown in Figure 8.6, where the influence is quantified as a normalized derivative of the RUL mean (accuracy) and standard deviation (precision) with respect to each selected input parameter (red indicates an inverse correlation) (Tang et al., 2008).

After the uncertainty sources and their contributions to final prognosis output have been analyzed, various uncertainty management techniques can be applied to manage and hopefully reduce the confidence bounds for the prognostics. Since uncertainty is typically caused by incomplete information or lack of knowledge, it can ideally be reduced by applying new data (or observations) as they become available (Tang et al., 2008).

8.3.1 Sources of Uncertainty

In many practical applications, it may be challenging to identify and quantify the different sources of uncertainty that affect prognostics. Researchers typically separate sources of uncertainty into categories to facilitate quantification and management. While it is customary to classify sources into aleatory (occurring because of physical variability) and epistemic (occurring because of a lack of knowledge), this classification may not be suitable for CBM purposes and RUL predictions (see

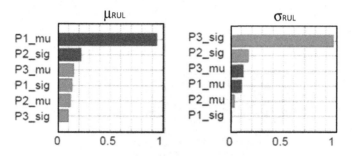

FIGURE 8.6   Normalized ranked Pareto chart (Tang et al., 2008).

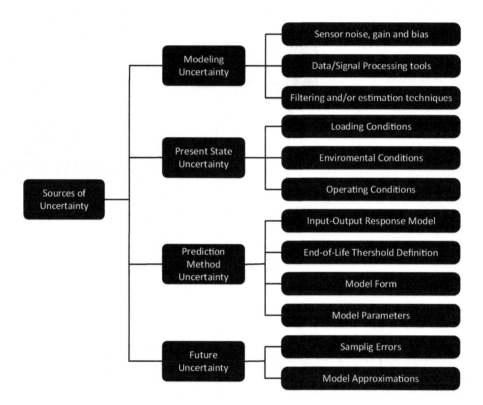

FIGURE 8.7   Sources of uncertainty (Sankararaman & Goebel, 2015).

Section 8.1.2). An approach more applicable to CBM is outlined below (Sankararaman & Goebel, 2013b) and shown graphically in Figure 8.7.

8.3.1.1   Present Uncertainty

Before prognosis can be performed, it is important to be able to estimate the state of the system at the time when RUL needs to be computed. This is related to state estimation commonly achieved using filtering. Output data (collected by sensors) are used to estimate the state, and a filtering approach (Kalman filtering and particle filtering) estimates the uncertainty in the state. In reality, it is possible to improve the estimate of the states and thereby reduce the uncertainty by using better sensors and improved filtering approaches (Sankararaman & Goebel, 2013b). Note that the system is at a particular state at any time instant, and uncertainty simply describes the lack of knowledge of the true state at that instant (Sankararaman & Goebel, 2015).

8.3.1.2 Future Uncertainty

The most important source of uncertainty in the context of prognostics is that the future is unknown, in that the loading and operating conditions are not known precisely. However, it is important to assess the uncertainty in these conditions before performing prognostics. If these quantities were known precisely (with no uncertainty), there would be no uncertainty about the true RUL of the system. Given the uncertainty, however, the RUL needs to be estimated using a model (Sankararaman & Goebel, 2013b).

The goal in prognostics is to individually characterize each of the above uncertainties and quantify their combined effect on prognostics. If we want to predict how long it may be possible to operate a system, the uncertainty in RUL can be quantified and expressed in terms of a probability distribution (Sankararaman & Goebel, 2014).

8.3.1.3 Modeling Uncertainty

A functional model must be used to predict future state behavior (Sankararaman & Goebel, 2013b) or, in other words, to model the system's response to anticipated loading, environmental, operational, and usage conditions (Sankararaman & Goebel, 2015). Meanwhile, EOL is defined using a Boolean threshold function which indicates EOL by checking whether failure has occurred or not. These two models are either physics-based or data-driven and are combined to predict the RUL.

In a practical sense, it may be impossible to develop models which accurately predict reality. Modeling uncertainty represents the difference between the predicted response and the true response and this cannot be known or measured accurately. The modeling comprises three main parts: model parameters, model form, and process noise. While it may be possible to quantify these terms until the time of prediction, knowing their values at future time instants is more challenging (Sankararaman & Goebel, 2013b).

8.3.1.4 Prediction Method Uncertainty

Even if all sources of uncertainty could be quantified accurately, it would still be necessary to quantify their combined effect on RUL prediction and thereby quantify the overall uncertainty in RUL prediction. It may not be possible to do this accurately in practice, adding further uncertainty (Sankararaman & Goebel, 2013b). For example, if a limited number of samples is used in a sampling-based approach to prediction, the estimated probability distribution will show uncertainty (Sankararaman & Goebel, 2015).

8.3.2 Sources of Uncertainty in Prognostics

Whether the diagnostic/prognostic algorithms are model-driven or data-driven, it is not feasible to eliminate all error factors. Prognostic predictions must deal with multiple sources of error, including modeling inconsistencies, system noise, and degraded sensor fidelity (Saha & Goebel, 2008).

8.3.2.1 Modeling Error

Modeling error stems from the inability to create an analytical model that represents the actual system exactly. The system dynamics are frequently too complex to be accurately modeled in a computationally feasible framework. Alternatively, there may be insufficient knowledge about the system itself or its response to various environmental stimuli.

8.3.2.2 Noise

Noise in a system can be generated by myriad sources. Some may be internal to the system while others infiltrate the system dynamics through noisy external inputs. In complex systems, the noise may be electrical, mechanical, or even thermal in nature.

- Electrical disturbances: Caused by relay chatter, faulty connectors, wiring crosstalk, electromagnetic interference, and electrostatic discharge.
- Mechanical disturbances: Caused by vibrations or component degradation due to aging, overloading, or corrosion.
- Thermal noise: Caused by insufficient cooling or uneven heat distribution. It can be highly disruptive to electrochemical systems like batteries.

8.3.2.3 Sensors

Sensor suites are increasingly used to monitor systems. Therefore, sensor fidelity has become critically important. Sensor noise can result from electrical interference, digitization error, dead band, sensor bias, backlash, and nonlinear response.

8.3.3 UNCERTAINTY SOURCE ANALYSIS

According to the filtering estimation principle, the prediction process of the stochastic filtering-based method can be summarized as shown in Figure 8.8 (Chen et al., 2016).

The prediction process consists of the following steps (Chen et al., 2016):

1. Data sampling and transferring: Different types of sensors are used to collect corresponding system parameters. The conversion of various physical signals to electrical signals is also required in this step.

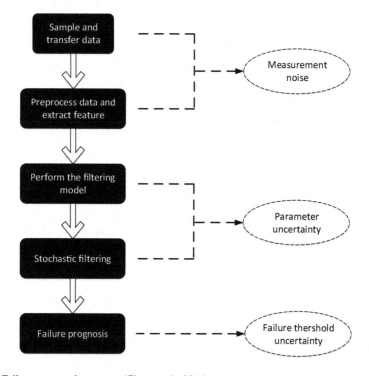

FIGURE 8.8   Failure prognosis process (Chen et al., 2016).

2. Data pre-processing and feature extraction: Data transmitted from sensors are pre-processed to a specified format for subsequent failure prognosis. Simplified, integrated, or compressed data are the output of this step.
3. Determining the filtering mode: As the different kinds of data have different characteristics, and the same data type at different system stages has different characteristics, the filtering model for the subsequent filtering calculations should be determined specifically for the data in question.
4. Stochastic filtering: The system state one-step estimation is completed by Kalman filtering.
5. Failure prognosis: If the system noise is assumed to be zero, and no new information is measured, the system state in the future can be calculated probabilistically.

In this process, different uncertainties are introduced and propagated step by step as shown in Figure 8.8 (Chen et al., 2016).

8.3.3.1 Model Parameter Uncertainty

A real system cannot be modeled accurately, and the error of a system's mathematical model cannot be known in advance. For system state estimation and prediction based on stochastic filtering, the system's model is also completely unknown. Such errors can be seen as system model parameter uncertainty (Chen et al., 2016).

8.3.3.2 Measurement Noise

Measurement noise or system output noise is the main uncertainty source during the failure prognosis process in a real environment. Environment noise, sensor inaccuracy, transmission noise, and a data reduction effect can be seen as measurement noise. The presence of these noises means the measurement result cannot reflect the true value of the measured item.

Mechanisms of measurement noise can be classified as system error, random error, and gross error. System error is the measurement tool's or system's inherent error, and gross error is error that is obviously inconsistent with the facts. Random error can be further divided into normal and abnormal distribution errors depending on the PDF. Generally speaking, system error and gross error can be removed using a mathematical method, and random error is assumed to follow a normal distribution. However, the statistical characteristics of dynamic noise in a real environment are very hard to acquire exactly; moreover, the initial "white" noise will be colored as it is transferred through signal channels and pre-processed. In most engineering applications, the measurement noise is only partly or approximately known or even totally unknown (Chen et al., 2016).

8.3.3.3 Failure Threshold Uncertainty

In traditional overhaul maintenance activities, the failure threshold, no matter whether the threshold value is given by human or intelligent monitoring systems, is based on experience and essentially fixed. However, the acceptance of traditional ideas may lead to the neglect of uncertainty in engineering practice. Such assumption or neglect may not be unreasonable in most cases, as the "point" at which the system performance degrades to failure (the failure threshold) will be changed for different operating environments, operators, and the system's individual differences. It is a challenging problem to consider the failure threshold's uncertainty and model the real failure zone (Chen et al., 2016).

8.3.4 OTHER SOURCES OF UNCERTAINTY

Uncertainty can enter mathematical models and experimental measurements in the following contexts (Wikipedia, 2020a).

8.3.4.1 Parameter Uncertainty

Parameter uncertainty occurs when model parameters inputted into a computer model (mathematical model) whose exact values are unknown and cannot be controlled in physical experiments or whose values cannot be exactly inferred using statistical methods. Examples include material properties in a finite element engineering analysis or multiplier uncertainty in the context of macroeconomic policy optimization.

8.3.4.2 Parametric Variability

Parametric variability stems from the variability of a model's inputted variables. For example, when the dimensions of an asset or system in a manufacturing process are not exactly as designed, this causes variability in its performance.

8.3.4.3 Structural Uncertainty

Structural uncertainty is also called model inadequacy, model bias, or model discrepancy. It stems from the lack of knowledge of the underlying physics of a problem and depends on how accurately a mathematical model describes the true object or system in a real-life situation, considering that models are generally just approximations of reality. For example, if we model the process of a falling object using the free-fall model, the model itself will be inaccurate, as air friction always exists. Thus, even if there are no unknown parameters in our model, we can expect a discrepancy between the model and true physics.

8.3.4.4 Algorithmic Uncertainty

Algorithmic uncertainty is also called numerical uncertainty or discrete uncertainty. It can be traced to numerical errors and numerical approximations per implementation of a computer model. The majority of models are too complicated to be solved exactly. For example, the finite element (or finite difference) method may be used to approximate the solution of a partial differential equation, but this introduces numerical errors. Two other examples are numerical integration and infinite sum truncation, both of which are necessary approximations in numerical implementation.

8.3.4.5 Experimental Uncertainty

Experimental uncertainty, also referred to as observation error, is based on the variability of experimental measurements. Experimental uncertainty is inevitable. We acknowledge its existence when we repeat a measurement many times using exactly the same settings for all inputs or variables.

8.3.4.6 Interpolation Uncertainty

Interpolation uncertainty stems from a lack of available data collected from experimental measurements or computer model simulations. For input settings without experimental measurements or simulation data, it is necessary to interpolate or extrapolate to predict responses.

8.3.5 Kinds of Uncertainty

Information science moves in a social environment whose future has to consider prediction and, as far as possible, to channel prescription. Uncertainty is an inherent feature of modern society. Uncertainty has its origins in the absence of information or the lack of knowledge; it also appears because of the strategic and institutional features of the networks that articulate and process problems. Types of uncertainty include (Bonome, 2011):

1. Substantive uncertainty.
2. Strategic uncertainty.
3. Institutional uncertainty.

Substantive uncertainty refers to the availability of information when we have to face problems of a complex entity. This uncertainty is related to the question of whether it is possible to have access to information. In addition, different actors have different perceptions of problems and view them from different frames of reference, so they also interpret the available information differently. The various interpretations of the meaning of information are an additional source of substantive uncertainty. More information does not necessarily lead to less uncertainty; in fact, it might lead to more uncertainty.

Strategic uncertainty appears when users who need to make decisions are cognitively insecure. Information users have to respond to the strategic actions of other agents, and to do this, they need to anticipate their behaviors. Those mechanisms in which several agents take part may make it difficult to predict the results of their interactions. This generates uncertainty to deal with the problem and its resolution.

Institutional uncertainty occurs when informative systems are part of an organized social environment. Together with the individual factors of the agents (different aims, interests, and perceptions of the users about the information), there are some contextual factors: different kinds of organizations, social networks, and administrative stratifications (local, regional, and national). Interaction between actors is difficult; their behavior will be guided by the tasks, opinions, rules, and language of their own organizations, administrative levels, and networks (Bonome, 2011).

Information science works jointly with computer sciences and information and communication technology to answer questions about uncertainty. With their help, information science deals with the following issues (Bonome, 2011):

- Lack of information.
- Interpretation of information.
- Excess information.
- Users' needs and desires.
- Interactions among different kinds of users.
- Links with surrounding cultural and organizational environments.
- Users' knowledge of the computational systems they are using.

8.3.6 UNCERTAINTY MODEL-BASED APPROACHES

Model-based prognostic approaches use a quantitative analytical model of component behavior to infer component state. These approaches also include component degradation dynamics, which can be modeled by an analytical description, given the methodology known as PoF or a stochastic model.

PoF is an approach that uses knowledge of a system's lifecycle loading conditions, geometry, and material properties to perform reliability modeling and identify potential failure mechanisms. Failure models require as input such information as material properties, geometry, environment, and operating loads. The loads are typically monitored *in situ*, and features (e.g., cyclic range, mean, and ramp rates) of the data are extracted and used in relevant PoF models to estimate damage and RUL. The uncertainty sources are included in these models and enable users to assess the impact of these uncertainties on the remaining life distribution and make risk-informed decisions. Because of the model complexity, the accumulated degradation is distributed using MCSs. The remaining life is then predicted from the accumulated damage distributions using confidence intervals.

Failure mechanisms for which physical models have been developed include low cycle and high cycle fatigue, overstress failure, corrosion, and ductile to brittle transitions.

When physical component degradation is unknown or difficult to model, statistical estimation techniques based on residuals and parity relations (the difference between the model predictions and system observations) are used to predict degradation. Predictions can be improved using state estimation techniques, such as extended Kalman filters and PFs. For example, this approach to prognostics has been used for lithium ion batteries, where a lumped parameter model was used along with an extended Kalman filter and PF algorithms to estimate RUL (Escobet et al., 2012).

8.3.7 Sources of Uncertainty in Prognostics and Health Management

Prognostics must be able to handle multiple sources of error, including system noise, modeling inconsistencies, or degraded sensor fidelity. The main sources of uncertainty in PHM are (Cristaldi et al., 2018):

1. Measurement uncertainty: Data collected through sensors are affected by a measurement uncertainty because of sensor inaccuracy. Uncertainty sources can be systematic or random.
2. Present uncertainty: A preliminary step in PHM is frequently the current state estimation of the system for which RUL prediction is required. The state estimation is typically performed by using filtering approaches (Kalman filtering, particle filtering, etc.) to the data collected through sensors. The measurement uncertainty of the data and the stochastic nature of the filtering approaches are unavoidably reflected in uncertainty in the definition of the system state, that is, the lack of knowledge of the "true" state of the system.
3. Future uncertainty: This is probably the most influential source of uncertainty in PHM as future operational conditions (loading, environmental, and usage conditions) cannot be precisely known in advance. Only assumptions can be made about them.
4. Model uncertainty: It is practically impossible to develop models that predict the underlying reality accurately. Model uncertainty includes model parameters, stochasticity, and process noise. Under-modeling issues are determined by different factors such as the ignorance of certain failure modes in the analysis or, in the case of the application of data-driven approaches, the lack of data describing possible failure scenarios.
5. Prediction method uncertainty: Once quantified, the above sources of uncertainty have to be accurately propagated on the final RUL forecast. Especially when dealing with different probability distribution families and/or non-parametric distributions, uncertainty propagation methods introduce some approximations which lead to additional uncertainty. A typical example is represented by sampling-based approaches, such as MCSs, widely used for RUL prediction; in this case, the use of a limited number of samples introduces uncertainty about the probability distribution of the RUL.

8.4 UNCERTAINTY IN TERMS OF PHYSICAL AND SUBJECTIVE PROBABILITIES

8.4.1 Physical Probabilities

Physical probabilities, also commonly called objective or frequentist probabilities, are related to random physical experiments such as rolling dice or tossing coins. Each trial of this type of experiment leads to an event (a subset of the sample space). Over the course of repeated trials, each event tends to occur at a persistent rate – that is, the relative frequency (see Section 8.1.1). Relative frequency is expressed and explained in terms of physical probabilities. Thus, physical probabilities are defined only in the context of random experiments.

There are two types of interpretation of physical probability: von Mises' frequentist interpretation (Von Mises, 1981) and Popper's propensity explanation (Popper, 1959). The frequentist interpretation is easier to understand and more widely used (Sankararaman & Goebel, 2013b).

Simply stated, randomness appears only with the presence of physical probabilities. If the true value of any particular quantity is deterministic, we cannot associate physical probabilities with it. To put it another way, when a quantity is not random but unknown, the tools of probability cannot be used to represent this type of uncertainty. To give one example, the mean of a random variable (i.e., the population mean) is deterministic, and we cannot talk about its probability distribution. In fact, in any type of parameter estimation, the underlying parameter is assumed to be deterministic; thus, we can only estimate this parameter.

Confidence intervals are used to address the uncertainty in the parameter estimate. However, interpreting confidence intervals can be confusing. Moreover, the uncertainty in the parameter estimate cannot be used in any further uncertainty quantification. To explain, consider the model of a battery: if the model's parameters are estimated under a particular loading condition, this uncertainty cannot be used to quantify the battery's response under a similar loading condition. It is not possible to propagate uncertainty after parameter estimation using confidence intervals – a serious limitation – but this is often necessary in system-level uncertainty quantification (Sankararaman, 2012).

Ultimately, the frequentist interpretation of probability has two limitations. First, a truly deterministic but unknown quantity cannot be assigned a probability distribution. Second, uncertainty represented using confidence intervals cannot be used for further uncertainty propagation. However, the subjective interpretation of probability overcomes these limitations (Sankararaman & Goebel, 2013b).

8.4.1.1 Frequentist View

In this interpretation, probability is defined as the fraction of times an event A occurs if the situation considered is repeated an infinite number of times. Taking a sample of repetitions of the situation, randomness causes the event A to occur a number of times and to not occur the rest of the time. Asymptotically, this process generates a fraction of successes, the "true" probability $P(A)$. This uncertainty (i.e., variation) is sometimes referred to as aleatory uncertainty (see Section 8.1.2).

In this context, let Ω be the sample space containing all the values that a given random variable Y of interest can assume. In the discrete case, a discrete probability distribution function $d_Y(y):\Omega \rightarrow$ [0, 1] exists such that $\sum_{y\in} d_Y(y) = 1$; in the continuous case, a PDF $p_Y(y)$ exists such that $\int P_Y(y)d_y$.

The number $d_Y(y)$ represents the (limit) frequency of observing y after many trials in the discrete case and the density of y in the continuous case. For any measurable subset A of Ω called event, the probability $P(A)$ of A is (Zio & Pedroni, 2013):

$$P(A) = \sum_{y\in A} d_Y(y) \qquad \text{(discrete case)}$$

$$P(A) = \int_A P_Y(y)dy \qquad \text{(continuous case)}$$

The probability $P(A)$ is required to have the following basic properties (Helton & Oberkampf, 2004:

1. If $A \in \Omega$, then $0 \le P(A) \le 1$.
2. $P(\Omega) = 1$.
3. If $A_1, A_2, \ldots, A_i, \ldots$ is a sequence of disjoint sets (events) from Ω, then $P(\cup_i A_i) = \sum i\, P(A_i)$.
4. $P(A) = 1 - P(\bar{A})$ (self-duality property). The probability of an event occurring, $P(A)$, and the probability of an event not occurring, $P(\bar{A})$, must sum to 1; thus, the specification of the likelihood of an event occurring in probability theory results in or implies a specification of the likelihood of that event not occurring.

Finally, in the continuous case, the CDF of Y is $F_Y: \Omega \rightarrow [0, 1]$, defined from the PDF $P_Y(y)$ as follows (Zio & Pedroni, 2013):

$$F_Y(y) = P((-\infty, y]) = P(Y \leq y) = \int_{-\infty}^{y} P_Y(t)dt, \quad \forall y \in \Omega \tag{9}$$

As an example, assume the random variable Y is normal, for example, $Y \sim N(5, 0.25)$: the corresponding PDF $P_Y(y)$ and CDF $F_Y(y)$ are shown in Figure 8.9, left and right, respectively. The probability that the variable Y is lower than or equal to $y_1 = 5.2$, that is, $P(Y \leq y_1 = 5.2) = \int_{-\infty}^{y_1=5.2} P_Y(y)dy = 0.79$, is pictorially shown in Figure 8.9 (left) as the shaded area between the PDF $P_Y(y)$ and the straight line $y_1 = 5.2$; notice that this probability is equal to the value of the CDF $F_Y(y)$ in correspondence of $y_1 = 5.2$; that is, $F_Y(5.2) = 0.79$ (Figure 8.9, right) (Zio & Pedroni, 2013).

Referring to the frequentist definition of probability given above, in practice, it is not possible to repeat the experiment an infinite number of times; thus, $P(A)$ needs to be estimated, for example, by the relative frequency of occurrence of A in the finite sample considered. The lack of knowledge of the true value of $P(A)$ is termed epistemic uncertainty (see Section 8.1.2). Epistemic uncertainty can be reduced by extending the size of the sample, but aleatory uncertainty cannot. For this reason, it is sometimes called irreducible uncertainty (Helton & Burmaster, 1996).

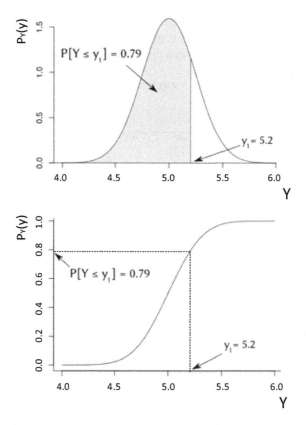

FIGURE 8.9 Probability density function, $P_Y(y)$ (left) and cumulative distribution function $F_Y(y)$ (right) of the normal random variable $Y \sim N(5, 0.25)$ (Zio & Pedroni, 2013).

8.4.1.2   Confidence Intervals: Frequentist Approach

Since R is Gaussian, estimating the parameters μ and σ is equivalent to estimating the PDF. In the context of physical probabilities (frequentist approach), the true underlying parameter μ is referred to as the population mean, and σ is the population standard deviation.

Now, let $\bar{x}$ denote the mean and s denote the standard deviation of the available n data. If data are limited, the sample parameters ($\bar{x}$ and s) will not be equal to the corresponding population parameters (μ and σ). The fundamental assumption is that, since there are true but unknown population parameters, it is meaningless to talk about the probability distribution of any population parameter. Instead, the sample parameters are treated as random variables; if another set of n data were available, we would obtain another realization of $\bar{x}$ and s. Using the sample parameters ($\bar{x}$ and s) and the number of data available (n), frequentists construct confidence intervals on the population parameters (μ and σ) (Sankararaman & Goebel, 2015).

Confidence intervals can be constructed for both μ and σ (Haldar & Mahadevan, 2000). For example, consider multiple nominally identical specimens of a component. Nominally identical implies inherent variability in these specimens' properties and behavior. Suppose they have been subjected to failure analysis, and we know their time-to-failure (TTF). If the true probability distribution of TTF across multiple specimens is assumed to be Gaussian, the $(1 - \alpha)\%$ confidence interval of the mean TTF time can be calculated as the following (Sankararaman & Goebel, 2015):

$$\left[\bar{x} - t_{\frac{\alpha}{2}} \frac{s}{\sqrt{n}}, \bar{x} + t_{\frac{\alpha}{2}} \frac{s}{\sqrt{n}} \right], \tag{10}$$

where $\bar{x}$ denotes the sample mean; s denotes the sample standard deviation; and n denotes number of samples. If the TTF times are given by {100, 105, 98, 110, 92, 97, 85, 120, 93, 101}, then $\bar{x} = 100.10$, $s = 9.87$, $n = 10$, and the 95% confidence interval on the mean TTF is given by [93.98, 106.22]. Using the properties of the chi-square distribution (χ^2), we can calculate the confidence interval on the variance as follows (Sankararaman & Goebel, 2015):

$$\left[\frac{(n-1)s^2}{\chi^2_{1-\frac{\alpha}{2}}}, \frac{(n-1)s^2}{\chi^2_{\frac{\alpha}{2}}} \right]. \tag{11}$$

For this numerical example, the corresponding confidence interval on the standard deviation is given by [6.79, 18.02]. While the above expressions for confidence intervals on mean and standard deviation are only applicable to Gaussian distributions, similar confidence intervals can be constructed for other types of distributions; it is generally easier to construct confidence intervals for the mean than for the standard deviation (or the variance).

It is important to interpret these confidence intervals correctly. The confidence intervals will decrease as more data are available; therefore, the width of these confidence intervals is simply related to the number of data. The actual uncertainty in the TTF times is given only by the estimate of the standard deviation, and this uncertainty is the result of variability in material properties, operating conditions, and so on across all the nominally identical specimens. But interpretation may be confusing or misleading. A 95% confidence interval on μ does not imply that the probability that μ lies in the interval is equal to 95%; μ is purely deterministic, and physical probabilities cannot be associated with it. The random variable in this case is $\bar{x}$, and the confidence interval is calculated using $\bar{x}$. Therefore, the probability that the estimated confidence interval contains the true population mean is equal to 95%. This is the correct interpretation. In any event, the width of the confidence intervals indicates the lack of infinite data, and the actual value of the standard deviation indicates the uncertainty in R (Sankararaman & Goebel, 2015).

In many applications, we may not know what type of probability distribution should be assumed to calculate confidence intervals; obviously, their calculation depends on the choice of distribution type (Gaussian, Weibull, and lognormal), and the presence of distribution type uncertainty adds more confusion to the interpretation of confidence intervals. As the sample size increases, the confidence intervals for mean and standard deviation may get narrower. This may be misleading, as the confidence intervals should be interpreted based on the underlying assumption of distribution type – and this assumption could be wrong. Researchers are working on computational methods to deal with distribution type uncertainty (Sankararaman & Mahadevan, 2013b), but they have not been implemented in PHM applications (Sankararaman & Goebel, 2015).

8.4.2 Subjective Probabilities

Subjective probabilities can be assigned to any statement. The statement does not need to be about an event which is a possible outcome of a random experiment. In fact, subjective probabilities can be assigned even in the absence of random experiments. The Bayesian methodology is based on subjective probabilities, considering the degrees of belief on the extent to which the statement is supported by existing knowledge and available evidence. The terms "subjectivist" and "Bayesian" are essentially synonymous.

Randomness in the context of physical probabilities is equivalent to a lack of information in the context of subjective probabilities (Calvetti & Somersalo, 2007). Even deterministic quantities can be represented using probability distributions which reflect the analyst's subjective beliefs about such quantities. As a result, probability distributions can be assigned to parameters that must be estimated; therefore, this interpretation facilitates uncertainty propagation after parameter estimation. Note that subjective probabilities can be applied in situations where physical probabilities are involved (Sankararaman, 2012).

The concept of likelihood and its use in Bayes' theorem are key to the theory of subjective probability. However, the numerical implementation of Bayes' theorem can be complicated, so several sampling techniques have been designed to address this issue.

Bayesian methods are used to solve various problems in engineering. For example, filtering techniques such as particle filtering or Kalman filtering are based on the use of Bayes' theorem and sequential sampling (Sankararaman & Goebel, 2013b).

8.4.2.1 Subjective (Bayesian) View

In this view, the probability of an event A represents the assigner's degree of belief in the occurrence of A. The probability can be assigned with reference to either betting or some standard event. If linked to betting, the probability of the event A, $P(A)$, is the price at which the assessor is neutral between buying and selling a ticket that is worth one unit of payment if the event occurs and is worthless otherwise (de Finetti, 1974; Singpurwalla, 2006). Following the reference to a standard, the assessor compares his/her uncertainty about the occurrence of the event A with some standard events, for example, drawing a favorable ball from an urn that contains $P(A) \times 100\%$ favorable balls (Lindley, 2000).

Given this, a subjective (Bayesian) interpretation of probability can be given where probability is a purely epistemic-based expression of uncertainty as seen by the assigner, based on his/her background knowledge (Zio & Pedroni, 2013).

Irrespective of reference, all subjective probabilities are conditioned on background knowledge K upon which the assignment is based. They are probabilities in the light of current knowledge (Lindley, 2006). To show the dependencies on K, it is common to write $P(A|K)$, but K is often omitted, as the background knowledge is tacitly understood to be a basis for the assignments. Elements of K may be uncertain and seen as unknown quantities, as pointed out by Mosleh and Bier (1996). However, the entire K cannot generally be treated as an unknown quantity and removed using the law of total probability, that is, by taking $E_K[P(A|K)]$ to obtain an unconditional $P(A)$ (Zio & Pedroni, 2013).

In this view, randomness is not seen as a type of uncertainty in itself but as a basis for expressing epistemic-based uncertainty. A relative frequency generated by random variation is referred to as a chance, to distinguish it from a probability, which is reserved for expressions of epistemic uncertainty based on belief (Singpurwalla, 2006; Lindley, 2006). Thus, the probability may be used to describe uncertainty about the unknown value of a chance. As an example, consider an experiment in which the event A of interest occurs $p \times 100\%$ of the times the experiment is performed. Suppose that the chance p is unknown. Then, the outcomes of the experiment are not seen as independent, as additional observations would provide more information about the value of p. On the contrary, if p was known, the outcomes would be judged as independent, since nothing more could be learned about p from additional observations of the experiment. Thus, conditional on p, the outcomes are independent, but unconditionally they are not; they are exchangeable. The probability of an event A for which p is known is simply p. In practice, p is generally not known, and the assessor expresses his/her (a priori) uncertainty about the value of p by a probability distribution $H(p)$. Then, the probability of A can be expressed as (Zio & Pedroni, 2013):

$$P(A) = \int P(A \mid p) dH(p) = \int p \, dH(p) \tag{12}$$

A common approach to risk analysis is to use epistemic-based probabilities to describe uncertainty about the true value of a relative frequency-interpreted probability (chance). This is called the probability of frequency approach (Kaplan & Garrick, 1981). Probability refers to the epistemic-based expressions of uncertainty and frequency to the limiting relative frequencies of events. By taking the expected value of the relative frequency-based probability with respect to the epistemic-based probabilities, both aleatory and epistemic uncertainties are reflected (Zio & Pedroni, 2013).

8.4.3 CHOICE OF INTERPRETATION

The frequentist and subjective approaches are well-established. They may yield similar results for a few standard problems involving Gaussian variables, but they will yield different interpretations. At times, both may be suitable for a given problem; for example, Kalman filtering has a frequentist interpretation based on least squares minimization. It also has a Bayesian interpretation that relies on continuously updating the uncertainty in the state estimates using Bayes' theorem. It is acceptable to interpret uncertainty using the frequentist approach or the Bayesian approach, provided the interpretation is applicable to the problem at hand (Sankararaman, 2015).

8.5 PROBABILITY DISTRIBUTION OF REMAINING USEFUL LIFE AS AN UNCERTAINTY PROPAGATION PROBLEM

8.5.1 UNCERTAINTY ASSOCIATED WITH RUL ESTIMATION

Appropriate maintenance actions can be taken depending on the estimated RUL. These actions may aim at eliminating the origin of a failure which can lead the system to evolve to any critical failure mode, delaying the instant of a failure by some maintenance actions or simply stopping the system if this is judged necessary.

As in any prediction work, a prediction error should be associated with the estimated value of the RUL (Figure 8.10). There are many possible sources of the prediction error: modeling hypotheses, non-significant data, the selected prediction tools, uncertainty in the thresholds' values, and so on. In addition, uncertainty is intrinsic to any prognostic work (Provan 2003). A prognosis involves a

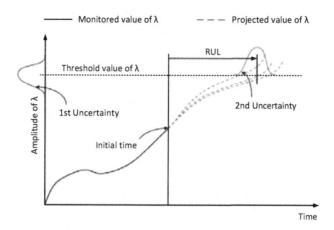

FIGURE 8.10 Uncertainty associated with RUL estimation (Medjaher & Zerhouni, 2009).

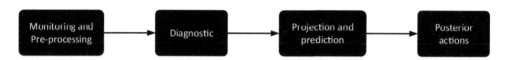

FIGURE 8.11 Summary of the ISO 13381-1: 2004 standard main steps.

projection into the future, and all future projections must contain some uncertainty, as the future cannot be predicted with certainty (Medjaher & Zerhouni, 2009).

The error associated with any RUL estimation should decrease as the time of the real failure approaches. This is exactly what happens in the case of a weather forecast: the predictions given at the beginning of a week for the next Sunday, for example, are less precise than those given for the same day but become more precise one or two days before because the predictions are adjusted each time new data are acquired.

As in weather forecasting, a confidence degree should be associated with any industrial prognostic work to render its conclusions more credible. Instead of telling an industrial worker that his/her machine will fail in *x* units of time, it would be more realistic to give an estimated RUL with a confidence value. When the uncertainty and confidence degree are included, the prognostic steps shown in Figure 8.11 become more detailed as shown in Figure 8.12 (Medjaher & Zerhouni, 2009).

The value of the estimated RUL is the output of a comparison of the projected state of the system and the predetermined threshold values. These values can be determined by using learning algorithms like those of neuro-fuzzy systems (Chinnam & Pundarikaksha, 2004). Note that, at the projection step, we do not necessarily need a value of a physical parameter; rather, we can use a desired performance, an achieved function, or an availability of a service, depending on the kind of system on which prognostics are performed (Medjaher & Zerhouni, 2009).

Three major challenges must be resolved before prognosis can be performed and the RUL can be calculated (Sankararaman et al., 2013):

1. State uncertainty: The first challenge is to estimate the current state of the system and the associated uncertainty. This problem is typically formulated as an estimation problem or a state identification problem, and filtering techniques such as particle filtering are used. The true state at any time instant is a deterministic quantity; that is, it is not precisely known and in most cases impossible to estimate with certainty due to:

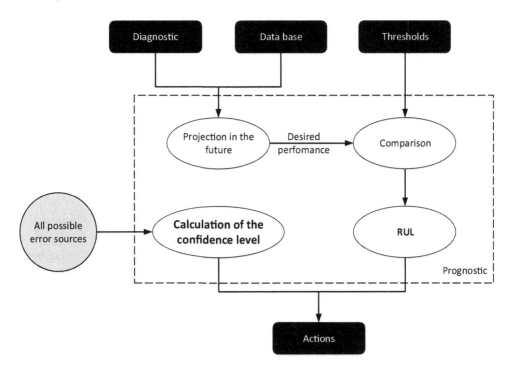

FIGURE 8.12   Prognostic and confidence degree (Medjaher & Zerhouni, 2009).

- Measurement errors and sensor noise (data uncertainty) in the data collected for estimation.
- Uncertainty in the initial state.
- Noise in the state space model used for estimation.
- Model uncertainty.
  Filtering techniques such as particle filtering are based on the Bayesian philosophy of uncertainty representation, according to which randomness can be perceived as a lack of information/precision and can be used to represent even an inherently deterministic quantity (which does not exhibit natural variability but is not known precisely) using a probability distribution.
2. Future loading uncertainty: In practical engineering systems, it is almost impossible to accurately predict the future loading and environmental conditions. Following the Bayesian approach, future loading needs to be represented as an uncertain quantity. An important challenge is to quantify the amount of uncertainty in future loading. After this uncertainty is characterized and quantified, it needs to be included in prognosis to quantify its effect on the uncertainty in RUL.
3. Process noise: Process noise is an important component of model uncertainty and commonly represented as a random variable which needs to be included while future predictions are made and the RUL is estimated. Although the process noise can be estimated using the data collected during the estimation stage, the future process noise may be significantly different from the estimated noise and cannot be known in advance. However, it is still necessary to assume statistical distributions for the process noise in the different states and include these distributions in the calculation of RUL uncertainty.

8.5.2   Uncertainty Propagation Methods for RUL Estimation

Researchers working in the areas of non-deterministic methods and uncertainty quantification techniques have developed several statistical methods for uncertainty propagation. The most general case of uncertainty propagation considers the mathematical function given by the following (Sankararaman & Goebel, 2013a):

$$Y = G\left(X_1, X_2, \ldots, X_n\right) \tag{13}$$

In this formulation, n inputs are given by X_i ($i = 1$ to n), and the uncertainty in each input is given by the PDF $f_{X_i}(x_i)$ or the CDF $F_{X_i}(x_i)$. The joint PDF of all inputs is denoted as $f_X(x)$.

The goal in uncertainty propagation is to compute uncertainty in Y, in terms of either the PDF $f_Y(y)$ or the CDF $F_Y(y)$. The entire CDF $f_Y(y)$ can be calculated as the following:

$$F_Y\left(y\right) = \int\limits_{g(X) < y} f_X\left(x\right) dx. \tag{14}$$

It is difficult to write a similar expression for PDF calculation, although the following equation makes a stab at it:

$$f_Y\left(y\right) = \int f_Y\left(y \mid x\right) f_X\left(x\right) dx. \tag{15}$$

In Equation (15), the domain of integration is such that $f_X(x) = 0$. Equation (15) is not really very meaningful: y is single-valued, given x, and hence, $f_Y(y|x)$ is nothing but a Dirac delta function. Alternatively, the PDF can be calculated by differentiating the CDF as:

$$F_Y\left(y\right) = \frac{dF_Y\left(y\right)}{dy}. \tag{16}$$

The various methods used to quantify uncertainty aim at solving the above equations in mathematically intelligent ways. The methods can be sampling-based or analytical. Some calculate the PDF of Y, while others calculate the CDF (Sankararaman & Goebel, 2013a).

8.5.2.1 Sampling-Based Methods

The most intuitive method for uncertainty propagation is MCS. Its basic underlying concept is to generate a pseudo-random number uniformly distributed on the interval [0, 1]. Then, the CDF of X is inverted to generate the corresponding realization of X. Next, several random realizations of X are generated, and the corresponding random realizations of Y are computed. The CDF $F_Y(y)$ is calculated as the proportion of the number of realizations where the output realization is less than a particular y_c. The generation of each realization requires 1 evaluation/simulation of G. Several thousands of realizations may be needed to calculate the entire CDF, especially with very high or very low values of y. Alternatively, the entire PDF $f_Y(y)$ can be computed by constructing a histogram based on the available samples of Y, using kernel density estimation (Sankararaman & Goebel, 2013a).

Variations of the basic Monte Carlo algorithm are listed below (Sankararaman & Goebel, 2013a):

1. Importance sampling: In this algorithm, random realizations are generated from a proposal density function; statistics of Y are estimated and then corrected based on the original density values and proposal density values.
2. Stratified sampling: The overall domain of X is divided into multiple sub-domains, and samples are drawn from each sub-domain independently. The process of dividing the overall

domain into sub-domains is called stratification. This method is applied when sub-populations within the overall population are significantly different.

3. Latin hypercube sampling: The method is commonly used in computer experiments. When sampling a function of N variables, the range of each variable is divided into M equally probable intervals, forming a rectangular grid. Sample positions are selected such that there is exactly 1 sample in each row and 1 sample in each column of the grid. Each resultant sample is used to compute a corresponding realization of Y and thus calculate the PDF $f_Y(y)$.

4. Unscented transform sampling: The approach focuses on estimating the mean and variance of Y accurately, not on the entire probability distribution of Y. Certain predetermined sigma points are selected in the X — space; these points are used to generate corresponding realizations of Y. The mean and variance of Y are calculated using weighted averaging principle.

8.5.2.2 Analytical Methods

A new class of methods developed by reliability engineers seeks to facilitate efficient, quick but approximate calculation of the CDF $F_Y(y)$; the focus is not on calculating the entire CDF function but on evaluating the CDF at a particular value (y_c) of the output; that is, $F_Y(Y = y_c)$ (Sankararaman & Goebel, 2013a).

The basic idea is to linearize a model G so that the output Y can be expressed as a linear combination of the random variables. Moreover, the random variables are transformed into uncorrelated standard normal space; the output Y is also a normal variable, as the linear combination of normal variables is normal. Therefore, the CDF value $F_Y(Y = y_c)$ can be computed using the standard normal distribution function. The transformation of random variable X into uncorrelated standard normal space (U) is denoted by $U = T(X)$ (Haldar & Mahadevan, 2000).

Since the model G is nonlinear, the calculated CDF value depends on the location of linearization. This linearization is done at the most probable point (MPP), which is the shortest distance from origin to the limit state, calculated in the U — space. CDF is calculated as follows:

$$F_Y(y_c) = \Phi(-\beta)$$

where Φ denotes the standard normal CDF function, and β denotes the shortest distance (see Figure 8.13).

The MPP and the shortest distance are estimated using a gradient-based optimization procedure solved using the Rackwitz−Fiessler algorithm (Rackwitz & Fiessler, 1978). The algorithm is based on repeated linear approximation of the nonlinear constraint $G(x) - y_c = 0$. This is commonly called the FORM. Several Forms are based on the quadratic approximation of the limit state (Sankararaman & Goebel, 2013a).

The entire CDF can be calculated using repeated FORM analyses and considering different values of y_c. Say we perform FORM for 10 different values of y_c; we calculate the corresponding CDF values and use an interpolation scheme to calculate the entire CDF. This can be differentiated to obtain the PDF. This approach is difficult, however; it is almost impossible to choose such multiple values of y_c, because the range of Y (i.e., the extent of uncertainty) is unknown. We can overcome this difficulty by using an inverse FORM method, choosing multiple CDF values and calculating the corresponding values of y_c. It is easier to choose multiple CDF values, as the range of CDF is known to be [0, 1], so this approach is simpler (Sankararaman & Goebel, 2013a).

8.5.2.3 Hybrid Methods

Some methods are hybrid, combining sampling with analytical methods. Surrogate modeling approaches include regression techniques, polynomial chaos expansion, and kriging. Each method uses different types of basic functions, so one may approximate G better than the other.

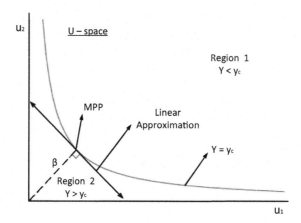

FIGURE 8.13 Most probable point concept (Sankararaman & Goebel, 2013a).

Several types of surrogate modeling techniques have been used for uncertainty propagation. The first step is to compute a few samples of X (i.e., training points) and the corresponding values of R (training values). Then different types of basic functions are constructed using this information, and interpolation is performed in multiple dimensions to facilitate the evaluation of G at untrained locations of X.

While surrogate modeling methods can be used to directly generate samples of R, researchers have developed a computational methodology to locally estimate the CDF (similar to the FORM) by emulating G, not globally, only near the curve represented by $G = r$ (Bichon et al., 2008).

8.5.2.4 Discussion

Sampling-based methods may require several thousands of samples to calculate the PDF or CDF. Thus, they are time consuming and may not be suitable for online prognostics and decision-making. Furthermore, sampling-based methods are not usually deterministic methods (with the exception of the unscented transform). In other words, every time a sampling-based algorithm is executed, it may result in a slightly different PDF or CDF. The ability to produce a deterministic solution can be an important criterion for existing verification, validation, and certification protocols, depending on the domain of interest.

Analytical methods are both computationally cheaper and usually deterministic, producing the same PDF or CDF every time the algorithm is executed. However, analytical methods are still based on approximations and, as such, can't really account for all types of uncertainty in prognosis.

For example, consider the FORM method discussed above, solved using gradient-based optimization equations. If $t_{EOL} \gg t_p$, then the number of elements in u and w may be of the order of a few hundreds or thousands, making it necessary to compute hundreds or thousands of derivatives of the equation. In that case, the computational efficiency of the analytical approach is about the same as a sampling-based approach. Although uncertainty propagation methods are available, their direct use for prognostics is challenging.

As discussed in the previous section, some researchers have advocated the use of surrogate models for uncertainty propagation. These models approximate the function $G(x)$ using different types of basic functions, such as radial basis, Gaussian basis, or Hermite polynomials. The models are inexpensive to evaluate and thus facilitate efficient uncertainty propagation. However, the use of surrogate models for uncertainty quantification in prognostics requires more research (Sankararaman & Goebel, 2013a).

8.5.3 Probability Distribution: Bayesian Approach

It is possible to address the problem of computing $f_R(r)$ from a subjective/Bayesian point of view. The Bayesian approach does not clearly differentiate between sample parameters and population parameters. The probability distribution of μ is directly computed using the available data; this uncertainty is referred to as the analyst's degree of belief about the underlying true parameter μ. Similarly, the probability distribution of σ can be computed using Bayes' theorem (Sankararaman & Goebel, 2015).

Now, consider a set of TTF times, given by r_i $(i = 1$ to $n)$. To compute the probability distribution of μ and σ, we construct their joint likelihood as the following (Sankararaman & Mahadevan, 2011):

$$L(\mu, \sigma) = \infty \prod_{i=1}^{m} f_R(r_i \mid \mu, \sigma). \tag{17}$$

For the most part, this approach is applicable to any type of parametric probability distribution, where the PDF can be expressed as $f_R(r|P)$. If R is Gaussian, then P represents the vector of the mean and standard deviation. If we let $f(P)$ denote the joint PDF of the distribution parameters P, it is easy to apply Bayes' theorem, choose uniform prior density $(f'(P) = h)$, and calculate the joint PDF as follows (Sankararaman & Goebel, 2015):

$$f(P) = \frac{hL(P)}{\int hL(P)dP} = \frac{L(P)}{\int L(P)dP}. \tag{18}$$

Note that the uniform prior density function can be defined over the entire admissible range of parameter P. For example, the mean of a normal distribution can vary in $(-\infty, \infty)$, while the standard deviation can vary in $(0, \infty)$, because the standard deviation is always greater than zero. Both are improper prior distributions because they do not have finite bounds.

If the TTF times are given by $\{100, 105, 98, 110, 92, 97, 85, 120, 93, 101\}$, the probability distribution of μ and σ can be calculated as shown in Figures 8.14 and 8.15 (Sankararaman & Goebel, 2015).

One realization of the parameters (μ and σ) uniquely defines the PDF $f_R(r)$. However, as the parameters are themselves uncertain, R is now represented by a family of distributions (Sankararaman & Mahadevan 2011, 2013a). This family of distributions will shrink to the true underlying PDF (denoted by $f_R^T(r)$) as the available data increase, and asymptotic PDF (as data increase) is simply reflective of the variability in material properties, operating conditions, and so on across all the nominally identical specimens.

In an alternative approach, a single unconditional PDF of R, which includes both the variability in X and the uncertainty in the distribution parameters P, can be expressed as follows (Sankararaman & Goebel, 2015):

$$f_R(r) = \int f_R(r \mid P) f(P) dP \tag{19}$$

The right-hand side of Equation (19) is no longer conditioned on P. Some researchers refer to this PDF $f_{R(}r)$ as the predictive PDF (Kiureghian, 1989) of R. For the predictive PDF for the above numerical example, see Figure 8.16 (Sankararaman & Goebel, 2015).

The predictive PDF $f_R(r)$ will indicate the presence of greater uncertainty in R than the original PDF $f_R^T(r)$, because it accounts for the lack of infinite data. As the data increase, $f_R(r)$ will tend

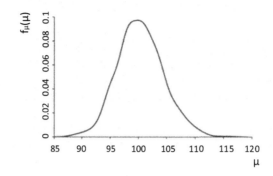

FIGURE 8.14 PDF of μ (Sankararaman & Goebel, 2015).

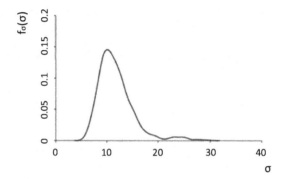

FIGURE 8.15 PDF of σ (Sankararaman & Goebel, 2015).

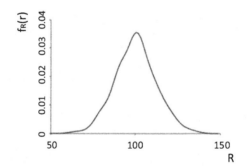

FIGURE 8.16 Predictive PDF of R (Sankararaman & Goebel, 2015).

toward $f_R^T(r)$. This is only true when the correct distribution type is assumed for R; in many cases, the choice of distribution type (i.e., the statistical model) is challenging and adds more uncertainty (Sankararaman & Mahadevan, 2013b).

REFERENCES

Baraldi P., Crenguta Popescu I., Zio E., 2012. Methods of uncertainty analysis in prognostics. Energy Department, Polytechnic of Milan. Via Ponzio 34/3, 20133 Milan, Italy. HAL ID: hal-00609156. https://hal-supelec.archives-ouvertes.fr/hal-00609156.

Bichon B., Eldred M., Swiler L., Mahadevan S., McFarland J., 2008. Efficient global reliability analysis for nonlinear implicit performance functions. AIAA Journal, 46(10), 2459–2468.

Bonome M. G., 2011. Prediction and prescription in the science of the artificial: information science and complexity. January 2011.

Calvetti D., Somersalo E., 2007. Introduction to Bayesian scientific computing: Ten lectures on subjective computing (Vol. 2). Springer.

Celaya J., Saxena A., Goebel K., 2012. Uncertainty representation and interpretation in model-based prognostics algorithms based on Kalman filter estimation. Proceedings of the Annual Conference of the PHM Society (pp. 23–27).

Chen J., Ma C., Song D., Xu B., 2016. Failure prognosis of multiple uncertainty system based on Kalman filter and its application to aircraft fuel system. Special Issue Article: Advances in Mechanical Engineering. 8(10), 1–13.

Chinnam R., Pundarikaksha B., 2004. A neuro fuzzy approach for estimating mean residual life in condition based maintenance systems. International Journal Materials and Product Technology, 20, 166–179.

Cristaldi L., Ferrero A., Leone G., Salicone S. 2018. A possibilistic approach for measurement uncertainty propagation in prognostics and health management. Dipartimento di Elettronica, Informazione e Bioingegneria. Politecnico di Milano. Milano, Italy.

de Finetti B., 1974. Theory of probability: A critical introductory treatment. Wiley.

Daigle M., Goebel K., 2011. A model-based prognostics approach applied to pneumatic valves. International Journal of Prognostics and Health Management. 2(2), 16.

Engel S. J., 2008. Prognosis requirements and v & v panel discussion on PHM capabilities: Verification, validation, and certification issues. International Conference on Prognostics and Health Management, Denver, CO. October 6–9, 2008.

Engel S. J., 2009. PHM engineering perspectives, challenges and "crossing the valley of death". Annual Conference of the Prognostics and Health Management Society. San Diego, CA.

Escobet T., Quevedo J., Puig V., 2012. A fault/anomaly system prognosis using a data driven approach considering uncertainty. WCCI 2012 IEEE World Congress on Computational Intelligence June, 10–15, 2012, Brisbane, Australia.

Goebel K., Eklund N. 2007. Prognostic fusion for uncertainty reduction. Proceedings of AIAA Infotech@ Aerospace Conference. Reston, VA: American Institute for Aeronautics and Astronautics, Inc.

Goebel K., Eklund N., Bonanni P., 2006. Fusing competing prediction algorithms for prognostics. Proceedings of 2006 IEEE Aerospace Conference.

González-Prida V., Zamora J., Guillén A., De La Fuente A., Viveros P., Martínez-Galán P., Candón E., Moreu P., 2019. Understanding the new context of uncertainty and risk under the 4th Industry Revolution. Proceedings of the 29th European Safety and Reliability Conference. M. Beer and E. Zio (Eds.), Research Publishing.

Haldar A., Mahadevan S., 2000. Probability, reliability, and statistical methods in engineering design, John Wiley & Sons.

Helton J. C., Burmaster D. E., 1996. Guest editorial: Treatment of aleatory and epistemic uncertainty in performance assessments for complex systems. Reliability Engineering & System Safety, 54, 91–94.

Helton J. C., Oberkampf W. L., 2004. An exploration of alternative approaches to the representation of uncertainty in model predictions. Reliability Engineering & System Safety, 85(1), 39–71. Special issue on Alternative Representations of Epistemic Uncertainty.

Kaplan S., Garrick B. J., 1981. On the quantitative definition of risk. Risk Analysis, 1(1), 11–27.

Kiureghian A. D., 1989. Measures of structural safety under imperfect states of knowledge. Journal of Structural Engineering, 115(5), 1119–1140.

Lindley D. V., 2000. The philosophy of statistics. The Statistician, 49(3), 293–337.

Lindley D. V., 2006. Understanding uncertainty. Wiley-Interscience.

Macheret Y., Teichman J., 2008. Effect of initial defect distribution on accuracy of predicting aircraft probability of failure. International Conference on Prognostics and Health Management, Denver, CO. October 6–9, 2008.

Medjaher K., Zerhouni N., 2009. Residual-based failure prognostic in dynamic systems. 7th IFAC International Symposium on Fault Detection, Supervision and Safety of Technical Processes, SAFE PROCESS'09, Barcelona, Spain.

Mosleh A., Bier V. M., 1996. Uncertainty about probability: A reconciliation with the subjectivist viewpoint. Special Issue of IEEE Transactions on Systems, Man, and Cybernetics, 26(3).

Orchard M., Kacprzynski G., Goebel K., Saha B., Vachtsevanos G., 2008. Advances in uncertainty representation and management for particle filtering applied to prognostics. International Conference on Prognostics and Health Management, Denver, CO. October 6–9, 2008. http://www.phmconf.org

Popper K., 1959. The propensity interpretation of probability. The British Journal for the Philosophy of Science, 10(37), 25–42.

Provan G., 2003. Prognosis and condition-based monitoring: An open systems architecture. 5th IFAC Symposium on Fault Detection, Supervision and Safety of Technical Processes, 57–62.

Rackwitz R., Fiessler B., 1978. Structural reliability under combined random load sequences. Computers & Structures, 9(5), 489–494.

Ratto M., Tarantola S., Saltelli A., 2001. Sensitivity analysis in model calibration: GSA-GLUE approach. Computer Physics Communications, 136, 212–224.

Saha B., Goebel K., 2008. Uncertainty management for diagnostics and prognostics of batteries using Bayesian techniques. IEEE Aerospace Conference, Big Sky, MT. March 1–8, 2008.

Sankararaman S., 2012. Uncertainty quantification and integration in engineering systems (Ph.D. Dissertation). Vanderbilt University.

Sankararaman S., 2015. Significance, interpretation, and quantification of uncertainty in prognostics and remaining useful life prediction. SGT Inc., NASA Ames Research Center, USA. Mechanical Systems and Signal Processing, 52–53, 228–247.

Sankararaman S., Goebel K., 2013a. Remaining useful life estimation in prognosis: an uncertainty propagation problem. American Institute of Aeronautics and Astronautics. NASA Ames Research Center, USA.

Sankararaman S., Goebel K., 2013b. Why is the remaining useful life prediction uncertain? Annual Conference of the Prognostics and Health Management Society 2013.

Sankararaman S., Goebel K., 2014. Uncertainty in prognostics: Computational methods and practical challenges. IEEEAC Paper # 2338, Version 1, 2014.

Sankararaman S., Goebel K., 2015. Uncertainty in prognostics and systems. Health Management. International Journal of Prognostics and Health Management.

Sankararaman S., Ling Y., Shantz C., Mahadevan S., 2011. Uncertainty quantification in fatigue crack growth prognosis. International Journal of Prognostics and Health Management, 2–1(1).

Sankararaman S., Mahadevan S., 2011. Likelihood based representation of epistemic uncertainty due to sparse point data and/or interval data. Reliability Engineering & System Safety, 96(7), 814–824.

Sankararaman S., Mahadevan S., 2013a. Separating the contributions of variability and parameter uncertainty in probability distributions. Reliability Engineering & System Safety, 112, 187–199.

Sankararaman S., Mahadevan S., 2013b. Distribution type uncertainty due to sparse and imprecise data. Mechanical Systems and Signal Processing, 37(1), 182–198.

Sankararaman S., Saxena A., Daigle M., Goebel K., 2013. Analytical algorithms to quantify the uncertainty in remaining useful life prediction. IEEEAC Paper #2336, Version 1.

Shafer G., 1976. A mathematical theory of evidence, Princeton University Press.

Singpurwalla N. D., 2006. Reliability and risk: A Bayesian perspective. Wiley.

Tang L., Kacprzynski G. J., Goebel K., Vachtsevanos G., 2008. Methodologies for Uncertainty Management in Prognostics. IEEEAC paper # 1263, Version 5.

Usynin A., Hines J. W., 2007. Uncertainty management in shock models applied to prognostic problems. AAAI Fall Symposium. November 8–11, 2007. Arlington, VA.

Villiers J. P., Laskey K., Jousselme A. L., Blasch E., Waal A., Pavlink G., Costa P., 2015. Uncertainty representation, quantification and evaluation for data and information fusion. Conference Paper: July 2015.

Von Mises R., 1981. Probability, statistics and truth. Dover Publications.

Wang H. F., 2011. Decision of prognostics and health management under uncertainty. International Journal of Computer Applications, 13(4), 1–5.

Wikipedia, 2020a. Uncertainty quantification. https://en.wikipedia.org/wiki/Uncertainty_quantification. Viewed: June 27, 2020.

Wikipedia, 2020b. Propagation of uncertainty. https://en.wikipedia.org/wiki/Propagation_of_uncertainty. Viewed: June 27, 2020.

Zio E., Pedroni N., 2013. Methods for representing uncertainty: A literature review. Risk analysis. No. 2013-03. Foundation for an Industrial Safety Culture (FonCSI).

9 RUL Estimation of Dynamic and Static Assets

9.1 PHYSICS OF FAILURE IN DYNAMIC AND NON-DYNAMIC ASSETS

The remaining useful life (RUL) of an asset or a system is defined as the time left from the current time to the end of its useful life (Si et al., 2011).

Figure 9.1 shows the evolution of the deterioration of a system. The RUL is the time between the current time, which corresponds to condition A, and the time when a maximum acceptable level of deterioration, condition B, is reached. The latter can correspond to the time of failure (i.e., for 100% of deterioration) or to a predefined lower threshold value. In general, a proper RUL prediction is useful in maintenance, saving resources and improving processes (Rodriguez Obando, 2018).

Prognostics deal with predicting the future behavior of engineering systems, and several sources of uncertainty will influence such predictions. A suitable prediction of RUL is accurate and precise, following certain criteria to evaluate a measure. However, unlike a physical magnitude, which can be quantified with an uncertainty level in the present time, the RUL prediction must include additional assumptions about the future operating conditions, as illustrated in Figure 9.1. Prognosis of RUL projects the diagnosis of current system condition (here, the deterioration in A) into the future using a model. Of course, in the absence of "future measurements", this necessarily entails propagated uncertainty.

Even if a given mechanical system model is widely known, several sources of uncertainty will affect the RUL prediction. In fact, it is not even meaningful to make such predictions without computing the uncertainty associated with the RUL. Some uncertainties can be considered endogenous to the system, for instance, the initial condition of its deterioration and its dynamic behavior; others can be considered exogenous to the system, such as measurement noise and process disturbances; still others can be considered as uncertainties in the future of the system, for instance, changes in the operating conditions. Therefore, accurately predicting the RUL is an open problem.

Several statistical and probabilistic approaches have been developed to tackle the problem. These include statistical data-driven approaches (Si et al., 2011). Model-based approaches, based on a model representing the physics of the degradation process built with specific experiments have been proposed (Jardine, Lin, & Banjevic, 2006), as well as data-based approaches using monitored data to construct the model (Rodriguez Obando, 2018).

For the last decade, the development of prognostics models has been a priority from both an academic and an operational point of view. The majority of prognostics research to date has focused on the prediction of the RUL of individual components and the propagation of a single mechanism leading to a single failure mode. However, industrial equipment is complex and can have concurrent multi-failure modes and multi-failure mechanisms involving various components and sub-components. The propagation of failure mechanisms may also involve several components, and various diagnostic tools can be used to detect and track them at different system scales.

Once predetermined degradation thresholds are reached, specific maintenance actions should be taken to avoid a system failure. Depending on the types of active failure mechanisms and

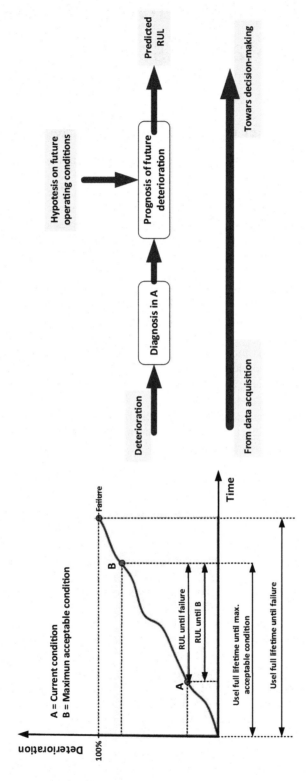

FIGURE 9.1   Remaining useful life, its concept and its use (Rodriguez Obando, 2018).

their progression, maintenance actions may not have the same effect to stop or slow down their propagation toward their related failure modes. It is therefore important to understand how the mechanisms propagate when we want to apply specific maintenance tasks to extend the RUL of complex equipment.

The approach to the predictive maintenance of complex equipment relies on a physics of failure (PoF) model. The model is based on expert knowledge and is also dynamic, as it uses diagnostic data (Blancke et al., 2018).

PoF is a science-based approach to reliability that uses modeling and simulation to design-in reliability. It helps to understand system performance and reduce decision risk during design and after the equipment is fielded. This approach models the root causes of failure such as fatigue, fracture, wear, and corrosion, leveraging the understanding of the processes and mechanisms that induce failure to predict reliability and improve product performance (Willems & Vandevelde, 2015).

The PoF concept is based on the relationships between requirements and the physical characteristics of the product and their variation in the manufacturing processes, and the reactions and interactions of product elements and materials under loads (stressors) and their influence on the fitness for use with respect to use conditions and time.

9.1.1 Physics of Failure Prognostic Models

Physics-based approaches focus on the equipment degradation process. They aim to model the propagation of equipment failure mechanisms by taking into account knowledge of the physics of degradation and feedback from domain experts. In such approaches, diagnostic data are often used to update initial conditions and to fine-tune the model. One of the main advantages of the PoF approach is that it is applicable even if the data are scarce, as it takes advantage of the knowledge gained. An adapted form of a generic methodology proposed by Gu and Pecht (2008) for PoF prognostic models is shown in Figure 9.2 (Blancke et al., 2018).

The methodology is based on the identification of failure modes, as in failure modes and effects analysis, but it also identifies the failure mechanisms that can lead to them. Once identified, prognostics models can be created for critical failure mechanisms. Of course, complex equipment may have several different failure modes and many failure mechanisms, requiring specific diagnostic tools to detect their state. For this purpose, Amyot et al. (2014) proposed an extension of failure modes, mechanisms, and effect analysis by discretizing the mechanisms using physical state of degradation. Each state can be detected by a unique combination of symptoms obtained using diagnostic tools. The proposed model consists of a causal graph where the nodes are physical states, and the edges represent all the identified failure mechanisms. Failure mechanisms propagate from a root cause to their related failure mode through a physical state succession, as shown in Figure 9.3. As a physical state can be present in different failure mechanisms, the causal graph enables failure mechanisms to share physical states. Amyot et al. (2014) also introduced an algorithm to detect active failure mechanisms based on a combination of active and inactive physical states (Blancke et al., 2018).

The dynamic causal graph model proposed by Amyot et al. (2014) enables various diagnostic data to be aggregated from different diagnostic tools at a system level. For this to evolve toward predictive maintenance, however, it is necessary to introduce the temporal aspect of causality (Blancke et al., 2018).

9.1.2 Physics of Failure Procedure

The application of the PoF approach requires the four main steps illustrated in Figure 9.4 and explained below (Matiü & Sruk, 2008).

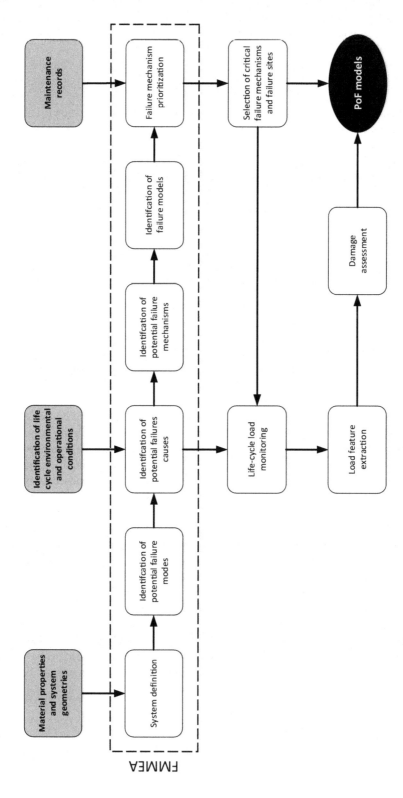

FIGURE 9.2   PoF-based proportional hazards model methodology (Gu & Pecht, 2008).

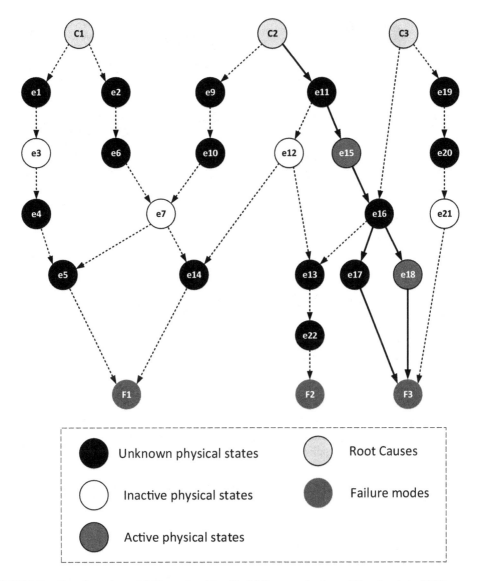

FIGURE 9.3 Causal graph model illustrating identified failure mechanisms (Blancke et al., 2018).

1. First, determine environmental factors by specifying or measuring them.
2. Second, isolate potential failure triads (site, mode, and mechanism). The sequence can differ depending on the available data. This step unites the identification of failure sites and the sites' corresponding failure modes with the mechanisms contributing to a potential failure mode. A comparative analysis of good and failed components, either in a field environment or a laboratory environment, could be helpful in this step.
3. Third, filter contributing environmental and/or operational factors.
4. Fourth, find the functional dependencies of all stresses and identify applicable models. Select the best fitting model for the specified operational and/or environmental conditions and determine the model's validity bounds. When the proper equations are known, it is possible to determine the life of components for any given operational/environmental condition.

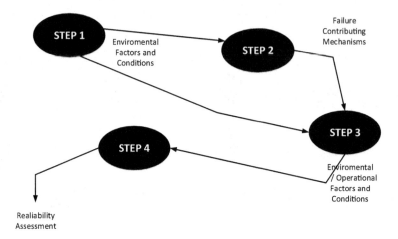

FIGURE 9.4 Generic PoF procedure (Matiü & Sruk, 2008).

9.1.3 Physics of Failure and Its Role in Maintenance

The majority of the challenges in the field of maintenance are associated with failures and their prevention. Understanding why, when and how components fail plays an important role in many aspects of maintenance management and engineering.

The key issue of maintenance is finding the optimal balance between minimizing the effects of failing (sub-) systems on the one hand, and minimizing the maintenance costs on the other. For critical systems, where failures have serious consequences in terms of costs, safety, environmental effects, or consequential damage, measures are generally taken to prevent such failures from occurring and spending a considerable budget on preventive maintenance is then justified.

However, for many critical applications in the military, aerospace, and nuclear power sectors, this risk avoidance has resulted in a very conservative maintenance approach. Many components are replaced long before they reach the end of their actual service life. Maintenance programs are thus effective (in preventing failures), but not very efficient (Tinga, 2013).

9.1.3.1 Predictive Maintenance and Prognostics

The main reason for the conservatism in the maintenance interval determination is the uncertainty in the expected service life of subsystems and components. Traditionally, researchers tried to capture this uncertainty in statistical distributions of failure times, for example, using the Weibull distribution function. However, this experience-based approach, where historical failure data are used to make predictions for future failures, has some important drawbacks.

First, a sufficiently large set of failure data is required to accurately determine the parameters of the distribution function. For critical systems, failures are prevented as much as possible, so failure datasets are, by definition, small. Second, the distribution functions are based on historical data and are thus associated with the usage profile in that period of time. When the present usage is significantly different, the distribution is not representative and thus cannot be used for predictive maintenance.

An alternative approach circumventing these drawbacks is the use of physical models for the prediction of failures, as illustrated in Figure 9.5. The purpose is to find out how the remaining life depends on the usage of the system as indicated in the upper half of the figure. The uncertainty in these relations, and the associated conservatism, can be reduced by zooming into the level of the physical failure mechanisms. By defining a failure model, the quantitative relations between the internal load (e.g., stress), which is directly related to the specific usage (e.g., rotational speed) of

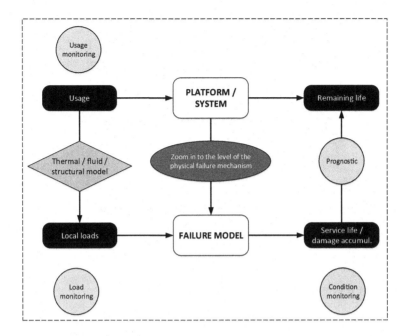

FIGURE 9.5 Model-based approach showing relations between usage, loads, condition, and life consumption (Tinga, 2013).

the system, and the resulting degradation rate is established. This gives the ability to accurately predict the expected time to failure (TTF), that is, if the usage of the system is monitored appropriately (Tinga, 2013).

The big challenge in this approach is, first, to assess the critical failure mechanism(s) (e.g., fatigue, corrosion, and wear) and the governing load. Second, a suitable model for the identified failure mechanism must be defined or developed.

The approach has been successfully applied to military systems in recent years, including gas turbines, helicopters (see Figure 9.6), navy frigates (Tinga & Janssen, 2012), TERs, and other military vehicles. These case studies have demonstrated that the application of the approach to real systems is feasible and provides benefits relative to the traditional (statistical) approach (Tinga, 2013).

9.1.3.2 Condition-Based Maintenance

The concept of condition-based maintenance (CBM) has been known for decades, but recent developments have considerably widened the applicability of this methodology. Many new (reliable) sensors have become available, enabling the monitoring of a wide range of load and condition parameters. Sensors detecting fatigue cracks, disbonded joints, erosion, impact damage, composite delamination, and corrosion are now available and enable structural health monitoring (SHM) of complex systems like aircraft or wind turbines. And the increased computational power of modern computers has made the analysis of all these collected data feasible.

With the boost in performance and availability of sensors and other hardware, condition monitoring systems are now offered by original equipment managers (OEMs) in several industries as a way to increase maintenance efficiency. However, many operators, and even some manufacturers, are only now starting to realize that CBM is not automatically possible when the condition monitoring system (i.e., sensors) is in place.

An additional requirement is that the collected raw data must be analyzed and interpreted to translate them into useful maintenance information. In this analysis step, PoF approaches are helpful, as

FIGURE 9.6   Navy frigate with several subsystems (Tinga, 2013).

the challenge of condition monitoring is to find features in the monitored signal that can be related to failure or degradation processes in the system.

For example, understanding the failure mechanisms in gearboxes helps to interpret the various features in the vibration monitoring signal obtained from the monitoring system. Similarly, features in the electro-chemical noise signals of corrosion processes can be attributed to certain corrosion mechanisms (Tinga, 2013).

In the phase of developing new condition monitoring systems, knowledge of failure mechanisms is also essential. When the critical failure mechanisms for a certain system are known, the suitable type of sensor can be selected and the appropriate location to monitor can be determined. This process was recently formalized in a CBM guideline (see Figure 9.7). In this decision diagram, a number of questions must be answered to identify whether a system is suitable for CBM. CBM is an appropriate concept for the considered system only when all questions can be answered positively (Tinga, 2013).

Finally, prognostics is important for CBM. The sensor data only provide information about the current state. Waiting until a monitored condition parameter exceeds a critical value means immediate action is required; this is difficult to plan (e.g., personnel and spare parts) and may have serious consequences for the system availability. Therefore, a prognostics method is required to determine when future maintenance activities are necessary.

In traditional condition monitoring systems, such as vibration analysis, trending methods and growth models are used to extrapolate trends in monitored condition parameters (e.g., vibration levels) to determine component replacement or repair intervals. However, these are experience-based approaches and, as such, have drawbacks. In this case, knowledge of the physical failure mechanisms can improve the maintenance efficiency.

By applying physics model-based prognostic methods, the effects of changes in use can be taken into account. The big difference with the prognostic methods used for fixed maintenance intervals is the availability of the monitoring data. First, this means the prognosis is not done for the complete service life, only for the fraction remaining after the last condition assessment. This limited scope makes the prediction generally more reliable and accurate. Second, the monitoring data can be used to validate the physical model used for the prediction. The consequence is that the model improves during operation, as more and more data can be used as feedback (Tinga, 2013).

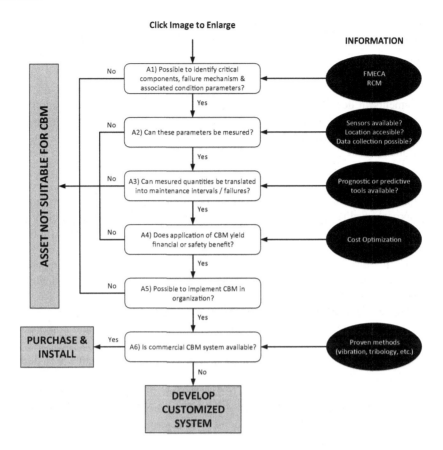

FIGURE 9.7 Decision diagram for condition-based maintenance (Tinga, 2013).

9.1.3.3 Root Cause Analysis

Despite the range of maintenance activities performed in industry, unexpected failures are unavoidable in practice. If the failure of a system has serious consequences, measures are generally taken to prevent it from happening again. But less critical failures can be extremely troublesome when they occur on a regular basis. In such cases, it is essential to identify the root cause of the failure.

The structured analysis of these types of real failures in industry can benefit from the knowledge of failure mechanisms. At the level of the physical failure mechanisms, the cause of a failure can be straightforward. Either the load on the system was too high, or the load-carrying capacity of the system was too low. The former could be caused by either by misusing the system or by using the system in a different way than it was designed for. The low capacity may be due to the application of wrong materials or a design error. The procedure consists of the following steps (Tinga, 2013):

1. Set-up fault tree to identify possible failure modes.
2. Prioritize failures (e.g., based on computer maintenance management systems data).
3. For critical failures, assess failure mechanism and governing load.
4. Solve the problem by either increasing capacity or reducing load.

9.1.4 ADVANTAGES OF THE PoF APPROACH IN RELIABILITY ENGINEERING

When the PoF approach is applied during development, the resulting advantages spread throughout the whole development process. They include the following (Matiü & Sruk, 2008):

- Compare design candidates: Potential failure modes that are specific for a certain environment can be analyzed before deciding whether the candidate component should be used in the design. The ability to compare design candidates' ability to support a specific requirement eliminates over-designing, contributing to cost-effectiveness.
- Identify design improvements: The PoF approach can issue a warning on the need for improvements in design or indicate the need for inspections during operation. As a result, test-analyze-fix (TAF) cycles, an essential part of the reliability improvement process, can be minimized, perhaps eliminated. The ability to minimize TAF cycles shortens the time-to-market and reduces the development cost. At the same time, the PoF approach can be used to analyze how design changes will affect reliability.
- Make realistic predictions: Because there are deeper insights into failures when failure mechanisms are known, overly optimistic or pessimistic predictions can be minimized. The PoF approach evaluates all elements for which reliability is a concern, including materials, structures, and technologies. Therefore, it can have a cost-effective impact on both maintenance and logistics.
- Estimate reliability more quickly: PoF allows an organization to check or compare the vendor-claimed reliability of a product in a specified environment. This could eliminate the need for long-lasting testing.
- Determine the life expectancy of components for different tasks: PoF can predict the life expectancy of a component in various mission profiles. An electronic component used on an automotive platform may have a very different life expectancy than the same component used on another platform. With globalization, it is a competitive advantage to design reliable systems for various environments.
- Optimize environmental stress screening (ESS): PoF models can take advantage of weak properties that are the consequences of either defects in materials or inappropriate manufacturing processes. Therefore, the approach can be used to determine the optimal number of burn-in/ESS cycles and their values. This optimization will help to avoid unnecessary product aging.
- Identify the optimal preventive maintenance interval: With environmental variations, the life expectancy of components will show considerable differences. More specifically, different environments will alter dominant failure modes. This, in turn, will lead to a different maintenance focus and the corresponding optimal preventive maintenance interval.

This non-exhaustive list suggests applying the PoF approach during the development process can identify and address a number of reliability concerns. In a best-case scenario, it will shorten the development cycle, reduce the development cost, and optimize maintenance (Matiü & Sruk, 2008).

9.2   RELIABILITY ESTIMATION AND PREDICTION

9.2.1   Reliability Estimation

Engineering systems are designed to perform sophisticated functions. For example, in the oil sands industry, heavy haulers carry tons of ore from one location to another, and slurry pumps contain a medium with high wear ingredients like sands. The components like gears and impellers in such systems are subject to cyclic stress and/or excessive wear which may lead to unexpected failures of the whole system, causing a loss in production and profit.

To alleviate such losses, companies can implement maintenance actions in which reliability plays an important role in making maintenance decisions. Reliability describes the ability of a system to work properly for a specified time period in a specified environment. The process of estimating reliability is referred to as reliability estimation; a proper execution of reliability estimation is crucial to the success of reliability based maintenance actions.

Reliability estimation uses event data and condition monitoring data. Reliability estimation methods can be differentiated by these two types of data.

Event data refer to information on what has happened to the system, such as installation, start-up, shutdown, and failure, as well as information on what has been done to the system, such as repair/replacement and overhaul. One type of event data widely used in reliability estimation is time-to-failure (TTF) data or lifetime data (data from the start to the failure of operation). For example, the TTF data of a light-emitting diode light may be 50,000 hours from the first time use to the ultimate failure.

Condition monitoring data refer to the versatile data measured by monitoring devices such as thermometers, pressure gauges, vibration sensors, and acoustic emission sensors, and they include value-type data, waveform data, and multidimensional data. All these types of data are time series data gathered at specified time intervals over a certain time period. Typical value-type data include oil analysis data, temperatures, pressures, and moistures. Waveform data display a waveform pattern and typically include vibration signals and acoustic signals. Ultrasonic data and visual images display images and are typical multidimensional data (Liu, 2016).

9.2.1.1   Reliability Estimation Based on Event Data

Reliability is defined as the probability that a system will perform its intended functions satisfactorily for a specified time period under specified operating conditions. Computationally, reliability is the probability that a system has a lifetime greater than a certain length of time. A probability density function (PDF) of system lifetime, also called the lifetime distribution, is usually adopted to estimate reliability. Several popular lifetime distributions are lognormal, exponential, and Weibull distributions.

In practical applications, lifetime distribution can be determined based on available information of event data, among which TTF data are the most widely reported. First, assume the lifetime of the system is a random variable and follows a certain type of lifetime distribution, for example, Weibull distribution. A goodness-of-fit test may be used to select the best distribution type. Then, estimate the parameters of the selected lifetime distribution based on available time-to-failure data using parameter estimation methods. Once the parameters of the lifetime distribution are determined, the reliability of the system can be estimated.

Reliability estimation based on event data has been studied for decades. Event data such as TTF data are collected from a large group of systems with the same characteristics (e.g., designed to achieve the same function, produced from the same batch, and operated under similar working and operating conditions). The TTF data of a large number of such systems are used to obtain the lifetime distribution for the entire group. The reliability of any individual system belonging to this group can thus be estimated.

However, there are some drawbacks. Reliability estimated using lifetime distribution based on event data cannot depict the specific condition of an individual system at a certain time point. For example, the reliability of a pump may be estimated as 80% based on event data, but it is impossible to know which specific pumps fall into the 20% of failure probability and fail at the time when the reliability is estimated. Moreover, TTF data are not always available, especially for expensive or highly reliable systems. If TTF data are limited or not available, reliability estimation based on event data is difficult to apply (Liu, 2016).

9.2.1.2   Reliability Estimation Based on Condition Monitoring Data

Because of the limitations of event data, condition monitoring data are attracting more interest in reliability estimation. With proper processing of the data, we could possibly obtain the reliability that reflects the ability of a specific system to fulfill its anticipated design function rather than a reliability estimated for a group of systems based on event data.

As mentioned in Section 9.2.1, there are two types of condition monitoring data. The value-type data can generally be directly used to represent system condition. For example, when the crack size on a gear tooth is greater than a certain value, the gearbox could be treated as having a failure. However, value-type data as such are difficult to obtain with non-intrusive means in practice, making online reliability estimation impossible. Some other value-type data, such as environmental temperatures, may not be sensitive to changes in a system's health condition.

Waveform data and multidimensional data are usually not directly used for reliability estimation. Data must be processed to extract quantitative measures, the so-called health indicator, to represent system health conditions. For example, the root mean square, standard deviation, and Kurtosis are used as indicators of the health of a gearbox. A large number of indicators have been studied. Current methods for reliability estimation usually comprise two key parts: thresholding and probability density estimation.

Thresholding refers to the process of determining a critical level of the indicator corresponding to the transition of system states. The critical level of the indicator is also called the threshold and is usually a pre-specified constant value. The performance of the system is regarded as deviating from the expected normal state if the indicator value exceeds the threshold.

Thresholds are traditionally determined with prior knowledge, for example, past experience, and design criteria. Event data have also been used to determine thresholds. For example, TTF data can be used to estimate the average availability of drill bits, and this can be treated as the threshold for making the maintenance decision for drill bits. If prior knowledge or TTF data are not available, thresholds can be determined directly from condition monitoring data. A boundary can be created between normal and abnormal data based on condition monitoring data and used as a threshold for reliability estimation.

Probability density estimation refers to the process of estimating the PDF of indicator values over a certain time period. It is often assumed that the indicator values over a certain time period follow an identical statistical distribution described by a PDF. As reported by Silverman (1986), probability density estimation methods are able to obtain the PDF of a given set of data. Therefore, it is applicable to use probability density estimation methods to obtain the PDF of an indicator value for reliability estimation. Methods for probability density estimation fall into two categories: parametric methods and non-parametric methods. Parametric methods estimate the PDF of data with the assumptions that the type of PDF is known and there are unknown finite parameters. Non-parametric methods estimate the PDF of data based on the data themselves, regardless of the prior information on the type of PDF.

It is also possible to use thresholding. With the acquired threshold and the PDF of the indicator values, the reliability is estimated as the probability that the indicator value does not exceed the threshold value.

Compared to reliability estimation based on event data, reliability estimation based on condition monitoring data can provide reliability estimates in accordance with the change of indicator values and reflect the degradation process of individual systems. It can also estimate the reliability for expensive or highly reliable systems for which the event data such as TTF data are not always available. However, reliability estimation based on condition monitoring data requires condition monitoring devices, and the threshold of indicator value is difficult to determine when there are not sufficient data (Liu, 2016).

9.2.2 Prognostics and Reliability

The need for the reliability, availability, and safety of a system is a determining factor in the effectiveness of industrial performance. As a consequence, the high costs of maintaining complex equipment are making it necessary to enhance maintenance support systems, and traditional concepts like preventive and corrective maintenance are progressively being replaced by new ones like predictive and proactive maintenance. Prognostics is now considered a key feature in maintenance strategies, as the estimation of the provisional reliability of an equipment and its RUL allows companies to avoid inopportune spending.

Considerable research supports the value of prognostics. However, in practice, choosing an efficient technique depends on classical constraints that limit the applicability of the tools: available data-knowledge-experiences, dynamic and complex systems, implementation requirements (precision, computation time, etc.), available monitoring devices, and so on. Moreover, implementing an adequate tool can be a non-trivial task, as it can be difficult to provide effective models of dynamic systems including the inherent uncertainty of prognostics. Ongoing developments are founded on two complementary assumptions (El-Koujok et al., 2008).

1. Real systems are increasing in complexity, and their behavior is often nonlinear, making modeling difficult, even impossible. Intelligent maintenance systems must take this into account.
2. In many cases, it is not too costly to equip dynamic systems with sensors, thus allowing real data to be gathered. Monitoring systems evolve in this way (El-Koujok et al., 2008).

9.2.2.1 From Maintenance to Prognostics

Maintenance activities use various methods, tools, and techniques with the goal of reducing maintenance costs while increasing reliability, availability, and safety. The steps of fault detection, failure diagnosis, and response development correspond to first, perceiving phenomena, second, understanding them, and third, acting in response to them. However, instead of simply understanding something like a failure (*a posteriori* comprehension) and fixing it, it may be a better idea to anticipate it and take proactive remedial action. This could be defined as prognostics.

The relative positioning of detection, diagnosis, prognostics, and decision-making/scheduling can be schematized as in Figure 9.8. In practice, prognostics are performed after a detection step: the monitoring system detects that the equipment passes an alarm limit, and this activates the prognostics process (El-Koujok et al., 2008).

9.2.2.2 From Prognostics to Predictions

The International Organization for Standardization says "prognostics is the estimation of TTF and risk for one or more existing and future failure modes" (ISO 13381-1 2004). Otherwise stated, it is the prediction of a system's lifetime given the current machine condition and past operating profile. Two salient characteristics of prognostics are the following (El-Koujok et al., 2008):

- Prognostics is mostly assimilated into prediction (a future situation must be caught).
- Prognostics is based on the failure notion, which implies a degree of acceptability.

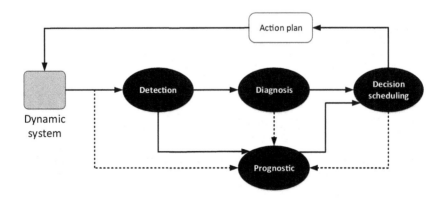

FIGURE 9.8 Prognostics within maintenance activity (El-Koujok et al., 2008).

A central problem is that the accuracy of a prognostics system is related to its ability to approximate and predict the degradation of equipment. In other words, starting from a current situation, a prognostics tool must be able to forecast future possible situations, and the prediction phase is thereby a critical one (El-Koujok et al., 2008).

9.2.2.3 From Prediction to Reliability

As mentioned earlier, an important task of prognostics is to predict the degradation of equipment. Thus, it is a process that allows *a priori* reliability modeling (El-Koujok et al., 2008).

Reliability ($R(t)$) is defined as the probability that a failure does not occur before time t. If the random variable ϑ denotes the TTF with a cumulative distribution function $F_\vartheta(t)= \mathrm{Prob}(\vartheta \leq t)$, then:

$$R(t) = 1 - F_\vartheta(t). \tag{1}$$

Assume the failure is not characterized by a random variable but by the fact that a degradation signal (y) passes a degradation limit (y_{lim}), and this degradation signal can be predicted $(\hat{y})$ with a degree of uncertainty (Figure 9.9) (El-Koujok et al., 2008).

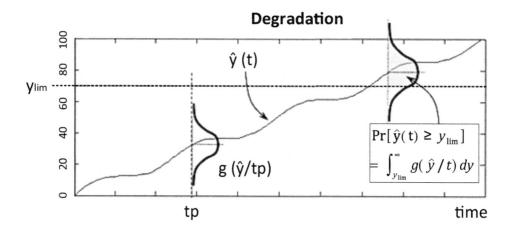

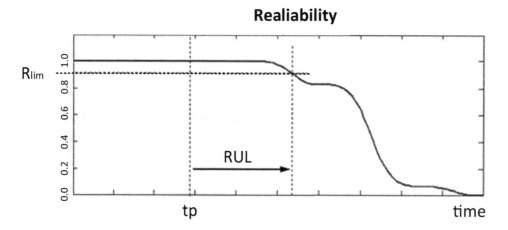

FIGURE 9.9   Prediction and reliability modeling (El-Koujok et al., 2008).

At any time t, the failure probability can be predicted as follows:

$$F(t) = Pr\left[\hat{y}(t) \geq y_{lim}\right].$$

(2)

Note that $g(\hat{y}/t)$ is the probability distribution function that denotes the prediction at time t. Thereby, by analogy with reliability theory, the reliability modeling can be expressed as follows:

$$R(t) = 1 - Pr\left[\hat{y}(t) \geq y_{lim}\right] = 1 - \int_{y_{lim}}^{\infty} g(\hat{y}/t)dy.$$

(3)

The RUL of the system can finally be expressed as the remaining time between the time of the prediction (tp) and the time to reach a reliability limit (R_{lim}) fixed by the practitioner (see Figure 9.2).

These explanations can be generalized with a multidimensional degradation signal. Finally, the *a priori* reliability analysis can be performed if an accurate prognostics tool is used to predict the degradation of equipment (El-Koujok et al., 2008).

9.2.3 RELIABILITY PREDICTION

Another way to approach reliability prediction is to define it as the process used to determine the mean time between failures (MTBF) of an item. Reliability prediction involves estimating the reliability of equipment or products prior to their production or modification. Successful reliability prediction generally requires developing a reliability model of the system. The level of detail of the model will depend on the level of design detail available at that time. Data required to quantify the model are obtained from such sources as company warranty records, customer maintenance records, component suppliers, or expert elicitation from design or field service engineers. Reliability prediction combines rigorous analysis procedures with expert judgment to develop a realistic estimate of product performance.

A reliability prediction is an important criterion in the new product development process. These predictions also play an important role in the process of equipment selection which can be carried out at any stage of a product development program to support the design process. When a reliability prediction is made, equipment designs can be improved, costly over-designs prevented, and development testing time optimized (Mirza Hyder, 2012).

These predictions are used to evaluate design feasibility, compare design alternatives, identify potential failure areas, trade-off system design factors, and track reliability improvement (ITEM Software, 2007).

9.2.3.1 Role of Reliability Prediction

Reliability prediction has many roles in the reliability engineering process. It is possible to determine the impact of proposed design changes on reliability by comparing the reliability predictions of the existing designs and the proposed designs.

The ability of the design to maintain an acceptable level of reliability under extreme environmental conditions can be assessed via reliability predictions. These predictions can also be used to evaluate the need for environmental control systems.

The effects of complexity on the designed system's operation can be evaluated by performing reliability prediction analysis. Results may suggest the need for redundant systems, back-up systems, subsystems, assemblies, or component parts. Some products on the market available to do this type of analysis are MIL-HDBK-217 (electronics reliability prediction), Bellcore/Telcordia (electronics reliability prediction), and Naval Surface Warfare Center (mechanical reliability prediction). These provide failure rate and MTBF predictions for electronic and mechanical parts and equipment.

A reliability prediction can evaluate the significance of reported failures and facilitate further analyses. For example, reliability predictions can be used to evaluate the probabilities of failure events described in alternate failure analysis models, such as modes, effects and criticality analysis, reliability block diagram analysis, or fault tree analysis (ITEM Software, 2007).

The reliability engineer has a difficult job in an age of electronic components, especially making solid predictions on component and system failure rates. Over time, reliability has increased, but paradoxically many of the tools used to make predictions have decreased in capability. To give an example, computers may have helped improve reliability through the Internet of Things, but they themselves suffer failure. Their components burn out, their circuits short, their solder joints fail, their pins bend, and their metals react with each other. Learning how to understand failure mechanisms and predict computer failure is a stand-alone profession.

Logistics, systems, and reliability engineers use failure rate predictions for many purposes: reliability analysis, cost trade studies, availability analysis, spares planning, redundancy modeling, scheduled maintenance planning, product warranties, and guarantees (Gipper, 2020). Reliability predictions are applied to the management of a product's life cycle for the following reasons (Gipper, 2020):

- Assessing the effect of product reliability on maintenance in order to predict maintenance frequency and establish the number of spares needed.
- Providing input into system-level reliability models to predict frequency of system outages in steady-state and during early life, expected downtime, and system availability.
- Providing input to unit and system-level lifecycle cost analyses to determine the cost over the lifecycle. This includes how often systems fail.
- Helping decide which product to purchase from a list of competing products.
- Setting standards for products requiring a reliability test by determining how often the system should fail; this will reveal if adequate testing is being done.
- Giving input into the analysis of complex control systems (e.g., weapon systems); in such cases, it is necessary to know how often various parts of the system, even redundant ones, are likely to fail.
- Giving input into designs for new products.
- Setting achievable in-service performance standards against which to judge actual performance, stimulate action, and adjust testing procedures if necessary.

9.2.3.2   Basic Concepts of Reliability Prediction

Even the best methodology of reliability prediction cannot be useful for practical engineers if it is not connected with techniques and equipment to obtain accurate initial information for prediction. The basic concept of accurate reliability prediction consists of the following basic steps (Klyatis & Klyatis, 2006):

1. Building an accurate model of real time performance.
2. Using the model to test the product and, as a result, to study the degradation mechanism over time and compare it with the real life degradation mechanism of this product. If these degradation mechanisms differ by more than a fixed limit, the model's real time performance must be improved.
3. Making real-time performance forecasts for reliability prediction using these testing results as initial information.

Each step can be performed in different ways, but reliability can be predicted accurately if researchers and engineers use the above concept.

Step 1 can be executed if we understand that in real life, the reliability of the product depends on a combination of different minimum numbers of input influences, such as temperature, vibration, and input voltage (for electronics), many of which connect and interact. The simulation of real life input influences will be complicated as in real life. For example, for a mobile product, we need to use multi-axis vibration.

To solve step 2, we must understand the degradation mechanism of the product and parameters of this mechanism. The results of the product degradation mechanism include data on the electrical, mechanical, chemical, thermal, and radiation effects. For example, the parameters of a mechanical degradation mechanism are deformation, crack, wear, and creep. In real life, different processes of degradation act simultaneously and in combination. Therefore, useful accelerated reliability testing includes simultaneous combinations of different types of testing (environmental, electrical, vibration, etc.), with the assumption that the failures are statistically independent. The degradation mechanism of the product in real life must be similar to this mechanism during accelerated testing.

To solve step 3, the reliability prediction technique has to be developed. The aspects that must be considered in this step involve specific manufacturing and field conditions (Klyatis & Klyatis, 2006).

9.2.3.3 Reliability Prediction Methods

To accurately predict reliability in any area, we need knowledge of a system's components, design, manufacturing process, and expected operating conditions. Once a prototype of a product is available, lab tests can make accurate reliability predictions. There are several different ways to predict reliability. Three main categories often used by government and industry are: empirical (standards based), PoF, and life testing. Each has advantages and disadvantages (Gipper, 2020).

1. Empirical reliability prediction methods

In this method, models for reliability prediction are developed from statistical curve fitting of historical failure data, either real data collected by the organization or data from manufacturers. The assumption is that causes of failure are inherently linked to components whose failures are independent of each other. Some parameters can be modified by integrating existing engineering knowledge. Empirical reliability predictions make fairly good estimates of reliability for similar or slightly modified parts.

Table 9.1 lists some of the commonly used empirical methods and the industry in which they are applied (Gipper, 2020).

2. Physics of failure (PoF)

TABLE 9.1
Empirical Prediction Methods in Common Use

Prediction Method	Applied Industry
MIL-HDBK-217F and Notice 1 and 2	Military
Bellcore/Telecordia	Telecom
IEC 62380 (RDF 2000)	Telecom
SAE Reliability Prediction Method	Automotive
PRISM	Military/Commercial

Source: Gipper (2020).

PoF analysis identifies and characterizes the physical processes and mechanisms that cause failures. It is based on an understanding of the failure mechanism and requires the application of a PoF model to the data. The foundation of the method is the use of computer models integrating deterministic formulae from chemistry and physics.

   3. Life testing

In life testing, a test is conducted on units operating under normal conditions. The sample must be sufficiently large for the method to be effective. TTFs are recorded and analyzed using a statistical distribution to estimate reliability. Since the real lifecycle of a system may be very long, its operating conditions can be accelerated and amplified to create a manageable test time of days or weeks. Some TTF data from life testing may be incorporated into empirical prediction methods or used to estimate the parameters for some of the (PoF) models. Life testing is sometimes called life data analysis, Weibull analysis, or highly accelerated life testing (Gipper, 2020).

9.2.3.4   Reliability Prediction Definitions
9.2.3.4.1   Failure Rates
Reliability predictions are based on failure rates.

   Conditional failure rate or failure intensity, $\lambda(t)$, can be defined as the anticipated number of times an item will fail in a specified time period, given that it was as good as new at time zero and is functioning at time t. It is a calculated value that provides a measure of reliability for a product. This value is normally expressed as failures per million hours (fpmh or 10^6 hours), but can also be expressed as failures per billion hours (fits or failures in time or 10^9 hours). For example, a component with a failure rate of 2 failures per million hours would be expected to fail 2 times in a million-hour time period.

   Failure rate calculations are based on complex models which include factors using specific component data, such as temperature, environment, and stress. In the prediction model, assembled components are structured serially. Thus, calculated failure rates for assemblies are a sum of the individual failure rates for components within the assembly (ITEM Software, 2007).

   There are three basic categories of failure rates (ITEM Software, 2007):

   1. Mean time between failures (MTBF)

MTBF is a basic measure of reliability for repairable items. MTBF can be described as the time passed before a component, assembly, or system fails, under the condition of a constant failure rate. Another way of stating MTBF is the expected value of time between two consecutive failures, for repairable systems. It is a commonly used variable in reliability and maintainability analyses.

   MTBF can be calculated as the inverse of the failure rate, λ, for constant failure rate systems.

$$MTBF = \frac{1}{\lambda}. \tag{4}$$

Note: Although MTBF was designed for use with repairable items, it is commonly used for both repairable and non-repairable items. For non-repairable items, MTBF is the time until the first (and only) failure after t_0 (ITEM Software, 2007).

   2. Mean time to failure (MTTF)

MTTF is a basic measure of reliability for non-repairable systems. It is the mean time expected until the first failure of a piece of equipment. MTTF is a statistical value and is intended to be the mean over a long period of time and with a large number of units. For constant failure rate systems, MTTF is the inverse of the failure rate, λ. If failure rate, λ, is in failures/million hours, MTTF = 1,000,000/ Failure Rate, λ, for components with exponential distributions. Or

$$MTTF = \frac{1}{\lambda \, \text{failures} / 10^6}. \tag{5}$$

For repairable systems, MTTF is the expected span of time from repair to the first or next failure (ITEM Software, 2007).

3. Mean time to repair (MTTR)

Mean time to repair (MTTR) is defined as the total amount of time spent performing all corrective or preventive maintenance repairs divided by the total number of those repairs. It is the expected span of time from a failure (or shut down) to the repair or maintenance completion. This term is typically only used with repairable systems (ITEM Software, 2007).

9.2.3.4.2 Failure Frequencies
Four failure frequencies are commonly used in reliability analyses (ITEM Software, 2007):

- Failure density $f(t)$: The failure density of a component or system, $f(t)$, is defined as the probability per unit time that the component or system experiences its first failure at time t, given that the component or system was operating at time zero.
- Failure rate $r(t)$: The failure rate of a component or system, $r(t)$, is defined as the probability per unit time that the component or system experiences a failure at time t, given that the component or system was operating at time zero and has survived to time t.
- Conditional failure intensity (or conditional failure rate) $\lambda(t)$: The conditional failure intensity (CFI) of a component or system, $\lambda(t)$, is defined as the probability per unit time that the component or system experiences a failure at time t, given that the component or system was operating, or was repaired to be as good as new, at time zero and is operating at time t.
- Unconditional failure intensity or failure frequency $\omega(t)$: The unconditional failure intensity of a component or system, $\omega(t)$, is defined as the probability per unit time that the component or system experiences a failure at time t, given that the component or system was operating at time zero.

9.2.3.4.3 Relationships between Failure Parameters
The following relations exist between failure parameters (ITEM Software, 2007):

$$R(t) + F(t) = 1 \tag{6}$$

$$f(t) = \frac{dF(t)}{dt} \tag{7}$$

$$F(t) = \int_0^t f(u) \, du \tag{8}$$

$$r(t) = \frac{f(t)}{1 - F(t)} \tag{9}$$

$$R(t) = e^{-\int_0^t r(u)\,du} \tag{10}$$

$$F(t) = 1 - e^{-\int_0^t r(u)\,du} \tag{11}$$

$$r(t) = r(t)e^{-\int_0^t r(u)\,du} \tag{12}$$

The definitions for failure rate $r(t)$ and CFI $\lambda(t)$ differ in that the failure rate definition addresses the first failure of the component or system rather than any failure of the component or system. If the failure rate is constant with respect to time or if the component is non-repairable, these two quantities are equal. In summary (ITEM Software, 2007):

$\lambda(t) = r(t)$ for non-repairable components
$\lambda(t) = r(t)$ for constant failure rates
$\lambda(t) \neq r(t)$ for the general case

The difference between the CFI $\lambda(t)$ and unconditional failure intensity $\omega(t)$ is that the CFI has an additional condition that the component or system has survived to time t. The relationship between these two quantities may be expressed mathematically as:

$$\omega(t) = \lambda(t)\big[1 - Q(t)\big]. \tag{13}$$

For most reliability and availability studies, the unavailability $Q(t)$ of components and systems is very much less than 1. In such cases $\omega(t) \approx \lambda(t)$ (ITEM Software, 2007).

9.2.3.4.4   Constant Failure Rates
If the failure rate is constant, then the following expressions apply:

$$R(t) = e^{-\lambda t} \tag{14}$$

$$F(t) = 1 - e^{-\lambda t} \tag{15}$$

$$f(t) = \lambda e^{-\lambda t}. \tag{16}$$

As can be seen from the equation above, a constant failure rate results in an exponential failure density distribution (ITEM Software, 2007).

9.2.3.5   Types of Reliability Prediction
9.2.3.5.1   MIL-HDBK-217 F
A military handbook for the reliability prediction of electronic equipment mentions two methods of reliability prediction:

1. The parts stress analysis method requires a greater amount of detailed information and is applicable during the later design phase when actual hardware and circuits are being designed.

2. The parts count method requires less information, generally part quantities, quality level, and the application environment. This method is applicable during the early design phase and during proposal formulation.

In general, the parts count method will result in a more conservative estimate (higher failure rate) of system reliability than the parts stress method (Mirza Hyder, 2012).

9.2.3.5.2 Bellcore

The Bellcore procedure provides three methods for predicting device and unit reliability. The method used is normally determined by the information that is available. The reliability prediction procedure for electronic equipment is the following (Mirza Hyder, 2012):

- Method I: The "parts count" or "black box" method is very similar to and was modeled on the MIL-217 standard (given in the previous section). This method assumes no reliability data are available on the devices and units included in the system. Predictions are based on generic reliability parameters. This is the simplest of the three methods and can be applied to devices or units.
- Method II: Use this method if laboratory failure rate data are available for some or all devices or units. This method allows the laboratory data to be combined with the generic data from Method I. The resulting prediction lies somewhere between Method I and Method II.
- Method III: Use Method III if field tracking reliability data are available for some or all devices or units. As above, this method will allow actual field reliability data to be combined with the generic data to obtain predictions. This is obviously the most accurate of the prediction methods but requires considerable actual field data.

9.2.3.5.3 Naval Surface Warfare Center

Individual mechanical components, such as valves and gearboxes, often perform more than one function at a time, and failure data for specific applications of non-standard equipment are rarely available. Failure rates of mechanical equipment are not usually described by a constant failure rate distribution because of wear, fatigue, and various stress-related failure mechanisms resulting in equipment degradation and are more sensitive to loading, operating modes, environmental conditions, and utilization rate (Mirza Hyder, 2012).

9.2.3.5.4 China 299B

This refers to Chinese military standards for electronic equipment, specifically to reliability prediction of various categories of electronic, electrical, and electromechanical elements to predict failure rates affected by environmental conditions, quality levels, various parameters and stress conditions (Mirza Hyder, 2012).

9.2.3.5.5 Non-Electronic Parts Reliability Data (NPRD)

Non-electronic parts reliability data (NPRD) provide failure rate data for a wide variety of component types, including mechanical, electromechanical, and electronic assemblies. They provide summary and detailed data sorted by part type, quality level, environment, and so on. A potential use for the NPRD method is to complement existing reliability prediction methodologies by providing failure rate data in a consistent format on various electrical, electromechanical, and mechanical parts and assemblies (Mirza Hyder, 2012).

9.2.3.5.6 RDF 2000

This refers to French telecom standards for electronics, specifically, reliability prediction of electronic components, printed circuit boards, and equipment (Mirza Hyder, 2012).

9.2.3.6 Need for an Effective Approach in Reliability Prediction

In the 1980s, the MIL-HDBK-217 (mentioned previously) was the epitome of prediction methodology in electronics. A more realistic approach was required, however, as these earlier methods did not account for design and manufacturing improvements. Knowles (1993) questioned the appropriateness of the reliability prediction concept in MIL-HDBK-217. He also thought the concept was often misunderstood. He highlighted that reliability is part of engineering – thus, any new approach must incorporate design integrity (Matiü & Sruk, 2008).

Pecht (1996) pointed to the PoF concept as an alternative approach by listing the main problems of classical prediction methods: out-of-date data are used for reliability prediction; they do not acknowledge the difference between removed and failed parts; they do not distinguish between design and manufacturing failures; failures can be due to overstress, damage due to incorrect manipulation of components, use of improper parts, and so on; they assume a constant failure rate; they use averaged values that are neither vendor specific nor device specific; they depend on inappropriate modeling; there are considerable discrepancies in predictions among the various classical approaches (Matiü & Sruk, 2008).

Reliability engineering communities in various countries had different visions of how to overcome these problems. In Europe, stakeholders worked to establish a standard based on the historical approach distinguishing suppliers' reliability practices, whereby predictions are generally based on vendor-component specifics. Others preferred the PoF approach (Matiü & Sruk, 2008).

9.3 SENSING TECHNOLOGIES IN DYNAMIC ASSETS AND FAILURE DIAGNOSIS

Recent advances in information and communication technologies, embedded systems, and sensor networks have generated significant research activity, including the development of cyber-physical systems (CPSs). A CPS consists of two components (Reppa et al., 2016):

1. Physical, biological, or engineered systems.
2. A cyber core, comprising communication networks and computational availability that monitors, coordinates, and controls the physical part.

Fault detection addresses the problem of determining the presence of faults in a system and estimating their instant of occurrence. Fault detection is followed by fault isolation, which deals with finding which components in the system are faulty or the type of fault.

Fault identification is described as the procedure of determining the size and the time variant behavior of the fault. In some cases, during the fault identification procedure, the extent of the fault and the risks associated with it are also assessed. The result of fault identification is essential to perform fault accommodation by either changing the control law or using virtual sensors or actuators in response to a fault, without switching off any system component.

Various methodologies have been developed for fault diagnosis, but the detection and isolation of sensor faults has become a challenging problem in recent years because of the large number of sensors and sensor networks used for (Reppa et al., 2016):

1. Monitoring and controlling large-scale CPS.
2. Providing rich and redundant information for executing safety-critical tasks.
3. Offering information to citizens and governmental agencies for resolving problems promptly in emergency situations.

For instance, in intelligent transportation, vehicles may be equipped with odometers, lasers, frontal camera video-sensors, a Global Positioning System (GPS), and speed or object tracking sensors to acquire and broadcast information relevant to performing tasks, such as cooperative or fully autonomous driving and avoidance of lane departure and collision. In smart buildings, multiple sensors are installed in different zones (measuring quantities such as temperature, humidity, CO_2, contaminant concentration, and occupancy), as well as in heating, ventilation, and air-conditioning (HVAC) systems to measure supply/return/mixed air temperature, supply/return air differential pressure, return air humidity, and so on. Such sensing information may be used to reduce the energy consumption of a building and maintain the desired living conditions, or to execute evacuation plans in safety-critical situations (e.g., fire). Undetected sensor faults can severely impact automation and supervision schemes, possibly leading to system instability, loss of information fidelity, incorrect decisions, and disorientation of remedial actions.

Sensor fault detection and isolation (FDI) methods are classified into physical redundancy based and model-based methods. In many applications, the physical redundancy approach is not used because of the high cost of installation and maintenance, as well as space restrictions. However, the evolution of micro technology has contributed to the reduction of the size and fabrication cost of sensors, making physical redundancy methods more cost effective. Current technological advances are geared toward the use of multiple, possibly heterogeneous, sensors, which are not necessarily co-located, but the measured variables may have redundant information, and this is useful for fault diagnosis purposes. For example, a smart building may have two sensors measuring the temperature in adjacent rooms; in such a case, the relationship (either known *a priori* or after operation) between the two measured quantities may be used to determine if one of the two sensors is faulty. With the trend toward larger numbers of sensors, there is a higher probability of multiple sensor faults occurring, an issue that has not been well studied in the fault diagnosis literature.

The majority of sensor FDI techniques rely on models only. These techniques are further categorized as quantitative or qualitative methods; the first relies on a nominal mathematical model describing the system, while the second uses symbolic and qualitative system representations. Some research work has been done on the combination of the two approaches. The majority of model-based sensor FDI methods are deployed in a centralized framework (see Figure 9.10(1)), but they are less suitable for large-scale and complex systems such as a network of interconnected CPS. In this context, centralized approaches have the following disadvantages (Reppa et al., 2016):

- Increased computational complexity of the FDI algorithms, as centralized architectures are tailored to handle (multiple) faults globally.
- Increased communication requirements because information is transmitted to a central point.
- Vulnerability to security threats, because the central cyber core in which the sensor FDI algorithm resides is a single-point of failure.
- Reduced scalability in case of system expansion, due to the use of a global physical model or black box.

A common design characteristic of non-centralized methods is that they handle a large-scale and complex system as a set of interconnected subsystems, and they employ local agents to perform diagnosis based on local subsystems' models. The local agents are commonly deployed in either a distributed (Figure 9.10(2)) or decentralized (Figure 9.10(3)) architecture. The classification of these architectures is based on the type of system interconnections, the cyber levels of diagnosis, the task of the local diagnosticians, and the type of communication and information exchanged between local and high-level diagnosticians (Reppa et al., 2016).

9.3.1 Fault Sensing and Diagnosis

Fault sensing and diagnosis are key components of many operations management automation systems. A "fault" is another word for a problem. A "root cause" fault is a fundamental, underlying

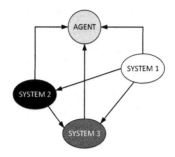

(1) Centralized architecture

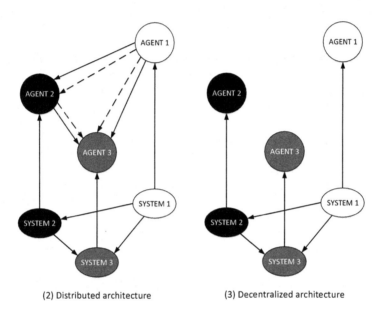

(2) Distributed architecture   (3) Decentralized architecture

FIGURE 9.10 Typical architectures for interconnected systems: (1) centralized approach, (2) distributed architecture, and (3) decentralized architecture (Reppa et al., 2016).

problem that may lead to other problems and observable symptoms (it might not be directly observable). A root cause is also generally associated with procedures for repair.

A fault or problem does not have to be the result of a complete failure of a piece of equipment, or even involve specific hardware. For instance, a problem might be defined as non-optimal operation or off-speculation product. In a process plant, root causes of non-optimal operation might be hardware failures, but problems might also be caused by poor choice of operating targets, sensor calibration errors, human error, and so on. A fault may be considered a binary variable ("okay" vs. "failed"), or there may be a numerical "extent", such as the amount of a leak or a measure of inefficiency.

A symptom is an observed event or variable value required to detect and isolate faults. If a symptom is the response to a question or an on-demand data request (when actively testing a system instead of just passively monitoring it), it is referred to as a test or test result.

Fault detection is recognizing that a problem has occurred, even if the root cause is as yet unknown. Faults may be detected by a variety of quantitative or qualitative means. This includes many multivariable, model-based approaches. It also includes simple, traditional techniques for single variables, such as alarms based on high, low, or deviation limits for process variables or rates of change, statistical process control measures, and alarms generated by packaged subsystems.

Fault diagnosis is pinpointing one or more root causes of problems, to the point where corrective action can be taken. This is also called "fault isolation", especially when emphasizing the distinction from fault detection. In common usage, "fault diagnosis" often includes fault detection, so "fault isolation" emphasizes the distinction.

Elements of operations management automation related to diagnosis include the associated system and user interfaces and workflow (procedural) support for the overall process. Workflow steps that might be manual or automated include notifications, online instructions, escalation procedures if problems are ignored, fault mitigation actions (what to do while waiting for repairs), direct corrective actions, and steps to return to normal once repairs are complete.

Automated fault detection and diagnosis depend on input from sensors or derived measures of performance. In many applications, such as those in the process industries, sensor failures are among the most common equipment failures. Therefore, those industries must focus on recognizing both sensor problems and process problems. Distinguishing between sensor problems and process problems is a major issue in these applications.

Sensors used in process industries include process monitoring instrumentation for flow, level, pressure, temperature, power, and so on. In fields such as network and systems management, they can include measures such as error rates, CPU utilization, queue lengths, dropped calls, and so on (Performity LLC, 2020).

9.3.1.1 Overview of Approaches to Fault Sensing and Diagnosis

9.3.1.1.1 Model-Based Reasoning

One of the major distinctions in approaches to fault detection and diagnosis is whether explicit models are used, and if so, what type. When models of the observed system are used as a basis for fault detection and diagnosis, this is often called model-based reasoning (Performity LLC, 2020).

9.3.1.1.2 Causal Models

An important example of model-based reasoning is the causal model. Causal models capture cause/effect information. In physical systems, causality is associated with time delay or lags between cause and effect. Mass or energy has to move, overcoming resistance by inertia, thermal inertia, inductance, or other physical phenomena (Performity LLC, 2020).

9.3.1.1.3 Fault Signatures, Pattern Recognition, and Classifiers

Pattern recognition is a general approach that directly uses the observed symptoms of a problem and compares them to a set of known symptoms for each possible problem, looking for the best match. The pattern, or fault signature, can be represented as a vector (one-dimensional array) of symptoms for each defined fault (Performity LLC, 2020).

9.3.1.1.4 Neural Networks

Neural networks are nonlinear, multivariable models built from a set of input/output data. They can be used to detect events and trends. They can also be used as diagnostic models in model-based reasoning or used directly as classifiers to recognize fault signatures (Performity LLC, 2020).

9.3.1.1.5 Procedural/Workflow Approaches – Modeling the Decision Process Rather than the Observed System

Some fault detection and diagnosis is handled by creating procedures for making decisions based on the observed data. This involves the direct modeling of the decision process rather than the modeling of the system being diagnosed (Performity LLC, 2020).

9.3.1.1.6   Event-Oriented Fault Detection, Diagnosis, and Correlation

An event represents a change of state of a monitored object. Alarms are examples of events. Diagnostics involving events can be significantly different from diagnostics involving a fixed set of variables (Performity LLC, 2020).

9.3.1.1.7   Passive System Monitoring vs. Active Testing

In the case of online monitoring systems, many diagnostic techniques assume routine scanning of every variable of interest. But it is often preferable to request non-routine tests. Diagnosis for maintenance purposes is based on testing (Performity LLC, 2020).

9.3.1.1.8   Rule-Based Approaches and Implementations

In most cases, rule-based systems just implement the approaches discussed above, more as a program control mechanism than a separate diagnostic technique (Performity LLC, 2020).

9.3.1.1.9   Hybrid Approaches

Pattern recognition by itself does not require a model. However, input to construct the signatures for known failures may be based on models; for instance, as residuals from models of normal behavior. This general technique applies to static or dynamic models.

For dynamic models, the patterns can be based on predicted measurement values vs. observed values in a Kalman filter. For example, SmartSignal Corporation offers products based on an empirical process model of normal operation used for fault detection, combined with other techniques for fault isolation.

Pattern recognition can also be combined with models of abnormal behavior. For instance, the SMARTS InCharge product uses a fault propagation model (qualitative cause/effect model of abnormal behavior). As part of the development process, this model was used to automatically construct fault signatures – a form of compiled knowledge. At runtime, diagnosis was based on matching observed data to the nearest fault signature. So at runtime, the product had the characteristics of a pattern matching solution.

In some cases, a qualitative model really just exists inside the application developer's head. That person directly constructs the signatures based on his or her knowledge, so the overall methodology is often a combination of pattern recognition with a model-based method.

Some tools, such as Graphical Diagnostic Assistant, are flexible enough to support multiple approaches to fault detection and diagnosis and also support the upfront filtering and event generation as well (Performity LLC, 2020).

9.3.2   Fault Management Mechanism for Wireless Sensor Networks

A sensor network is a collection of sensor nodes which coordinate with each other to perform a specific function. There are usually many sensor nodes, and they are densely deployed either inside the phenomenon of interest or very close to it. They can be used for various application areas (e.g., health, military, and home). Failures are inevitable in wireless sensor networks (WSNs) because of inhospitable environments and unattended deployment. Therefore, it is necessary to detect network failures in advance and take appropriate measures to sustain network operation (Asim et al., 2010).

9.3.2.1   Fault Detection

Since sensor network conditions constantly change, network monitoring alone may not be sufficient to identify network faults. Therefore, fault detection techniques need to be in place to detect potential faults. There are two types of fault detection in WSNs: explicit and implicit detection. The first is performed directly by the sensing devices and their sensing applications. In the second, anomalistic phenomena might disable a sensor node from communication or behave improperly and must be identified by the network itself. Implicit detection is normally achieved in two ways: active or passive models. Active detection is carried out by the central controller of the sensor network. Sensor nodes continuously send keep-alive messages to the central controller to confirm their existence. If the central controller does not receive the update message from a sensor node after a pre-specified period of time, it may believe the sensor is dead. A passive detection model (event-driven model) triggers the alarm only when failure has been detected. However, this model will not work properly if a sensor is disabled from communication due to intrusion, tampering, or movement out of range.

Fault detection mainly depends on the type of application and the type of failures. Existing failure detection approaches are classified into two primary types: centralized and distributed (Asim et al., 2010).

9.3.2.1.1 Centralized Approaches

In centralized fault management systems, a geographical or logical centralized sensor node usually identifies failed or misbehaving nodes in the whole network. This centralized node can be a base station, a central controller, or a manager. It usually has unlimited resources and performs a wide range of fault management tasks. Some common centralized fault management approaches are as follows (Asim et al., 2010).

Sympathy is a debugging system used to identify and localize the cause of failures in a sensor network application. The sympathy algorithm does not provide automatic bug detection. It depends on historical data and metrics analysis to isolate the cause of failure. Sympathy may require nodes to exchange neighborhood lists, which is expensive in terms of energy. A sympathy flooding approach means the knowledge of global network states is imprecise, possibly leading to incorrect analysis (Ramanathan et al., 2005).

Staddon, Balfanz, and Durfee (2002) enabled the base station to construct an overview of the network by integrating each piece of network topology information (i.e., node neighbor list) embedded in the node's usual routing message. This approach uses a simple divide-and-conquer rule to identify faulty nodes. It assumes the base station is able to directly transmit messages to any node in the network and rely on other nodes to route measurements to the base station. It also assumes each node has a unique identification number. This method enables the base station to know the network topology via route-discovery protocols. Once the base station knows the node topology, it detects the faulty node by using a simple divide-and-conquer strategy based on adaptive route update messages (Asim et al., 2010).

The centralized approach is suitable for certain applications, but it has limitations. It is not scalable and cannot be used for large networks. In addition, because of the centralized mechanism, all traffic is directed to and from the central point. This creates communication overhead and fast energy depletion. Moreover, the central point is a single point of data traffic concentration and potential failure. Finally, if a network is portioned, nodes that are unable to reach the central server are left without any management functionality (Asim et al., 2010).

9.3.2.1.2 Distributed Approaches

This is an efficient way of deploying fault management. In a distributed approach, each manager controls a sub-network and may communicate directly with other managers to perform management functions. Distributed management provides better reliability and energy efficiency and has lower communication cost than centralized management (Asim et al., 2010).

The algorithm proposed for faulty sensor identification in Ding et al. (2005) is purely localized. Nodes in the network coordinate with their neighbor nodes to detect faulty nodes before contacting the central point. In the scheme, a sensor's reading is compared with its neighbor's reading; if the difference is large or large but negative, the sensor is likely faulty. This algorithm can easily be scaled for large networks, but the probability of sensor faults needs to be small. In addition, if half of the sensor neighbors are faulty and the number of neighbors is even, the algorithm cannot detect the fault.

The algorithm developed by Chen, Kher, and Somani (2006) tries to overcome the limitations of this approach by identifying good sensor nodes in the network. The results are used to diagnose faulty nodes and then propagated in the network to diagnose all other sensor nodes. This approach performs well with an even number of sensor nodes and does not require the sensors' physical location. It is not fully dynamic, however, and must be preconfigured. Furthermore, each node should have a unique identifier (ID) and the center node should know the existence and ID of each node.

In a scheme proposed by Marti et al. (2000), sensor nodes police each other to detect faults and misbehavior. A node listens-in on the neighbor it is currently routing to and can determine whether the message it sent was forwarded. If the message was not forwarded, it concludes its neighbor is a faulty node and chooses a new neighbor to route to (Asim et al., 2010).

9.3.2.2 Fault Diagnosis

In the diagnosis stage, detected faults are properly identified by the network system and distinguished from irrelevant or spurious alarms. Fault diagnosis includes fault isolation (where is the fault located), fault identification (the type of detected fault), and root cause analysis (what caused the fault). However, no comprehensive descriptive model is available to identify or distinguish various faults in WSNs and thereby support the network system through accurate fault diagnosis or action-taken in the fault recovery stage. Existing approaches are based on hardware faults and consider hardware components' malfunctioning only. Some assume system software is already fault tolerant.

Koushanfar, Potkonjak, and Sangiovanni-Vincentelli (2002) describe two fault models. The first corresponds to sensors with binary outputs, and the second is based on sensors with continuous (analog) or multilevel digital outputs. Clouqueur, Saluja, and Ramanathan (2004) only consider faulty nodes caused by a harsh environment. Thus, there is a need to design a generic fault model that is not based on individual node level, but also considers the network and management aspects (Asim et al., 2010).

9.3.3 Multi-Sensor Measurement and Data Fusion Technology

Automated workplaces are widely regarded as important factories of the future because of their flexibility and efficiency. For example, in industries relying on machining processes, such as turning, drilling, and milling, the demand for machined parts with tighter surface finishes and higher machining precision has led to the need for better, more precise control of automated machining processes, mainly because of intensified global competition, shortened product life cycles, and diversified product demand. To meet the growing demand, manufacturers are increasingly turning to automated processing systems, which can reduce dependence on operators during production. The realization of automated processing systems requires reliable and robust monitoring systems for the online and offline monitoring of critical machining processes.

Sensors are widely used physical devices that respond to a physical stimulus by giving an electrical output. As shown in Figure 9.11, modern instrumentation systems are equipped with many different sensors, each with its own function. Each sensor in the monitoring system can measure a certain parameter independently and use a special signal-processing algorithm to combine all of the independent measurements into a complete measurement value. This kind of system is called a multi-sensor system. For instance, when industries using machining processes implement a multi-sensor system to monitor and control metal cutting operations and machine tools, they see a significant reduction in machining errors and better surface finishes. This method is increasingly used for lower cost, powerful microprocessors. Meanwhile, sensing instruments make the use of signal-processing systems and digital closed-loop control quite economical. In other words, a multi-sensor system is an important component for achieving higher machining accuracy (Kong et al., 2020).

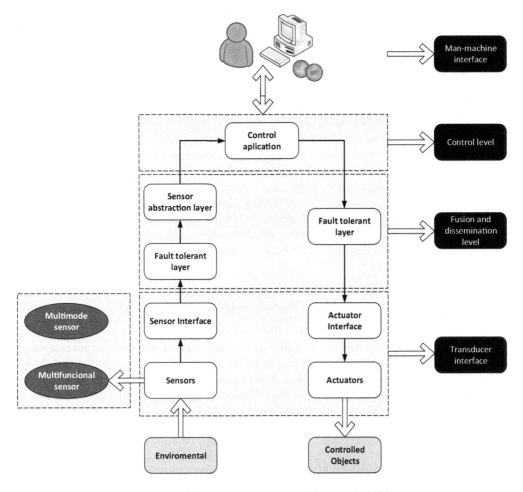

FIGURE 9.11   Framework for intelligent multi-sensor system (Kong et al., 2020).

In a multi-sensor system, the key to achieving the system's function is the coordination of all sensors. Figure 9.11 illustrates the general framework of an intelligent system with multi-sensor capabilities. In a multi-sensor system, the sensors provide measurement data to the data fusion layer, and it uses advanced signal processing algorithms to combine the information into a whole. Meanwhile, the control application provides sufficient control signals to the actuator.

The process of combining the measurement data of all sensors and using special algorithms to provide a complete overview of the measurement is called multi-sensor data fusion. Data fusion takes full advantage of multiple information sources and combines complementary or redundant information in space or time. Using specific standards, data fusion can obtain consistent interpretation of a tested object, so that the information system performs better than a system composed of subsets. Data fusion uses mathematical methods and technical tools to synthesize different sources of information, obtaining high-quality, useful measurement data. Compared with the results achieved by the independent processing of data from a single source, the advantages of data fusion include improved detectability and reliability, expanded range of spatial-temporal perception, reduced degree of ambiguity of inference, improved detection accuracy, increased dimension of target features, improved resolution of spatial questions, and enhanced fault-tolerant ability of the system (Kong et al., 2020).

9.4 PROPORTIONAL HAZARDS MODEL AND PHYSICAL STRESSORS

Proportional hazards models (PHMs) proposed by Cox (1972) are a class of survival models in statistics. Survival models relate the time that passes before some event occurs to one or more covariates that may be associated with that quantity of time. In a PHM, the unique effect of a unit increase in a covariate is multiplicative with respect to the hazard rate. For example, taking a drug may halve an individual's hazard rate for having a stroke or changing the material from which a manufactured component is constructed may double its hazard rate for failure. Other types of survival models such as accelerated failure time models do not exhibit proportional hazards. The accelerated failure time model describes a situation where the biological or mechanical life history of an event is accelerated (or decelerated) (Wikipedia, 2020b).

The model has been used primarily in medical testing analysis to model the effect of secondary variables on survival. It is more like an acceleration model than a specific life distribution model, and its strength lies in its ability to model and test many inferences about survival without making any specific assumptions about the form of the life distribution model (NIST/SEMATECH, 2012).

The PHM is also used in the analysis of reliability data, especially for small electromechanical appliances (Mendes, 2014).

9.4.1 PROPORTIONAL HAZARDS MODEL WITH TIME-DEPENDENT COVARIATES

The inclusion of time-dependent covariates allows the statistical modeling to enlarge its applicability quite significantly and achieve the full potential of the PHM. PHMs have advantages over accelerated failure time models, for example, as the latter are parametric in nature and show an inability to handle time-dependent covariates. Most applications in reliability can benefit from the use of modeling that solves the problem of modeling time-dependent covariates, as classical reliability theory is limited when varying stress levels are used to estimate product reliability from accelerated reliability testing (Mendes, 2014).

9.4.2 PROPORTIONAL HAZARDS MODEL

PHMs generally assume the product of an unspecified hazard rate $\lambda_0(t)$ that depends on time, known as the baseline hazard rate, and a positive functional term $\psi(X, \beta)$ as shown in Equation (17) that incorporates the time-independent effects of covariates. Hence (Mendes, 2014):

$$\lambda\left(t\middle|X\right) = \lambda_0\left(t\right)\psi\left(X,\beta\right), \tag{17}$$

where X is a row vector consisting of the covariates, and β is a column vector consisting of the regression parameters. The covariate vector is associated with the system, and β is the unknown parameter of the model, defining the effects of the covariates.

The baseline hazard rate is the failure rate under the standard condition, $X = 0$, and requires $\psi(X \beta) = 1$ when there is no influence of the covariates on the model failure time.

The assumption of the multiplicative effect of the covariates and the baseline hazard rate implies that the ratio of the hazard rates of any two process variables observed at any time t associated with covariate sets x_1 and x_2, respectively, will be a constant with respect to time and proportional to each other. This explains why the model is called the PHM.

Different parameterization forms of $\psi(X \beta)$ can be used in Equation (17) such as:

- Exponential form: $exp\left(X\beta\right)$
- Logistic form: $log\left(1 + exp\left(X\beta\right)\right)$

- Inverse linear form: $\dfrac{1}{(1+X\beta)}$
- Linear form: $(1+X\beta)$.

The logistic, inverse linear, and linear forms are not very convenient because it is difficult to choose the regression parameters β for the possible values of X that make $\psi(X\beta) > 0$.

If the exponential form is assumed, the mathematical calculation is less complex, and the hazard rate can be written as in Equation (18):

$$\lambda(t,X) = \lambda_0(t)\exp(X\beta) = \lambda_0(t)\exp\sum_{j=1}^{q} x_j\beta_j \tag{18}$$

where $x_j = 1, 2,\ldots, q,\ldots, x_p$ are covariates, and $\beta_j = 1, 2,\ldots, q$ are the covariates associated with the system and are the unknown parameters of the model (independent variables), defining the effects of each of the q covariates. The multiplicative factor $\exp(X\beta)$ addresses the risk of failure because of the presence of a covariate x.

The related reliability functions are given by Equations (19) and (20):

$$R_0(t) = exp\left(-\int_0^t \lambda_0(x)dx\right) = exp(-H_0(t)) \tag{19}$$

$$R(t,x) = exp(-H_0(t)) * exp\left(\sum_{j=1}^{q} x_j\beta_j\right), \tag{20}$$

where $R_0(t) = exp\left(-\int_0^t \lambda_0(x)dx\right) = exp(-H_0(t))$ is the baseline reliability function dependent only on time, $H_0(t)$ is the cumulative hazard rate, and $R(t,x)$ is the total reliability function dependent on time and covariates.

The baseline hazard rate $\lambda_0(t)$ can be modeled by parametric models such as Weibull, exponential, and lognormal distributions.

A key assumption is proportionality, which means the failure rates for the two observations and with different covariate values are independent of time and can be expressed in Equation (21) as:

$$\frac{\lambda_i(t)}{\lambda_j(T)} = e^{\beta_1\left(x_{i1}-x_{j1}\right)+\beta_2\left(x_{i2}-x_{j2}\right)+\ldots+\beta_k\left(x_{ik}-x_{jk}\right)} \tag{21}$$

Various graphical and analytical numerical methods can be used to evaluate the applicability of the proportional hazard rate for the analysis of TTF datasets (Mendes, 2014). Section 9.4.6 explains the Cox regression model in detail, using the model to show the relations between the hazard rate and a set of covariates.

9.4.3 PROPORTIONAL HAZARDS MODEL ASSUMPTION

Let $z = \{x, y,\ldots\}$ be a vector of one or more explanatory variables believed to affect lifetime. These variables may be continuous (like temperature in engineering studies, or dosage level of a particular drug in medical studies) or they may be indicator variables with the value 1 if a given factor or condition is present, and 0 otherwise.

Let the hazard rate for a nominal (or baseline) set $z_0 = \{x_0, y_0, \ldots\}$ of these variables be given by $h_0(t)$, with $h_0(t)$ denoting a legitimate hazard function (failure rate) for some unspecified life distribution model.

The PHM assumes changed hazard function can be written for a new value of z (NIST/SEMATECH, 2012):

$$h_z(t) = g(z) * h_0(t) \qquad (22)$$

In other words, changing z, the explanatory variable vector, results in a new hazard function that is proportional to the nominal hazard function, and the proportionality constant is a function of z, $g(z)$ independent of the time variable t.

A common and useful form for $g(z)$ is the log linear model which has the equation $g(x, y) = e^{ax}$ for one variable and $g(x, y) = e^{ax+by}$ for two variables, and so on (NIST/SEMATECH, 2012).

9.4.4 Properties and Applications of the Proportional Hazards Model

1. The PHM is equivalent to the acceleration factor concept if and only if the life distribution model is a Weibull (this includes the exponential model, as a special case). A Weibull with shape parameter γ and an acceleration factor AF between nominal use fail time t_0 and high stress fail time t_s (with $t_0 = AFt_s$) has to $g(s) = AF^\gamma$. In other words, $h_s(0) = AF^\gamma * h_0(t)$.

2. Under a log-linear model assumption for $g(z)$, without any further assumptions about the life distribution model, it is possible to analyze experimental data and compute maximum likelihood estimates and use likelihood ratio tests to determine which explanatory variables are highly significant. However, special software is needed to do this kind of analysis (NIST/SEMATECH, 2012).

9.4.5 Cox Regression

In survival analysis, Cox regression and PHMs belong to a class of models used to model the risks affecting the survival of a population of subjects (Wikipedia, 2020a).

Cox regression analysis models relations between the hazard rate and a set of one or more discrete or continuous covariates. The model is solved using the method of marginal likelihood (Kalbfleisch & Prentice, 1980).

Survival analysis analyzes elapsed time. The response variable is the time between a starting point and an ending point. The latter is either the occurrence of the event of interest, that is, failure, or the end of the subject's participation in analysis.

Two properties of these elapsed times invalidate standard statistical techniques (e.g., t-tests, analysis of variance, multiple regression). The first is that the time values are frequently positively skewed, and standard statistical techniques require normally distributed data. Although this could be corrected using transformation, it is easier to adopt a more realistic data distribution. The second is that some data are censored; that is, the end point has not yet been reached when the subject leaves the analysis. The analysis may have ended before the subject's response occurred; alternatively, the subject withdrew from active participation for one reason or another.

Two functions are of interest in the analysis of survival data: the survivor and the hazard functions.

If T is the survival time, that is, the elapsed time from the starting point to the ending point, the values of T will have a probability distribution. If we take an example from human health, we can let T denote the elapsed time between the diagnosis of cancer and the patient's death (NCSS,

2020). Now, suppose the PDF of the random variable T is given by $f(T)$. The probability distribution function of T is then given by

$$F(T) = Pr(t < T) = \int_0^T f(t) dt. \tag{23}$$

Survivor function $S(T)$ is the probability that the subject (the cancer patient) survives past T. This leads to

$$S(T) = Pr(T \geq t) = 1 - F(T). \tag{24}$$

The hazard function is the probability that our subject (the cancer patient) experiences the event of interest (relapse, death) during a short time interval, as he or she has survived to the start of that interval. The hazard function is expressed mathematically as:

$$h(T) = \lim_{\Delta T \to 0} \frac{Pr\left(T \leq t \langle (T + \Delta T) \big| T \leq t\right)}{\Delta T} = \lim_{\Delta T \to 0} \frac{F(T + \Delta T) - F(T)}{\Delta T} = \frac{f(T)}{S(T)} \tag{25}$$

The cumulative hazard function $H(T)$, the sum of the individual hazard rates from time zero to time T, is expressed as:

$$h(T) = \int_0^T h(u) \, du \tag{26}$$

The hazard function is the derivative or slope of the cumulative hazard function. The latter is related to the cumulative survival function by either

$$S(T) = e^{-H(T)} \tag{27}$$

or

$$H(T) = -\ln\left(S(T)\right). \tag{28}$$

It is obvious that the distribution, hazard, and survival functions are mathematically related. For reasons of practicality and convenience, the hazard function is used in the basic regression model.

Cox's (1972) expression of the relations between a set of covariates and the hazard rate is the following:

$$\ln\left(h(T)\right) = \ln\left[h_0(T)\right] + \sum_{i=1}^{p} x_i \beta_i \tag{29}$$

or

$$h(T) = h_0(T) * e^{\sum_{i=1}^{p} x_i \beta_i}, \tag{30}$$

where $x_1, x_2, \ldots, x_p$ are covariates, $\beta_1, \beta_2, \ldots, \beta_p$ are regression coefficients to be estimated, T is elapsed time, and $h_0(T)$ is baseline hazard rate when all covariates are equal to zero. Thus, the linear form of the regression model can be written as:

$$ln\left[\frac{h(T)}{h_0(T)}\right] = \sum_{i=1}^{p} x_i \beta_i. \tag{31}$$

The exponential of both sides of this equation is the ratio of the actual and baseline hazard rates, or the relative risk. We can write this as:

$$\frac{h(T)}{h_0(T)} = exp\left(\sum_{i=1}^{p} x_i \beta_i\right) = e^{x_1\beta_1 + x_2\beta_2 + \ldots + x_p\beta_p}. \tag{32}$$

The regression coefficients can thus be interpreted as the relative risk when the value of the covariate is increased by one unit.

Most regression models include an intercept term, but this one does not; if one were to be included, it would be part of $h_0(T)$. Note that the model does not include T on the right-hand side. That is, the relative risk is constant for all time values, thus explaining why we use this as a PHM.

Also note that the model only needs to use the ranks of the failure times to estimate the regression coefficients. The actual failure times are only used to generate the ranks. Thus, we will arrive at the same regression coefficient estimates regardless of whether our time values are days, months, or years (NCSS, 2020).

9.5 HYBRID MODELS FOR DYNAMIC AND NON-DYNAMIC ASSETS

9.5.1 Hybrid Models

By integrating model-based (or physics-based) and data-driven approaches, hybrid models attempt to benefit from both. The main idea of a hybrid approach is to create a prognosis model able to manage uncertainties and accurately estimate RUL. Javed, Gouriveau, and Zerhouni (2014) divide this into two parts: a series approach and a parallel approach.

9.5.1.1 Series Approach
The series approach, also known as the systematic modeling approach, integrates the model-based (physics-based) approach that has prior knowledge of the process and the data-driven approach that is suitable for estimating unmeasurable parameters of the process (see Figure 9.12) (Ramezania et al., 2019).

The series approach cannot be considered model-based because parameters are adjusted using a data-driven method (Ramezania et al., 2019).

9.5.1.2 Parallel Approach
Some researchers advocate using model-based and data-driven models in parallel. More precisely, the parallel approach uses a data-driven method to estimate the residuals that cannot be explained by the model-based approach (see Figure 9.13) (Ramezania et al., 2019).

9.5.2 Hybrid Approach

Figure 9.14 shows the different categories of prognosis approaches (Ramezania et al., 2019). Each technique has some limitations.

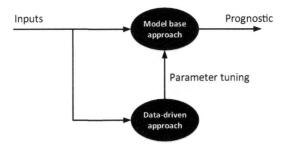

FIGURE 9.12 Hybrid series approach (Ramezania et al., 2019).

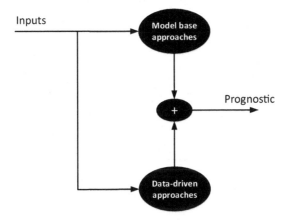

FIGURE 9.13 Hybrid parallel approach (Ramezania et al., 2019).

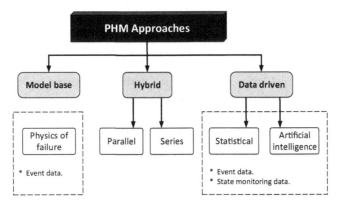

FIGURE 9.14 Categorization of proportional hazards model approaches (Ramezania et al., 2019).

A hybrid (or fusion) approach combines data-driven and model-based (or physics-based) approaches to get the best from each. For example, model-based approaches can compensate for a lack of data, and data-driven approaches compensate for the lack of knowledge about system physics. Fusion can take place either before RUL estimation (pre-estimate) when physics-based and data-driven models are fused to perform RUL estimation or after RUL estimation by fusing the results from each approach to get the final RUL (post-estimate).

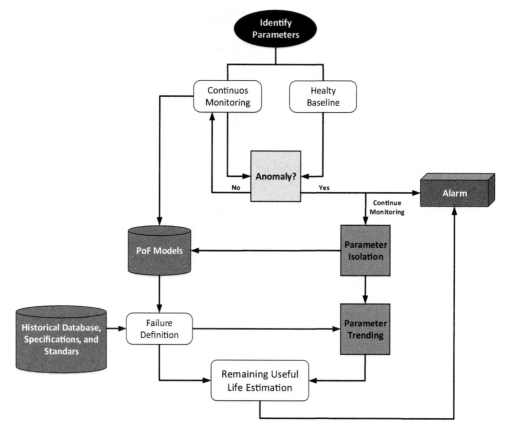

FIGURE 9.15 Fusion approach (Cheng & Pecht, 2009).

A nine-step fusion approach to the RUL estimation of electronic products is shown in Figure 9.15 (Cheng & Pecht, 2009). These steps can be used to develop a fusion approach for any other application (Elattar et al., 2016).

For example, Cheng and Pecht (2009) studied RUL estimation for ceramic capacitors using the fusion approach. Another application was for the prognostics of a lithium ion battery (Goebel et al., 2008). In still another application, researchers used a fusion approach to study aircraft engines' bearings (Goebel et al., 2006); they reported that the method gave a more accurate and robust outcome than using either data-driven or physics-based models alone.

Although the fusion approach is used to eliminate the drawbacks of physics-based and data-driven methods while reaping the benefits, it still has the disadvantages of both methods to a certain extent, albeit not at the same level as if each method were used individually. The Kalman filter and particle filter are used to implement this fusion methodology.

Figure 9.16 shows a flowchart explaining how to select the appropriate prognostics approach for certain applications (Goebel, 2007).

9.5.3 COMBINATION MODELS

In real-world prognostic processes, the trends of the various parameters are diverse; thus, it is difficult to make predictions using a single prediction method. A combination prediction method is better suited for prognostics. A well-designed method combining two or more prognostic approaches for data extraction, analysis, and modeling has the following advantages (Peng et al., 2010):

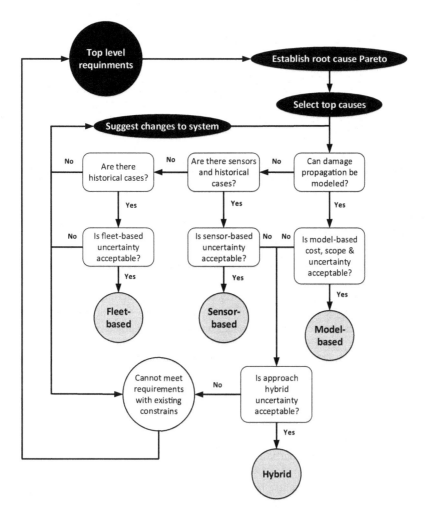

FIGURE 9.16   Prognostics approach selection (Elattar et al., 2016).

1. Problems of individual theories will be offset, and their merits capitalized on.
2. Complexity of computation may be reduced.
3. Prediction precision could be improved.

The application of artificial neural network (ANN) analysis is usually incorporated with knowledge-based techniques such as fuzzy logic (FL) and expert systems. For example, Brotherton et al. (2000) combined a dynamically linked ellipsoidal basis function neural network with rule extractors and applied it to gas turbine engine prognostics. Garga et al. (2001) introduced a hybrid reasoning method integrating machinery data into a feedforward neural network, trained based on a simple representation of explicit domain knowledge, and used this for gearbox health prognosis. The proposed neuro-fuzzy algorithm is a combination of an ANN and fuzzy inference system. More specifically, they combined the linguistic description of a typical fuzzy inference system with learning procedures inspired by neural networks. These algorithms are particularly adaptive, lucid, robust, and highly flexible (Peng et al., 2010).

Wang (2004) compared the results of recurrent neural network (RNN) and neural fuzzy (NF) inference systems for predicting fault damage propagation trends. A properly trained NF system will perform better than RNNs in forecasting accuracy and training efficiency. Chinnam and Baruah

(2004) presented an NF approach to estimate RUL when failure data and a specific failure definition model are not available, but domain experts with relevant knowledge and experience are available. Satish and Sarma (2005) combined neural networks and FL to form a fuzzy backpropagation (BP) network for identifying the present condition of a bearing and estimating the RUL of a motor. Xue et al. (2005) developed a fuzzy mathematical model with a radial basis function neural network to predict the potential faults of a coal-fired boiler.

Kothamasu and Huang (2007) presented a NF modeling approach based on an adaptive learning Mamdani fuzzy model for system diagnosis and prognosis. Their robust and lucid modeling system can assume the role of a decision-making aid in the process of CBM. The comprehensibility of the system is emphasized so it can effectively serve as a decision aid for domain experts and be adaptive to some ordinary modifications and even continuous improvement by interacting with users. The comprehensibility of an NF system usually deteriorates once rules are tuned, so these researchers introduced Kullback–Leibler mean information into the system to solve the problem. It can evaluate and refine tuned rules to make the system easily interpretable.

Since the Gray model, GM (1, 1), is good at smoothing time series data, and ANN has stronger ability in nonlinear time series prediction than others, Dong et al. (2004) designed a multi-parameter condition prediction model based on the combination of GM (1, 1) and BP neural network to predict the conditions of equipment in a power plant. By fully using (extraction and analysis) the operating data, condition monitoring data, and operation statistical data, they were able to predict the conditions of equipment, and the results were more reasonable than single characteristic parameter prediction.

Shetty, Mylaraswamy, and Ekambaram (2008) modeled a degrading system for aircraft auxiliary power units as a collection of prognostic states (health vectors) that evolve continuously over time. The proposed multivariate state space model includes an age-dependent deterioration distribution, component interactions, and effects of discrete events arising from line maintenance actions and/or abrupt faults. Mathematically, the proposed model can be summarized as a continuously evolving dynamic model, driven by non-Gaussian input and switches according to the discrete events in the system. The system identification and recursive state estimation scheme for the developed non-Gaussian model is derived from a partially specified distribution framework.

Mohanty et al. (2008) developed a hybrid prognosis model for real-time RUL estimation of metallic aircraft structural components. Their prognosis framework combines information from off-line physics-based, off-line data-driven, and online system identification-based predictive models. These components can be explicitly expressed by Gaussian processes, which are based on a data-driven approach within a Bayesian framework. The Gaussian process model projects the input space to an output space by probabilistically inferring an underlying nonlinear function for relating input and output. For the off-line prediction, the input space of the model is trained with parameters that affect fatigue crack growth. For online prediction, the model input space is trained using features found in piezoelectric sensor signals rather than training input space with loading parameters, as these are difficult to measure in a real flight-worthy structure. A new output space for corresponding unknown crack length or damage state can be predicted using the trained Gaussian process model (Peng et al., 2010).

REFERENCES

Amyot N., Hudon C., Lévesque M., Bélec M., Brabant F., St-Louis C., 2014. Development of a hydrogenerator prognosis approach. CIGRE paper A1-202.

Asim M., Mokhtar H., Merabti M., 2010. A self-managing fault management mechanism for wireless sensor networks. International Journal of Wireless & Mobile Networks, 2(4), 184.

Blancke O., Combette A., Amyot N., Komljenovic D., Lévesque M., Hudon C., Tahan A., Zerhouni N., 2018. A predictive maintenance approach for complex equipment based on petri net failure mechanism propagation model. European Conference of the Prognostics and Health Management Society 2018.

Brotherton T., Jahns J., Jacobs J., Wroblewski D., 2000. Prognosis of faults in gas turbine engines. Proceedings of the IEEE Aerospace Conf., 18–25 Mar. 2000, Big Sky, MT (USA), 6, 163–171.

Chen J., Kher S., Somani A. K., 2006. Distributed fault detection of wireless sensor networks, Proceedings of DIWANS 06, 2006.

Cheng S., Pecht M., 2009. A fusion prognostics method for remaining useful life prediction of electronic products. In: 5th Annual IEEE Conference on Automation Science and Engineering, Bangalore, Karnataka, India, 102–107).

Chinnam R. B., Baruah P., 2004. A neuro-fuzzy approach for estimating mean residual life in condition-based maintenance systems. International Journal of Materials and Product Technology, 20, 166–179.

Clouqueur T., Saluja K., Ramanathan P., 2004. Fault tolerance in collaborative sensor networks for target detection. IEEE Transactions on Computers, 320–333.

Cox D. R., 1972. Regression models and life tables. Journal of the Royal Statistical Society, B 34, 187–220.

Ding M., Chen D., Xing K., Cheng X., 2005. Localized fault-tolerant event boundary detection in sensor networks, Proceedings of the 24th Annual Joint Conference of the IEEE Computer and Communications Societies, 2, 902–913.

Dong Y. L., Gu Y. J, Yang K., Zhang W. K., 2004. A combining condition prediction model and its application in power plant. Proceedings of the Int Conf on Machine Learning and Cybernetics, August 26–29, 2004, Shanghai (China), 6, 3474–3478.

El-Koujok M., Gouriveau R., Zerhouni N., 2008. Development of a prognostic tool to perform reliability analysis. FEMTO-ST Institute, UMR CNRS 6174 – UFC/ENSMM/UTBM. Automatic Control and Micro-Mechatronic Systems Department, Besançon, France. Proceedings of the European Safety and Reliability and Risk Analysis Conference, ESREL'08, and 17th SRA-EUROPE, Valencia: España.

Elattar H. M., Elminir H. K., Riad A. M., 2016. Prognostics: A literature review. Complex & Intelligent System, 2, 125–154.

Garga A. K., McClintic K. T., Campbell R. L., Yang C. C., Lebold M. S., Hay T. A., Byington C. S., 2001. Hybrid reasoning for prognostic learning in CBM systems. Proceedings of the IEEE Aerospace Conference, 6, 2957–2969.

Gipper J., 2020. Choice of reliability prediction methods. Vita Technologies. http://vita.mil-embedded.com/articles/choice-reliability-prediction-methods. Viewed: July 3, 2020.

Goebel K., 2007. Prognostics and health management. Guest lecture, ENME 808A University of Maryland.

Goebel K., Eklund N., Bonanni P., 2006. Fusing competing prediction algorithms for prognostics. In: Proceedings of 2006 IEEE aerospace conference, Big Sky, MT, USA, 4–11 March 2006.

Goebel K., Saha B., Saxena A., Celaya J., Christophersen J., 2008. Prognostics in battery health management, instrumentation and measurement magazine. IEEE, 11(4), 33–40.

Gu J., Pecht M., 2008. Prognostics and health management using physics-of-failure. 2008 Annual Reliability and Maintainability Symposium, January 28–31, 2008, 481–487.

ISO 13381-1, 2004. Condition monitoring and diagnostics of machines – prognostics – Part1: General guidelines. Int. Standard, ISO.

ITEM Software, 2007. Reliability prediction basics. www.reliabilityeducation.com/reliabilityeducation/ReliabilityPredictionBasics.pdf. Viewed: July 3, 2020.

Jardine A. K. S., Lin D., Banjevic D., 2006. A review on machinery diagnostics and prognostics implementing condition-based maintenance. Mechanical Systems and Signal Processing, 20(7), 1483–1510.

Javed K., Gouriveau R., Zerhouni, N., 2014. SW-ELM: A summation wavelet extreme learning machine algorithm with a priori parameter initialization. Neurocomputing, 123, 299–307.

Kalbfleisch J. D., Prentice R. L., 1980. The statistical analysis of failure time data. John Wiley and Sons.

Klyatis L. M., Klyatis E. L., 2006. Accelerated quality and reliability solutions. Elsevier Science.

Kong L., Peng X., Chen Y., Wang P., Xu M., 2020. Multi-sensor measurement and data fusion technology for manufacturing process monitoring: A literature review. International Journal of Extreme Manufacturing, 2 (2).

Koushanfar F., Potkonjak M., Sangiovanni-Vincentelli A., 2002. Fault tolerance techniques in wireless ad-hoc sensor networks. UC Berkeley Technical reports.

Kothamasu R., Huang S. H., 2007. Adaptive Mamdani fuzzy model for condition-based maintenance. Fuzzy Sets Systems, 158, 2715–2733.

Knowles I., 1993. Is it time for a new approach? IEEE Transactions on Reliability 1993, 42(1), 2–3.

Liu Y., 2016. Reliability it time for a new approach. A thesis submitted in partial fulfillment of the requirements for the degree of Master of Science. Department of Mechanical Engineering. University of Alberta.

Marti S., Giuli T. J., Lai K., Baker M., 2000. Mitigating routing misbehavior in mobile ad hoc networks. ACM Mobicom, 2000, 255–265.

Matiü Z., Sruk V., 2008. The physics-of-failure approach in reliability engineering. Proceedings of the ITI 2008 30th International Conference on Information Technology Interfaces, June 23–26, 2008, Cavtat, Croatia.

Mendes A. C., 2014. Proportional hazard model applications in reliability. Northeastern University, Boston, MA. April 2014.

Mirza Hyder A. B., 2012. Reliability prediction and analysis of a mechanical valve. A Thesis Submitted in partial fulfillment of the Requirements for the award of the Degree of Master of Technology in Mechanical Engineering (Product Design). Departments of Mechanical Engineering. Jawaharlal Nehru Technological University Anantapur College of Engineering.

Mohanty S., Chattopadhyay A., Peralta P., Das S., Willhauck C., 2008. Fatigue life prediction using multivariate Gaussian process. Collection of Technical Papers AIAA/ASME/ASCE/AHS/ASC Structures, Structural Dynamics and Materials Conf, 7–10 Apr 2008, Schaumburg, IL (USA), Art. No. 2008-1837.

NCSS, 2020. Chapter 565: Cox regression. NCSS Statistical Software. https://ncss-wpengine.netdna-ssl.com/wp-content/themes/ncss/pdf/Procedures/NCSS/Cox_Regression.pdf. Viewed: July 2, 2020.

NIST/SEMATECH, 2012. Engineering statistic handbook. Chapter 8. Assessing product reliability. Handbook April, 2012. www.itl.nist.gov/div898/handbook/apr/section1/apr167.htm. Viewed: July 3, 2020.

Pecht M., 1996. Why the traditional reliability prediction models do not work: Is there an alternative. Electronic Cooling, 2(1): 10–12.

Peng Y., Dong M., Jian Zuo M., 2010. Current status of machine prognostics in condition-based maintenance: A review. International Journal of Advanced Manufacturing Technology, 50, 297–313.

Performity LLC, 2020. A guide to fault detection and diagnosis. https://gregstanleyandassociates.com/whitepapers/FaultDiagnosis/faultdiagnosis.htm. Viewed: July 10, 2020.

Ramanathan N., Chang K., Kohler E., Estrin D., 2005. Sympathy for the sensor network debugger. In Proceedings of 3rd ACM Conference on Embedded Networked Sensor Systems (SenSys '05), San Diego, California (pp. 255–267).

Ramezania S., Moini A., Riahic M., 2019. Prognostics and health management in machinery: a review of methodologies for RUL prediction and roadmap. International Journal of Industrial Engineering & Supply Chain Management, 6(1), 38–61.

Reppa V., Polycarpou M. M., Panayiotou C. G., 2016. Sensor fault diagnosis. Foundations and Trends R in Systems and Control, 3(1–2), 1–247.

Rodriguez Obando D. J., 2018. From deterioration modelling to remaining useful life control: A comprehensive framework for post-prognosis decision-making applied to friction drive systems. These pour obtenir le grade de: Docteur de la Communaute Universite Grenoble Alpes. Spécialité: Automatique-Productique. Thèse soutenue publiquement le 13 Novembre 2018.

Satish B., Sarma N. D. R., 2005. A fuzzy BP approach for diagnosis and prognosis of bearing faults in induction motors. IEEE Power Engineering Society General Meeting, 12–16 Jun 2005, San Francisco, CA (USA), 3, 2291–2294.

Shetty P., Mylaraswamy D., Ekambaram T., 2008. A hybrid prognostic model formulation and health estimation of auxiliary power units. Journal of Engineering for Gas Turbines and Power, 130(2), 021601.

Si X-S., Wang W., Hu C-H., Zhou D-H., 2011. Remaining useful life estimation: A review on the statistical data driven approaches. European Journal of Operational Research, 213(1), 1–14.

Silverman B. W., 1986. Density estimation for statistics and data analysis. CRC Press.

Staddon J., Balfanz D., Durfee G., 2002. Efficient tracing of failed nodes in sensor networks. In First ACM International Workshop on Wireless Sensor Networks and Applications USA, 2002.

Tinga T., 2013. Physics of failure and its role in maintenance. Netherlands Defence, Academy & University of Twente. February 4, 2023. www.maintworld.com/R-D/Physics-of-Failure-and-its-Role-in-Maintenance. Viewed: July 6, 2020.

Tinga T., Janssen R. H. P., 2012. The interplay between deployment and optimal maintenance intervals for a navy frigate. In: European Safety and Reliability Conference, Helsinki.

Wang W. Q., Golnaraghi M. F., Ismail F., 2004. Prognosis of machine health condition using neuro-fuzzy systems. Mechanical Systems and Signal Processing, 18, 813–831.

Wikipedia, 2020a. Cox regression. https://es.wikipedia.org/wiki/Regresi%C3%B3n_de_Cox. Viewed: July 2, 2020.

Wikipedia, 2020b. Proportional Hazards Model. https://en.wikipedia.org/wiki/Proportional_hazards_model. Viewed: July 2, 2020.

Willems G., Vandevelde B., 2015. Physics-of-failure based reliability-by-design. IMEC-CEDM. 29 June 2015.

Xue G. X., Xiao L. C., Bie M. H., Lu S. W., 2005. Fault prediction of boilers with fuzzy mathematics and RBF neural network. Proceedings of International Conference on Communications, Circuits and Systems, 27–30 May 2005, Hong Kong (China), 2, 1012–1016.

10 Principles of Digital Twin

10.1 PRINCIPLES OF DIGITAL TWIN

10.1.1 INTRODUCTION

Since the emergence of the idea of Industry 4.0, industry has started to use digital twins. Simply stated, a digital twin is a realistic digital representation of a physical thing (Bolton et al., 2018). It is used to describe the behavior of a real system and to derive solutions for problems in that system.

The digital twin incorporates a system's engineering data, operation data, and behavior descriptions into simulation models. The simulation models are specific for their intended use and for the problem to be solved. The digital twin evolves along with the real system along the whole lifecycle and integrates all currently available knowledge about it (Boschert & Rosen, 2016). The goal of the digital twin is to support all stakeholders in the system during all lifecycle phases to increase productivity (nexDT).

Today, modeling and simulation are standard processes in system development, used, for example, to support design tasks or to validate system properties. Simulation-based solutions are used to optimize operations and predict failure. In general, simulation technology can integrate data-driven and physics-based approaches to reach the next level of merging the real and virtual world in all lifecycle phases. The digital twin for mechatronic and cyber-physical systems (CPS) ensures information created during design and engineering is available and ready for evaluation during the operation of the system. This is often neglected, as design and operation are generally disconnected.

By using simulation models, it is possible to interpret measurements and operational and fleet data in a different way than just looking for deviations from the norm. Several modes of failure can be simulated for the current situation to reproduce the actual measurement signals. The comparison of the simulated signals with the measured ones can help to identify failure modes. The integrated use of data-driven and physics-based models promises an efficient approach to the optimized operation of CPS and analytics and diagnostic solutions for their complex components.

A mathematical model is a description of a system using mathematical concepts and language. A mathematical model may help to explain a system and to study the effects of different components, and to make predictions about behavior (Krishna et al., 2019).

The next generation digital twin will transport data, information, and executable models of all system elements from development – delivered by the component supplier – to operation. This information will be used for the cost-effective development of assist systems during operation, for example, autopilot solutions for production systems and service applications. Applications will include improved maintenance solutions which can explain anomalies and identify potential failure causes. This networking of information will also allow the increased use of field data in the development of variants and follow-up products. The value creation process will be closed by a feedback loop (Boschert et al., 2018).

10.1.2 WHY DIGITAL TWINS MATTER

With a digital twin, that is, an up-to-date representation of an operating asset, an organization can control and/or optimize the asset and the larger system. The digital twin representation captures

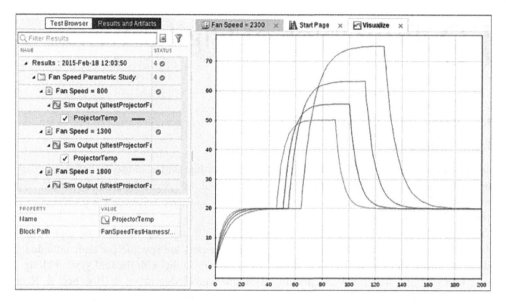

FIGURE 10.1 Monte Carlo simulations to evaluate possible behavior (Mathworks, 2020).

both the current state and the operating history. Digital twins enable optimization, improve efficiency, lead to automation, and permit the evaluation of future performance. The models can be employed for other purposes, for example, virtual commissioning. They can also be used to influence next-generation designs.

Digital twins are used in the following areas:

1. Operations optimization: Using myriad variables (e.g., weather, fleet size, energy costs, and performance factors), digital twins can run almost unlimited numbers of simulations to evaluate readiness or suggest adjustments to current system set-points. In this way, system operations can be optimized to reduce risk, decrease cost, and improve efficiency (Figure 10.1) (Mathworks, 2020).
2. Predictive maintenance: Digital twins can calculate the remaining useful life (RUL) of equipment to determine the best time to service or replace (see Figure 10.2) (Mathworks, 2020).
3. Anomaly detection: The digital twin runs parallel to the real asset. It immediately flags any behavior deviating from the expected behavior. Depending on the industry, by detecting anomalies, the twin is able to avert catastrophic damage, for example, in an off-shore oil rig (see Figure 10.3) (Mathworks, 2020).
4. Fault isolation: Anomalies detected by the model may trigger further simulations to isolate the fault and identify the root cause, allowing maintainers to take the appropriate actions (see Figure 10.4).

10.1.3 HOW DIGITAL TWINS WORK

A digital twin models an Internet of Things (IoT) asset. Thus, it will include the components, behaviors, and dynamics of the IoT asset. Modeling methods include physics-based or first principles methods (e.g., mechanical modeling) and data-driven methods (e.g., deep learning). A digital twin can also combine various behaviors and methods and can be expanded over time (see Figure 10.5) (Mathworks, 2020).

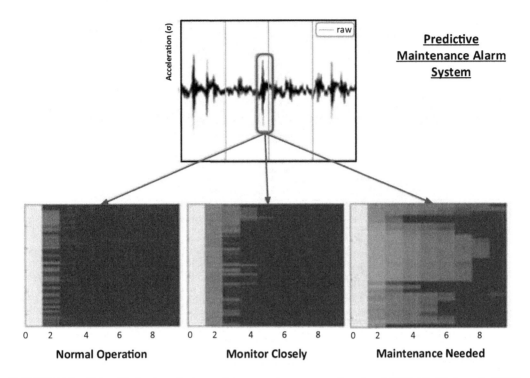

FIGURE 10.2   Baker Hughes' predictive maintenance alarm system, based on MATLAB (Mathworks 2020).

The models must be kept up-to-date with the IoT assets in operation; for example, data are dir-ectly streamed from the assets into algorithms that tune the digital twin. This allows organizations to consider such asset aspects as environment, age, and configuration.

If the digital twin is up-to-date, it can be used to predict the asset's future behavior, refine the control of the asset, or optimize its operation. For example, the model could simulate sensors that are not on the real asset or simulate various future scenarios to inform current and future operations. The digital twin could also determine current operational state if real current inputs are used in the model (Mathworks, 2020).

10.1.4   VALUE OF DIGITAL TWINS

Digital twins offer the following (Oracle, 2017; Rasheed et al., 2020):

1. Real-time remote monitoring and control: It is almost impossible to gain an in-depth view of a very large system physically in real time. Because of its nature, a digital twin can be accessible anywhere. The performance of the system can be monitored and controlled remotely using feedback mechanisms.
2. Greater efficiency and safety: Digital twinning will enable greater autonomy, with humans in the loop only when required. The dangerous, dull, and dirty jobs will be allocated to robots, with humans controlling them remotely. Humans will be able to focus on more creative and innovative jobs.
3. Predictive maintenance and scheduling: A comprehensive digital twinning will ensure mul-tiple sensors monitoring the physical assets will be generating Big Data in real time. Through a smart analysis of data, faults in the system will be detected in advance, enabling better scheduling of maintenance.

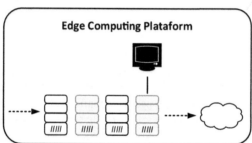

FIGURE 10.3 Prototype of IIoT deployment on oil rig using Simulink Real-Time (Mathworks, 2020).

4. Scenario and risk assessment: A digital twin will enable what-if analyses, resulting in better risk assessment. It will be possible to synthesize unexpected scenarios and study the response of the system and the corresponding mitigation strategies.
5. Better intra- and inter-team synergy and collaborations: With greater autonomy and all the information immediately available, teams can better utilize their time, leading to greater productivity.
6. More efficient and informed decision support system: Availability of quantitative data and advanced analytics in real time will assist in more informed and faster decision-making.
7. Personalization of products and services: With evolving market trends, the demand for customized products and services will increase. A digital twin in the context of factories of the future will enable faster and smoother shifts in the face of changing needs.
8. Better documentation and communication: Readily available information in real time, combined with automated reporting, will keep stakeholders well informed, thereby improving transparency.

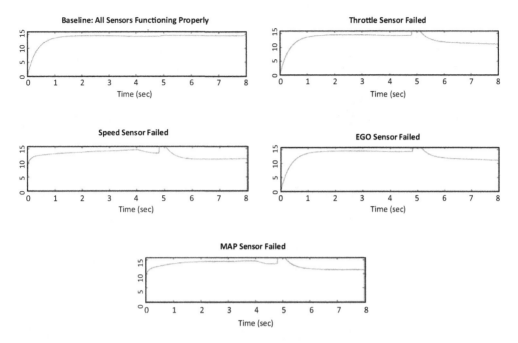

FIGURE 10.4 Fault isolation of a fuel control system (Mathworks, 2020).

10.1.5 INTRINSIC CHARACTERISTICS OF DIGITAL TWINS

A digital twin has the following five core characteristics (Ove Erikstad, 2017):

1. Identity: The digital twin connects to a single, real, and unique physical asset, such as a ship, a semisubmersible, a riser, or a wind turbine. A state on the digital twin corresponds one-to-one with a potential observation on a particular physical asset.
2. Representation: The digital twin captures an essential physical manifestation of the real asset in a digital format, such as computer-aided design (CAD) or engineering models with corresponding metadata.
3. State: A digital twin differs from a traditional CAD/computer-aided engineering model, by having the capability to render quantifiable measures of the asset's state in (close to) real time.
4. Behavior: The digital twin reflects basic responses to external stimuli (forces, temperatures, chemical processes, etc.) in the present context.
5. Context: The digital twin describes external operating contexts, such as wind, waves, and temperature, in which the asset exists or operates within.

10.1.6 DIGITAL TWINS: WHAT, WHY, AND HOW?

The "what" perspective identifies the intrinsic characteristics of digital twins, as explained above. The "why" perspective focuses on their purpose and role, to make them a useful part of a business or engineering landscape. The "how" perspective concerns the architecture and technology platform needed for their realization (Ove Erikstad, 2017).

10.1.6.1 The "Why" Perspective

"Why" is linked to the opportunities afforded by a continuously updated digital model, as opposed to assessing the real asset directly. It is possible to do monitoring and inspection on the digital twin

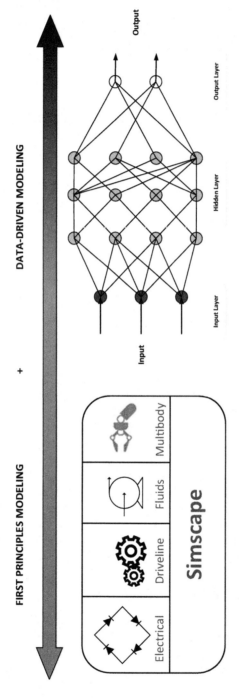

FIGURE 10.5 Digital twin modeling methods: first principles and data-driven (Mathworks, 2020).

instead of the asset itself, thus saving (part of) the effort. This is particularly important when access is a challenge, such as offshore structures or subsea installations. Data can be aggregated at relatively high fidelity, such as stress cycle counting in fatigue life utilization calculations.

Examples of useful applications of a digital twin for a high value, complex asset include the following (Ove Erikstad, 2017):

- Remaining life assessment of structure.
- Inspection/maintenance planning based on true load history.
- Determination of relationship between loads and power production for control system policies.
- Early damage detection for preemptive maintenance and shutdown prevention.
- Hindsight to foresight: Access to (aggregated) time series for design feedback.
- Virtual inspection support.
- Prediction of consequences of (adverse) future operating conditions.
- Inspection/monitoring process support (cost reduction).
- Visualization and inspection of stresses at inaccessible/hidden locations.

10.1.6.2 The "How" Perspective

Implementation will depend on the asset type and will be linked to the needs for accuracy, quality, availability, feasibility, and similar parameters, traded off with cost and technology readiness. A minimum digital twin implementation must at least comprise the following parts:

- Edge capabilities for observing key aspects of the real asset's state and behavior: This typically implies sensors with corresponding edge processing capabilities for data quality enhancements, such as calibration, filtering, and time synchronization.
- Digital twin core runtime, using the input stream from the edge to render a (near) real-time digital reflection of the asset's state.

The application layer subscribes to selected data streams from the digital twin as an integral part of various business/industry processes. This can be specific end-user applications for monitoring and control or legacy applications for maintenance and asset management, or the data stream from the twin might feed into data analytics and machine learning (ML) stacks for pattern recognition and decision support (Ove Erikstad, 2017).

10.1.7 A Digital Twin Example: SAP Digital Twin for Wind Power

An example from the offshore renewables domain can illustrate what a digital twin solution looks like. Traditional condition monitoring solutions for wind power systems are developed to report the operational state with the objective of understanding changes which may occur over time. The main purpose of the monitoring is to prevent damage to the system and to ensure efficient operation of the asset. Control systems are traditionally based on generic algorithms, and maintenance actions often follow predefined plans. The concept of real-time structural integrity management has not been implemented in condition monitoring solutions.

SAP's (SAP is a company specializing in enterprise application software) digital twin for wind power monitoring enables operators to implement adaptive control strategies, as well as improved predictive maintenance tactics based on the physical condition of the system at any time, using a digital representation of the real asset. Control actions such as the yawing of a wind turbine can be simulated in advance of a control decision, considering consequences for both power production and structural lifetime. A yaw misalignment, for example, to the incoming wind direction can lead to a decreased power production. At the same time, the blade root can become sensitive to the yawing direction by increasing or decreasing the overall fatigue loading of the system. Engineering simulation also allows operators to record structural loading throughout the asset lifecycle; they can increase the uptime because they have a better overview of the structural integrity and, thus, the option to react preventively.

Already implemented in a pilot project with Arctic Wind, this wind turbine digital twin solution supports maintenance operations and structural capability utilization (Ove Erikstad, 2017).

10.1.8   DIGITAL TWIN ORIGIN: PHYSICS AND SIMULATION

Though the concept of a physics-based digital twin is relatively new, its foundation can be traced to widely used and well-known constructs in the engineering community, as illustrated in Figure 10.6. It is based on engineering analysis and Newtonian physics. When time is added as a dimension, asset behavior can be assessed, allowing the analysis of operations based on anticipated load cases. The further evolution of the digital twin is based on two fundamental contributions; switching from (anticipated) load cases to sensor-based observations as the main input for the model, and switching from simulated time to (close to) real time (Ove Erikstad, 2017).

10.1.9   USE OF DIGITAL TWIN IN OPERATIONS

10.1.9.1   Point Machine for Train Switches

Points are part of railway switches and, as such, are a key element of the rail network infrastructure. They are distributed all over the network, and their maintenance is crucial to guarantee safety and undisturbed operation. The points are responsible for a large amount of the railway's operational costs, as monitoring and maintenance are mainly manual. Points diagnostics systems like the Sidis W compact (Sidis, 2010) from Siemens are used to monitor the current condition of the point machine by analyzing the electrical power demand of the drive and point machine operation module. However, a prediction of the future behavior remains difficult. The interaction of a railway switch and its drive (point machine) is complex. Both subsystems have many different parameters, making it difficult to predict their behavior (Boschert et al., 2018).

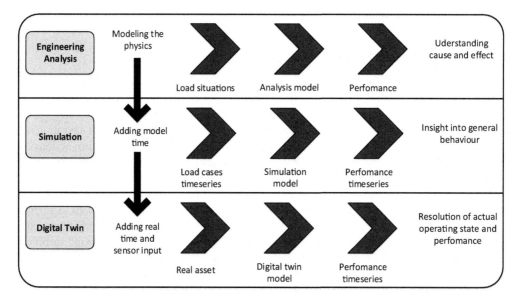

FIGURE 10.6   Digital twin foundations in simulation and engineering analysis (Ove Erikstad, 2017).

10.1.9.2 Planning the Digital Twin

Physics-based simulation models, as well as live data from the points' diagnostics system, can be used to identify a possible malfunction. Therefore, a "template" of this digital twin feature is specified during the conceptual design phase. This template describes how the different components are linked together and interact. During the ongoing design and construction process, it is filled with (sub-)models as soon as they are created, resulting in a complete system model that fulfils the specified product functionalities. Defining the template during early planning helps to ensure the simulation models are created at the time when the lowest effort is needed: during the component development. It is also possible to create the models afterwards, but at a higher cost, as implicit knowledge may have been lost (Boschert et al., 2018).

10.1.9.3 Digital Twin during Operation Phase

The system model allows the operation of the points' machine to be simulated for a customer/site specific configuration. Depending on the complexity of the underlying system, the implementation can be done at different locations, ranging from embedded logic in the control unit of the drive or included as an assistance module for the switch operator as service on demand, for example, for maintenance workers. In the suggested set-up (assistance system), the failure identification module compares the measurement data from the live system with results from the physics-based simulation. This direct comparison is possible as, by design, the results of the two input sources (simulation model and points diagnostics system) are comparable.

As soon as the failure identification module detects a significant deviation of the two models, it raises a notification in the dashboard and analyzes possible root causes of the failure, also displayed to the operator.

As usually every point has an individual configuration, the failure identification module has to be configured individually. However, in several situations, different points show a similar behavior. For example, those close to each other are subject to similar environmental conditions, like temperature or rainfall (Boschert et al., 2018).

10.1.9.4 Hybrid Analysis and Fleet Data

The connection between the digital twin and operational data offers a wide range of new services, from failure detection to diagnosis.

The connection of the digital twin and a physical instance is established via sensors. Sensor data are received and processed in real time. But if data are submitted to a cloud and processed there, synchronization and delay management are important. Depending on the application, sensors measure physical quantities like accelerations, displacements, strains, temperatures, or current signature and power in the case of points' machines. In this situation, it is relevant not only to monitor data, but also to detect and diagnose failure states, as well as give recommendations for future operation or maintenance.

A second approach is to rely on data and use artificial intelligence (AI) algorithms on those data. However, new services cannot be assessed because there is a lack of data on them. In this situation, simulation models help to test the system virtually in different environments or failure modes. Fleet data are generated using simulation models and variants of them. This approach even applies for unique items. Once products are in operation, these data can be gradually enriched by data from the field (Boschert et al., 2018).

10.1.10 Digital Twin Reference Model

Figure 10.7 shows a model based on digital twin methodologies to reduce risk in process plants. An organization can employ the model in the following ways (Bevilacqua et al., 2020):

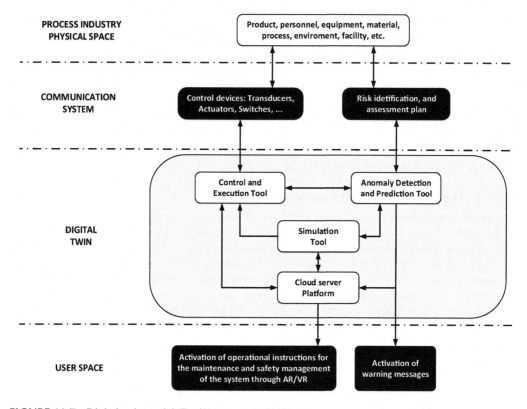

FIGURE 10.7   Digital twin model (Bevilacqua et al., 2020).

1. Creating a virtual process that parallels the physical one, thus permitting both static and dynamic analysis of the physical process.
2. Sending information to other interconnected digital objects, thus increasing safety.
3. Intercepting anomalies quickly, thus permitting prompt intervention to minimize possible damage and/or support preventive/predictive maintenance.

For ease of interpretation, Figure 10.7 should be read from top to bottom. The sections are as follows:

1. Process industry physical space: Physical space refers to all observable plant elements in production that are monitored and sensed and may be actuated and controlled, including, for example, products, workers, equipment, materials, processes, and environment. This is the primary environment that the company aims to control via the digital twin.
2. Communication system: Communication refers to transferring data or information between the digital twin and the plant. This system connects observable plant elements to digital entities, and vice versa, so they can be synchronized. Data are collected using sensors, cameras, actuators, and other composite devices. A three-dimensional (3D) model representation and a risk identification and assessment plan are included in this layer so the simulation system and the anomaly prediction and detection tool can be set up.
3. Digital twin: The digital twin layer must be able to do the following:
   - Acquire data from sensors, actuators, devices installed on the physical system.
   - Visualize data coming from the field.
   - Analyze these data to detect anomalies and make predictions.

- Integrate anomaly detection algorithm data to compare trends.
- Develop digital twin that simulates behavior (i.e., what-if scenarios).
- Integrate digital twin and compare it with the physical space to determine anomalous behaviors.
- Identify dangerous situations.
- Activate and manage alerts to all concerned stakeholders
- Support maintenance following anomaly detection (Bevilacqua et al., 2020).

A digital twin system relies on four main tools:

1. Control and execution tool: This tool is dedicated to the management or control of industrial processes. It executes a program to elaborate on the digital and analog signals coming from sensors and directed to the actuators in an industrial plant. Otherwise stated, it permits the physical system to communicate with the cyber system at the output through sensors, transducers, and so on, and at the input through the control of actuators, switches, and so on.
2. Simulation tool: This tool permits the company to virtually model the processes. It includes a simulation model, as well as a behavioral and functional model. The simulation tool can work online (inputs come from sensors) or off-line (inputs are entered manually). In the off-line case, the tool provides a virtual representation of the physical assets, thus allowing managers to analyze what-if scenarios without needing to physically realize them. In one example, the tool could commission a new plant in a virtual way to identify operator risk before the plant is activated. In another example, it could be used to simulate a maintenance activity and identify the related risks. In the online case, the tool receives information from sensors on physical assets; it modifies its parameters when the assets change their conditions. The online application can allow a company to compare the various data from the simulation system and activate a warning signal if a discrepancy between two values is beyond defined thresholds (Bevilacqua et al., 2020).
3. Anomaly detection and prediction tool: This tool uses ML algorithms to predict why faults happen, that is, the causes (anomaly detection), and to predict how long the system can keep working before it breaks down or goes beyond operating parameters (anomaly prediction and residual life assessment) (see Figure 10.8). An example of an anomaly predicted through ML in a process plant is the following (Bevilacqua et al., 2020). An operator erroneously closes a shut-off valve. As a result, the plant goes into overpressure, thus risking the integrity of the piping and the safety of the workers. Yet the tool is able to detect an increase in pressure at certain points, identify the problem, and warn the operator about the anomaly.
4. Cloud server platform: Real-time data are required but in today's operating environment, huge amounts of data are available. A normal server architecture is not enough, because in the normal operating environment, the enormous amount of data would not make its operation stable. By the same token, classic relational databases are not able to withstand an excessive number of requests for simultaneous access to reading and writing. A cloud solution is therefore called for. The platform provides application programming interfaces (APIs) or external calls with related authentication. The platform handles the following:
   - Data coming from a programmable logic controller (PLC).
   - Analysis of input data (i.e., sensor readings).
   - Elaborations coming from data analysis engines and visualized for comparison.
   - Data coming from sensors to the simulated model of the plant.
   - Data coming from the simulated model (what-if scenarios) and their comparison to real data.
   - Data coming from the sensors to the 3D model of the plant.
   - Data coming from the 3D model and the comparison of these data with the real data.

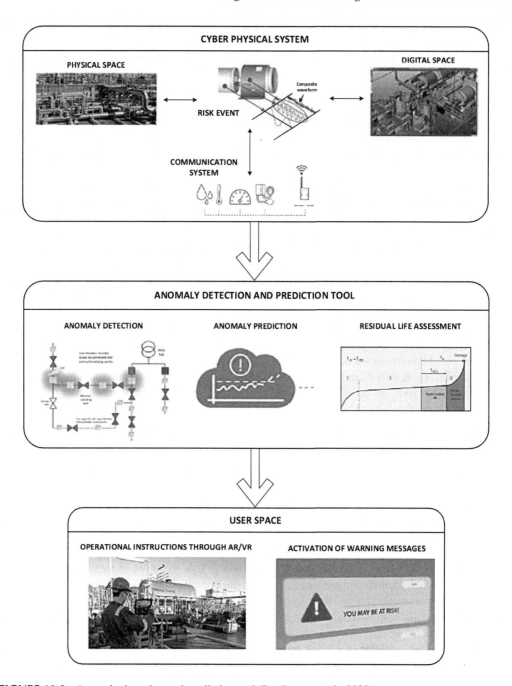

FIGURE 10.8 Anomaly detection and prediction tool (Bevilacqua et al., 2020).

The cloud platform must also manage historical data related to both field sensors and analysis, as well as user permissions to avoid inappropriate deletions in non-applicable areas
(Bevilacqua et al., 2020).

5. User space: In this context, "user" refers to a human, a device, or a system, such as a manufacturing execution system (MES) or enterprise resource planning (ERP). It includes:

- Activation of operational instructions for the maintenance and safety management of the system through augmented reality (AR)/virtual reality (VR).
- Activation of alarms and warnings.

This layer includes classes of advanced services; for example, if the ML system predicts risk situations, operators may be warned of anomalies through wearable systems (Bevilacqua et al., 2020).

10.1.11 PHYSICAL MODEL-BASED DIGITAL TWINS

As part of Industry 4.0, digitalization will play an important role in producing accurate components made in multi-stage automated, autonomous processes, thus saving money and development time and increasing manufacturing flexibility. CPSs will play a key role in this concept of Industry 4.0 architecture. A CPS can be described as a set of physical devices or processes, objects that interact with a virtual cyberspace. This digital representative or digital twin of the real world can be used, for example, to monitor and control the physical entity and to compare the virtual digital twin with reality. The model itself can be based on data-based or physics-based simulations or on both.

Figure 10.9 gives a brief illustration of the history of modeling up to the digital twin (Post et al., 2017).

As noted in previous sections, the digital twin is a representative of the real world and must be able to fully simulate all relevant behavior of the product/process during the lifecycle, or a subset of this. It can be a part of the product lifecycle management (PLM) system. The digital twin can be based on different types of information, for example, real data of the process or product stored in statistical models using AI, like neural network-based deep learning algorithms. It could also be based on physics-based models; it could be analytical, phenomenological, or even very complex micro-mechanical-based nonlinear finite element method (FEM) models. Of course, the digital twin can also be based on the combination of different model concepts, and even the physics-based models can be calibrated based on real time production data.

In all cases, these models have to represent all relevant behavior of the real system in all conditions and must be very mature, robust, validated, and usable as a part of a real-time control system or real-time virtual reality.

In the domain of zero defect manufacturing of metal precision parts, one of the issues in the digital twin is the interoperability in material data over a chain when different metal processes are

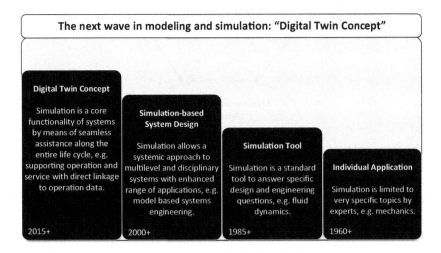

FIGURE 10.9 Digital twin: next wave in simulation technology (Post et al., 2017).

interacting with each other, and the material data have to be mapped over the production chain. This difficulty increases if different commercial solvers are involved. A digital twin has to be very robust in finding a solution in the whole of the production space, because it will be a part of a real running production system. The worldwide research on improving material models, both macroscopic and microscopic, leads to more complexity and new models which have to be implemented with the same robustness as the old ones, over a number of different solvers (Post et al., 2017).

10.1.11.1   Model-Based Control

A digital process twin can be used as a part of an industrial control system: it can be combined with real data from an industrial process, and the real data of the process can even update or improve the digital twin. This digital twin can be based on a physics-based model of the process, also called the FEM. For it to be of value, it must solve the following issues (Post et al., 2017):

- The physical FEM model must be able to describe the relevant behavior of the process.
- The model must be validated.
- The solution must be robust and mature.
- The model must be in real time.
- The model must represent all industrial processes in the production chain.

In most cases, these multi-stage FEM models cannot be used directly as part of the control because of the long calculation times. To solve this problem, a meta-model can be created based on the physics-based model using statistical techniques like design and analyses of computer experiments or other data mining techniques, leading to a good and fast representative model of the complex FEM model. This meta-model must be able to calculate and find a good solution in the whole domain of the industrial environment. This also means the FEM model must be very robust and mature, with the ability to calculate a solution in the whole design space of the meta-model. If successful, this digital twin can be used for much more than only model based control, as suggested in Figure 10.10 (Post et al., 2017).

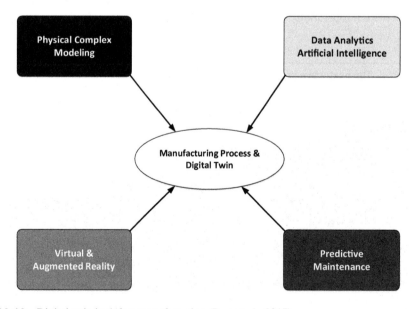

FIGURE 10.10   Digital twin in defect manufacturing (Post et al., 2017).

10.2 FUNCTIONAL MOCK-UP FOR COMPLEX SYSTEM ASSEMBLY

10.2.1 DEFINING COMPLEX SYSTEMS

Systems can be categorized into three types: simple, complicated, and complex. Simple systems are just that. The outside observer has no problem discerning the operation of the system. The system is completely predictable, the inputs are highly visible, the actions performed on those inputs are obvious and transparent, and the outputs are easily predictable (Sargut & McGrath, 2011).

Complicated systems are also completely predictable, and the system follows well defined patterns. The difference between simple systems and complicated systems is the component count. Complicated systems have many more components. Complicated systems are often described as intricate, but the inputs are well known, as are the resulting outputs. The connection between components is linear and straightforward. A mechanical watch would be a representative example of a complicated system (Nature, 2008).

Complex systems are in a different class of systems entirely. There is little agreement as to how to even describe a complex system, let alone the definition of the term "complexity". Complex systems have been characterized as being a large network of components, with many-to-many communication channels, and sophisticated information processing that makes prediction of system states difficult (Mitchell, 2009).

10.2.2 COMPLEX SYSTEMS AND ASSOCIATED PROBLEMS

While man-made complex systems can be considered relatively new phenomena, the issue of complex systems as it relates to serious and catastrophic problems goes back decades. The first major discussion of this issue is often considered to be Perrow's seminal work on the inherent dangers of complex systems, *Normal Accidents* (1984), in which he defines the difference between complex and complicated systems.

Perrow defines complexity in terms of interactions. He defines linear interactions as "those in expected and familiar production or maintenance sequences, and those that are quite visible even if unplanned". He defines complex interactions as "those of unfamiliar sequences, or unplanned and unexpected sequences, and either not visible or not immediately comprehensible". Perrow also uses "tightly coupled" to describe complex systems and "loosely coupled" to describe complicated systems. Perrow's claim, which he supports with numerous examples, is that complex systems lead quite naturally to the idea of "normal accidents".

Perrow's list of normal accidents, better described as disasters, include the Three-Mile Island nuclear reactor meltdown, the 1979 DC-10 crash in Chicago, and maritime disasters. The common thread linking these examples is the human element in interacting with complex systems. In other words, these are socio-technical systems. The human element takes two forms: human inconsistency, both deliberate and accidental, in following rules, processes, and procedures, and a lack of sense-making, that is, the inability to make sense out of the inputs and stimuli being presented.

Of course, Perrow was writing before the ubiquitous presence of computers. These can prevent human inconsistency that is accidental or even intentional. Computers forget nothing and perform processes over and over again without deviation. Computers can even go a long way in preventing the deliberate, error-causing human acts, by sensing what the system's state should be and comparing it against what it is, and raising an alarm if the two do not match. One of the examples Perrow uses is a fire door being propped open when it should not have been. In today's environment, a sensor would have been placed on that door and triggered an alarm in the computer that the door was open when it should be closed. In other words, the use of computers dramatically decreases the numbers of even deliberate human intent to not follow procedures and processes.

The other source of human interaction problems, sense-making, has played a role in major system disasters. The core issue here is that humans often do not do a good job in making sense of the information streaming at them, especially in stressful situations. As a result, they sometimes do not make the right decisions and not infrequently make exactly the wrong decision in the heat of a crisis. Weick (1990, 2005) has a list of disasters where this has happened, including the NASA disasters, Challenger and Columbia, and what can be classified as a system-of-systems failure, the air accident involving two 747s colliding on Tenerife in the Canary Islands.

This is a much more difficult issue to address, as computers do not do any sense-making. They simply execute what they have been programmed to do. However, computers can perform simulations before the system is deployed to determine how the system reacts in a wide variety of conditions and train users of the system under abnormal conditions they might face. Simulated systems might "front-run" in real-time developing conditions to assist humans in making the correct sense of developing situations and overcoming the biases that can negatively affect human decision-making (Troyer, 2015).

10.2.3   FUNCTIONAL MOCK-UP INTERFACE

The functional mock-up interface (FMI) defines a standardized interface to be used in computer simulations to develop complex CPSs. It is based on the following premise: if a product is to be assembled from a range of parts interacting in complex ways, each controlled by a complex set of physical laws, it should be possible to create a virtual product that can be assembled from a set of models, each representing a combination of parts, each a model of the physical laws, as well as a model of the control systems (using electronics, hydraulics, and digital software) assembled digitally.

To create the FMI standard, numerous software companies and research centers worked together in the MODELISAR project. The FMI standard is used in the model-based development of systems, for example, to design the electronic devices used inside vehicles, including, among others, active safety systems, electronic stability program controllers, and combustion controllers (Wikipedia, 2020).

FMI is implemented in the following:

1. Model exchange.
2. Co-simulation.
3. Applications.
4. PLM.

FMI implementation by a software modeling tool enables the creation of a simulation model that can be interconnected or a software library called a functional mock-up unit (FMU) (Wikipedia, 2020).

10.2.3.1   Functional Mock-Up Interface for Model Exchange and Co-Simulation

The FMI defines an interface to be implemented by a FMU. The FMI functions are used by a simulation environment to create one or more instances of the FMU and to simulate them, typically together with other models. An FMU may either have its own solvers or require the simulation environment to perform numerical integration. The goal of this interface is that the use of an FMU in a simulation environment should be reasonably simple (Modelica Association, 2019). There are basically two types of FMI interfaces:

• FMI for model exchange interface: This is an interface to the model of a dynamic system described by differential, algebraic, and discrete-time equations. It provides an interface to evaluate these equations as needed in different simulation environments, as well as in embedded control systems, with explicit or implicit integrators, and fixed or variable step size. The interface is designed to allow the description of large models.

- FMI for co-simulation interface: This interface is designed for the coupling of simulation tools (simulator coupling and tool coupling) and for coupling with subsystem models, exported by their simulators together with their solvers as runnable code. The goal is to compute the solution of time-dependent coupled systems consisting of subsystems continuous in time (model components that are described by differential-algebraic equations) or time-discrete (model components described by difference equations, for example, discrete controllers). In a block representation of the coupled system, the subsystems are represented by blocks with (internal) state variables $x(t)$ connected to other subsystems (blocks) of the coupled problem by subsystem inputs $u(t)$ and subsystem outputs $y(t)$.

In tool coupling, the modular structure of coupled problems is exploited in all stages of the simulation process, beginning with the separate model setup and pre-processing of the individual subsystems in different simulation tools. During time integration, the simulation is again performed independently for all subsystems restricting the data exchange between subsystems to discrete communication points. Finally, the visualization and post-processing of simulation data is done individually for each subsystem in its own native simulation tool (Modelica Association, 2019).

The two interfaces have many things in common, in particular, an FMI 2.0 "C" API. All required equations or tool coupling computations are evaluated using standardized C functions. C is used because it is the most portable programming language today and is the only programming language that can be used in all embedded control systems.

An FMI description schema (XML) defines the structure and content of an XML file generated by a modeling environment. This XML file contains the definition of all variables of the FMU in a standardized way, making it possible to run the C code in an embedded system without the overhead of the variable definition (the alternative would be to store this information in the C code and access it via function calls, but this is not practical for embedded systems or for large models). Furthermore, the variable definition is a complex data structure, and tools should be free to determine how to represent this data structure in their programs. This approach allows a tool to store and access the variable definitions (without any memory or efficiency overhead of standardized access functions) in the programming language of the simulation environment, such as C++, C#, Java, or Python. Note that there are many free and commercial libraries in different programming languages to read XML files into an appropriate data structure (Modelica Association, 2019).

An FMU (in other words, a model without integrators, a runnable model with integrators, or a tool coupling interface) is distributed in one ZIP file. The ZIP file contains:

- FMI description file (in XML format).
- C sources of the FMU, including the needed run-time libraries used in the model, and/or binaries for one or several target machines, such as Windows dynamic link libraries (.dll) or Linux shared object libraries (.so). The latter solution is commonly used if the FMU provider wants to hide the source code to secure the contained know-how or to allow a fully automatic import of the FMU in another simulation environment. An FMU may contain physical parameters or geometrical dimensions, which should not be open. However, some functionalities require source code.
- Additional FMU data (such as tables or maps) in FMU specific file formats.

A schematic view of an FMU is shown in Figure 10.11.

10.2.4 THE USE OF FMUs FOR THE DIGITAL TWIN

As noted above, a digital twin is a coupled model of a real system; it is stored in a cloud platform and represents its status based on data analysis and physical sensory information (Lee & Lapira, 2013).

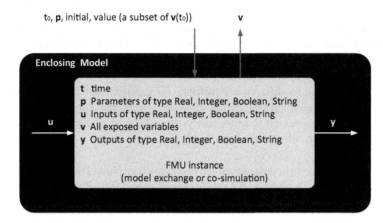

FIGURE 10.11 Data flow between the environment and an FMU. Blue arrows: information provided by the FMU. Red arrows: information provided to the FMU (Modelica Association, 2019).

Different simulation modules can simulate different behaviors of the system that are replicated in the digital environment. The core simulation model is built with the representation of the pieces of equipment of the production system. Each piece is connected to one or more modules representing the specific behavior of interest. The behavior modules are black-boxes that take standard input data from both the simulated and field environments and output computed data also in a standard format, as in Figure 10.10.

FMI is used for the black-box modules, creating FMUs. This allows the function to be exported to different simulation environments, thus granting the independence of the FMU modules from the single simulation tool and allowing them to be reused with little change (Figure 10.11). This has already been successfully used in CPS-based systems (Yun et al., 2017; Wang & Baras, 2013).

Recall from the previous section: the FMU modules are used for discrete event simulation (DES) and constituted by a unique FMU zip-file composed of three main parts: an XML file to define the FMU variables; the model's equations (a set of C functions); other data, for example, tables or comments about the model.

The FMU shown in Figure 10.12) is created for a DES model for energy consumption computation.

Energy consumption depends on the actual state of each equipment piece at each moment, as shown in Figure 10.13. The FMU is connected to the field to get information on the updated machine states (Seow et al., 2013).

10.2.5 OBJECTIVES OF FMI APPLIED TO PRODUCT LIFE METHOD

The FMI product life method (PLM) interface standard specifications define simulation and co-simulation integration methods, and requirements for the supporting PLM system to guarantee that the other components implementing FMI, such as simulation and co-simulation servers, editors, or other applications, may properly store and retrieve the relevant data for the PLM. The PLM server provides services to manage the lifecycle of (Itea, 2011):

- Executable units and the associated files (.fmu).
- Co-simulation configuration files.
- Results files.
- Executable unit's data for scenarios: Parameters, input, and initialization.

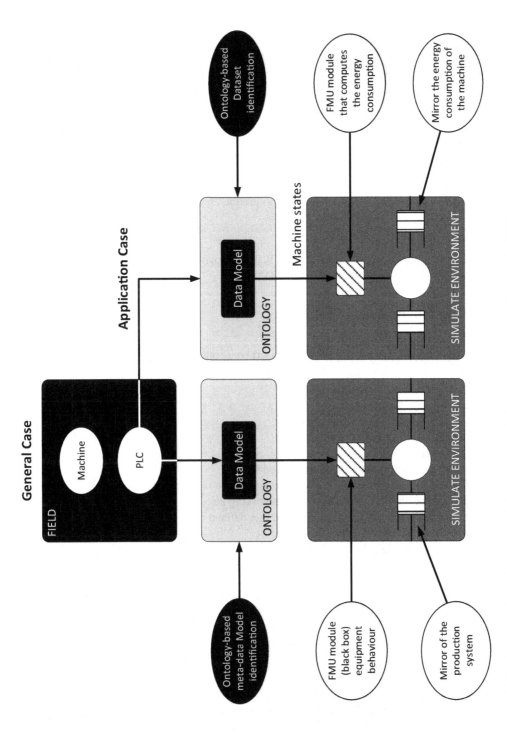

FIGURE 10.12 Relationship between field, ontology, and simulated environment (Seow et al., 2013).

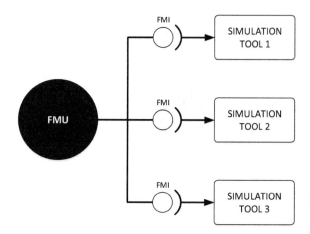

FIGURE 10.13 FMU module reuse in different simulation environments through FMI standard (Seow et al., 2013).

Also recall from the previous discussion that there are two types of interfaces: FMI for model exchange, and FMI for co-simulation (Itea, 2011).

10.2.6 Functions of Product Life Method

10.2.6.1 Summary of PLM Functions

The PLM capabilities needed to implement the FMI PLM interface standard are (Itea, 2011):

- Navigation & search functions:
  - Access to FMU properties.
  - Search of simulation content attributes.
  - Navigate on links between PLM items.
- Basic PLM functions:
  - Deploy and collect FMUs and associated files.
  - Check-out or download the working directory of a target on the network.
  - Check-in or import into PLM from the working directory of a target on the network.
  - Launch remotely the application on a target defined on the network (see Figure 10.14).
- Administrative functions (see Figure 10.15):
  - Manage access for users to simulation objects.
  - Create a new simulation content.
  - Edit/delete existing simulation content.
  - Associate simulation content with simulation result or co-simulation configuration.

In this generic approach, tools can be launched by PLM. In reality, it is a matter of implementation. It also depends on the connection possibility. Generally, for co-simulation, the tools are cascade-launched by a co-simulation engine. The FMI PLM interface doesn't prescribe how the functionality should be implemented by the PLM. The approach works without imposing the development of a specific API on top of the PLM. Basic PLM functions are adequate to support the FMI PLM interface, in a fully open approach (Itea, 2011).

10.2.6.2 Network Description

The topology of the available resources (machines, simulation tools, etc.) in the enterprise network includes the following.

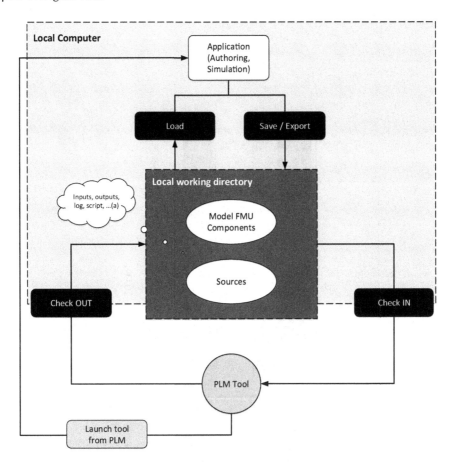

FIGURE 10.14 Services between external tools and PLM; (a) Additional information such as input simulation data, calculated data, acquired data, generated reports, log files etc. (Itea, 2011).

The files stored in the PLM system have to be extracted and deployed on specific hosts, as defined by the PLM user. Before this stage, as a prerequisite, the network's available hosts and applications must be declared in the PLM system.

Each host is identified and associated with a list of available applications. This information will help users choose the target (Host + Application) for editing, simulation, or post-processing activities, for example, according to their hardware and software specifications and versions (see Figure 10.16).

10.2.6.3 Deployment Description

The configuration deployed in the network and available for the authoring tools (simulation engine, modeling tool, post-processing tool, etc.) includes the following.

To prepare a simulation (or any other activity as defined in the use cases), the PLM has the ability to associate the FMUs and the available means (host and applications). This is shown in Figure 10.17 (Itea, 2011).

Figure 10.17 indicates the following (Itea, 2011):

- The same FMU can be used once or several times; thus, several instances may be executed.
- Resources files can be used in the same way, although simulation parameters should probably be different.
- From different instances of the same FMU, different results will be collected.

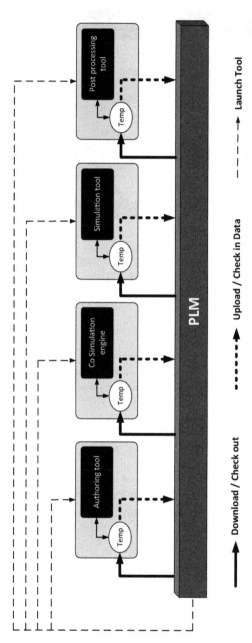

FIGURE 10.15   List of interactions between PLM and "outside PLM" applications (Itea, 2011).

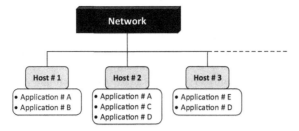

FIGURE 10.16   Available hosts and applications on the network (Itea, 2011).

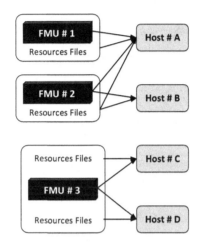

FIGURE 10.17   FMU deployment on hosts (Itea, 2011).

As a consequence, additional PLM information has to be found in the deployment directories and vice versa when checking-in the information produced by the simulation. This is necessary for the PLM system, but not for the deployment description. This information includes:

- Tree structure identification of the FMU and the configuration to be simulated.
- FMU identification and version.

Table 10.1 shows how various scenarios are mapped onto PLM services and the functional capability required for PLM to implement the FMI methodology.

10.3   INTEGRATION OF LOW-LEVEL DIGITAL TWINS

10.3.1   DIGITAL TWIN: TOWARD AN EVALUATION FRAMEWORK

The objective of the framework is to evaluate the current state of digital twins across five key levels. The levels help the industry use common language when describing a digital twin and its capabilities. The aim of the framework is to enable clients and collaborators to participate at all stages of development.

Figure 10.18 presents the evaluation framework. The metrics of the framework are shown in Table 10.2. The four metrics are: autonomy, intelligence, learning, and fidelity. While these four metrics are conceptually correlated, they should be treated independently (ARUP, 2019).

TABLE 10.1
Mapping of Scenario to PLM Services

FMI Required Capability	Scenario §	FMI for PLM Mandatory Format	Functional Capability Needed for PLM Engine to Implement Methodology	Comment
Search and navigate	§3.23.2 §3.33.3 §3.4.13.4.1 §3.4.33.4.3 §3.53.500 §3.73.7	NO	• Search object in PLM. • Navigate on PLM links.	
Create and Edit PLM objects	§3.1.23.1.2 §3.23.2 §3.33.3 §3.4.13.4.1 §3.4.33.4.3 §3.53.5 §3.73.7	YES	• Create new executable unit. • Create new Result object. • Create new co-simulation configuration. • Edit Object's attribute.	
Check-out	§3.23.2 §3.4.13.4.1 §3.53.5	YES	• PLM search. • Check out. • PUT on network location.	Files are extract to new directory
Launch	§3.23.2 §3.4.23.4.2 §3.53.5	NO	• Remote application start. • Wait end of Process activity.	May be limited to launch application Graphic user interface
Check-in	§3.1.23.1.2 §3.23.2 §3.4.33.4.3 §3.53.5	YES	• Check in file. • Check in directory. • Get form network location. • Import metadata.	
Simulation Control	§3.4.23.4.2	NO	• This is out of PLM scope.	No PLM Action requested during simulation. Main process can be launched by PLM.

Source: Itea (2011).

In Table 10.3, the framework moves through five levels, beginning with a simple digital model. As the model evolves, feedback and prediction increase in importance. At higher levels, ML capacity, domain-generality, and scaling potential all come into play. By the highest levels, the twin is able to reason and act autonomously and to operate at a network scale (incorporating lower-level twins, for example) (ARUP, 2019).

10.3.2 Current Technologies Deployed in Digital Twin

A digital twin integrates different technologies with several functions. The following sections describe these technologies and explain the problems connected to their implementation and use in the digital twin (Bevilacqua et al., 2020).

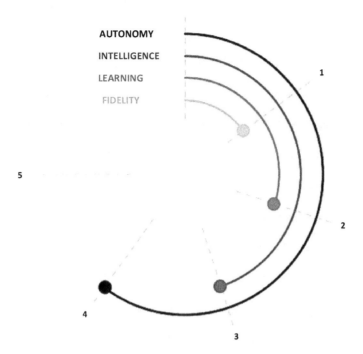

FIGURE 10.18   Digital twin metrics framework: autonomy, intelligence, learning, and fidelity (ARUP, 2019).

10.3.2.1   Industrial Internet of Things (IIoT) and Digital Twin

The increased use of IoT devices in real applications is a direct result of the proliferation of embedded sensors, low-power wireless communications, and signal processing algorithms. IoT connects the real world to cyberspace, enabling the sensing of objects and processes, data gathering, ML, and real-time feedback.

Simply stated, the digital twin is a dynamic digital replica of a real physical process. Its backbone technology is the use of IoT for real-time and multi-source data gathering. The ability of a digital twin to duplicate a physical object in terms of the amount of information acquired via IoT sensors is extremely advantageous for many reasons. However, there are still some remaining research challenges to be solved (Bevilacqua et al., 2020):

1. Sensing methods that monitor physical metrics in an industrial context and use less resources are more practical in Industrial IoT (IIoT).
2. Although wireless and battery-free sensors can support lightweight and robust monitoring, remaining concerns include how to extend the capabilities of wireless signals and how to increase the battery life.
3. Processing data from networked sensors will require computing architectures to be upgraded, for example, collaborative edge computing. To reduce the workload of cloud architecture and save bandwidth, we need to develop a resource-constrained IoT device with modern analysis techniques, for example, deep learning.
4. Efficient data transmission methods need to be integrated into the IoT radio chip to correctly implement the digital twin. These include, for example, low-power wide-area networks, low power communications, parallel backscatter for high throughput, and software-defined low-power wireless (Bevilacqua et al., 2020).

TABLE 10.2
Working Levels of Digital Twins

Levels

Level 1

A digital model linked to the real-world system but lacking intelligence, learning or autonomy; limited functionality, e.g., a basic model of a map.

Level 2

A digital model with some capacity for feedback and control, often limited to the modeling of small-scale systems, e.g., building temperature sensors which feed information back to a human operator.

Level 3

A digital model able to provide predictive maintenance, analytics and insights, e.g., predicting the life expectancy of rail infrastructure, enabling repairs or replacements before asset failure.

Level 4

A digital model with the capacity to learn efficiently from various sources of data, including the surrounding environment. The model will have the ability to use that learning for autonomous decision-making within a given domain, e.g., the model can automatically communicate real-time route recommendations through various modalities (app, signage, radio), allowing drivers to better plan their journey.

Level 5

A digital model with a wider range of capacities and responsibilities, ultimately approaching the ability to autonomously reason and to act on behalf of users (artificial general intelligence). Intuitively, a level 5 model, such as a model of a neighborhood in a smart city, would take responsibility for the tasks one would presently expect a human operator to manage, as well as to react to previously unseen scenarios. Another hallmark of this level would be the interconnected incorporation of lower-level twins, e.g., take the level 4 example of traffic updates across a network. In a smart city scenario, numerous independent systems work in parallel to provide feedback to a central decision-making network to deliver value to city-level leaders.

Source: ARUP (2019).

IoT sensors are increasingly part of the industrial environment, thus making digital twins more viable. As time goes on, IoT devices will continue to be refined; consequently, digital-twin scenarios will be able to include smaller and less complex objects, yielding additional benefits to companies.

Digital twins can be used to predict different outcomes based on variable data. Therefore, with the right software and data analytics, digital twins can optimize IoT deployment in a company for maximum efficiency and help the designers figure out where things should go or how they operate before they are physically deployed.

TABLE 10.3
Definitions of Digital Twin Metrics

Metric	Definition
Autonomy	Autonomy represents the ability of a system to act without human input. There are five levels of autonomy. At level 1, there is complete absence of autonomy, with the user controlling all aspects of the digital twin. At level 2, the twin is user-assisted. At this level, prompts and notifications of system activity are expected, but autonomy is limited. At level 3, the twin has partial autonomy; it has the ability to alert and to control the system in certain ways. At level 4, the twin has high autonomy; it is able to perform critical tasks and to monitor conditions with little to no human intervention. Finally, at level 5, the twin operates safely in the absence of human intervention.
Intelligence	Intelligence represents the ability of digital twins to replicate human cognitive processes and perform tasks. There are five levels of intelligence. At level 1, the twin has no intelligence. At level 2, the twin has reactive intelligence (the twin only responds to stimuli and cannot use previously gained experiences to inform present actions). At level 3, the twin uses learning to improve its response and is capable of learning from historical data to make decisions. At level 4, the twin understands the needs of other intelligent systems. Finally, at level 5, the twin is self-aware with human-like intelligence and self-awareness.
Learning	Learning represents the ability of a twin to automatically learn from data to improve performance without being explicitly programmed to do so. Through machine learning, a twin classifies aspects of the systems (objects, behaviors) using reinforced learning. There are five levels of learning. At level 1, the twin has no learning component. At level 2, the twin is programmed using a long list of commands. At level 3, the twin is trained using a supervised learning approach (using labeled data able to provide feedback and prediction performance). At level 4, the twin is trained using unsupervised learning (the twin uses no labels and tries to make sense of the environment on its own). At level 5, the twin uses reinforcement learning by interacting with its environment. With reinforcement learning, the twin learns from past feedback and experiences to find the optimal way to improve performance; it uses a reward system for good performance.
Fidelity	Fidelity represents the level of detail of a system, or the degree to which measurements, calculations, or specifications approach the true value or desired standard. There are five levels of fidelity. At level 1, the twin has low accuracy and can be considered a conceptual model. At level 2, the twin has a low to medium range of accuracy and can be used to extract measurements. At level 3, the twin has a medium range of accuracy and can be used as a reliable representation of the physical world. At level 4, the twin can provide precise measurements, and at level 5, the twin has a high degree of accuracy and can be used in for life safety and critical operational decisions. Fidelity, therefore, depends on the requirements of a given asset operator, rather than constituting an absolute property of a digital twin.

Source: ARUP (2019).

The better able a digital twin is to duplicate the physical object, the more likely it is that efficiencies and other benefits will accrue. For instance, in manufacturing, where we find more highly instrumented devices, digital twins might more accurately simulate how these devices have performed over time. This, in turn, could help companies predict future performance and possible failure (Shaw & Fruhlinger, 2019).

10.3.2.2 Digital Twin, Cyber-Physical System, and Internet of Things

Although digital twins, CPS, and IoT all use networking and sensors, the digital twin is a different (albeit related) concept, as shown in Figure 10.19 (Lua et al., 2019).

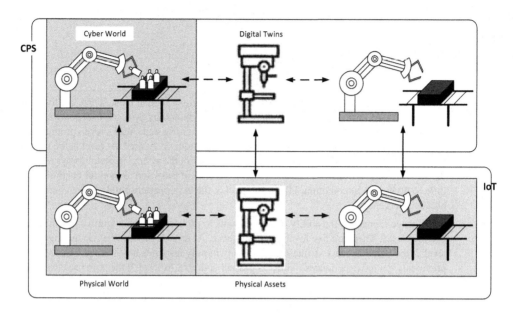

FIGURE 10.19   Relationship between digital twin, CPS, and IoT (Lua et al., 2019).

A CPS is characterized by a physical asset and its digital twin. By contrast, a digital twin is limited to the digital model, not the twinning physical asset, although a digital twin cannot live without its twinning asset in the physical space. In other words, the digital twin represents the prerequisite for the development of a CPS.

IoT refers to connections between a network of physical assets through which data can flow between themselves. The connections are made possible by the secure implementation of computer networks, the Internet, and communication protocols. However, despite the connectivity, IoT does not include the idea of digital models in cyberspace. IoT is the infrastructure in the physical space for connecting physical assets (Lua et al., 2019).

10.3.2.3   Enabling Technologies

Digital twin enabling technologies can be slotted into five categories: physics-based modeling, data-driven modeling, Big Data cybernetics, infrastructure and platforms, and human-machine interface. This section addresses the challenges of using these various enabling technologies; Table 10.4 gives an overview (Rasheed et al., 2020).

Physics-based modeling: To this point, the engineering community has been driven mostly by a physics-based modeling approach (see Figure 10.20 for the hierarchical stages). This approach consists of observing a physical phenomenon of interest, developing a partial understanding of it, putting the understanding in the form of mathematical equations, and ultimately solving them. Because of partial understandings and numerous assumptions along the line, from observation of a phenomenon to solution of the equations, a large portion of the physics can potentially be ignored. The physics-based approach can be subdivided broadly into experimental and numerical modeling.

Data-driven modeling: While physics-based models are the workhorse in the design phase, with the abundant supply of data in a digital twin context, open source cutting edge and easy-to-use libraries (tensorflow, torch, and openAI), cheap computational infrastructure (CPU, GPU, and TPU), and high-quality, readily available training resources, data-driven modeling is becoming very popular. Compared to the physics-based modeling approach, as illustrated in Figure 10.21, this approach is based on the assumption that since data are a manifestation of both known and unknown physics, by developing a data-driven model, we can account for the full physics.

TABLE 10.4

Mapping between Common Challenges and Enabling Technologies

Challenges	Enabling Technologies
Data management, data privacy and security data quality.	Digital platforms, cryptography and block chain technologies, Big Data technologies.
Real-time communication of data and latency.	Data compression, communication technologies like 5G and Internet of Things technologies.
Physical realism and future projections.	Sensor technologies, high fidelity physics-based simulators, data-driven models.
Real time modeling.	Hybrid analysis and modeling, reduced order modeling, multivariate data-driven models.
Continuous model updates, modeling the unknown.	Big Data cybernetics, hybrid analysis and modeling, data assimilation, compressed sensing and symbolic regression.
Transparency and interpretability.	Hybrid analysis and modeling, explainable artificial intelligence.
Large scale computation.	Computational infrastructure, edge, fog and cloud computing.
Interaction with physical asset.	Human machine interface, natural language processing, visualization augmented reality and virtual reality.

Source: Rasheed et al. (2020).

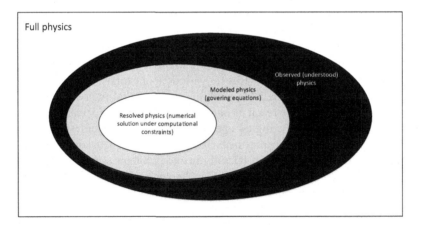

FIGURE 10.20 Physics-based modeling: based on first principles but can only model part of known physics due to assumptions at different stages (Rasheed et al., 2020).

Big Data cybernetics: The objective of cybernetics is to steer a system toward a reference point. To achieve this, the output of the system is continuously monitored and compared against a reference point. The difference, called the error signal, is then applied as feedback to the controller which, in turn, generates a system input that can direct the system toward the reference point. At times, the quantity of interest that is required to compare against the reference cannot be measured directly and hence has to be inferred from other quantities that are easier and cheaper to measure. With an increase in computational power and availability of Big Data, there are two major improvements

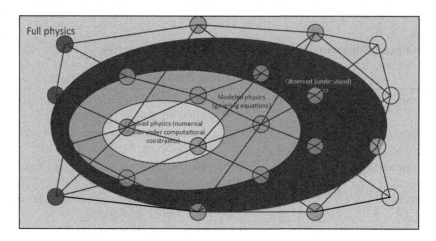

FIGURE 10.21 Data-driven modeling: data are a manifestation of both known and unknown physics (Rasheed et al., 2020).

possible within the cybernetics framework. First, the controller can be improved by increasing its complexity to incorporate more physics. Second, the Big Data can be used for a better estimation of the quantity of interest.

Big Data technologies: The infrastructure for storing and processing high volume data has advanced considerably. Many available platforms are available to handle Big Data projects in terms of blending, integration, storage, centralized management, interactive analysis, visualization, accessibility, and security. Many information technology (IT) vendors benefit from Hadoop technology, which allows us to execute tasks directly from its hosting place without copying to local memory.

The IoT is becoming increasingly popular to develop smart technologies in sectors ranging from healthcare to agriculture, from transportation to energy. The availability of high-speed Wi-Fi Internet connection, digital machines, low-cost sensors, and the development of ML algorithms to perform real-time analysis has contributed to enabling IoT. The leading IoT vendors provide reliable and innovative platforms to set up IoT devices for various applications.

Communication technologies: A reliable working of any digital twin will require information arising from different components to reach its intended target on time. For example, during a robotic surgery, the action of a surgeon in a digital operating room should manifest into action in reality without any latency. With so many sensors, there is a problem of fast data communication. The state-of-the-art communication technology like 4G will run into problems as more and more devices start sharing the limited radio-frequency spectrum. 5G technology with a wider frequency range might accommodate many more devices but that requires communication at much higher frequencies (30–300GHz) than those currently used by our mobile networks.

Computational infrastructures: According to Moore's Law, the performance and functionality of computers/processors can be expected to double every two years with the advances in digital electronics. In scientific computing applications, the speeds of both computation and communication have substantially increased over the years. However, communication becomes more challenging as the number of processing elements increases and/or the number of grid points decreases within processing elements; this can constitute a major bottleneck in exascale computing systems.

Digital twin platforms: Kongsberg Digital is a company delivering platform as a Service (PaaS) and software as a service (SaaS) services for energy, oil and gas, and maritime industries. Kongsberg partners with cloud vendors, including Microsoft Azure, to provide enabling IT services (Ullevik, 2017). In 2017, Kongsberg launched its open digital ecosystem platform, called KognifAI. It combines a group of applications on the cloud focusing on optimal data accessibility and processing. KognifAI is built on cybersecurity, identity, encryption, and data integrity. Its main infrastructure can be used to easily scale applications and services. KognifAI offers digital twin solutions in maritime, drilling and wells, renewable energies, etc. Kongsberg's dynamic digital twin combines safety with fast prototyping and implementation, and connects offshore and onshore users in oil and gas industry. It can provide a model to represent an existing physical system, a planned installation (e.g., greenfield digital twin), or maintenance and repair (e.g., brownfield).

Human-machine interface: As the demarcation between humans and machines fades in the context of digital twin, there will be a need for more effective and fast communication and interaction. While augmented/virtual reality will be required to create a detailed visualization of the assets, natural language processing and gesture control will be a common mode of interaction.

10.3.2.4 Digital Twin and Simulation

In very simple terms, a digital twin is a digital replica of a real-world object. It is something like simulation but is much more. A digital twin is a high-fidelity representation of the operational dynamics of its physical counterpart, enabled by near real-time synchronization between cyberspace and physical space. The operational dynamics are critical elements of a digital twin because a twin's behavior is based on near real-time data coming from the actual physical counterpart. By contrast, simulation focuses on what could happen in the real world (what-if scenario), not what is currently happening. In the manufacturing context, a digital twin can be used for monitoring, control, diagnostics, and prediction, other than just simulation (Lua et al., 2019).

Although simulation is used in the engineering field to investigate well-defined problems, it has certain limits. For example, most simulations solve a specific request, such as design or online optimization (Schluse & Rossmann, 2016). Because the structured and complicated process of model simulation alters only one aspect, more complicated data exchange processes may call for a new simulation setting phase. By contrast, a digital twin can be used for the entire lifecycle of an object in real time, as it is a digital representation of an existing physical object. Thanks to sensors connecting the physical object with its digital copy, a digital twin can represent the object's state and show its changes in real time.

When simulation techniques are applied to digital twins, the result is a proper controlling and monitoring system. A digital twin simulation interface (DTSI) responds to the needs related to the product or the asset lifecycle simulation. The DTSI can be used to understand what is happening on the shop floor in real time and to update the real system with improvements suggested by the digital model.

To sum up: the digital twin follows the real system in the physical world in real time to be able to monitor, adjust, and optimize real processes, anticipate failures, and increase efficiency using simulation (Bevilacqua et al. 2020).

10.3.2.5 Machine Learning and Digital Twin

ML and digital twins work together in today's smart factories, Digital twins make use of real data to simulate the real production plant. ML models require real data to acquire knowledge, to be trained, and then to be tested to validate their effectiveness. Digital twins work in parallel with real production plants, using the validated ML models and simulating new production situations to identify available improvements.

The digital twin verifies the concreteness of the ML model in unusual situations; for example, it can be used to predict risk when it is not possible to test the ML model with real data. In these cases, the digital twin enables operators to test the model with simulated data without any real risk to workers. The validated ML models then provide predictions for recognized situations, which are useful for the digital twin to simulate change. On the one hand, the digital twin improves the capacities of the ML model; on the other hand, the ML model automates the digital twin.

ML is more complex in the context of digital twins for the following reasons (Bevilacqua et al., 2020):

- Data availability: ML requires information. Without good quality data, it is impossible to develop and implement a ML model. Therefore, data must be checked. They must be analyzed, visualized, validated, and understood. Many organizations still lack digitalized data, relying on analogic PLCs and a few digital monitoring systems.
- Environment complexity: Industries are socio-technical systems difficult to simulate and predict because they are variable and irregular. Processes, resources, and people interact. Some interactions are linear, cause-effect, and predictable, while others are nonlinear, complex, and unpredictable.

10.3.2.6   Augmented and Virtual Reality and Digital Twin

A digital twin can be useful for AR and VR, because the twin provides a virtual and realistic view of the environment in which historical and real-time data are integrated with people. However, given the enormous amount of data and information generated by the digital twin in real time, it is difficult to provide users with the information they need in an easy and intuitive manner. Note that AR does not replace the physical world; rather, it allows users to see the physical world with overlapping virtual objects. It also gives them the opportunity to interact with the physical world to perform specific tasks, or it alerts them to risks.

An architecture integrating the digital twin and AR/VR comprises the following (Bevilacqua et al., 2020):

1. Calibration: To obtain a clear and intuitive data visualization, the 3D or two-dimensional (2D) models must be perfectly aligned with the physical part. This is called calibration.
2. Control process: The control process allows users to interact with both the physical and virtual part of the digital twin. Users can apply this information to support decision-making and directly control the physical part through the AR device.
3. Augmented process: The AR device receives data from the virtual part, and after calibration, it correctly presents these data to users. The process must provide users with an intuitive and clear AR view of the information coming from the digital twin.

There are some remaining challenges to solve before AR is more widely used in manufacturing (Bevilacqua et al., 2020):

1. Managing real-time data: An enormous amount of real-time data are exchanged between the manufacturing process, the cloud, and the AR device. These must be correctly managed to support potential users.
2. Developing 3D and 2D modeling: The ability to recognize, track, and follow the target object through 3D and 2D modeling will improve the quality of AR utilization.
3. Improving reliability: An AR device used in manufacturing must be reliable and robust because the work environment is dirty, noisy, and dangerous.
4. Ensuring user cooperation: At any one time, there will be multiple users and operators, and this requires cooperation. The AR infrastructure must be flexible enough to permit data exchange between multiple devices.

10.3.2.7   Cloud Technology and Digital Twin

The digital twin can be understood as a group of complex systems composed of mathematical models, computational methods, and software services permitting real-time synchronization between a virtual system and a real process. In other words, it takes advantage of various ways of analyzing data coming from different types of sensors installed on the real system.

Creating a digital twin of a process or a part of an industrial plant requires a set of computational services that represent models of their interactions. Each requires specific computational resources.

One solution is to use a cloud infrastructure that offers both flexibility and high processing performance. For example, a cloud system that uses the "Containers as a Service" model can manage a high amount of data while supporting the algorithm execution container. A few cloud vendors offer this model for digital twin implementation, including IBM, Amazon, and Microsoft Azure (Bevilacqua et al., 2020).

The digital twin is not yet widely used in the industrial context. Companies thinking about developing a digital twin mention concerns about data privacy, security, and connectivity (Bevilacqua et al., 2020).

10.3.2.8 Extending the Relevance of Predictive Maintenance

Predictive maintenance concepts such as condition-based monitoring are not new, but asset digital twins and predictive analytics are encouraging more widespread adoption by providing more visually meaningful predictive maintenance foresight.

Interest in predictive maintenance strategies has increased in line with other predictive analytics technologies, so automation, PLM, and original equipment manufacturer vendors are focusing their marketing efforts on asset digital twins. Asset digital twins can add substantially more value in situations where multiple asset populations are deployed across different geographies and/or operating conditions. Twins enable companies to harvest even greater insights to enable cost reductions through more effective organization and scheduling of field service and engineering resources (Aitken, 2020).

10.3.2.9 Operational Process Digital Twin: Diagnostic and Control Capability

The digital twin gives dynamic diagnostic capabilities, allowing companies to analyze end-to-end process performance in real time. The twin links sensor data from the physical world to analytical and data-mining algorithms to better understand and manage process performance. The use of real-time PLC devices and proportional–integral–derivative (PID) process control loop systems is not new, as these automation concepts were implemented across many manufacturing sites during Industry 3.0. However, Industry 4.0 is unlocking a wider opportunity to apply these technologies at scale in a cost-effective way – which is essential given the increasing market urgency.

10.3.2.9.1 The Importance of Diagnostics and Control

At recent technology exhibitions, Microsoft has been demonstrating its internal manufacturing IIoT capabilities, showcasing the business and operational performance tracking implemented across their global manufacturing network. They can drill down into performance detail and problem root cause level in real time using live data feeds connected across Microsoft Azure and a customized suite of Microsoft Power BI analytics.

Using this dynamic diagnostic ability, they can more rapidly determine the appropriate corrective action and reduce costs associated with sub-optimal performance. While analytical functionality is in-built within Microsoft's type of diagnostic digital twin, automated corrective control actions aren't yet. That will require additional functionality linking analytical results to algorithms driving appropriate methods of process control actuation (e.g., triggering machine alerts or changing conveyor speed). ML and AI methods will progressively offer enhanced algorithmic intelligence to support such decisions.

Digital twins delivering diagnostics together with control capability can deliver impressive results by supporting the optimization of real-world asset and process performance in the following ways (Aitken, 2020):

- Capture and analyze operational data from assets and processes in real time.
- Use those data to present and chart operational statistics.
- Trigger alerts to prompt closer monitoring.

- Diagnose issues and identify performance improvement actions.
- Link physical assets and control systems containing control logic and algorithms to automatically adjust or optimize process parameters via an actuation loop.

A diagnostic and control digital twin enables more effective execution of an operational plan. It supports increased productivity, improves control, and can help drive more stable and reliable operational performance (Aitken, 2020).

10.3.2.10   Operational Process Digital Twin: Predictive Capability

A predictive digital twin goes a step further by unlocking opportunities beyond using current state data to drive asset and operational process performance. It is created using specially designed predictive simulation software and can be used to evaluate planned or potential future scenarios. This allows decision-makers to test and understand the impact of each scenario, identifying opportunities and risks, without incurring any cost.

Such foresight is invaluable in de-risking and optimizing key business and operational decisions, for example, evaluating digital technology investments, refining resource planning, testing new business process models, and improving schedule performance.

Predictive digital twins work by modeling individual events using a time-based engine, considering resources, constraints, and interactions with other events. They reflect the process rules and variability affecting the behavior of the real-life systems and complex operating environments. In this way, they mirror the dynamic processes experienced in actual businesses, whether manufacturing facilities, airports, or contact centers.

The level of visualization deployed within a predictive digital twin should match the intended function. In some cases, a fully immersive virtual reality model of a complete factory is required, for example, getting investor buy-in for an innovative new factory design concept. In other cases, a high-level, animated ARIS 2D process model may suffice, for example, when creating an enterprise digital twin (EDT) for users of an administrative business process who are mainly interested in analytical outcomes (Aitken, 2020).

10.3.2.11   Gamify Decision-Making with Prescriptive Analytics

The predictive digital twin enables decision navigation at a business level by connecting key business questions to the right data sources. Questions like "What if?", "What's best?" and "How do we?" can be answered using dynamic models of real business and operational processes.

Essentially, decision-making can be gamified when different scenarios can be run across the short, medium and long term. This helps build an informed investment case for technology implementations because they are driven by the business strategy. As a result, the business is designed to meet future challenges.

Most people would like a simple solution for complex problems – automated decision-making. The hope is that digital twins and AI will provide it. This is not yet the case, but it is likely to be a reality in the not-so-distant future. Several advanced simulation software packages (like Lanner's WITNESS Horizon) provide optimization capabilities together with scenario experimentation functionality. This enables a predictive digital twin to hunt for the optimum answer, such as balancing the expenditure and resources required to achieve a targeted outcome. This is known as prescriptive rather than predictive analytics (Aitken, 2020).

This type of digital twin analytics can provide the following:

- Ensure strategic business questions drive technology innovation plans.
- Facilitate and de-risk new business model testing and planning.
- Get answers to questions like "What should our business look like in 5 years?" and "How do we best manage ourselves to meet predicted demand?"

- Validate plans and process changes.
- Make better decisions earlier by anticipating future opportunities and risks.
- Link machine, system, and process data to wider business analysis.
- Connect operational technology and IT to the rest of the value chain, thus facilitating planning and digital transformation.
- Justify new digital technology investment plans.
- Identify the specific data that must be captured to optimize operational and business performance (rather than collecting and analyzing all available data).
- Connect to other systems, for example, ERP and MES, to improve scheduling quality.

This digital twin supports the testing and optimization of business and operational planning as opposed to controlling operational execution. It can also link the business level to operational processes and key asset data feeds to maximize end-to-end business value. This ensures the Industry 4.0 technology and data strategies are strategically aligned with the business, thus turning Big Data into smart data (Aitken, 2020).

10.3.2.12   Enterprise Digital Twins

The IoT is driving the creation of physical assets and product digital twins that add value for their users, but business managers are left with a requirement for digital twins that represent and support the control of their areas of responsibility, that is, the processes and resources that allow their business to function.

These higher level, more abstract, business digital twins can use lower level asset digital twins to pull data they require to represent the real-world system. They require an understanding of process and decision logic to add context and insight for those using them. These are known as EDTs. Their main objective is to capture the holistic business operating model for control and management purposes.

Many organizations are now using EDTs to coordinate the critical interdependencies between their resources (people, processes, and technology) so that they can understand, manage, and optimize complex digital business transformations.

Importantly, EDT can be linked to Level 1 and Level 2 digital twins to maximize their relevance and decision-making value. Businesses will increasingly want to harness the insights and foresights (via predictive digital twins) provided by each level of digital twin within a connected digital twin ecosystem (Aitken, 2020).

This type of digital twin gives companies the following capabilities:

- Understand, manage, and adapt to business transformation in a coordinated manner.
- Design and implement digital business transformation roadmaps.
- Better manage digital business transformation.
- Understand the impact of changes on customer experience.
- Test and de-risk scenarios before changes are implemented across the business.
- Track performance against benchmarked or target key performance indicators (KPIs).
- Optimize the business by balancing resources and performance across end-to-end processes.

EDTs should be used to take control of a business when performance is inconsistent with expectations, yet the root causes are either unknown or too challenging to resolve. If too much time is being spent managing exceptions rather than planning strategic improvements, or the level of risk is blocking transformational action, the EDT can resolve this and unblock the route forward (Aitken, 2020).

10.3.2.12.1   Predictive Twin

Once device virtualization is implemented, a functional abstraction is obtained for interacting with the device. For example, the device can be interrogated or controlled through the virtualization abstraction. The model can react to the current status of the device. However, merely reacting to a situation is neither sufficient nor optimal. For example, just knowing that a machine has developed a problem is good. But knowing that a machine is likely to develop a problem in future is even more important as it gives the user time to deal with the problems before they occur (Oracle, 2017).

Behavioral and predictive modeling can be done in two ways (Oracle, 2017):

Physics-based approach: A model can be created using physics-based approach utilizing the knowledge of the exact design as well as manufacturing parameters of the asset (see Section 10.1.13.1). Techniques such as finite element analysis are often used to create fairly accurate models that typically answer the "what if" questions. For example, using such models, users can estimate the stress patterns on various parts of a machine for a given set of loading conditions. In practice, building these models takes significant effort by the team that designed the products to create models with reasonable fidelity. The math involved in creating the FEM models tends to be fairly complex; hence, these models tend to be fairly static and they do not adapt to a complex and continuously changing environment. The biggest drawbacks of these models include:

- It often takes the original designers of the machines to create these models; customers who buy an assembled product cannot put together a model that suits their needs.
- While these models can answer questions about performance on various load conditions, they do not provide guidance for fixing the issues.
  Analytical/statistics-based models: A predictive model can be built using ML techniques without necessarily involving the original designers. A data analyst can create a predictive model merely based on external observations of the machine. This option is far more practical because it enables the creation of various models based on the end customer's needs. Another significant aspect is that they take the whole system into account, that is, the contextual data. This allows the creation of models that are far superior in their effectiveness and usability than physics-based models.

Not all predictive models are created equal. There is a spectrum of complexity based on the user's goals.

10.3.2.12.2 *Predictive Digital Twins in Use*
Predictive digital twins make it possible to evaluate and justify future-state scenarios, and many leading firms in the automotive and manufacturing sectors already mandate the use of this technology for stress-testing business and operational change plans.

As Industry 4.0 momentum increases, companies across many sectors and geographies are increasingly active in using this technology. In Japan, Mitsubishi has created WITNESS Horizon models to interface with MES data feeds to optimize process design. Meanwhile, companies like Hayward Tyler and Meggit are linking predictive twins with their business and operational planning systems, such as ERP and MES. These businesses are connecting their value chains to predictive twins to enable decision navigation at a business level (Aitken, 2020).

10.3.2.13 Digital Twin Approach to Predictive Maintenance
Predictive maintenance solutions relying on digital twins can more precisely monitor equipment health and recognize anomalies in a timely manner. However, there are some limitations to their deployment.

The level of digitalization in industry is rapidly increasing, especially with the current focus on predictive maintenance. Given the obvious value of IoT-driven predictive maintenance solutions, many organizations are venturing into the realm of predictive maintenance. However, digital twin technology is not yet part of this trend.

A simple definition of digital twin technology is the following: it involves creating a virtual representation of a physical asset or a system (e.g., an industrial machine, a production line, or a factory) in order to model its state and simulate its current and future performance. Digital twins continuously learn and are powered by ML algorithms. This allows them to adapt to changes in the state of the physical object.

In an industrial setting, digital twins have many practical uses. They can be used to improve product design, monitor equipment health, or simulate operations.

The following provides a simplified explanation of how digital twins enable predictive maintenance (Shiklo, 2020).

1. Building a model: Creating a digital twin requires building an accurate 3D model of the physical asset. Modeling experts collaborate with mechanical, electrical, and process engineers to describe and virtually present physical properties of the asset and its components.
2. Inputting real data from the physical asset: The 3D model is powered with IoT data from sensors attached to the asset, including records about performance, condition, and environment (e.g., temperature).
3. Adding contextual data: To improve functionality, the digital twin software is integrated with enterprise and shop floor management systems. To give one example, contextual data (e.g., regulatory, financial, operational data) from ERP systems will allow the digital twin to predict how a pump will function given a variety of external conditions.
4. Adding historical data: The model also requires historical data on the asset's failure modes and criticality.
5. Activating the digital twin: The digital twin-based predictive maintenance software takes in real-time sensor data on the health and working conditions of an asset and analyzes them against historical data on the asset's failure modes and their criticality, and contextual data fetched from enterprise and shop floor management systems. A neural network detects abnormal patterns in the incoming sensor data and reflects the patterns in predictive models, which are then used to predict failures. If an asset's current configuration is likely to lead to a failure, the digital twin localizes the issue, assesses the criticality, notifies the technicians, and recommends an action.

Digital twins can predict failure but they have other maintenance-related uses as well, including:

- Calculating maintenance-related KPIs: By combining historical data on failures, risk factors, machine configuration, and operating scenarios, a digital twin can calculate mean time between failures (MTBF), RUL, end of life (EoL), and so on.
- Forecasting the behavior of equipment under a variety of circumstances: The digital twin is an accurate real-time model of equipment condition and performance and, as such, can be used to run simulations and predict how equipment will respond to certain factors, including runtime or severe operating conditions.
- Simulating maintenance scenarios: Maintainers can use digital twins to test various solutions for different maintenance scenarios and see if they work before applying them.

10.3.2.13.1 Remaining Challenges
Although digital twin-enabled predictive maintenance stands to benefit an organization, it can be challenging to make the switch.

First, there is a need for accuracy and this can be problematic. To be effective, a model must precisely reflect the physical object's properties, including mechanical and electrical ones. This requires input from a variety of stakeholders – facility managers, process engineers, electrical engineers, equipment vendors, and others – adding another layer of complexity to deployment.

Second, detailed data on equipment failures are required. These data should be collected over an extended period (e.g., a year) so that the degradation process can be observed. In many cases, such data are not available.

Third, if there is any change in equipment configuration or state, the digital twin must be remodeled; this includes modifying its underlying algorithms. Such modifications, whether at the machine level (replacing original parts with made-to-order ones) or the organizational level (changes to operating policy), are not always reflected in factory specifications and, thus, cannot be precisely simulated.

Fourth, deploying digital twin-based predictive maintenance is time-consuming and labor-intensive.

Nevertheless, once it is in place, the technology will help organizations recognize disruptions in asset performance in a timely manner. They will be able to forecast potential problems and simulate various maintenance scenarios to find the best possible solutions. Ultimately, it will help them eliminate machine downtime, reduce equipment maintenance costs, improve equipment reliability, and extend equipment life (Shiklo, 2020).

10.4   FAILURE FORECASTING AT THE SYSTEM LEVEL

10.4.1   FAILURE FORECASTING

When building equipment, manufacturers either build it to avoid failures, even though this increases the original capital costs, or build it and sell it at a low or break-even cost, expecting to make profits with the sale of replacement parts (Reliabilityweb.com 2020). Thus, end-users need to know the forecasted failures. One way to do this is to look at spare parts sold for similar equipment and estimate the number of units they will have working.

A failure is an event which renders an asset unavailable for its intended purpose during a certain time interval. The failure can be partial or total, or even catastrophic; it can be sudden, intermittent, or gradual (Reliabilityweb.com, 2020).

Failures cost money and often cause safety and/or environmental problems. With accurate failure forecasting, organizations can save time and money. They can anticipate expected failures now and in the future. Failure forecasting analysis can be performed during equipment design. Not all failures can be caught ahead of time: unexpected failure modes arise during operation, causing loss of service and high costs. This type of event calls for follow-up analysis (Reliabilityweb.com, 2020).

10.4.1.1   Anomaly Detection and Analysis

In complex communication systems, such as core routers, data are collected in the form of time-series. Three kinds of techniques are used to detect anomalies in time-series data.

1. Distance based anomaly detection uses a distance measure between a pair of time-series instances to represent the similarity between these two time-series. The smaller the overall distance is, the closer this pair of time-series instances will be. Instances far away from others will be identified as abnormal.
2. Window-based anomaly detection divides time series instances into overlapping windows. Anomaly scores are first calculated per window and then aggregated to be compared with a predefined threshold. When the overall anomaly score of a single time-series instance significantly exceeds a predefined threshold, the instance is identified as abnormal.
3. Prediction-based anomaly detection is the third technique. First, a machine-learning-based predictive model is learned from historical logs. Next, predicted values are obtained by feeding test data to this predictive model. These predicted values are then compared with the actual measured data points. The accumulated difference between these predicted and the actual observations is defined as the anomaly score for each test time series instance.

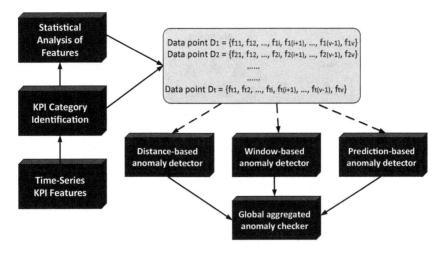

FIGURE 10.22   Depiction of feature-categorization-based hybrid anomaly detection (Jin et al., 2016).

However, a single class of anomaly detection methods is effective for a limited number of time-series types. A feature-categorization-based hybrid method may be more appropriate. Figure 10.22 illustrates the proposed technique. First, time-series data of different features extracted from the core router system are fed to a KPI-category identification component. As features belonging to different KPI categories often exhibit significantly different statistical characteristics across the timeline, natural language processing techniques are used to ensure different KPI categories, such as configuration, traffic, resource type, and hardware, can be identified. Features belonging to different KPI categories may have similar trends or distributions across time intervals; therefore, a statistical analysis component is needed to ensure all features exhibiting similar statistical characteristics are placed in the same class. After these steps, a data point, Dt, with v features can be divided into different groups, $C_a, C_b,\ldots, C_k,\ldots, C_r$, where each group has different statistical characteristics. Next, each group of features is fed to the anomaly detector that is most suitable for this type of feature. Finally, the results provided by different anomaly detectors are aggregated, so an anomaly can be detected in terms of the entire feature space (Jin et al., 2016).

Although the proposed feature-categorization-based hybrid method can help to detect a wide range of anomalies, not all anomalies are useful and necessary for predicting system failures. First, the temporal and spatial localities of neighboring components lead to co-occurrences of similar anomalies. Second, some anomalies are caused by workload variations or temporary external noise, making them irrelevant for predicting system failures. Since the number of possible anomalies will increase from hundreds to tens of thousands when more new features are identified and extracted from the raw log data, anomaly analysis is needed to remove irrelevant and redundant anomalies before predicting failures (Jin et al., 2016).

10.4.1.2   Failure Prediction

Figure 10.23 shows the temporal relationship between faults, anomalies, and failures. Assume a fault occurs in the system at time point t_r. After a period of time, a wide range of anomalies begin to appear at time point t_{as}. Finally, at time point t_f, the system encounters a fatal failure and crashes. Two important time intervals are defined here. First, δt_1, referred to as the lead time, is the time interval between the occurrence of the last anomaly and the occurrence of the predicted failure. It is defined as $\delta t_1 = t_f - t_{ae}$. Only if this lead time is larger than the time required to take preventive actions can a prediction become useful in reality. Second, the parameter δt_d is defined as the time interval between the occurrence of the first and last anomaly, that is, $\delta t_d = t_{ae} - t_{as}$. Since the failure

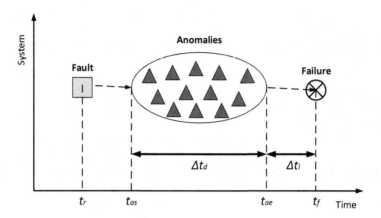

FIGURE 10.23   Temporal relationship between faults, anomalies, and failures (Jin et al., 2016).

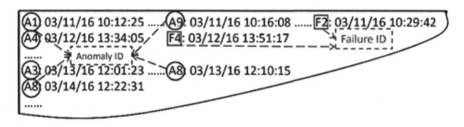

FIGURE 10.24   An example of anomaly event sets (Jin et al., 2016).

prediction is based on the detected anomalies in the system, δt_d can be considered the temporal length of the dataset (Jin et al., 2016).

Using the proposed anomaly detector and analyzer, representative anomalies can be identified and recorded. When they are correlated with logs of system failures, two types of anomaly event sets can be formed: failure-related anomaly event sets and non-failure-related anomaly event sets. An example of these two types of anomaly event sets is shown in Figure 10.24. Here, A_i represents the ID of each anomaly, and F_j represents the ID of each failure. The failure-related anomaly event set consists of records that always end with a failure event F_j, while the non-failure-related anomaly event set consists of records that do not have any failure events (Jin et al., 2016).

An efficient failure predictor should not only predict whether failures will occur, but also predict the type/category and occurrence time of those failures. Therefore, as shown in Figure 10.25, the proposed failure predictor consists of two main components, the classifier and the regressor, so that both the category and the lead time of failures can be predicted. First, the historical logs including both failure-related and non-failure-related anomaly events are fed as training data to both the classifier and the regressor to build corresponding learning models. Second, a set of newly detected anomalies is fed to these learning models. Finally, the learned classifier outputs which type of system failures will be triggered by the current anomalies, and the learned regressor outputs the predicted lead time for this type of system failure (Jin et al., 2016).

One key step implicit in Figure 10.25 is building training datasets from historical anomaly event sets for both the classification component and the regression component. Suppose a set of anomalies $A = \{A_1, A_2, \ldots, A_N\}$ and a set of system failures $F = \{F_1, F_2, \ldots, F_M\}$ have been identified from historical log H. The training dataset D for the classification component can then be built. Each record H_i in the historical log can contain one or more anomalies and either no failure or one failure. If the

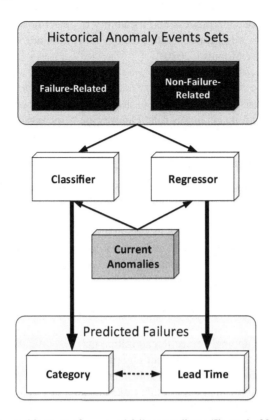

FIGURE 10.25 Overview architecture of proposed failure predictor (Jin et al., 2016).

anomaly A_j appears in the record H_i, $D_{ij} = 1$; otherwise $D_{ij} = 0$. Note that $D_{i(N+1)}$ represents the failure category of the record H_i. If the failure F_k appears in the record, H_i, $D_{i(N+1)} = k$. If no failures occur in the record, $D_i(N + 1) = 0$. The process of building the training dataset T for the regression component is similar. The only difference is that the occurrence times of anomalies and failures needs to be included. If the anomaly A_j appears at time t_j in the record, H_i, $T_{ij} = t_j$; otherwise, $T_{ij} = 0$. If the failure F_k appears at time t_k in the record H_i, $T_{i(N+1)} = t_k$. If no failures occur in the record, $T_{i(N+1)} = 0$.

Different machine-learning techniques can be applied for classification and regression in a failure predictor. The support vector machine (SVM) algorithm offers several advantages, such as overfitting control through regularization parameters and performance improvement via custom kernels. Specifically, multiclass SVM can be applied for the classification component and support vector regression for the regression component (Jin et al., 2016).

10.4.1.3 Anomaly Detection Solutions for Predictive Maintenance of Industrial Equipment

The future lies in data or, more specifically, the ability to operationalize data in real time and translate those data into insights that can be acted on. Real-time data are especially important in the maintenance context. Predictive maintenance, powered by Big Data analytics and ML, is gaining popularity; however, most predictive maintenance techniques lack the following:

- Historical data showing a range of failure types.
- Normal equipment performance thresholds.
- Condition-based data suggesting what foreshadows failure.

A good equipment data collection process and the ability to detect normal and abnormal behavior in the collected data are the basis for accurate predictions. If both are in place, algorithms can be trained to make predictions (infopulse, 2020).

10.4.1.3.1   Key Technologies Underpinning Predictive Maintenance

In the past, organizations used regular equipment inspections to collect performance data, but IIoT solutions are now replacing humans in this role. Digitization is increasingly popular in industry for the following reasons:

- Increased data storage and improved data processing capacities are available on the cloud.
- Sensors, devices, and asset-level connectivity cost less and have better quality.
- Big Data analytics, ML, and AI technologies are commoditized.

Organizations can better capture deviations in system and equipment behavior. By combining historical and real-time sensor data, along with additional information on failure rates, they can predict anomalies and take proactive steps to mitigate them. Through predictive analytics, organizations can uncover previously hidden occurrences, boost the performance of individual assets, and maximize the performance of the entire supply chain (infopulse, 2020).

Depending on the type of data, anomaly detection algorithms can be developed using the methods shown in Table 10.5.

The Donut unsupervised anomaly detection algorithm may be particularly effective for analyzing data obtained from a single sensor. The Donut algorithm is based on the variational autoencoder (VAE). A VAE is a deep learning-based model able to produce new, unseen data, a distinctive feature making it particularly useful for image rendering tasks and anomaly detection. However, the Donut algorithm is more accurate than standard VAEs for unsupervised learning, and it detects a wider array of anomalies in the given data.

If multi-sensor data are being operationalized, the isolation forest technique, paired with visualizations, works even better than the Donut. In this technique, unsupervised models cumulatively assess the data across several metrics to find an anomaly. Then, they analyze the individual metrics to detect granular anomalies. With this two-pronged approach, organizations can capture

---

TABLE 10.5
Anomaly Detection Algorithms

Category	Method
Statistical anomaly detection	• Univariate. • Multivariate. • Time series model.
Machine learning anomaly detection	• Bayesian networks. • Markov models. • Neural networks including generative adversarial networks (GANs) and variational autoencoders (VAEs). • Isolation forest with visualizations. • Genetic algorithms. • K-nearest neighbor. • Decision trees. • Clustering and outlier detection.

Source: infopulse (2020).

very small deviations in performance and investigate them. Isolation forest has fewer memory requirements than other methods, the time to train is faster, and there is less linear time complexity than other methods, so results are fast, possibly in real time (infopulse, 2020).

10.4.1.3.2 Anomaly Detection Algorithms and Alert Fatigue

Merely installing sensors to gauge equipment performance can lead to alert fatigue. Not every deviation from the threshold value means a failure is imminent. Nor does the threshold-based analysis of individual sensors paint the full picture of equipment performance. And more sensors lead to more monitoring. Human operators may be overwhelmed by the volume of information coming their way and fail to recognize the important patterns hidden among the less significant ones. Here, anomaly detection algorithms are particularly useful: they help process the enormous amount of data. Most notably, they translate the data into actionable insights (infopulse, 2020).

10.4.1.3.3 Unsupervised Algorithms and Limited Data

It can be difficult (or prohibitively expensive) to measure some equipment parameters, such as extremely high temperatures or performance across a fleet of similar equipment. Standard predictive maintenance technologies don't solve the problem, but AI-driven anomaly detection does. For example, intelligent systems can run simulations. They can also perform highly accurate data interpolation based on other parameters to compensate for the lack of historical data (infopulse, 2020).

10.4.1.3.4 Anomaly Detection Algorithms and Real-Time Intelligence

State-of-the-art anomaly detection algorithms can gauge the performance of an almost unlimited number of machines and report on equipment anomalies in only a few minutes. Whereas predictive maintenance systems offer a long-term outlook on performance, anomaly detection algorithms provide real-time information (infopulse, 2020).

REFERENCES

Aitken A., 2020. Industry 4.0: Demystifying digital twins. Lanner, Future Proof.

ARUP, 2019. Digital Twin. Towards a meaningful framework. London.

Bevilacqua M., Bottani E., Ciarapica F. E., Costantino F., Di Donato L., Ferraro A., Mazzuto G., et al., 2020. Digital twin reference model development to prevent operators' risk in process plants. Sustainability 2020, 12, 1088.

Bolton A., Butler L., Dabson I., Enzer M., Evans M., Fenemore T., Harradence F. et al., 2018. The Gemini Principles: Guiding values for the national digital twin and information management framework. Centre for Digital Built Britain and Digital Framework Task Group.

Boschert S., Heinrich C., Rosen R. 2018. Next Generation Digital Twin, Proceedings of TMCE 2018, 7–11 May, 2018, Las Palmas de Gran Canaria, Spain Edited by: I. Horváth, J.P. Suárez Rivero and P.M. Hernández Castellano, Organizing Committee of TMCE 2018, ISBN 978-94-6186-910-4.

Boschert, S., Rosen, R. 2016. Digital twin: The simulation aspect. In P. Hehenberger, D. Bradley (Eds.), Mechatronic futures challenges and solutions for mechatronic systems and their designers (pp. 59–74). Springer-Verlag.

infopulse, 2020. Anomaly Detection Solutions for Predictive Maintenance of Industrial Equipment. January 17, 2020. www.infopulse.com/blog/anomaly-detection-solutions-for-predictive-maintenance-of-industrial-equipment-and-systems/#:~:text=With%20the%20help%20of%20real,merely%20the%20onset%20of%20it. Viewed: June 22, 2020.

Itea, 2011. FMI PLM interface specification for product lifecycle management (PLM) of modeling, simulation and validation information, MODELISAR (ITEA 2 – 07006), V1.0, 31 March 2011.

Jin S., Chakrabarty K., Zhang Z., Chen G., Gu X., 2016. Anomaly- detection-based failure prediction in a core router system. VALID 2016: The Eighth International Conference on Advances in System Testing and Validation Lifecycle.

Krishna, Addepalli V., Balamurugan M., 2019. Security mechanisms in cloud computing-based Big Data. Handbook of Research on the IoT, Cloud Computing, and Wireless Network Optimization. IGI Global, 2019. 165–195.

Lee J., Lapira E., 2013, Predictive factories: The next transformation, Manufacturing Leadership Journal, 20(1), 13–24.

Lua Y., Liub C., I-Kai Wang K., Huanga H., Xua X., 2019. Digital twin-driven smart manufacturing: Connotation, reference model, applications and research issues. Robotics and Computer Integrated Manufacturing, 61, 101837.

Mathworks, 2020. What is a Digital Twin? 3 things you need to know, www.mathworks.com/discovery/digital-twin.html. Viewed: May 28, 2020.

Mitchell, M., 2009. Complexity: A guided tour. Oxford University Press.

Modelica Association, 2019. FMI Functional Mock-up Interface 2.0.1 October 2, 2019.

Nature, 2008. Language: Disputed definitions. Nature, 455, 1023–1028.

Oracle, 2017. Digital twins for IoT applications: A comprehensive approach to implementing IoT digital twins. Oracle White Paper. January 2017.

Ove Erikstad S., 2017. Merging physics, Big Data analytics and simulation for the next-generation digital twins. Norwegian University of Science and Technology. HIPER 2017, High-Performance Marine Vehicles, Zevenwacht, South-Africa, 11–13 September 2017.

Perrow C., 1984. Normal accidents: Living with high-risk technologies. Basic Books.

Post J., Groen M., Klaseboer G., 2017. Physical model based digital twins in manufacturing processes. Forming Technology Forum 2017. October 12 & 13, 2017, Enschede, The Netherlands. Jan Post J., Digital Fabrication.ENTEG, PO box 221, 9700 AE Groningen, University of Groningen, The Netherlands.

Rasheed A., San O., Kvamsdal T., 2020. Digital twin: Values, challenges and enablers from a modeling perspective. January 28, 2020. Digital Object Identifier.

Reliabilityweb.com, 2020. Failure forecast. Reliabilityweb.com "A Culture of Reliability". https://reliability web.com/articles/entry/failure_forecast. Viewed: May 29, 2020.

Sargut, G., McGrath, R. G., 2011. Learning to live with complexity. Harvard Business Review, 89(9), 68–76.

Schluse M., Rossmann J., 2016. From simulation to experimentable digital twins: Simulation-based development and operation of complex technical systems. In Proceedings of the 2016 IEEE International Symposium on Systems Engineering (ISSE), Edinburgh, UK, October 3–5, 2016.

Seow Y., Rahimifard S., Woolley E., 2013. Simulation of energy consumption in the manufacture of a product, International Journal of Computer Integrated Manufacturing, 26(7), 663–680.

Shaw K., Fruhlinger J., 2019. What is a digital twin and why it's important to IoT. Network World. January 31, 2019. www.networkworld.com/article/3280225/what-is-digital-twin-technology-and-why-it-matters. html. Viewed: May 29, 2020.

Shiklo B., 2020. A digital twin approach to predictive maintenance. Informationweek IT Network. www. informationweek.com/big-data/ai-machine-learning/a-digital-twin-approach-to-predictive-main-tenance/a/d-id/1333331#:~:text=Along%20with%20the%20prediction%20of%20failures%2C%20 digital%20twin%20technology%20provides%3A&text=Combining%20historical%20data%20 about%20failures,EoL%2C%20MTBF%2C%20and%20more. Viewed: June 22, 2020.

Sidis W., 2010. Compact points diagnostics system. Siemens AG 2010 Order No: A19100-V100- B927-X-7600.

Troyer, L., 2015. Expanding sociotechnical systems theory through the trans-disciplinary lens of complexity theory. In F.-J. Kahlen, S. Flumerfelt, A. Alves (Eds.), Trans-disciplinary perspectives on system complexity. Springer.

Ullevik C. D. L. R., 2017. The impact of digital disruption in the maintenance service industry, in the oil and gas sector. M.S. thesis, Dept. Mech. Struct. Eng. Mater. Sci., Univ. Stavanger, Norway, U.K., 2017.

Wang B., Baras J. S., 2013. HybridSim: A modelling and co-simulation toolchain for cyber-physical systems. In 17th IEEE/ACM International Symposium on Distributed Simulation and Real Time Applications, (pp. 33–40).

Weick, K. E., 1990. The vulnerable system: An analysis of the Tenerife air disaster. Journal of Management, 16(3), 571–593.

Weick, K. E., 2005. Making sense of blurred images: Mindful organizing in mission STS-107. In M. Farjoun, W. Starbuck (Eds.), Organization at the limit: Lessons from the Columbia disaster, Blackwell (pp. 159–177).

Wikipedia, 2020. Functional Mock-up Interface. https://en.wikipedia.org/wiki/Functional_Mock-up_Interface#cite_note-modelica_Jan10-2.

Yun S., Park J., Kim W., 2017. Data-centric middleware based digital twin platform for dependable cyber-physical systems. In 2017 Ninth International Conference on Ubiquitous and Future Networks (pp. 922–926).

11 Application of Prognosis in Industry, Energy, and Transportation

11.1 MECHANICAL SYSTEMS: MAINTENANCE ACTIVITIES IN AUTOMOTIVE AND RAILWAY SECTORS, AIRCRAFT APPLICATIONS, ROTATING EQUIPMENT (BEARINGS, PUMPS, GEARBOXES, MOTORS)

11.1.1 Mechanical Systems: Maintenance Activities in Automotive Sector

The automotive industry is competitive worldwide. Automobile manufacturers and their suppliers are recognized for their competitiveness and results. The ability to generate employment and develop technology or advanced production processes has made the industry an example for other industrial sectors to follow. The sector is an example of success for its dynamism and its ability to generate growth in an increasingly complex environment. However, to consolidate what has been achieved so far and to evolve and to be comfortably positioned in the future, decisive actions have to be taken. It is increasingly necessary to analyze the current situation and identify elements which can be enhanced. The industry must reduce waste, optimize processes and the flow of information, and reduce overall costs to retain a competitive advantage (Dos Reis et al., 2019).

11.1.1.1 Current Composition of the Automotive Industry

A diverse variety of sectors and activities comprise the automotive industry, including the following (Australian Industrial Systems Institute, 2020):

- Motor vehicle and motor vehicle parts manufacturing.
- Motor vehicle and motor vehicle parts wholesaling.
- Motor vehicle, parts, and tire retailing.
- Automotive repair and maintenance.
- Agricultural, mining, and lifting machinery.
- Fuel retailing
- Motor vehicle rental.
- Motorsports.
- Outdoor power equipment.
- Motorcycles.
- Marine.

11.1.1.2 Technological Change, Skills, and Changing Job Roles

The increasing complexity of motor vehicles – as evidenced in the merging of electronic and mechanical technologies, intelligent transport systems, navigation, tracking, and infotainment systems, and the embedded network of computerized controls that manage these technologies – is placing greater demands on the skills base of the workforce.

DOI: 10.1201/9781003097242-11

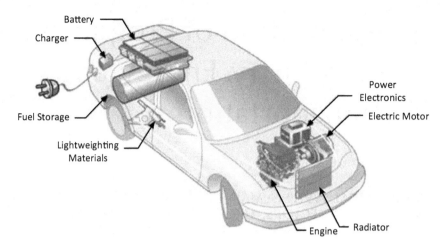

FIGURE 11.1 Vehicle technologies: hybrid and battery electric vehicles (Australian Industrial Systems Institute, 2020).

Mechanics used to work with less complex vehicle technology and could fix almost every problem across a broad range of motor vehicles; today's automotive technician is more likely to be a specialist. With the rate of technological change, it is difficult for even an experienced technician to keep up with the required technical knowledge without constant upskilling and training.

A key problem area within the current skills base that is often raised by industry is the absence of effective practical skills in vehicle diagnostics. This involves troubleshooting or faultfinding skills, along with the appropriate action to repair the problem. Even with the use of diagnostic scan tools in modern vehicle servicing that identify diagnostic trouble codes for particular vehicle faults, there is still a large element of misdiagnosis or failure to pinpoint the real source of particular vehicle problems.

There has traditionally been a divide between mechanical and electrical technicians, with a general reluctance to cross over into each other's space. The integration of mechanical and electronic technology in modern vehicles has changed this balance. Many vehicle service and repair workshops now expect a mechanical technician to have competency in all but the most complex electrical and mechanical tasks, particularly in independent workshops.

Staff in these workshops are required to work on a variety of vehicle brands, as distinct from the service department of an original equipment manufacturer (OEM) affiliated dealership that may specialize in only one or a limited number of brands.

As vehicle technologies evolve, and with the increased adoption of hybrid and battery electric vehicles over time, there will be a greater segmentation of skills within the automotive industry, with narrower and deeper specializations in vehicle brands or technologies being the norm (see Figure 11.1) (Australian Industrial Systems Institute, 2020).

11.1.1.3 Achievement of Reliability by Maintenance Activities and Tools in the Automotive Sector

Many maintenance activities and tools can be employed to achieve optimal reliability; some of these are briefly described in the following subsections (Schmidt & Schmidt, 2018).

11.1.1.3.1 Total Productive Maintenance

Total productive maintenance (TPM) has helped raise uptime and overall equipment effectiveness (OEE) by taking a team approach to maintenance and making a concerted effort to improve reliability and serviceability. TPM activities vary depending on the unit and production area concerned. In the body shop, for example, equipment can be shut down for a brief period during every work shift for cleaning and adjustment. In other areas, shutdowns can occur more frequently (a few minutes every couple of hours) or less frequently (a longer time period once a week). In some cases, workers may be given full-time TPM duties.

Admittedly, TPM has some limits. Some tasks belong entirely to maintenance. In addition, workers performing TPM must be properly trained, so the equipment is correctly maintained. In some cases, both basic and advanced TPM activities are outsourced.

The goal of TPM is to improve reliability through increased worker communication, responsiveness, and ownership (Schmidt & Schmidt, 2018).

11.1.1.3.2   Long-Term Planning

Long-term planning addresses the maintenance needs of critical systems over a timeframe of several years. It identifies the major and minor maintenance activities required to ensure the assets' long-term sustainability. This plays an instrumental role in highlighting maintenance issues for which resources are needed. Some assets may be used more than others and, as such, may require regular overhaul. The long-term plan considers the useful life of equipment and determines when it must be returned to as-good-as-new condition (Schmidt & Schmidt, 2018).

11.1.1.3.3   Design for Maintainability

Plants have been substantially extended, for example, changing from a two-line assembly system to a one-line assembly system with all vehicles assembled on the same line. After this type of extension and conversion, plants can take a new approach to a number of things, including maintenance, setting up new equipment, and installing it in a manner that increases accessibility and maintainability (Schmidt & Schmidt, 2018).

11.1.1.3.4   Standardization and Partnering for Maintainability

Plants are generally big supporters of equipment standardization. For example, some plants have converted their programmable logic controllers (PLCs) so they share a PLC platform with the company. This type of standardization enables corporate and cross-plant partnering.

Most companies have more than one site, and most of the equipment is the same at all company sites. This makes it possible to receive solutions to problems from plants around the world via a corporate production network, thus increasing OEE and uptime. Ultimately, maintenance best practices are determined globally not locally.

All information gathered and lessons learned can be accessed and used by all sites. Sharing these insights from the experiences of the different plants on a company-wide platform ensures the reliability of both equipment and maintenance processes. Partnering/idea-sharing between plants and their maintenance, repair, and operations suppliers is another way to increase reliability (Schmidt & Schmidt, 2018).

11.1.1.3.5   Condition Monitoring

When plants are extended and/or converted, most take the opportunity to install sensors on highly critical equipment components. The initial system likely monitored a plant in zones, but now a problem can be identified in individual equipment. When degradation or other issues are discovered, action is taken immediately, and issues are addressed before an incident can occur (Figure 11.2) (Schmidt & Schmidt, 2018).

The sensors and the PLCs interface with the plant's computerized maintenance software and systems applications and processing (SAP) system to submit current status information. After comparing these data to standard deviations, SAP can determine whether action is required and what kind of action. Depending on various factors, it may produce a work order, make a notification in

FIGURE 11.2 High-tech maintenance: passion for perfect production processes (Schmidt & Schmidt, 2018).

the computerized maintenance management system (CMMS), make a phone call, or send an email. Condition monitoring equipment is used for reasons of cost efficiency (Schmidt & Schmidt, 2018).

11.1.1.3.6 Root Cause Focus

Today, the plant and its maintenance organizations are dedicated to identifying the source of problems and preventing their reoccurrence. For example, in the body shop, if a line breakdown takes 15 or more minutes to resolve, maintainers may perform root cause analysis (RCA), using the "five-whys" method.

In this method, the question "why" is asked five times. All possibilities are considered, and the associates stay tuned until the problem is solved. RCA can't be done quickly, and it requires manpower to be implemented, ranging from people on the shop floor to the planning group. An appropriate course of action must be drawn up for simple tasks, that is, in cases when it is unreasonable to carry out the "five-whys" RCA with the effort and expenditure it entails. In such cases, a full RCA is not necessary. Furthermore, if new equipment malfunctions, it will take maintainers longer to solve the problem and rectify it, because they are still learning about its special features.

The recognition of problems and their solutions are transferable to all similar pieces of equipment in the plant, and the groups concerned ensure that the information they have gained is shared with senior departmental managers. This way, the managers are continually kept up to date with what maintenance is doing, as well as why it is important and what is being done to ensure the production processes remain up and running. In other words, a spotlight is constantly trained on equipment performance and maintenance.

Plants can also determine the root cause of problems using Lean Six Sigma projects. Lean Six Sigma is data-driven, making it different from other problem-solving measures (Schmidt & Schmidt, 2018).

11.1.1.3.7 Constant Change

Today's plants function on intervals. Each new car line that is introduced requires new equipment to produce the vehicles, providing some OEE benefits. The equipment is often cutting-edge technology. This means the maintenance staff is constantly learning.

The changes and staff turnover mean maintenance must consistently re-evaluate its preventive maintenance (PM) processes and activities. For example, if a machine undergoes maintenance eight times per year and shows no problems, the annual PM may be cut in half. The aim is to find the right balance, that is, not to perform too much, not to perform too little, and to perform the right thing. PM is rapidly changing from time-based to condition- or cycle-based. PM is based on how hard the equipment is running. Digitalization and demographical change will lead to new challenges and new technology for maintenance (Schmidt & Schmidt, 2018).

11.1.1.4 Total Productive Maintenance in Automotive Industry

The methodology of TPM originated in Japan and consists of optimizing industrial processes by maximizing equipment performance to generate greater productivity. The implementation of the process requires time and expertise, as the length of time required for TPM depends on the dimensions of each industrial unit. TPM requires investment and effort, but it can have very beneficial results over time, such as increasing the satisfaction of employees of the industrial unit at all levels and achieving higher productivity goals.

TPM is oriented toward organizations where capacity depends on the machines and is based, among other things, on autonomous maintenance. This requires the creation of a motivating and stimulating culture that encourages teamwork and coordination between production and maintenance, as well as the training of personnel. With TPM, it is possible to identify losses which decrease efficiency by interfering with production, including defects in the process, reduced operating speed, loss of time, adjustment and setup of machines or downtime, minor stoppage losses, and equipment failures. TPM yields several benefits to an industrial unit, such as increased control over tools and equipment, reduction of equipment failure time by improving the response time, and reinforced coordination between production and maintenance (Dos Reis et al., 2019). In other words, TPM is an enabling tool to maximize the effectiveness of equipment by setting and maintaining the optimum relationship between people and their machines.

11.1.1.4.1 Methodology

Measurement is an important requirement of continuous improvement process, and it is necessary to establish the appropriate metrics for TPM to work well. From a generic perspective, TPM can be defined in terms of OEE which, in turn, can be considered a combination of operation maintenance, equipment management, and available resources. The goal of TPM is to maximize equipment effectiveness, and OEE is used as a measure.

OEE can be calculated using the formulas shown below (Talib Bon & Lim, 2015):

$$OEE = Availability\,(A) \times Perfomance\ Efficiency\,(P) \times Rate\ of\ Quality\,(Q) \tag{1}$$

$$Availability\,(A) = \frac{Operating\ Time}{Planned\ Production\ Time} \tag{2}$$

$$Operating\ Time = Planned\ Production\ Time - Down\ Time \tag{3}$$

$$Planned\ Production\ Time = Shift\ Length - Breaks \tag{4}$$

$$Perfomance\ Efficiency\,(P) = \frac{Ideal\ Cycle\ Time}{\left(\dfrac{Operating\ Time}{Total\ Pieces}\right)} \tag{5}$$

$$Ideal\ Cycle\ Time = \frac{Shift\ Length}{Scheduled\ Number\ of\ Products} \tag{6}$$

$$Rate\ of\ Quality\,(Q) = \frac{Good\ Pieces}{Total\ Pieces} \tag{7}$$

$$Good\ Pieces = Total\ Pieces - Reject\ Pieces. \tag{8}$$

To explain how this works, consider the Manufacturing Operation and Engineering Department of Company X. The department wants to determine its OEE and implement TPM. In this case, the company can use non-probability sampling to collect data as the data are from one organization – Company X – only. In non-probability sampling, each element of the population does not have an equal probability of selection. For non-probability sampling, purposive or judgmental sampling is commonly selected as the method of choosing samples. The subject to be studied is fixed; in this case, the company is looking at the machines in the production line. Thus, the research sample comprises the machines in the Manufacturing Operation and Engineering Department.

The instrument used for data collection in this case can be observation. Observation is carried out by closely observing the activities of machines. Data are recorded based on observations. These data will become the input of OEE calculation (Talib Bon & Lim, 2015).

Company X uses Microsoft Excel to analyze its data and to calculate OEE. The company then uses a graphic method to display the findings to get a clearer picture of implementation of TPM and to identify elements that affect OEE the most. After its analysis of the OEE, the company makes improvements (Talib Bon & Lim, 2015).

11.1.2 Mechanical Systems: Maintenance Activities in Railway Sector

Railway traffic is steadily increasing, and there is a growing need for efficient operation and maintenance of rolling stock systems. The increased operation of articulated trains has also challenged maintenance organization and planning.

The selection of optimal maintenance strategies for each component will lead to increased availability, operational performance, and profitability. To this end, suitable tools are needed to analyze, compare, and optimize maintenance strategies (Eisenberger & Fink, 2017).

11.1.2.1 Scheduling Preventive Railway Maintenance Activities

A railway system needs a substantial amount of maintenance. To prevent unexpected breakdowns as much as possible, PM is required.

Reliability, that is, punctuality and safety, are important aspects of railway transport. The quality of the railway infrastructure has a major influence on the reliability of the railway system as a whole. Therefore, it is important to have enough PM of the infrastructure (e.g., rail, ballast, sleepers, switches, and fasteners). However, maintenance is very expensive, and budgets for maintenance are always under pressure. For instance, some governments who subsidize rail drastically reduced the amount of money spent on maintenance at the end of the 1990s, with major consequences on the punctuality of the railway system a couple of years later.

It is important to reduce maintenance costs without reducing the maintenance itself (Budai et al., 2004).

11.1.2.1.1 Railway Maintenance Planning

Preventive railway maintenance is performed to reduce the probability of the occurrence of a failure in the components of the railway infrastructure and/or to maximize the operational benefit. The frequency of PM may be based on calendar time, operating time, or the actual condition of the infrastructure components.

PM on railways can be subdivided into small routine activities and larger projects. Routine (spot) maintenance activities consist of inspections or small repairs, for example, inspection of rail, switches, level crossings, overhead wires, and signaling systems, as well as switch lubrication. These do not take much time and are done frequently, a few times a year. Larger maintenance projects include renewal work, such as ballast cleaning, rail grinding, and tamping. They are carried out once/twice every few years (Budai et al., 2004).

Preventive railway maintenance is carried out in most countries during train service (Budai & Dekker, 2002). In the actual train timetable, possible possession allocations are scheduled for maintenance so that the maintenance does not affect the regular train service very much. Many countries use timetabling software (e.g., Viriato and Opentrack) to find free intervals or periods with less impact on the train operators. Carrying out maintenance during train service might be unsafe for the maintenance crew. Therefore, maintenance may be done at night (when there are fewer trains) or during the day with an interruption of the train service.

Because of the safety requirements, train cancellation is required when maintenance is done during the day, so companies must arrange alternative transport (e.g., using buses) during the track possession time (Budai et al., 2004). If the maintenance is done at night, companies can design a cyclic static schedule (e.g., Den Hertog et al., 2001; Van Zante-de Fokkert et al., 2001). Miwa, Ishikawa, and Oyama (2001) present an optimal schedule for a specific maintenance work, namely an annual schedule for the tie (sleeper) tamping.

Every time a track is maintained, it is blocked for train service. Furthermore, in these track possessions, the maintenance activities (small routine work and larger projects) are clustered as much as possible to disturb the railway operation as little as possible (Budai et al., 2004).

Since rail is an important transportation mode, proper maintenance of the lines, repairs, and replacements must occur in a timely fashion to ensure efficient operation. Moreover, since some failures will have a strong impact on the safety of the passengers, it is important to prevent them by carrying out PM in time and according to some predefined schedule. Since the infrastructure maintenance costs represent a huge part of the total operating costs, operations research tools are required to help maintenance planners to come up with optimal maintenance plans.

An optimization model to improve rail maintenance decisions by creating a schedule for carrying out PM activities is normally presented. Maintenance works are assigned to different time periods (months/weeks), minimizing the track possession cost or the track possession time. Routine maintenance works and projects are planned together. Furthermore, since the maintenance scheduling problem is a complex optimization problem, and for a large set of instances it is difficult and time consuming to solve the problem to optimality, it is necessary to develop some approximation methods, which still give solutions close to the optimal ones (Budai et al., 2004).

11.1.2.2 Current Maintenance Challenges in Railway Industry

Regular maintenance is essential to keep vehicles and components in an operational state and remedy any faults and failures. Poor maintenance can result in train delays, cancellations, and hazardous events – even fatalities. Such events affect competitiveness and profitability and, as such, are extremely important.

Maintenance has become more complex in the railway industry. For one thing, articulated trains are replacing traditional locomotive and passenger coaches, and they require more complex maintenance. For another, relatively simple mechanical systems are being replaced by complex mechatronic devices and systems that combine mechanical, electronic and information technology, and these obviously have quite different degradation and failure characteristics.

It is necessary to have an optimal maintenance strategy in place, but the increased complexity of railway systems, accompanied by the increased requirements for reliability, availability, and safety, has made this more difficult.

Modular maintenance could be one way to cope with the increased complexity of railway vehicles while accommodating the need for availability. In modular maintenance, line units can be replaced by another unit and then repaired in the workshop, thus increasing the vehicle's availability.

Although there is a possibility of doing things differently, and condition monitoring devices can make data on the system's condition available to maintainers, railway maintenance is dominated by traditional methods, that is, corrective or periodically scheduled maintenance (Eisenberger & Fink, 2017).

11.1.2.3 How to Implement Efficient Railway Maintenance through Digitalization

Digitalization allows railway companies to collect and use data in new ways. They can rethink maintenance, starting with the business model. Through digitalization, they can reduce costs, increase reliability, and optimize availability.

Digitalization improves efficiency in both daily activities and long-term planning. Areas of application include spare parts planning, maintenance planning, obsolescence monitoring, failure prediction, technical documentation, access to remote experts, and e-learning, among others.

For instance, Alstom has taken an innovative approach to condition-based and predictive maintenance though its HealthHubTM. This integrated range of decision-support tools, including cameras, sensors, data loggers, and lasers, continuously monitors asset health. Alstom's next proposed step is dynamic maintenance planning (DMP), the next phase of digitalization in maintenance. DMP transforms collected data into action to allow more flexible, responsive, and fluid maintenance (see Figure 11.3) (Global Railway Review, 2019).

DMP works as follows: The first phase of DMP is concerned with monitoring assets, that is, infrastructure, rolling stock, and signaling systems. In this phase, sensors, cameras, lasers, data loggers, and so on continuously monitor the state of assets and identify changes in performance. In the second phase, the collected data are transformed into smart data using algorithms capable of automatically identifying failures and sending alerts. Data are bundled and sent to a web platform to predict the remaining useful life (RUL) and to anticipate failures. The ability to transmit and interpret these data is essential for system reliability. The third phase involves concrete actions based on the findings.

DMP's benefits include the following: information flows are automated; all activities are visible and can be easily traced; maintenance is optimized; and efficiency is improved (Global Railway Review, 2019).

11.1.3 MECHANICAL SYSTEMS: MAINTENANCE ACTIVITIES IN AIRCRAFT APPLICATIONS

11.1.3.1 Aircraft Maintenance and Repair

The successful use of new materials and structural concepts relies on maintenance programs that cost-effectively ensure passenger safety. This section gives an overview of current aircraft maintenance programs, including inspection and repair processes and future needs (National Research Council, 1996).

11.1.3.1.1 Aircraft Maintenance

Maintenance programs evolve and are developed for each new type of aircraft based on previous experience with similar materials, engines, components, or structures. New materials or structures, for which experience is limited, are observed more frequently until a basic level of confidence is established. Time extensions to inspection intervals are based on observations made during routine service checks. A typical airline maintenance and service plan is outlined in Table 11.1. The objectives of an effective maintenance program are as follows (National Research Council, 1996):

- Ensure, through maintenance activity, that the inherent safety and reliability imparted to an aircraft by its design are sustained.
- Provide opportunities to restore levels of safety and reliability when deterioration occurs.
- Obtain information for design modification when inherent reliability is not adequate.
- Accomplish the above at the lowest possible cost.

11.1.3.1.2 Structural Maintenance

Any new aircraft program is based on assessing structural design information, fatigue and damage tolerance evaluations, service experience with similar aircraft structures, and relevant test results.

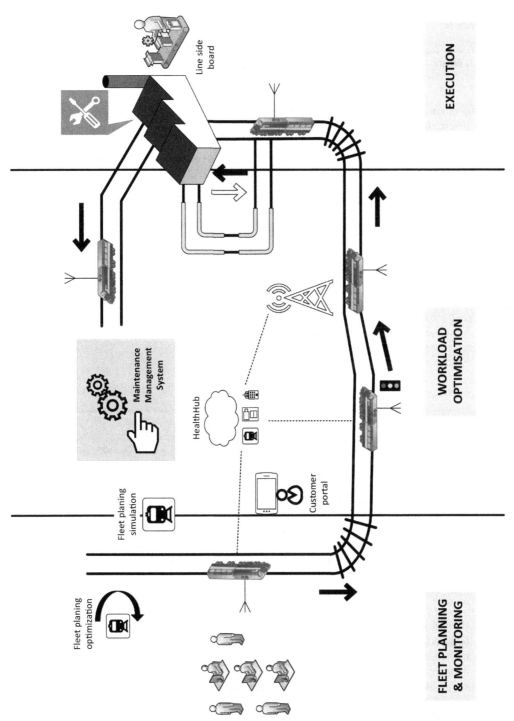

FIGURE 11.3   Dynamic maintenance planning (DMP) (Global Railway Review, 2019).

TABLE 11.1
Typical Airline Maintenance and Service Plan

When Service Is Performed	Type of Service Performed	Impact on Airline Service
Prior to each flight	"Walk-around" – visual check of aircraft exterior and engines for damage, leakage, and brake and tire wear.	None
Every 45 hours (domestic) or 65 hours (international) flight time	Specific checks on engine oils, hydraulics, oxygen, and specified unique aircraft requirements.	Overnight layover service
Every 200–450 hours (22–37 days) flight time	"A" check-detailed check of aircraft and engine interior, services and lubrication of systems such as ignition, generators, cabin, air conditioning, hydraulics, structure, and landing gear.	Overnight layover service
Every 400–900 hours (45–75 days) flight time	"B" check (or "L" check) – torque tests, internal checks, and flight controls.	Overnight layover service
Every 13–15 months	"C" check-detailed inspection and repair program on aircraft engines and systems.	Out of service for 3–5 days
Every 2 years (narrow-body aircraft)	Inspection and reapplication of corrosion protective coatings.	Out of service up to 30 days
Every 3–5 years	Major structural inspection with attention to fatigue damage, corrosion, etc. Aircraft is dismantled, repaired, and rebuilt. Aircraft is repainted as needed.	Out of service up to 30 days

Source: National Research Council (1996).

The maintenance task generally evaluates sources of structural deterioration, including the following: accidental damage, environmental deterioration, and fatigue damage; susceptibility of the structure to each source of deterioration; the consequences of structural deterioration to continuing airworthiness, including the effect on the aircraft (e.g., loss of function and reduction of residual strength, multiple-site or multiple-element fatigue damage, the effect on aircraft flight or response characteristics caused by the interaction of structural damage or failure with systems or power plant items, or in-flight loss of structural items); and the applicability and effectiveness of various methods of detecting structural deterioration, taking into account inspection thresholds and repeat intervals (National Research Council, 1996).

11.1.3.1.3   Component Maintenance

The application of new materials will not cause undue maintenance difficulties or hardship for the airlines if the aircraft designer is familiar with component experience. Airline experience indicates that hardware items wear out, but statistical old-age wear-out in complex mechanical, electrical, and avionic components is not a dominant pattern of failure. In fact, over 90% of generic part types show either random distribution of failure or gradually increasing probability of failure with age (National Research Council, 1996).

The reliability of a part or component of aircraft hardware is only as good as its inherent design (supported by adequate maintenance) allows it to be. Hence, it is generally accepted that (National Research Council, 1996):

1. Good maintenance allows parts to reach their potential reliability.
2. Over-maintaining does not improve reliability, but does waste money.
3. Under-maintaining can degrade reliability. In general, fundamental design changes are required to correct inherent component reliability problems.

Three approaches to PM have proven to be effective (National Research Council, 1996).

- Hard time involves removing a unit from service when it reaches a pre-ordained parameter value.
- Functional check or inspection involves monitoring a characteristic dimension or usage/operating parameter of a piece of hardware to determine if it is still suitable for continued operation, or if it should be removed to prevent an in-service failure.
- Functional verification requires performing an operational check of hardware functions to determine each function's availability if it is normally hidden from the scrutiny of the flight and operating crew.

There are many components for which measurement of deterioration, periodic removal for maintenance, and hidden function verification are not economically feasible or beneficial. Such parts require routine performance or reliability monitoring, and no PM is required or desirable. Modern aircraft are more tolerant of failures than older aircraft designs because of the increased redundancy provided in the design (National Research Council, 1996).

Most airlines classify specific component maintenance tasks as follows (National Research Council, 1996):

- Lubrication or servicing: The replenishment of the consumable reduces the rate of functional deterioration.
- Operational or visual check: Identification of the failure must be possible.
- Inspection or function check: Reduced resistance to failure must be detectable, and the rate of reduction in failure resistance must be predictable.
- Restoration: The item must show functional degradation characteristics at an identifiable age, have a large proportion of units survive to that age, and be able to be restored to a specific standard of failure resistance.
- Discard: The item must show functional degradation characteristics at an identifiable age, and a large proportion of units are expected to survive to that age.

Malfunctions of components should be evident to the operating crew, have no direct adverse effect on safety (whether they occur as a single or multiple event), and minimize the effect on the operation of the aircraft itself (National Research Council, 1996).

11.1.3.2 Aircraft Maintenance Operations

Throughout the years, the understanding and planning of maintenance checks for airplanes have changed substantially. In the very beginning, the structure of airplanes was quite basic. Therefore, maintenance was usually straightforward and often planned manually. When necessary, it was performed after a short period of flying time. Even more comprehensive activities such as repairs and overhauls, which take place on a frequent basis, used to be performed *ad hoc*. However, the manual planning of maintenance became increasingly impracticable in a more dynamic environment where both costs and complexity of the airplanes kept rising. As a result, systematic planning of maintenance became necessary to save costs and achieve greater efficiency.

The airline industry is not comparable to any other transportation industry. Flights consist of more than just the take-off and the landing: all has to be put in place, and authority and maintenance requirements have to be met. Maintenance checks have to be performed with care to make sure that every plane leaving the ground is reliable, safe, and airworthy, of course, at the lowest possible cost. It is obvious that maintaining fleets properly is of key importance to stay one of the safest transportation options. Airline companies are trying to achieve the right balance between three key values: profit, safety, and optimal planning of their activities (Van den Bergh et al., 2013).

11.1.3.2.1   Type of Maintenance

Aircraft maintenance can be scheduled and unscheduled maintenance, line and hangar maintenance, routine and non-routine checks, and so on. Various terms are used to refer to the same type of maintenance. For example, there are four major types of maintenance checks (A, B, C, and D) which each airplane has to undergo after a certain number of flying hours, as regulated by the Federal Aviation Administration (FAA). These four types can also be considered scheduled or PM, and overlaps are common. Table 11.2 indicates the types of maintenance (Van den Bergh et al., 2013).

Each of the four significant types of maintenance checks (A, B, C, and D) varies in scope, duration and frequency. The first one, the A-check, occurs most frequently and has to be performed about every 65 flight-hours, or approximately once a week. This check comprises the inspection of landing gear, engines, and control surfaces. The second substantial check, the B-check, is performed slightly less frequently, about every 300 to 600 hours. This involves a more extensive visual inspection and lubrication of all moving parts, for example, of the horizontal stabilizers of the plane. The two largest checks, types C and D, are called the heavy maintenance checks. A C-check is an inspection that takes about one to two weeks, once every year. A D-check, which includes, among others,

---

TABLE 11.2
Type of Maintenance

A-check

B-check

C-check

D-check

Scheduled maintenance

Unscheduled/emergency maintenance

Routine/non-routine maintenance

Short/mid/long maintenance

Light or heavy/base/hangar maintenance

Line maintenance

Preventive maintenance

Corrective maintenance

Predictive/on-condition maintenance:
• Layover maintenance
   ✓ Short-term layover
      o Pre/(post) flight inspection.
      o Transit check.
      o Daily check (night stop check/service check).
   ✓ Regular checks

OR, IN, DE level maintenance

Turnaround inspection

Other

Source: Van den Bergh et al. (2013).

TABLE 11.3
Taxonomy of Aircraft Maintenance

	Lay-over Maintenance or Light Maintenance		Heavy Maintenance
	Line maintenance	Line or hangar maintenance	Hangar maintenance
SCHEDULED or PREVENTIVE or ROUTINE	Short-term	Mid-term or regular checks	Long-term
	Pre-flight, transit, daily checks	A-check B-check Av-check M-check	C-check D-check Balance checks
UNSCHEDULED or NON-ROUTINE	*Predictive or on-condition maintenance*	*Predictive or on-condition maintenance*	*Predictive or on-condition maintenance*
	Corrective or emergency maintenance	*Corrective or emergency maintenance*	*Corrective or emergency maintenance*

Source: Van den Bergh et al. (2013).

stripping, painting, and cabin refurbishment, varies from a three-week to two-month inspection and is done once every four years.

Within these types of checks, sometimes an extra division is made. Clarke et al. (1996) divided the A-check into an M- and Av-check. The M-check is an inspection carried out every two to three days on every aircraft; the Av-check consists of the M-check in addition to an avionics inspection and is carried out every four to five days. Some airline companies break the C-check into four quarter C-checks, which Talluri (1998) calls balance-checks (Van den Bergh et al., 2013).

Since there is a great variety of terminology for the different types of maintenance, it is not always easy to know what exactly someone means by each of these checks. Some describe A- and B-checks as short-term checks and C- and D-checks as long-term checks. Others say an A-check is short or light, and a B-check is long or heavy. There is no real consistency.

The terms scheduled, preventive, line, and corrective maintenance are also commonly used types. Again, the subdivisions are not clearly separated. Only preventive and corrective maintenance are noticeably different, while preventive and scheduled maintenance are often used as synonyms. Table 11.3 gives the most common definitions in a comparative, tabular form. With respect to uncertainty, aircraft maintenance can be divided into scheduled (preventive or routine) maintenance and unscheduled or non-routine maintenance, indicated by the rows in the table. The columns refer to the intensity of the workload, starting from short-term (i.e., frequent and light) and going to long-term (heavy) maintenance. Most agree that line maintenance consists of both scheduled and unscheduled maintenance. Confusion arises, however, when the types of tasks need to be specified. Some see line maintenance as only short-term maintenance, while others also add the A- (and even B-) checks. Line maintenance got its definition from "on line" maintenance, referring to all the maintenance that can be done at the gate or on the apron. All other maintenance is categorized as "hangar" maintenance. The table categorizes both classification fields as midterm maintenance (i.e., A- and B-checks), whereas the heavy maintenance (i.e., C- and D-checks) is categorized as hangar maintenance only (Van den Bergh et al., 2013).

11.1.3.3 Aircraft Servicing, Maintenance, Repair, and Overhaul: Changed Scenarios through Outsourcing

Aircraft maintenance consists of a number of complex activities, typically referred to as servicing, maintenance, repair, and overhaul (SMRO). The associated maintenance problems require strict oversight. Outsourcing such complicated areas is risky, but it can be done through close monitoring with the outsourcing agencies with comprehensive service-level agreements to attain high-level proficiency. For example, the Indian aircraft industry (IAI), on average, spends more of its financial resources, time, and effort on SMRO activities annually than on manufacturing, design, and development activities. Many international companies are collaborating with IAI to promote and build better products quickly and cost-effectively.

Successful outsourcing depends on developing the right framework to outsource non-core activities, while maintaining constructive oversight. This allows a company to concentrate on its core competencies. As global delivery and outsourcing become key strategies for the aerospace industry, a company's policies should allow its partner companies to achieve product service excellence in their own areas with cost savings, accelerated time to market, and increased competitive advantage. Precise decision-making in terms of either "outsource" or "in-source" at the right time is another key factor (Varaprasada Rao & Chaitanya, 2017).

Aircraft maintenance is a complex, comprehensive, and continuous process. The whole aircraft needs to be inspected and maintained; modifications are carried out, and the necessary parts are replaced to meet the international standards set by oversight authorities like the Civil Aviation Authority, FAA, Joint Aviation Authorities, and Centre for Military Airworthiness and Certification. In general, an aircraft must be maintained with proper condition monitoring of all assemblies, subassemblies, rotatables, shelf-life-parts, and line replacement units after a stipulated period of time or flight hours/flight cycles. Some parts will have a specific life (flight cycle/shelf life) limit and must be replaced immediately upon reaching the threshold limits. Many other parts must be examined for problems to ensure flight safety. In general, the following are routine maintenance tasks (Varaprasada Rao & Chaitanya, 2017):

- Cleaning the aircraft and its main avionic components.
- Cleaning and checking shelf life parts as per standard of preparation .
- Performing avionics maintenance.
- Applying corrosion prevention compound.
- Lubricating and overhauling parts.
- Draining leaked fluids and checking fuel systems.
- Servicing hydraulic and pneumatic systems.
- Replacing rotatables and LRUs as per guidelines.
- Inspecting and checking for general wear and tear.

Avionics maintenance is a crucial field of aircraft maintenance, as it deals with electrical and electronic systems. These parts are vital for the safety of the aircraft. The navigation and communication systems in avionics, including radar, instrument, computer system, radio communication, and Global Position System (GPS) maintenance, require close examination. A strong knowledge of electrical wiring, looming, and technical skills is required to work in avionics maintenance (Varaprasada Rao & Chaitanya, 2017).

Aircraft maintenance is crucial to ensure the safety of passengers. The International Aircraft Advisory Circular explains that the aircraft maintenance programming needs to have specific objectives for the safety of the aircraft and the passengers. The specific objectives listed in the Circular are (Varaprasada Rao & Chaitanya, 2017):

1. Each aircraft released to service is airworthy and has been properly maintained for operations in air transportation.
2. Maintenance and alterations that are performed, or that other persons perform on the aircraft, should be in accordance with the maintenance manual.
3. Only competent personnel with adequate facilities and equipment can perform maintenance and alterations on the aircraft.

TABLE 11.4
Ten Elements of Air-Carrier Maintenance Program

1. Airworthiness responsibility;
2. Air carrier maintenance manual;
3. Air carrier maintenance organization;
4. Accomplishment and approval of maintenance and alterations;
5. Maintenance schedule;
6. Required Inspection Items;
7. Maintenance recordkeeping system;
8. Contract maintenance;
9. Personnel training;
10. Continuing Analysis and Surveillance System (CASS).

Source: Varaprasada Rao and Chaitanya (2017).

The Circular also lists ten elements that are a must for an air carrier's maintenance program (Varaprasada Rao & Chaitanya, 2017). These are shown in Table 11.4.

11.1.3.3.1 Aircraft Maintenance Checks
Aircraft maintenance checks are done on all aircraft (commercial and civil) after a specified amount of time and/or usage. Military aircraft normally follow specific maintenance programs; these are generally similar to the programs of commercial/civil operators in terms of flight safety. Commercial airlines and other commercial operators of large turbine-powered aircraft use continuous inspection programs approved by inspection agencies: FAA in the USA, Civil Aviation Safety Authority in Australia, European Aviation Safety Agency in Europe, Transport Canada in Canada, Director General of Aeronautical Quality Assurance in India and so on.

Each operator follows its respective country's guidelines for both routine and detailed inspections. As mentioned previously, these include A-, B-, C-, and D-checks. A- and B-checks are lighter checks. Table 11.5 reviews these checks (Varaprasada Rao & Chaitanya, 2017).

11.1.3.3.2 Aircraft Servicing: Maintenance, Repair, and Overhaul Framework
Figure 11.4 is a simple illustration of an aircraft in a hangar being inspected by the inspection team (Varaprasada Rao & Chaitanya, 2017).

Figure 11.5 shows an aircraft servicing framework. An airline operator's in-house maintenance framework, including the maintenance center's responsibilities, material and part supplies, and component part supplies, along with bought-out-items inspection and their coordination requirements, is shown in a flowchart in Figure 11.5 (Varaprasada Rao & Chaitanya, 2017).

Three major types of problems can be attributable to maintenance (Varaprasada Rao & Chaitanya, 2017):

• Problems related to maintenance control: Ineffective maintenance control systems, such as ineffective rules, regulations, and standards.
• Problems related to incomplete maintenance: A specific prescribed maintenance activity is prematurely terminated. In such circumstances, the correct maintenance procedures are followed but not done completely, as per regulatory measures, such as a rotatable or LRU not removed, not fitted, or not set correctly.
• Problems due to incorrect maintenance action: An incorrect procedure is followed, and work is completed which did not achieve its aim. This can be traced to omissions of the maintainer, and it causes further problems.

TABLE 11.5
Aircraft Maintenance Checks – Simple View

CHECK	LOCATION	DESCRIPTION	DURATION *
LINE/TRANSIT	At Gate	Daily (before first flight or each stop when in transit). Visual Inspection, fluid levels, wheels and Brakes, emergency equipment.	1 Hour
At Gate	Routine light Maintenance, Engine Inspection	10 Hours (One Shift/ Overnight)	At Gate
CHECK	At Gate	If carried out, similar to A Check but with different tasks (May occur between consecutive A Checks)	10 Hours to 1 Day
CHECK	Hanger	Structural Inspection of Airframe, opening access panels, routine and non-routine maintenance, run-in tests.	3 Days to 1 Week
D CHECK	Hanger	Major Structural Inspection of Airframe after paint removal, engines, landing gear and flaps removed, instruments, electronic and electrical equipment removed, interior fittings (seats & panels) removed, Hydraulic and pneumatic components removed.	1 Month and above

* Duration generally depends on type of defects found and remedial measures needed.
Source: Varaprasada Rao and Chaitanya (2017).

FIGURE 11.4   Checking defense aircraft at a hangar (Varaprasada Rao & Chaitanya, 2017).

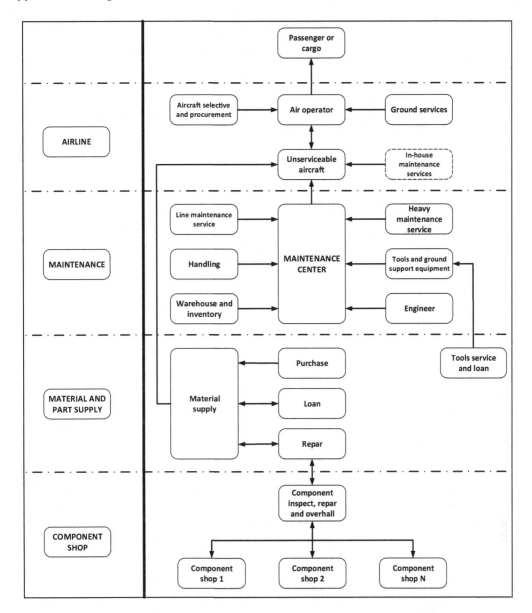

FIGURE 11.5 Aircraft servicing: maintenance, repair, and overhaul framework (Varaprasada Rao & Chaitanya, 2017).

Problems reported during inspections by supervisory authorities are shown in Figure 11.6 (Varaprasada Rao & Chaitanya, 2017).

11.1.4 MECHANICAL SYSTEMS: MAINTENANCE ACTIVITIES IN ROTATING EQUIPMENT (BEARINGS, PUMPS, GEARBOXES, MOTORS)

11.1.4.1 Maintenance Activities in Bearings

The rolling-contact bearing is an element of machinery with a very important role, as it dominates the performance of the machine. If one of the bearings breaks or seizes, not only the machine but

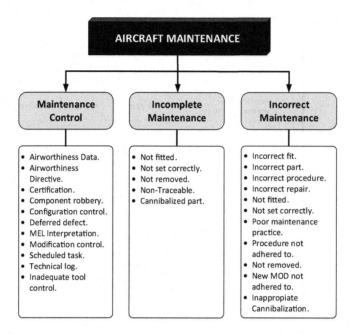

FIGURE 11.6 Aircraft maintenance: snag analysis main groups (Varaprasada Rao & Chaitanya, 2017).

also the assembly line may stop. If one of the axle bearings of an automobile or a railway car breaks down, a serious accident could occur. To avoid trouble, every bearing manufacturer should make efforts to assure the highest quality for each bearing and should emphasize that the user must carefully handle and maintain all bearings.

Every bearing becomes unserviceable in the course of time even if it is installed correctly and operated properly. The raceway surfaces and the rolling contact surfaces of the rolling elements are repeatedly subjected to compressive loads, and the surfaces eventually flake. The life of a rolling-contact bearing is defined as the total number of revolutions before flaking occurs. Alternatively, it is the number of operating hours at a certain constant speed before flaking.

The bearing may become unserviceable because of seizing, breakage, wear, false brinelling, corrosion, and so on. These problems are caused by improper selection or handling of the bearing. The problems are avoidable by correct selection, proper handling, and maintenance, and are distinguished from the fatigue life of the bearing.

However, breakdowns due to improper application, bearing design, and maintenance are more frequent than flaking due to rolling fatigue in the field (New Technology Network. 2017).

11.1.4.1.1 Inspection of Bearings

Inspection of a machine's bearings during operation is important to prevent unnecessary bearing failure. The following methods are generally adopted to inspect the bearing (New Technology Network, 2017):

1. Check of bearings in operation: Included are the check of bearing temperature, noise, and vibration, and the examination of the properties of the lubricant to determine when it should be replenished or exchanged.
2. Inspection of bearings after operation: Any change in the bearing is carefully examined after operation and during periodic inspections so as to take measures to prevent recurrence.

It is important for proper bearing maintenance to determine inspection requirements and intervals, based on the importance of the system or machine, and adhere to the established schedule (New Technology Network, 2017).

11.1.4.1.2 Inspection When Machine Is Running
When the machine is running, the following bearing inspections are relevant (New Technology Network, 2017):

1. Bearing temperature.
2. Operating sound of bearing.
3. Vibration of bearing.
4. Lubricant selection.
   - Grease lubrication.
   - Oil lubrication.
5. Relubrication.

11.1.4.1.3 Check of Bearings after Operation
Bearings after operation and those removed during periodic inspection should be carefully checked visually for symptoms on each component to evaluate whether the bearings' operating conditions are satisfactory. If any abnormality is detected, find the cause and apply a remedy by checking the abnormality against possible failure cases.

The bearing is generally usable up to the end of the rolling fatigue life if handled properly. If it fails earlier, it may be due to some fault in the selection, handling, lubrication, and/or mounting of the bearing. It is sometimes difficult to determine the real cause of bearing failure because many interrelated factors are possible.

It is possible to prevent the recurrence of similar problems by considering possible causes according to the situation and condition of the machine on which the bearings failed. Installation location, operating conditions, and surrounding structure of the bearings should be taken into consideration as well.

Figure 11.7 shows the bearing parts that may fail (New Technology Network, 2017).

11.1.4.2 Maintenance Activities in Pumps
The importance of pumps to the daily operation of buildings and processes necessitates a proactive maintenance program, which incorporates a preventive and predictive maintenance schedule. Most pump maintenance activities center on checking packing and mechanical seals for leakage, performing maintenance activities on bearings, assuring proper alignment, and validating proper motor condition and function without consideration for pump efficiency. Improving efficiency will decrease both maintenance and operating costs. The activities and basic measures to improve pump efficiency are the following (Metro Pumps & Systems, 2020):

1. Shut down unnecessary pumps.
2. Restore internal clearances if performance has changed.
3. Trim or change impellers if the head is higher than necessary.
4. Control by throttle instead of running wide-open or bypassing flow.
5. Replace oversized pumps.
6. Use multiple pumps instead of one large one.
7. Use a small booster pump.
8. Change the speed of a pump for the most efficient match of horsepower requirements.

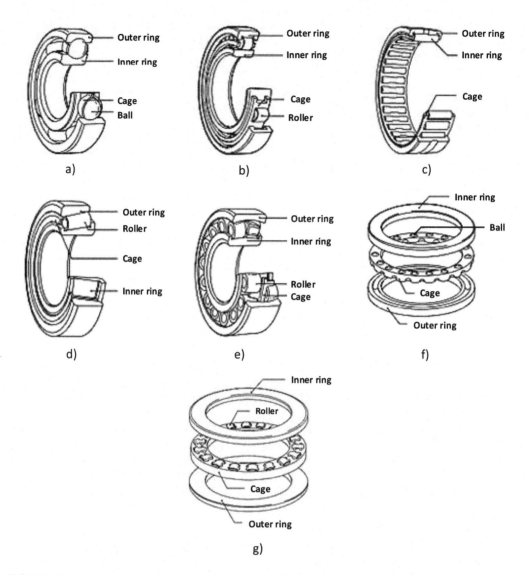

FIGURE 11.7 (a) Deep groove ball bearing; (b) cylindrical roller bearing; (c) needle roller bearing; (d) tapered roller bearing; (e) spherical roller bearing; (f) thrust ball bearing; (g) thrust roller bearing (New Technology Network, 2017).

11.1.4.2.1 *Preventive Maintenance Checklist for Centrifugal Pumps*

To guarantee an efficiently performing centrifugal pump that provides service with minimal repairs and shutdowns, institute a sound PM program.

Whether operating under harsh or mild external conditions, environmental surroundings can put a great deal of stress on the pumping equipment that is required to run 365 days a year. To achieve optimum performance and reliability in a centrifugal pump, it must operate close to its best efficiency point (BEP) – the point at which the hydrodynamic unbalanced load of the centrifugal pump is at its minimum.

When a pump operates at a point some distance from the actual BEP, the result is an overall increase in hydrodynamic unbalanced load. This, in turn, affects the performance, reliability, and efficiency of the centrifugal pump. (Based on experience and experiments, the unbalanced load is at its peak at the shutoff point.)

In any operational atmosphere, a routine maintenance program will extend the life of a pump, as well-maintained equipment lasts longer and requires fewer and less-expensive repairs. The following is a basic checklist for the PM of centrifugal pumps (Holland & Burg, 2017).

Annual maintenance: Once a year, the pump's performance should be checked and recorded in detail. Performance benchmarks should be established for a new pump during the early stages of a pump's operation, when the installation adjustments are correct. These benchmarking data should include:

- The head pressure developed, as measured by the suction and discharge gauges.
- Centrifugal pump flow rate.
- Motor amp draw.
- Pump vibration signature.

During the annual assessment of a pump's performance, any changes in the benchmarks should be recorded and used in determining the level of maintenance that may be required to get the pump back to operating at its BEP.

Quarterly maintenance:

- Check integrity of pump foundation.
- Check tightness of hold-down bolts.
- Change oil of oil-lubricated pumps after 200 hours of operation for a new pump, and every three months or 2,000 operating hours thereafter, whichever comes first.
- Grease bearings of grease-lubricated pumps every three months or 2,000 operating hours, whichever comes first.
- Check shaft alignment.
- Grease motor bearings according to the manufacturer's instructions.

Bearing lubrication tips:

Any centrifugal pump operation and maintenance program must emphasize bearing lubrication. Realistically, all pump bearings will eventually fail. When two surfaces rub together, whether lubricated or not, friction will win in the end:

- Oil-lubricated bearings require the use of non-foaming and non-detergent oils. Avoid over-lubrication which can be just as damaging as under-lubrication.
- Excess oil will cause a slightly higher horsepower draw and generate additional heat, which can cause frothing. If the lubricating oil is cloudy, it may indicate excessive water content from condensation. If this is the case, the oil must be changed immediately.
- For pumps equipped with re-greaseable bearings, avoid mixing greases of differing consistencies or types. Note that the shields must be located toward the interior of the bearing frame. When re-greasing, confirm the bearing fittings are absolutely clean to prevent contamination, as this can decrease bearing life.
- Over-greasing must be avoided, as this can cause localized high temperatures in the bearing races and create caked solids.

Additional parts examination:

If a part on a malfunctioning pump requires replacement, this is an ideal time to examine the pump's other parts for signs of fatigue, excessive wear, and cracks. Any worn parts should be replaced if they do not meet the following part-specific tolerance standards:

- Bearing frame and foot: Inspect for rust, scale, cracks, and roughness. Check machined surfaces for pitting or erosion.
- Bearing frame: Inspect tapped connections for dirt. Clean and chase threads. Remove loose and foreign material. Inspect lubrication passages for blockage.
- Shaft and sleeve: Inspect for grooves or pitting. Check bearing fits and shaft runout. Replace shaft and sleeve if worn or shaft runout exceeds 0.002 inches.
- Casing: Inspect for wear, corrosion, and pitting. Replace casing if wear exceeds a depth of 1/8-inch. Check gasket surfaces irregularities.
- Impeller: Inspect for wear, erosion, and corrosion. Replace impeller if vanes are bent or show wear exceeding 1/8-inch.
- Frame adapter: Inspect for cracks, warpage, and corrosion. Replace adapter if any is present.
- Bearing housing: Inspect for wear, corrosion, cracks, and pits. Replace housings if worn or out of tolerance.
- Seal chamber/stuffing box cover: Check for pitting, cracks, erosion, and corrosion. Inspect chamber face for wear, scoring, and grooves. Replace if worn more than 1/8-inch deep.
- Shaft: Check shaft for corrosion or wear and straightness. Maximum total indicator reading at sleeve journal and coupling journal should not exceed 0.002 inches.
- Severe service conditions: If the pump is used in severe service conditions, such as in highly corrosive liquids, maintenance/monitoring intervals should be shortened (Holland & Burg, 2017).

11.1.4.3 Maintenance Activities in Gearboxes

Gearboxes are essential devices found everywhere in industrial manufacturing facilities, providing an even distribution of power and torque wherever needed to fuel manufacturing productivity and profits. Despite how ubiquitous gearboxes may be, it's easy to overlook their repair needs. Forgoing routine maintenance on any gearbox could spell disaster for the entire production system. To keep everything running smoothly, it is necessary to take the time to perform basic upkeep and repairs on gearboxes to get the most out of their lifespan (Broadwind, 2015).

11.1.4.3.1 Preventive Maintenance Tips

To prevent the need for any costly repairs or downtime, the first and most important part of any maintenance plan is fixing problems before they happen. Gearbox oil can be a key indicator of gearbox health and can prolong gearbox life span (Broadwind, 2015):

1. The gearbox's oil should be regularly changed according to the needs of the specific system. Just like in a car's engine, oil lubricates the gears and prevents them from grinding against each other.
2. In the process of changing oil, a sample of the old oil can be collected and sent to a lab, where an analysis of particles inside it can reveal potential underlying issues in the gearbox.
3. Taking note of any excessive surface heat, odd vibrations, or unusual noises coming from a gearbox can help identify most issues that may need a closer look.

Keeping up with these simple, routine tasks can extend the life of a gearbox by several years and reduce unplanned operations and maintenance expenses, and this means increased production and profits for an organization (Broadwind, 2015).

11.1.4.3.2 Inspection and Maintenance Activities

Table 11.6 shows some common inspection and maintenance activities and the suggested intervals of execution for a gearbox. They are explained at more length below (Tos Znojmo, 2010).

TABLE 11.6
Inspection and Maintenance Intervals

Interval	Inspection and Maintenance
At least once a month	• Visual inspection for cleanliness of the gearbox or variator surface (dust, other impurities).
At least every 6 months	• Visual inspection. • Check noise during operation. • Check oil level. • Check temperature rise. • Fill re-lubrication devices with grease (at gearboxes equipped with re-lubrication devices).
Every X operation hours or every 24 months	• Oil change. • This interval is to be shortened if gearbox is operating under extreme. • conditions (high humidity, aggressive environment, extreme temperature changes).
Every 10 years	• Complete overhaul. • This interval is to be shortened if gearbox is operating under extreme conditions (high humidity, aggressive environment, extreme temperature changes).

Source: Tos Znojmo (2010).

Remember that inspection and maintenance must be performed by qualified personnel only. Furthermore, when it is being dismantled and serviced, equipment must not be in operation. The equipment must be isolated from the electrical supply, and measures must be taken to prevent accidental connection during maintenance (Tos Znojmo, 2010).

1. Visual inspection
   Check the gearbox surface for cleanliness. Dust/dirt must not be thicker than 1 mm. Check the gearbox for mechanical damage. If resilient mounting blocks are used, they must be checked. The gearbox must be overhauled if an oil leak or mechanical damage is detected. If this is the case, the supplier service agent or the manufacturer should be contacted.
   Oil around the lips of oil seals is not a sign of damage, however, as shafts are oiled during mounting. In fact, the shafts/oil seals should not be operated unless the surface under the lips is lubricated (Tos Znojmo, 2010).
2. Noise inspection during operation
   Excessive vibrations, excessive noise, and excessive temperature rise are signs that the gearbox may be damaged. It must be dismantled and overhauled (Tos Znojmo, 2010).
3. Oil level inspection
   Gearboxes which are not equipped with oil check plugs do not need to have their oil level inspected. Oil can be inspected only when the gearbox is not in operation. For gearboxes equipped with an oil gauge, the oil level should be maintained at the central position. Oil used for topping up must be the same as the original oil (Tos Znojmo, 2010).
4. Re-greasing
   Some gearboxes have bearings requiring re-greasing (Tos Znojmo, 2010).
5. Oil change
   Synthetic and mineral lubricants must not be mixed. If the type or make of oil is changed, rinse and clean the gearbox. Gearboxes are usually filled with oil suitable for long-term operation (synthetic oil), precluding the need for frequent oil changes. However, gearboxes can be filled

with oil requiring frequent changes if specified by the customer. Oil can be changed only when a gearbox is not in operation, and measures must be taken to prevent its accidental introduction into operation. The following tasks are carried out in an oil change (Tos Znojmo, 2010):

- Place a suitable vessel under the drain hole.
- Unscrew the inspection or drain plug.
- Drain the entire volume of oil.
- Rinse the gearbox with rinsing oil.
- Put the inspection or drain plug back in and tighten it using recommended torque.
- Put the required amount of oil into the gearbox through the filling hole.

6. Overhauling

A gearbox overhaul should be carried out by a specialized service agent who has necessary tooling or by qualified personnel. The overhaul could also be done by suppliers or manufacturers if they perform warranty repairs. Before an overhaul, the gearbox is dismantled to establish the wear of its parts. In an overhaul, all bearings are always changed, as are all seals, shaft seals, covering rigs, and lids (Tos Znojmo, 2010).

11.1.4.4 Maintenance Activities in Motors

A well-designed motor maintenance program, when correctly used, can include PM, predictive maintenance, and reactive maintenance. Inspection cycles depend upon the type of motor and the conditions under which it operates.

Motors need regular maintenance to avoid failure and prolong their lifespan. Generally speaking, motors and motor parts should be maintained and tested at least every six months. Only then is it possible to maintain a motor's life and its efficiency. Some factors affecting the life span of an improperly maintained motor are shown in Figure 11.8 (Csanyi, 2016).

Preventive motor maintenance: This kind of maintenance aims to prevent operating problems and to ensure the motor operates continuously and reliably. PM is usually intended to maintain a whole system (see Figure 11.9) (Csanyi, 2016).

Predictive motor maintenance: The objective of this kind of motor maintenance is to ensure that the right kind of maintenance is carried out at the right time. To define these two parameters, it is

FIGURE 11.8   Effect of lack of maintenance on life span of the motor (Csanyi, 2016).

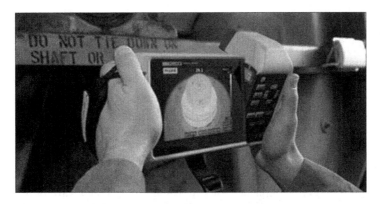

FIGURE 11.9   Motor preventive maintenance (Csanyi, 2016).

FIGURE 11.10   Motor preventive testing (Csanyi, 2016).

necessary to monitor the motor operation regularly and thereby detect problems before they actually occur. Keeping a log-book may help maintainers to compare historical data on a wide range of parameters and anticipate potential problems (Csanyi, 2016).

Reactive motor maintenance: The goal of reactive or breakdown maintenance is to repair and replace a motor when it fails. It does not imply regular service or tests.

Unexpected downtimes are costly, as an entire production process can stop until repairs are made. Regular PM can help prevent failures and thus avoid unexpected production stoppages. Figure 11.10 shows a test of PM on the motor (Csanyi, 2016).

11.1.4.4.1   Electric Motors: Maintenance
Electric motors are used in industrial, commercial, and residential settings and have many important applications: they propel trains, automobiles, compressors, pumping systems, and so on. Their components deteriorate over time and with increased volume of work, so their maintenance is crucial for operating continuity. The maintenance of an electric motor can be preventive and/or corrective.

PM is performed at set intervals to verify the motor's ongoing operation, perform such procedures as lubrication, and replace parts as recommended based on the workload of the motor. The maintenance should be planned to interfere as little as possible in operations.

Corrective maintenance is performed after a failure and thus interrupts production. Downtime can be minimized if the spare parts needed for repair are on hand and qualified personnel are available.

Electrical and mechanical components are both subject to continuous use, and they wear out. To conserve the life of the motor, both component types must be maintained.

The maintenance of the electrical components (windings, brushes, etc.) includes revising the connections, changing the carbon brushes, and making periodic measurements. The instruments used include megohmmeter, micro-ohmmeter, multimeter, oscilloscope, and others. They obtain values of insulation, continuity, frequency, current, voltage, power, and waveforms.

The mechanical components (bearings, shaft, housing, etc.) have a conservation program that consists of cleaning the mechanical components, reviewing the bases and nuts of the motor, and taking measurements that allow the speed, balance of the rotor, torque applied by the load, and temperature to be evaluated (Electrical Equipment, 2020).

PM of electric motors includes the following:

- Inspect the motor at regular intervals.
- Remove dust, oil, and dirt in the fan cover to maintain good ventilation and allow the motor to cool.
- Check the seals.
- Check mechanical and electrical connections and fix bolts.
- Check the bearings, especially for noise and vibration.
- Replace worn bearings with new ones, connect them in cold or hot (oil baths), and lubricate them.
- Use a megohmmeter to test insulation resistance when the motor is turned off and has been off for a while.
- Use only qualified personnel to disassemble motors, if disassembly is required.
- Clean windings with dielectric solvent and bake in infrared light when there is a lot of dirt in the coils.
- Perform start-up tests after maintenance to verify the motor's operation.

Corrective maintenance activities for electric motors include the following scenarios:

- Motor does not start: Check voltages of grid, contact of brushes, fuses, contacts, correct connections, voltages in rotor, circuit of starting resistors, and insulation of windings.
- Motor overheats: Check load; clean grills, ventilation slots, terminal board connections, and stator winding.
- Motor smokes and burns: Check windings; keep ventilation circuits clean to repair or rewind as necessary.
- Excessive current is absorbed during operation: Check the load and replace the motor; otherwise, check rings, brushes, insulation, resistance circuit, and rotor winding to repair or rewind.

PM aims to reduce interruptions due to corrective maintenance. Thus, adequate planning of the maintenance interval is essential (Electrical Equipment, 2020).

11.2 INDUSTRIAL ENTERPRISES: CHEMICAL, CONTINUOUS-TIME PRODUCTION PROCESSES

11.2.1 CONTINUOUS PRODUCTION

Continuous production processes are flow production methods used to manufacture, produce, or process materials without interruption. As the term suggests, in continuous production, the materials being processed are continuously in motion. Continuous generally means operating 24-7 with infrequent shutdowns for maintenance, for example, semi-annual or annual maintenance. This interval depends on the industry. Some can operate for years without a shutdown (Wikipedia, 2020).

A continuous production system produces items for stock supplies, not specific orders. Companies using continuous production systems first make a sales forecast to estimate the demand for the product. They then prepare a master schedule to adjust the sales forecast based on past orders and present level of inventory. The inputs are standardized, and a standard set of processes and sequence of processes is adopted. Routing and scheduling for the whole process can also be standardized. After they have designed a master production schedule, the next step for companies is to prepare a detailed plan, considering basic production information, a bill of materials, machine load charts, equipment, personnel, and material needs.

In continuous production, items are manufactured in lot sizes, and the production process follows a clearly defined sequence in a predetermined order. Storage is not needed, thus reducing material handling and transportation facilities. The method has been defined as first-in-first-out (Money Matters, 2020).

11.2.1.1 Characteristics of Continuous Production

1. Standard products with ongoing, large demand throughout the year.
2. Standardized inputs and standardized sequence of operations.
3. Efficient division of labor.
4. Minimum but constant material handling.
5. Minimum flow of work at any point.
6. Little work-in-progress.
7. Good use of productivity techniques.
8. Minimum cost of production per unit.
9. Rigid quality control.
10. More maintenance (Money Matters, 2020).

11.2.1.2 Types of Continuous Production

The three types of continuous production are: mass, process, and assembly production (Money Matters, 2020).

11.3.1.2.1 Mass Production

In mass production, one type of product or a maximum of two or three is manufactured in large quantities, with little emphasis on consumers' orders. The main characteristics of this system are: standardization of products, processes, materials, and machines, and uninterrupted flow of materials.

Mass production systems are popular in industries where production is not interrupted, for example, electronics, electrical, automobiles, bicycles, and container industries.

These systems offer economies of scale, as the volume of output is large. The quality of products tends to be uniform and high thanks to standardization and mechanization. Individual expertise is less important; the quality level depends on the plant's quality control systems and management policy (Money Matters, 2020).

11.2.1.2.2 Process Production

In process production, a single raw material can be transformed into different kinds of products at different stages of the production process. The flow of this type of production may be classified as analytical or synthetic (Money Matters, 2020):

- Analytical processes: The transformation of a raw material into different products. For example, crude oil becomes gas, naphtha, or petrol, and coal is processed to obtain coke, coal, gas, or coal-tar.
- Synthetic processes: The mixing of two or more materials to manufacture a product. For instance, lauric acid, myristic acid, plasmatic acid, stearic acid, and linoleic acid are synthesized to manufacture soap (Money Matters, 2020).

11.2.1.2.3 Assembly Production

In this type of production, two or more components are combined in a finished product. Manufactured parts are joined into sub-assemblies or final assemblies. Assembly production is used for cars, radios, televisions, bicycles, watches, cameras, and the like.

An assembly line is a type of flow production developed in the automobile industry in the USA. The assembly line improves production efficiency and leads to cost reductions. Assembly lines are especially useful when producing a limited variety of similar products on a mass scale or in large batches on a regular or continuous basis.

A key decision in assembly production is the layout of the assembly line. The design depends on the product design and location of production. In addition, technology must be balanced with other manufacturing facilities to optimize results. Design decisions consider the following (Money Matters, 2020):

1. Work flow rate.
2. Direction of manufacturing operations.
3. Workers' convenience and comfort.
4. Availability of service facilities, for example, water, electricity, compressed air, oxygen.
5. Materials' supply and demand.

In an assembly line, each machine receives material from the previous machine and passes it on to the next. Therefore, the location of machines is automatically regulated by the sequence of operations.

In addition, every operator needs to have free and safe access to each machine. There should be space for forklifts and trucks to move freely when delivering materials or collecting the finished products. Passageways should not be blocked, and workers should not be in danger of being hit by moving vehicles. The floor space may also be used commercially (Money Matters, 2020).

11.2.1.3 When Is Continuous Production Suitable?

A continuous production system is best suited to organizations intending to produce a limited variety of products on a large scale. In this way, the heavy fixed costs of specialized equipment used to operate at low cost per unit can be distributed over a high volume of output.

Industries who meet the following requirements can make good use of continuous production (Money Matters, 2020):

1. Uniform demand.
2. High volume of production.
3. Product standardization.
4. Process balancing.

11.2.2   FUTURE PRODUCTION CONCEPTS IN THE CHEMICAL INDUSTRY

In the last few years, chemical and pharmaceutical industry companies have been working on two major production concepts to improve their production of chemicals, drugs, materials, or biotechnology products: continuous-flow and modularized production. The general goal is to more quickly produce at a higher quality, while being less wasteful.

With today's globalized and volatile markets, the reduction of time to market is as essential as safe, resource-efficient, and flexible production. The chemical industry is facing an increasing demand from fast growing and vibrant markets, such as China, India, or Brazil, as well as a trend to customize specialty and fine chemicals. This leads to high product varieties, produced either in small amounts or over a hundred tons per year.

The question is whether continuous production processes and modularized production systems are helpful in this context (CHEManager, 2016).

11.2.2.1   Traditional Batch Processing vs. Continuous Production Methods

A batch process is long-term run and requires chemical engineering know-how and calculations, as well as experimental results from the lab and pilot plant prototyping. Every step, from planning to production, is difficult, requires a high investment, and increases time to market. In addition, the market has a potentially high deviation rate, as time to market is too long.

Continuous-flow and modularized process approaches are now being used to overcome the disadvantages of the batch process and reduce development time of a chemical or biotechnological production process from initial idea to market operation with simultaneous energy and resource efficiency. Continuous production methods have a smaller ecological footprint, the needed equipment is much smaller and easier to handle, process cycle times and operating costs are lower, and they have maximized quality control and a higher level of automation, which leads to less human interaction and allows smarter and digitized process control for new trends like the Internet of Things (IoT). Another key component of continuous-flow production is that the process is fully integrated, which means the products of one section flow into the next (CHEManager, 2016).

11.2.2.2   Continuous Manufacturing vs. Modularized Plant Systems

Modularized plant systems, working with continuous manufacturing methods as a key enabler, allow fast reactions to increasing or decreasing market demands. The goal is to use standard modules for continuous manufacturing. Therefore, modules and components must be integrated and multi-scalable to significantly accelerate modeling and process design. By using continuous manufacturing laboratory equipment with high similarity to the final process equipment, the detailed engineering of the final production facility can be realized with the chosen laboratory plant structure. The production facility is then built by preconfigured modules. The combination of these components into modules and the associated integrated information modeling from the process design to the initial operation are essential cross-cutting activities. They reduce throughput times, while optimizing the energy efficiency of the process.

The development of scalable components supports the concurrent development of appropriate planning and hardware modules for recurring process steps and frequently used components, such as pumps, columns, reactors, and infrastructure. These modules must be integrated into a planning tool that supports the entire design process from early process development in the laboratory up to the three-dimensional plant model.

The modularization of key components, as well as the data integration and data management through various phases in the plant design cycle, contributes significantly to an increased efficiency and reduced time to market, and allows industry-wide use. Models from automotive industry supply chains can be adapted, as they offer great potential for specialty and fine chemicals companies to develop synergy and competitive advantages. Ultimately, the continuous manufacturing and modularized plant system approach could lead to cost-effective production over the long-term right from the start, by offering an optimal balance between investments and operating costs, as well as future updates (CHEManager, 2016).

11.2.3 CONTINUOUS MANUFACTURING IN PHARMACEUTICAL AND CHEMICAL INDUSTRIES

In the pharmaceutical and chemical industries, the shift from the conventional batch manufacturing system to the continuous manufacturing method is gaining momentum. As the previous section suggests, the advantages of continuous manufacturing include an increase in efficiency (productivity and economy), enabled through the maximized automation by the unit operations interconnection, enhanced product quality and safety by a continual automated monitoring of processes, environmental impact reduction by decreasing waste, through a high rate of reaction efficiency, and space saving because of the compact size of the equipment.

To realize continuous manufacturing, it is essential to develop new catalysts and processes to replace batch reactions with continuous ones and to develop sensor technologies to achieve advanced continual monitoring. The development of virtual measurement technologies, called soft sensors, is particularly interesting, as they fully leverage simulation technology to estimate data that are difficult to obtain through actual measurements (Inada, 2019).

11.2.3.1 Factors behind the Rising Momentum toward Continuous Manufacturing

In the pharmaceutical and chemical fields, the basic approach to production has been the batch manufacturing method, characterized by the sequential arrangement of large tanks that are managed individually for each process. Continuous manufacturing is a relatively new production method to produce or process products without interruption by constantly supplying raw materials while the manufacturing process is underway.

In the pharmaceutical industry, the transition from batch to continuous manufacturing is not popular because in the high value-added pharmaceuticals domain, sufficient revenues can be ensured with batch methods. Switching production methods would require additional capital investment, making the change unattractive even if there were no problems with the product quality or manufacturing process.

However, there are signs that this type of change is required. In the mid-1990s, for example, a problem with product quality came to light in the USA. Defective products of batch-produced pharmaceuticals were released into the market, and this became a public issue. In addition, the lack of production volume adjustment function has been pointed out as a disadvantage of batch manufacturing, driving the move toward a continuous manufacturing system, as it allows the production of the required volume when needed. Finally, the US Food and Drug Administration's (FDA) industry guidance issued in 2004 included recommendations for the use of continuous manufacturing for pharmaceuticals, and in 2018, the adoption of continuous manufacturing by the pharmaceutical and fine chemicals industries was mentioned as an important issue in its national strategy (Strategy for American Leadership in Advanced Manufacturing) (Inada, 2019).

11.2.3.2 What Is Continuous Manufacturing?

11.3.3.2.1 Advantages of Continuous Manufacturing
The conventional batch manufacturing method is one in which raw materials are input into a piece of production equipment, and the produced output is collected after the completion of each specific unit production operation. In this method, since operations are stopped for each separate process, such as raw material input, manufacturing (chemical reactions, refining), and product discharge, operations tend to be complicated and labor intensive (Figure 11.11) (Inada, 2019).

Continuous manufacturing is a production method in which raw materials are continuously injected into a manufacturing facility, and products are discharged continuously during the period

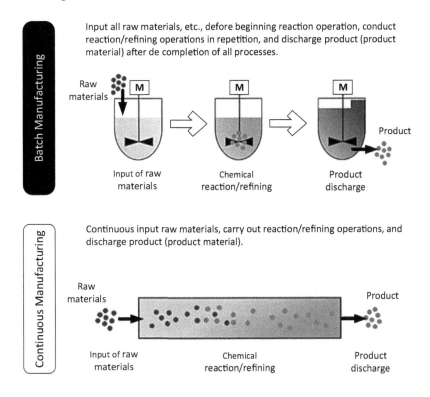

Input all raw materials, etc., defore beginning reaction operation, conduct reaction/refining operations in repetition, and discharge product (product material) after de completion of all processes.

Batch Manufacturing

Raw materials

Input of raw materials

Chemical reaction/refining

Product discharge

Product

Continuous input raw materials, carry out reaction/refining operations, and discharge product (product material).

Continuous Manufacturing

Raw materials

Product

Input of raw materials

Chemical reaction/refining

Product discharge

FIGURE 11.11 Schematic of batch and continuous manufacturing (Inada, 2019).

in which the production processes are in operation. In this method, multiple processes are automatically controlled, contributing to simplifying the overall operation and reducing the workload requirements for human operators.

There are many advantages of continuous production (see Section 11.3.3). Ultimately, although the amount of capital investment will be higher than batch manufacturing, the operating cost reduction is expected to improve the overall economic efficiency of the business enterprise (see Table 11.7) (Inada, 2019).

In addition, by adjusting the production time, the required amount can be produced on demand without waste. The method thus offers significant advantages to the overall pharmaceutical product manufacturing operation, including product storage and distribution.

As the production volume can be adjusted, the same equipment can be used both for the development and production, meaning that, in general, scale-up processes are unnecessary when moving on to the commercial production phase. This translates into a reduction of the time and costs involved in equipment development for new drugs (Inada, 2019).

11.2.3.2.2 Elemental Technologies for the Realization of Continuous Manufacturing
Given that batch and continuous manufacturing have different reaction mechanisms, the first step in achieving continuous manufacturing in pharmaceutical manufacturing is exploring process parameters, such as optimal catalysts for continuous manufacturing, temperature and pressure levels, and reaction vessel shapes that can maximize production efficiency.

Because production cannot be stopped in a continuous manufacturing system, measurement technologies for real-time monitoring of the reaction state during the processes play an important role. In the early 2000s, when the pharmaceutical industry initially began to turn its

TABLE 11.7
Features of Batch Production and Continuous Production

	Batch Manufacturing	Continuous Production
Raw material input/ Product output	Raw materials are injected into the process operation no continuously, and the product (product material) is discharged collectively after the operation is completed.	Raw materials are injected into the process operation continuously, and the product (product material) is discharged continuously and sequentially after a certain time.
Production processes	Each operation is started and stopped repeatedly by operator handling.	Production is continuous through interconnected unit operations and automation without operator management.
Production facility area	Large space needed	Space saving
Scaling-up	Individual verification processes and dedicated equipment are needed for each scale at the stage of development and validation, and different equipment is needed for commercial production.	Equipment required for development can be designed in line with actual production, and a quick transition to commercial production is possible by simply adjusting the production time.

Source: Inada (2019).

attention to continuous manufacturing, there were still some issues with measurement technologies, but recent technological advancements have made it more feasible to implement continuous manufacturing.

Process analytical technology (PAT) is a key technology for solving issues of real-time monitoring. PAT is a general term for technologies used for real-time monitoring of reaction behaviors during production processes, including the progress of reactions, temperature, and pressure. The development of virtual measurement technology, called soft sensors, is of special interest (see Figure 11.12). Soft sensors enable the estimation of information that is difficult to measure in real time through a simulation technique. It is attracting a great deal of interest as an alternative to real measurements using actual sensors, and various manufacturers are promoting the development of soft sensor technologies (Inada, 2019).

Some variables, such as temperature, pressure, and flow rate in the chemical reaction process, can be measured in real time. But other variables, such as concentration and density, cannot be measured unless a sample is taken and analyzed in the laboratory. Since such analysis takes time, its real-time property is lost, and a time delay problem can occur in the process control.

With a soft sensor, a numerical model can be constructed, for example, based on previously obtained data on inputs (temperature and pressure) and outputs (concentration and density), and target output values can be estimated in real time. By collecting actual temperature and pressure information that changes every moment on-site using a real sensor and inputting the obtained data into a soft sensor, it becomes possible to monitor real-time changes in concentration and density without going through time-consuming analysis. In addition, the technology can contribute to reducing the work hours required for sample collection and analysis, as well as the cost of analytical equipment.

The development of soft sensors for monitoring processes with complex reactions is ongoing, as they represent a vital technology in the implementation of continuous manufacturing (Inada, 2019).

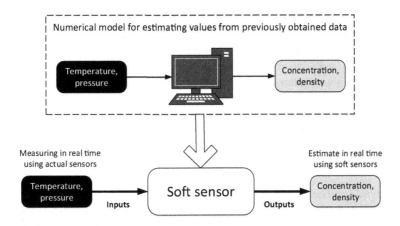

FIGURE 11.12 Overview of soft sensor technology (Inada, 2019).

11.3 MARINE SYSTEMS: SHIPBOARD MACHINERY AND LOGISTICS MAINTENANCE

11.3.1 SHIPBOARD MACHINERY MAINTENANCE

Shipboard machinery maintenance is key to the accurate operation of a ship. Therefore, the ship's machinery plant should be kept well maintained and clean at all times to assure the safe and smooth operation and functionality of the ship. Furthermore, ample new spare parts and equipment must be on board to perform maintenance in a timely manner under the supervision of fully competent staff according to the manufacturer's instructions. For example, crankshaft deflections need to be taken frequently at regular intervals. Fuel oil and lubricant oil must be analyzed to ensure that they do not contain debris or contamination. Main and auxiliary engines could break down, causing severe damage, such as fire. Consequently, the ship may have to terminate its services until lengthy and costly repairs and maintenance measures are accomplished.

Unfortunately, maintenance requirements are not standardized, and requirements will differ substantially from one ship to another. Some ship owners employ a fundamental PM policy, while others operate more sophisticated systems based on condition monitoring.

An approach based on operating at scheduled intervals offers a manageable and controlled maintenance program. Although it may not be most efficient, it is tailored to operational performance trends and individual elements of machinery. Furthermore, by evaluating the condition monitoring data, the extent to which machinery is withdrawn from use can be reduced. This approach may be applied to a vast variety of reciprocating and rotating shipboard machinery. For instance, tail-shafts can be controlled and monitored on a frequent basis, thereby allowing lengthy periods between withdrawals for accurate inspection. The time interval between raising main stream turbine casings may be extended in a similar fashion. As a result, unproductive repair, as well as dry dock downtime, may be kept to a minimum level.

Reliability centered maintenance can further optimize a maintenance program. This technique employs a structured assessment of functions to derive the most efficient maintenance strategy. Operational performance can be improved by maintaining ship availability and keeping downtime minimal.

Proper shipboard machinery maintenance and loss prevention measures include the following checks (UKEssays, 2017):

1. Lubricating oil: Lubricating or lube oil sampling should be done at frequent levels and regularly recorded as a recommended practice. Lube oil purifiers must be operated continuously

and adequately. The gravity disc should be chosen so as to achieve the desired oil-water interface at a maximum level of temperature nearing 90°C according to specifications. Furthermore, the lube oil feed system applied to the purifiers must be evaluated to ensure the optimum flow rate between the feed pump and the purifier's capacity. Old lube oil feed system designs use a direct drive pump with a larger capacity flow than recommended through the purifiers. If the system shows heavy water contamination, the lube oil in the sump tank should be transported to a settling tank, the sump tank must be cleaned, and new fresh oil must be filled to the minimum level as recommended by the engine manufacturer. The contaminated lube oil is drained and circulated by the purifiers, and after analysis, the future use of the oil is assessed. If solid particles are found in the system, the piping system should be cleaned, and the entire engine flushed.

2. Crankshaft safety: The engine should be stopped as soon as the oil mist detector alarm rings or engine overheating is detected. To avoid further damage, the main cause of overheating must be identified and correction measures taken before the engine is restarted. The lube oil must be kept as clean as possible by continuously using the lube oil purifiers at the recommended temperature over 90°C. Lube oil filters must be frequently and routinely maintained and placed in a clean and suitable environment. Crankshaft deflections should be regularly taken to ensure the operation of the engine is within the limit permitted by the manufacturer.

3. Bunkers: Bunkering processes, including fuel-testing procedures, must be carefully reviewed to ensure accuracy in procedures when dealing with off-specification bunkers. Every precaution must be taken to ensure adequate bunker supplies are available to allow proper testing before any new bunkers are used. Water, high ash, and total sediment potential content must also be considered, as well as high sodium and water content indicating the presence of seawater within the bunkers.

At sea, loss of structural integrity due to grounding and collision are the most significant contributors to accidental pollution. Hulls must be designed by considering two important factors:

1. Improvement in the hull strength as well as energy absorption.
2. Sufficient residual strength subsequent to damage for allowing salvage operations.

To this point, the major focus has been on preventing pollution from tankers, but there is now greater concern about the potential consequences of damage and accidents to bunker tanks and other types of ships, some which may be carrying thousands of tons of fuel oil.

11.3.2 MAINTENANCE AND REPAIR OF SHIPBOARD MACHINERY AND EQUIPMENT

Table 11.8 expands on the previous section's information about the maintenance and repair of shipboard machinery and equipment. The table includes the maintenance of marine pumps, valves, air compressors, heat exchangers, diesel engines, turbochargers, marine lubricating systems, deck machinery, boilers, and refrigeration units. It can be considered a basic maintenance checklist (Australian Industry Standards, 2018).

11.3.3 LOGISTICS MAINTENANCE

Simply stated, maintenance keeps mechanical equipment or machinery working. Regardless of the equipment's size, efficient maintenance can prolong its life and lead to more favorable outcomes. This is true of maintenance in all industries. As elsewhere, on board a ship, each piece of equipment requires maintenance at regular intervals.

TABLE 11.8
Elements and Performance Criteria

Elements of Essential Outcomes	Performance Criteria
Follow safe work practices	Work Health and Safety (WHS)/Occupational Health and Safety (OHS) procedures relevant to maintaining shipboard machinery and equipment are complied with.
	Safety hazards are identified and reported according to safety and vessel procedures.
	Before use, tools, equipment, and testing devices needed to carry out maintenance activities for correct operation and safety are checked according to safety and vessel procedures.
	Before beginning maintenance activities, isolation precautions are implemented according to safety and vessel procedures.
Marine pump maintenance	Maintenance requirements for pump are determined according to safety, manufacturer, and vessel procedures and documentation.
	Appropriate procedures, materials, tools and equipment for maintaining pump selected according to safety, manufacturer and vessel procedures.
	Relevant information is extracted from drawings and technical specifications required to perform maintenance activities.
	Pump is disassembled, inspected and serviced according to safety, manufacturer, and vessel procedures.
	Pump is reassembled and tested according to safety, manufacturer and vessel procedures.
	Performance of pump is confirmed against recommended performance specifications according to safety, manufacturer, and vessel procedures.
Valve maintenance	Maintenance requirements for valve are determined according to safety, manufacturer, and vessel procedures and documentation.
	Appropriate procedures, materials, tools and equipment for maintaining valve are selected according to safety, manufacturer, and vessel procedures.
	Relevant information is extracted from drawings and technical specifications required to perform maintenance activities.
	Valves is removed for maintenance according to safety, manufacturer, and vessel procedures and documentation.
	Valve is disassembled and valve maintenance is performed according to safety, manufacturer, and vessel procedures and documentation.
	Valve is reassembled and tested according to safety, manufacturer, and vessel procedures and documentation.
Air compressor maintenance	Maintenance requirements for air compressor are determined according to safety, manufacturer, and vessel procedures and documentation.
	Appropriate procedures, materials, tools and equipment for maintaining air compressor are selected according to safety, manufacturer, and vessel procedures.
	Relevant information is extracted from drawings and technical specifications required to perform maintenance activities.
	Air compressor is disassembled and inspected according to safety, manufacturer, and vessel procedures.
	Air compressor is reassembled, tested and adjusted according to safety, manufacturer, and vessel procedures.
	Performance of air compressor is confirmed against recommended performance specifications according to safety, manufacturer, and vessel procedures.

(continued)

TABLE 11.8 (Continued)
Elements and Performance Criteria

Elements of Essential Outcomes	Performance Criteria
Heat exchanger maintenance	Maintenance requirements for heat exchanger are determined according to safety, manufacturer, and vessel procedures and documentation.
	Appropriate procedures, materials, tools and equipment for maintaining heat exchanger are selected according to safety, manufacturer, and vessel procedures.
	Relevant information is extracted from drawings and technical specifications required to perform maintenance activities.
	Heat exchanger is disassembled and inspected according to safety, manufacturer, and vessel procedures.
	Heat exchanger is reassembled, tested and adjusted according to safety, manufacturer, and vessel procedures.
	Performance of heat exchanger is confirmed against recommended performance specifications according to safety, manufacturer, and vessel procedures.
Diesel engine maintenance	Maintenance requirements for diesel engine are determined according to safety, manufacturer and vessel procedures and documentation.
	Appropriate procedures, materials, tools, measuring instruments and equipment for maintaining diesel engine are selected according to safety, manufacturer, and vessel procedures.
	Relevant information is extracted from drawings and technical specifications required to perform maintenance activities.
	Diesel engine components are disassembled and inspected for wear and deterioration according to safety, manufacturer, and vessel procedures.
	Routine maintenance on diesel engines is performed according to manufacturer and vessel procedures.
	Diesel engine components are refurbished, as required, according to manufacturer and vessel procedures.
	Specialized tools and measuring instruments are used to maintain and refurbish diesel engines/components according to safety, manufacturer, and vessel procedures.
	Diesel engine is reassembled, tested and adjusted according to safety, manufacturer, and vessel procedures.
	Performance of diesel engine is confirmed against recommended performance specifications according to safety, manufacturer, and vessel procedures.
Turbocharger maintenance	Maintenance requirements for turbocharger are determined according to safety, manufacturer, and vessel procedures and documentation.
	Appropriate procedures, materials, tools and equipment for maintaining turbocharger are selected according to safety, manufacturer, and vessel procedures.
	Relevant information is extracted from drawings and technical specifications required to perform maintenance activities.
	All components of turbocharger are disassembled and inspected for wear and deterioration according to safety, manufacturer, and vessel procedures.
	Turbocharger is reassembled, tested and adjusted according to safety, manufacturer, and vessel procedures.
	Performance of turbocharger is confirmed against recommended performance specifications according to safety, manufacturer, and vessel procedures.

(continued)

TABLE 11.8 (Continued)
Elements and Performance Criteria

Elements of Essential Outcomes	Performance Criteria
Marine boiler inspection	Inspection requirements for marine boiler are determined according to safety, manufacturer, and vessel procedures and documentation.
	Appropriate procedures for inspecting marine boiler are selected according to safety, manufacturer, and vessel procedures.
	Relevant information is extracted from drawings and technical specifications required to perform inspection activities.
	Marine boiler is inspected for repair or general maintenance according to safety, manufacturer, and vessel procedures.
	Performance of marine boiler is confirmed against recommended performance specifications according to safety, manufacturer, and vessel procedures.
Marine refrigeration unit inspection	Inspection requirements for marine refrigeration unit are determined according to safety, manufacturer, and vessel procedures and documentation.
	Appropriate procedures for inspecting marine refrigeration unit are selected according to safety, manufacturer, and vessel procedures.
	Relevant information is extracted from drawings and technical specifications required to perform inspection activities.
	Marine refrigeration unit is inspected for repair or general maintenance according to safety, manufacturer, and vessel procedures.
	Performance of marine refrigeration unit is confirmed against recommended performance specifications according to safety, manufacturer, and vessel procedures.
Marine lubricating system maintenance	Inspection and maintenance requirements for lubricating systems are determined according to safety, manufacturer, and vessel procedures and documentation.
	Relevant information is extracted from drawings and technical specifications required to perform inspection and maintenance activities.
	Purifier maintenance procedures are applied according to safety, manufacturer, and vessel procedures.
	Components of lubricating system are inspected according to safety, manufacturer, and vessel procedures.
Maintenance and repair of deck machinery	Maintenance and/or repair requirements for deck machinery are determined according to safety, manufacturer, and vessel procedures and documentation.
	Appropriate procedures, materials, tools and equipment for maintaining and/or repairing deck machinery are selected according to safety, manufacturer, and vessel procedures.
	Relevant information is extracted from drawings and technical specifications required to perform maintenance activities.
	Deck machinery maintenance and/or repair procedures are implemented according to safety, manufacturer, and vessel procedures.
	Deck machinery is tested and adjusted according to safety, manufacturer, and vessel procedures.
	Performance of deck machinery is confirmed against recommended performance specifications according to safety, manufacturer, and vessel procedures.

Source: Australian Industry Standards (2018).

Historically, a ship had many crew members and engineers, and this allowed maintenance to be done quickly and easily. Today, there are fewer crew and more machinery. This makes it extremely important to plan maintenance in advance. Efficient planning is the key to productive maintenance.

Three types of maintenance are used on ships (Mohit, 2020):

1. Preventive or scheduled maintenance: Maintenance is based on running hours (e.g., 4000 hrs and 8000 hrs) or calendar intervals (e.g., yearly, monthly). The maintenance is carried out irrespective of machinery condition. If specified in the schedule, parts must be replaced, even if they can still be used.
2. Corrective or breakdown maintenance: Maintenance is performed when machinery breaks down. This is not a good method, as the machinery may be required in the event of an emergency. In addition, it may incur more expenses, as during a breakdown, other parts may be damaged. However, parts are used to the end of life.
3. Condition maintenance: Machinery parts are checked regularly using sensors. Their condition is accessed and maintenance done accordingly. This system requires experience and knowledge, as erroneous interpretations may damage the machinery and lead to costly repairs.

11.4　MEDICAL: HOSPITAL 4.0

Hospital 4.0 (or Smart Hospitals) is the application of the concept of Industry 4.0, or the fourth industrial revolution, to the hospital setting. Cyber-physical systems (CPS), the IoT, and cloud computing are essential components of Hospital 4.0.

CPS communicate and cooperate with each other in real time over IoT. They also communicate and cooperate with humans through IoT and Internet of Services. This is already a reality in modern hospitals. In fact, hospitals are leading the way in the adoption of Industry 4.0 concepts.

Hospital 4.0 has four design principles (Hospital International Congress, 2017):

- Interoperability: Equipment, devices, sensors, and people can connect and communicate via IoT or Internet of People.
- Virtual hospital: Information systems can create a virtual image of real hospital models using data sensors.
- Maintenance: Systems' availability and the ability of people to maintain them are supported, including the use of more recent tools like augmented reality and holography.
- Decentralized decisions: Tasks are performed as autonomously as possible at each level using cyber systems.

Hospital 4.0 aims to optimize resources and processes through better organization of information, patients, clinical, and administrative staff. In the same way, it is possible to speak of Smart Patients, that is, ordinary people who purchase special devices capable of monitoring their own health conditions and who constitute the so-called Health IoT (HIoT) or Smart Health.

As healthcare continues to become more competitive, the ability to assess tradeoffs between resource utilization, service, and operating costs is increasingly important, for example, with respect to appointment access, waiting room delays, or telephone service. In addition, stochastic simulation analysis can be used to study and improve health processes and other complex systems. Simply stated, Smart Health and, consequently, Hospital 4.0, apply the IoT to medicine. In addition to the IoT, the application of predictive algorithms and artificial intelligence (AI), capable of processing large amounts of data, plays a fundamental role in this context. Hospital 4.0 is a new way of conceiving the hospital, where the process of delivery of healthcare, medical monitoring, waiting rooms, and operation scheduling is radically changed to adapt and better serve patients and clinicians.

Some remaining healthcare challenges are the following (Cassettari et al., 2019):

- Self-diagnosis systems for patients: With wearable devices, patients could monitor their health status and thus self-diagnose a number of diseases and health conditions without having to go to a clinic or hospital. We are already seeing some of these devices hit the market.
- Patient monitoring: In today's hospitals, it is common to have continuous or intermittent life parameter assessments available. The additional requirement would be to make the system as space-saving as possible and, from a Hospital 4.0 perspective, to guarantee remote access by doctors. The future prospect is to build automated hospitals, in which a complete imaging system is set up as soon as the hospital is checked into the system.
- Digital data archive: The aim is to integrate the devices with the digital medical records to guarantee a constant and automatic update of the patient's vital conditions and of the care and therapies to which he or she is subjected.
- Acceptance of AI: AI is beginning to play an increasingly important role in diagnoses and therapies.
- Coordination and collaboration: The same patient can often be treated and examined by different professionals; constant interaction is therefore necessary to enable the exchange of useful information. For this reason, sharing information is fundamental.

The strengths of a principle of hospital modernization, based more and more on the constant integration of the IoT in daily hospital tasks, come with some risks, mainly associated with the problem of safety but also related to the need for a generational change, so that the technology is something easy to use (Cassettari et al., 2019).

11.4.1 Hospital 4.0: Digital Transformation in Hospitals

The interconnection of different devices through the IoT, the use of CPS and AI are salient features of the era of Industry 4.0. Digitization is key to Industry 4.0. Not only equipment is digitized, but also products, supplies, and practically everything that surrounds us. All digitized "things" are called "Smart" or "4.0", such as smart cities, smart factories, smart hospitals, and also smart drugs. In simple terms, it could be said that digitizing is making those things self-identify, providing information about themselves to every other thing through the Internet.

Digitization enables activities not imagined a few years ago. As we write this book, the Covid-19 pandemic is ongoing. Many medical consultations are virtual. This is likely to continue post-pandemic. The same is true for hospitals. Beds will be occupied by critically ill patients, as the least critically ill will be treated at home. The development of new technologies may even decrease the cost of healthcare.

Hospital facilities were in a crisis phase even before the pandemic struck because of changing health care needs. The existing structures, based on poly-specialist structures, are facing new challenges caused by the aging of the population, the progress of medical science, the increased demand for well-being, and the increased costs of medicine. The general approach is not integrated and not focused on the patient but on the hospital. Quality management is mainly based on centralized control, methodologies are lacking, and patient information is incomplete or unreliable.

Hospitals exist to care for patients, who sometimes feel like part of a production line. This must be changed. Digital transformation is no longer an option simply to add value to organizations. It must also answer users' demands. However, transforming the hospital organization from a multifunctioning center where the patient is treated by different therapeutic units to an integrated center capable of providing a personal care service that involves him/her as a subject represents an innovative challenge for the coming years.

The concept of Hospital 4.0 rests on three pillars (Salimbeni., 2020):

1. Sensors for patient monitoring.
2. Fast and efficient interconnected system, allowing massive data exchange in real time.
3. Ability to retain, retrieve, and analyze large amounts of data.

These technologies allow the construction of connected health systems, predictive diagnostics, and the revolution of "mHealth", or mobile health. Hospital 4.0 promises to yield the following benefits:

- Hospital 4.0 represents a new decentralized, patient-focused, and personalized hospital model. It is linked to the so-called "Theranostic" concept, where therapy and diagnosis come together and are carried out in real time. This form of personalized medicine combines diagnosis and therapy at the molecular level. It is more efficient and safer than classic medical practice.
- Hospital 4.0 increases efficiency by adding the ability to collect data across the board and communicate and analyze these data, thus obtaining useful information. Robotics and CPS will manage processes independently and proactively.
- Hospital 4.0 will improve accessibility to medical and pharmacological treatments, increase efficiency, and convey significant personalization of the health system and medications. Successful hospitals will not be those with better economic ratios, but those who know how to better interpret the present social, economic, and technological moment.

Those hospitals who have already started the digital transformation are much better prepared to face the Covid-19 pandemic, and their patients are better off (Salimbeni, 2020).

REFERENCES

Australian Industry Standards, 2018. MARB014 Maintain and repair shipboard machinery and equipment. Australian Government. July 9, 2018.

Australian Industrial Systems Institute, 2020. Training for the automotive service and repair sector. Automotive Pathway Program. Melbourne VIC 3000, Australia.

Broadwind, 2015. Gearbox maintenance tips. Keep Your Gearbox Humming Along. Broadwind GearBox Ebook-edit-lowres-2015.

Budai G., Dekker R., 2002. An overview of techniques used in planning railway infrastructure maintenance and its effect on capacity. In Geraerds, W.M.J., & Sherwin, D. (eds), Proceeding of the IFRIMmmm (maintenance management and modelling conference), Växjö, Sweden.

Budai G., Huisman D., Rommert Dekker R., 2004. Scheduling preventive railway maintenance activities. Econometric Institute, Erasmus University Rotterdam. September 15, 2004.

Cassettari L., Patrone C., Saccaro S., 2019. Industry 4.0 and its applications in the Healthcare Sector: A systematic review. XXIV Summer School "Francesco Turco" – Industrial Systems Engineering. September 2019.

CHEManager, 2016. Future Production Concepts in the Chemical Industry. April 04, 2016. www.chemanager-online.com/en/topics/production/future-production-concepts-chemical-industry. Viewed: July 29, 2020.

Clarke L. W., Hane C. A., Johnson E. L., Nemhauser G. L., 1996. Maintenance and crew considerations in fleet assignment. Transportation Science, 30, 249–260.

Csanyi E., 2016. Regular motor maintenance to avoid failure (and prolong its lifespan). Electrical Engineering Portal. February, 29 2016. https://electrical-engineering-portal.com/regular-motor-maintenance-to-avoid-failure-and-prolong-its-lifespan. Viewed: July 28, 2020.

Den Hertog D., Van Zante-de Fokkert J. I., Sjamaar S.A., Beusmans R., 2001. Safe track maintenance for the Dutch railways, Part I: Optimal working zone division. Tech. rept. Tilburg University, The Netherlands.

Dos Reis M. D. O., Godina R., Pimentela C., Silva F. J. G., Matias J. C. O., 2019. A TPM strategy implementation in an automotive production line through loss reduction. 29th International Conference on Flexible Automation and Intelligent Manufacturing (FAIM2019), June 24–28, 2019, Limerick, Ireland. Procedia Manufacturing, 38, 908–915.

Eisenberger D., Fink O., 2017. Assessment of maintenance strategies for railway vehicles using Petri-nets. 20th EURO Working Group on Transportation Meeting, EWGT 2017, 4–6 September 2017, Budapest, Hungary. Transportation Research Procedia 27 (2017) 205–214.

Electrical Equipment, 2020. Maintenance in electric motors. http://engineering.electrical-equipment.org/ene rgy-efficiency-motors/maintenance-in-electric-motors.html. Viewed: July 27, 2020.

Global Railway Review, 2019. How to implement efficient railway maintenance through digitalisation. April 16, 2019. www.globalrailwayreview.com/article/80517/implementation-railway-maintenance. Reviewed: July 18, 2020.

Holland D., Burg R., 2017. Preventive maintenance checklist for centrifugal pumps. Mahan's Thermal Products. Jul 17, 2017. www.plantservices.com/articles/2017/fh-preventive-maintenance-checklist-centrifugal-pumps. Reviewed: July 18, 2020.

Hospital International Congress, 2017. Hospital 4.0 Conference. September 6–8, 2017. University of Coimbra. https://cemuc.dem.uc.pt/hospital40. Reviewed: July 18, 2020.

Inada Y., 2019. Continuous manufacturing development in pharmaceutical and fine chemicals industries. Mitsui & Co. Global Strategic Studies Institute Monthly Report. December 2019.

Metro Pumps & Systems., 2020. Pump and motor preventive maintenance program. Pumps, Parts, Systems, Engineering Services, Field & Shop Repairs, Upgrades.

Miwa M., Ishikawa T., Oyama, T., 2001. Modeling the optimal decision-making for multiple tie tamper operations. Proceedings of the 5th World Congress on Railway Research, Cologne, Germany.

Mohit K., 2020. How maintenance work is done onboard a ship? Marine Insight. August 6, 2020. www.marineinsight.com/guidelines/how-maintenance-work-is-done-onboard-a-ship. Viewed: August 8, 2020.

Money Matters, 2020. Continuous production system|Characteristics|Types|Merits| Demerits|Suitability. https://accountlearning.com/continuous-production-system-characteristics-types-merits-demerits-suit ability. Viewed: July 29, 2020.

National Research Council, 1996. New materials for next-generation commercial transports. Chapter 7: Aircraft maintenance and repair. The National Academies Press.

New Technology Network (NTN), 2017. Care and maintenance of bearings. NTN Corporation. CAT. No. 3017-II/E.

Salimbeni S., 2020. Hospital 4.0: La transformación digital en los hospitales. April 4, 2020.

Schmidt S., Schmidt B. S. G., 2018. Maintainability, reliability and serviceability: Industrial examples automotive industry. Bulletin of Engineering, 11 (January–March).

Talib Bon A., Lim M., 2015. Total productive maintenance in automotive industry: Issues and effectiveness. Proceedings of the 2015 International Conference on Industrial Engineering and Operations Management Dubai, United Arab Emirates (UAE), March 3–5, 2015.

Talluri K. T., 1998. The four-day aircraft maintenance routing problem. Transportation Science, 32.

Van den Bergh J., De Bruecker P., Beliën J., Peeters J., 2013. Aircraft maintenance operations: State of the art. Research Paper. Faculteit Economie en Bedrijfswetenschappen. Campus Brussel. November 2013. Nr. 2013/09.

Tos Znojmo, 2010. Installation, operation and maintenance manual for gearboxes and variators. January 2010. https://d2.tos-znojmo.cz/wp-content/uploads/2018/11/tos_navod_a4_univerzalni_uk.pdf. Viewed: July 20, 2020.

UKEssays, 2017. Ship machinery and equipment maintenance. April 24, 2017. www.ukessays.com/essays/engineering/ship-machinery-and-equipment-maintenance-engineering-essay.php. Viewed: July 21, 2020.

Van Zante-de Fokkert J. I., Den Hertog D., Van den Berg, F. J., Verhoeven, J. H. M., 2001. Safe track maintenance for the Dutch railways, part ii: maintenance schedule. Tech. rept. Tilburg University, The Netherlands.

Varaprasada Rao M., Chaitanya M., 2017. Aircraft servicing, maintenance, repair & overhaul – the changed scenarios through outsourcing. International Journal of Research in Engineering and Applied Sciences. 7(5), 249–270.

Wikipedia, 2020. Continuous production. https://en.wikipedia.org/wiki/Continuous_production#:~:text=Con tinuous%20production%20is%20. Viewed: July 29, 2020.

Index

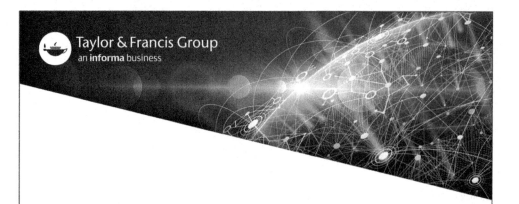

Printed in the United States
by Baker & Taylor Publisher Services